ERNST MACH – A DEEPER LOOK

BOSTON STUDIES IN THE PHILOSOPHY OF SCIENCE

VOLUME 143

ERNST MACH – A DEEPER LOOK

Documents and New Perspectives

Edited by

JOHN BLACKMORE

KLUWER ACADEMIC PUBLISHERS
DORDRECHT / BOSTON / LONDON

Library of Congress Cataloging-in-Publication Data

Ernst Mach-a deeper look : documents and new perspectives / edited by
John Blackmore.
p. cm. -- (Boston studies in the philosophy of science ; v.
143)
Includes bibliographical references and index.
ISBN 0-7923-1853-6 (HB : alk. paper)
1. Mach, Ernst, 1838-1916. 2. Physicists--Austrian--Biography.
3. Philosophers--Austrian--Biograhy. I. Blackmore, John T.
II. Series.
Q174.B67 vol. 143
[QC16.M32]
001'.01 s--dc20
[530'.092] 92-20024

ISBN 0-7923-1853-6

Published by Kluwer Academic Publishers,
P.O. Box 17, 3300 AA Dordrecht, The Netherlands.

Kluwer Academic Publishers incorporates
the publishing programmes of
D. Reidel, Martinus Nijhoff, Dr W. Junk and MTP Press.

Sold and distributed in the U.S.A. and Canada
by Kluwer Academic Publishers,
101 Philip Drive, Norwell, MA 02061, U.S.A.

In all other countries, sold and distributed
by Kluwer Academic Publishers Group,
P.O. Box 322, 3300 AH Dordrecht, The Netherlands.

Printed on acid-free paper

Printed in the Netherlands

This book is dedicated to my kind and loving wife:

SETSUKO TANAKA

TABLE OF CONTENTS

SECTION I: HISTORICAL DOCUMENTS

CHAPTER 1: SOME BIOGRAPHICAL DOCUMENTS 1

A. 'Johann Mach 1840–1862' by Leo Fürst 1
B. 'Early Memories' by Marie Mach 3
C. 'Ernst Mach' by Marie Mach 10
D. 'Ernst Mach' by Ernst Mach 17

CHAPTER 2: ERNST MACH IN PRAGUE 29

First Study: Ernst Mach & the Conflict of Nations by Dieter Hoffmann 29
Second Study: Prague Data on Ernst Mach (1867-1895) by Dieter Hoffman & Ivo Tretera 47

A. Mach's *Lotus* Lectures 47
B. Mach's Assistants 48
C. Mach's Doctoral Students 48
D. Mach's *Habilitanten* 50
E. Correspondence & Notes 50
F. Lectures by Topic 55

CHAPTER 3: ERNST MACH AS A HISTORIAN OF SCIENCE AS SEEN IN BOOK REVIEWS AND NOTES (1896-1932) 57

A. On Mach's *Principles of the Theory of Heat:*
E. Mach (1896) 57
J.E. Trevor (1897) 57
B. On Mach's *Science of Mechanics:*
E. Lampe (1901 & 1905) 59
W. Ostwald (1905) 61
C.D. Broad (1916) 62
W.V. Lenzen (1932) 63
C. On Mach's *The Principles of Physical Optics:*
M. von Laue (1921) 65
E. Bloch (1922) 66
J. Stark (1922) 67
S. Brodetsky (1922) 68
L.C. Martin (1926) 69

CHAPTER 4: MACH CORRESPONDENCE 71

A. Ernst Mach and Wilhelm Ostwald (1889-1913) 71
B. Ernst Mach and Hugo Dingler (1910-1915) 92

SECTION II: PHILOSOPHICAL DOCUMENTS

CHAPTER 5: SOME PHILOSOPHICAL DOCUMENTS 111

A. 'Lectures on Psycho-Physics - Conclusion' (1963) by Ernst Mach 111
B. A Draft Foreword to the Russian Translation of *Die Analyse der Empfindungen* (1906/1907) by Ernst Mach 115
C. 'Sensory Elements & Scientific Concepts' (1910) by Ernst Mach 118

CHAPTER 6: THE MACH-PLANCK POLEMICS 127

A. 'The Unity of the Physical World Picture - Section 4' (1908/1909) by Max Planck 127
B. 'The Leading Thoughts of My Scientific Epistemology and Its Acceptance by Contemporaries' (1910) by Ernst Mach 133
C. 'On Mach's Theory of Physical Knowledge' (1910) by M. Planck 141
D. Mach departs from the Austrian Academy of Sciences (1913) 147

CHAPTER 7: MACH AND EINSTEIN 151

A. 'Mach Looks Through a Sprinthariscope' by Stefan Meyer 151
B. 'Einstein's Visit to Mach' by Philipp Frank 152
C. 'Ernst Mach' by A. Einstein 154
D. 'F. Adler & L. Mach on Einstein' 159
E. 'Einstein and the Philosophies of Kant and Mach' in *Nature* 162
F. 'Einstein's Kyoto Talk' by B. Brauner 163
G. 'Mach & Einstein' by B. Brauner 164

CHAPTER 8: MACH'S RELATIVITY VS. EINSTEIN'S RELATIVITY 165

A. 'The Relativity Concept in Physics' by Emil Wiechert 165
B. 'Ernst Mach and Relativity' by Hugo Dingler 170
C. 'Objective Relativism' by Arthur O. Lovejoy 174

SECTION III: ANTHOLOGY

CHAPTER 9: URSULA BAATZ / The Scientist as a Buddhist 183

CHAPTER 10: MIODRAG CEKIC / Mach's Phenomenalism as a Link Between Physics and Psychology 201

CHAPTER 11: RUDOLF HALLER / Poetic Imagination and Economy: Ernst Mach as Theorist of Science 215

CHAPTER 12: KEIICHI NOÉ / Mach's Relativism vs. Apriorism and the Mechanistic World View 229

CHAPTER 13: MARIO BUNGE / Mach's Critique of Newtonian Mechanics 243

CHAPTER 14: GERALD HOLTON / More on Mach and Einstein 263

CHAPTER 15: RYOICHI ITAGAKI / Three Batches of Reasons for Mach's Rejection of Einstein's Theory of Relativity 277

CHAPTER 16: SETSUKO TANAKA / Mach, Einstein, and Kuwaki 297

CHAPTER 17: MICHAEL A. SANTONE / Ernst Mach's Influence on Four Japanese Thinkers 333

CHAPTER 18: FRIEDRICH STADLER / The 'Verein Ernst Mach' - What was It Really? 363

CHAPTER 19: MIODRAG CEKIC / The Influence of Mach's Philosophy in Yugoslavia 379

CHAPTER 20: HENK VISSER / Mach, Utrecht, and Dutch Philosophy 403

Bibliography 431

Index 453

ERNST MACH

INTRODUCTION

1.

One purpose of this book is to present previously obscure biographical information and make it readily available in English. A second purpose is to make the contributions of recent European and Asiatic scholars about Mach more widely known. A third goal is to provide English-speaking readers with ready access to enough basic documents and source material to be able to follow and contribute to current controversies over Ernst Mach, the Austrian physicist, physiologist, psychologist, philosopher, and historian.

On the physical side, many scientists have been drawn to Mach's *Science of Mechanics,* which originally came out in 1883 and which by the year 1901 was widely considered a classic (See Chapter Three). Mach's book appealed to a number of ideological trends within physics which had already begun to attack Newton's mechanics and theory of absolute space, time, and motion. These trends were already using the distinction between "classical" and "modern" physics even *before* the work of Planck and Einstein. Four of the best-known ideologies were Wilhelm Ostwald's "energetics" which he held from approximately 1892 to 1908, the "electro-magnetic world picture" associated with Heinrich Hertz and some of the work of Hendrik Antoon Lorentz, which still had defenders as late as 1910, "positivism" which is still prominent in philosophy of science in some of its avatars, and Ernst Mach's closely related "physical phenomenology" and theory of "Denkökonomie" which also retain a number of supporters. What united these movements was opposition to atomism and the epistemological representism which they believed underlay Newtonian mechanics. During the last decades of the 19th century they attempted to treat thermodynamics, electromagnetism, or phenomenalism as more basic than atomism, discontinuity theory, and what we now call statistical mechanics.

In opposition, Ludwig Boltzmann pointed out that thermodynamics rested on assumptions from statistical mechanics and atomism rather than the reverse, a fact which threatened Ostwald's "energeticism". After the turn of the century, Einstein and Perrin helped prove the reality of atoms such that discontinuity theory could not be abolished, and gradually, Einstein's special theory of relativity was distinguished from the work of Lorentz and considered much more fundamental and reliable than the vain effort to reduce everything to electromagnetism. The collapse of the anti-atomistic and anti-representationalist ideologies within physics did not mean that Mach's book on mechanics lost admirers. Both Planck and Einstein themselves, regardless of how disenchanted they became with aspects of Mach's philosophy (and both men had initially been strong admirers of it), nevertheless, continued to consider Mach's *Science of Mechanics* as an outstanding work. Why? Because by challenging Sir Isaac Newton's doctrines of Absolute space, time, and motion during the 1880's, even if the challenge was basically epistemological rather than physical, enabled physicists twenty years later to be more receptive to the truly revolutionary physical work of Max Planck and Albert Einstein. But one should add that in order to understand the Mach-Planck polemics which came to a climax in 1910 (See Chapter Six in our book) it is necessary to keep in mind that Max Planck, a philosophical "conservative", wanted to disentangle the physical contributions of both Einstein and himself from some of the earlier ideological movements which had long considered Mach to be one of their leaders.

Ernst Mach often claimed that the most decisive event in his life concerning philosophy was when he read Immanuel Kant's *Prolegomena to a Future Metaphysics* at the age of fifteen. This book encouraged him to oppose what he called "metaphysics", that is, indirect and representist world views, such as those espoused by Descartes and Locke and which have been strongly represented in science by means of many assumptions held by Galileo, Newton, Darwin, and more recently the mature Planck and Einstein.

Mach as a philosopher of science, however, justified his presentism, his desire to identify the physical world with what can be directly noticed, with the need to reconcile physics with psychology, such that scientists did not have to shift world views in going from one discipline to the other. On the other hand, Descartes and Locke had thought that one of the greatest advantages of representism and mind-matter dualism were their ability to reconcile science with rationally-held religion, and common sense.

If one agrees with Mach that reconciling physics with psychology is more important - at least for practising scientists - than bringing science into closer conjunction with rationally-held religion and common sense, then one may be attracted to Mach's world view, or with one of its closely related successors. But needless to say, such matters are very controversial.

In short, Mach's presentism, his tendency to identify the physical world with sensations, conscious experience, or what can be directly noticed is still a very strong and live tradition. Nor has the issue of whether presentism is compatible with physical science, especially theoretical physics, ceased to be important and controversial, and if compatible, whether it is better for science than representism or another type of indirect world view. But some philosophers have maintained that even could it be proved that phenomenalism was incompatible with physical science, that a case might still be made that important aspects of Mach's philosophy of science may have been held independently from his phenomenalism, and hence still have value. Critics, on the other hand, allege that while it may always be possible to salvage inconsistent parts of a world view, that this scarcely makes the world view or philosophy of science more impressive. A more neutral view might be that Mach's philosophy has had a remarkable influence, first, on the early work of Einstein, and later, on the Vienna Circle and Logical Positivism such that to understand scientific and intellectual currents of the 20th century one has to comprehend the role of Mach's way of thinking.

Many people have admired Mach's philosophy because it seemed to reject the notions of force and cause, but this observer has been intrigued by Mach's approach for the opposite reason. By refusing to accept an inflated conception of natural laws, as if mere conditional patterns were causes, determined causes, or replaced causes, Mach opened the way for future philosophers of science to develop a more realistic idea of what makes, pressures, or influences things to happen, a notion which takes into account the energetic foundation of all forces as causes and the capacity of unforseen natural and human variables to trump so many law-based predictions.

2.

As one can easily verify by looking at the bibliography in this book, there has been a remarkable resurgence of interest in Ernst Mach over the last few years, especially in German-speaking countries. Numerous books by Mach have been republished, often after an interval of fifty or sixty years. In addition, numerous new books about Mach have recently come out plus several times as many articles. But the renaissance has been uneven, touching some countries much more strongly than others.

Ernst Mach often thought of himself as a physicist with a strong interest in the epistemology of science. He apparently wanted to base what he considered real on the epistemological position that only sensations or what he called "elements" could be absolutely certain. But for people who deny that anything can be absolutely certain for finite and fallible human beings, that is, who reject epistemology or theory of knowledge in this extreme sense, there is a tendency to identify Mach's ontology or world view not just with his phenomenalism but also with the other things he *normally* treated as real, even if they could not be known with absolute, infallible, or unconditional certainty. An advantage of using this approach concerning Mach is that his theory of economy now makes much more sense and becomes much more compatible with the rest of his philosophy. What is "Denkökonomie"? A slightly disguised form of traditional means-end, volitional logic, something quite essential in both practical life and many aspects of doing science. Though the recent renewel of interest in Mach has many sources, perhaps the one with the most philosophical potential focuses on what might be called a "post-epistemological" investigation of the "new"

Mach, that is, of the positions he seems to have assumed or held, but with less - often considerably less - than absolute certainty. Below is a diagram of the five types of things Mach often or normally assumed were real:

CONSCIOUS EXPERIENCE

RELATIONS ELEMENTS NOT-YET RELATED PHENOMENA

THE PRACTICAL WORLD

Some of the different types of relations which Mach usually treated as if they were real include the following:

OBSERVATION is vital in science and potentially indubitable.

MATHEMATICAL FUNCTIONS are an end-goal of science & very certain.

MEANS-END LOGIC gives direction to human activity but is not certain.

ECONOMY IN SCIENCE[1] as if science were an end-in-itself is not certain.

ECONOMY IN SCIENCE[2] as if science were a means to an end is uncertain

MEMORY is generally useful in science and life but is not certain.

ANALOGY gives information about other minds but is also not certain.

In terms of Mach's Darwinian position that all human activity including science and epistemology are tools for helping the human race survive on a Civilized level and his Lamarkian corollary that acquired characteristics such as economical habits may be subject to inheritance, it seems difficult to give primacy either to his phenomenalistic or epistemological claims about knowing sensations or elements with absolute certainty. they too are merely means or tools to help further a probably valid biological end. Concerning his theory of economy, in the larger sense it seems to have been focused on biological purposes. This suggests that the criticisms of Planck, Einstein, and others directed at Mach's theory while largely valid when treating science as an end in itself may have overlooked Mach's larger message. Mach and his colleague Ludwig Boltzmann were contemporaries of Darwin and thought very largely in Darwinian or Lamarkian terms, but Planck and Einstein, men from a later generation, tended to treat physics as an autonomous discipline and not as a means to an end outside of itself. This significantly restricted how they understood Mach's theory of economy and the use of economical methods in science. They could appreciate the use of economy as logical simplicity, but resisted psychological and biological notions, such as saving time and effort, since good scientists must be willing to spend as much time and effort to do the best work possible. But right as they probably were in a narrow sense about eventually successful investigation, and much of their own work tended to be disproportionately successful and important at least for a number of years, they were clearly not thinking of science itself as a means to an end, much less a biological end, or even of the general utility within science of not wasting time and effort, something which the determined pursuit of false trails can also result in. Did Einstein himself perhaps waste a certain amount of both time and effort during the second half of his life? Persist in "uneconomical science"? Let us refrain from unkind or premature response. The point is, that Mach's theory of economy may well deserve the second look it is receiving from more and more scientists, philosophers, administrators, and perhaps in the future, financial experts. Any economizing may seem to go against the wish of scientists to be given all of the resources needed to make progress, regardless of how it impoverishes other disciplines and areas of life, but to prevent possible over-reaction in the future against "wasteful" science, perhaps even physicists could use a bit more psychological and biological "Denkökonomie".

Furthermore, a little noticed but quite important difference between Mach's approach toward science and that of subsequent Logical Positivists has been their respective attitudes toward the metaphysical as something "meaningless". For the latter, that has put it beyond science, but for Mach it was precisely what seemed metaphysical or meaningless as the not-yet-related which *invited* scientific investigation in order to make it meaningful by finding relations to what was sensory or empirical. The metaphysical or meaningless as the not-yet-experienced or not-yet-understood or even as what appeared incapable of either was an opportunity or challenge for science for Mach, but outside science for the Positivists. In this sense, Paul Feyerabend is probably right in distinguishing Mach so sharply from his successors.

Nevertheless, it is surely true that Mach, like most philosophers, especially unwilling ones, was not always consistent. For example, sometimes he regarded space and time as elements, sometimes as relations, sometimes as unsolved problems, and sometimes he carefully distinguished between metrical, physical, physiological, and psychological types of spaces and times. Furthermore, he occasionally made efforts to reduce the above relations to their sensory components, but most of the time he treated them like most people do, namely, as useful ideas or abstractions. Such inconsistencies may be related to his reluctance to consider himself a philosopher or to have a philosophy. Also, to admit the possible or probable existence of other than elements would have conspicuously compromised his ostensible ontological monism, something his theory of economy was reluctant to allow him to do. But from the perspective of the many recent thinkers who have been even more reluctant to claim absolute certainty about anything, discovering the pluralistic aspects of the non-certaintistic side of Mach's ontology and world view has merely meant helping to make his overall philosophical position more compatible with his general opposition to dogmatism and overbelief.

3.

Concerning the form and content of this book, the work has been divided into three sections. The first part is primarily biographical and presents a number of basic papers or previously unpublished documents. Section two performs the same service for philosophical material and controversies. The third and main section is an anthology with contributions by eleven scholars. Of the twelve articles, seven are new, three are English versions of foreign articles, and two are previously published classics.

In section three, the first four contributions are rather friendly toward Mach's philosophy and defend him on many points. The next three papers are critical, particularly of Mach's philosophy of science and his views on physics and relativity theory. The remaining papers are primarily historical and present much new or little-known information about Mach's life, work, and influence in different cities, countries, and periods of history. One might add that scholars of Planck, Einstein, Ostwald, Dingler, and a number of other contemporaries of Mach will also find material which may interest them. Several Japanese researchers have also honored the editor by contributing valuable work to this anthology.

While several articles in this book direct attention almost exclusively toward philosophical issues, numerous others focus on historical relationships between Mach and his contemporaries making it easier to link his contributions to his subsequent influence on scientists and philosophers. Concerning topical debates, the book touches on Mach's position with respect to atomism (chapters 1, 4 - 7, 15, & 16), briefly on Ludwig Mach's alleged forgery of the preface to his father's *Optik* (chapters 1, 7, 8, 14, & 15), and goes into extensive detail on differences between Mach's and Einstein's respective theories of relativity and on Mach's attitude toward the younger man's approach (chapters 7, 8, 14, 15, 16, & 20).

Concerning inclusiveness, first, since earlier anthologies have focused largely on Mach's "positivism" and laboratory contributions to physics and psychology, we have directed our attention more toward history and other aspects of his philosophy, and second, while the contributors to this book have taken a wide variety of positions on Mach's phenomenalism, physics, and contributions to history and philosophy of science, variety which when combined to create a synthesis may significantly increase our understanding, this does not mean that this or any anthology can substitute for reading his own work, especially with respect what he actually wrote about controversial subjects. Thus, we have also included much original material for the first time or for the first time in English to help this volume become *more* than an anthology. *It is - as the reader can determine for himself - a source book plus an anthology.* If it can help stimulate more research into Mach and his *oevre* and increase appreciation for the central role his ideas have exercized in the history of early 20th century physics and philosophy, while making some progress toward the solution of lasting, festering, or controversial issues, especially about the compatibility of presentist perspectives with science, then this editor will be content.

Concerning editing and translating, it has been a challenge and a pleasure to help prepare a number of foreign articles for English-speaking readers. In general, if there is praise, let the authors receive it; if blame, put it on the editor, who in somewhat more than half of the chapters did the translating and in all cases made the final revisions. What most pleases the editor is the amount of serious research which historically-oriented philosophers have carried out. Many of them have broken new ground, opened up new realms for future study.

Above all, the editor would like to acknowledge and thank the unfailing cooperation of the Ernst-Mach-Institut in Freiburg im Breisgau together with Robert S. Cohen, Annie Kuipers, numerous Mach scholars, and the many patient publishers of previous books and articles. Without their help this book would not have been possible.

He also wants to thank several publishers for allowing a number of articles, or as a rule, shortened or revised versions of them, to be republished. These sources have been cited in the appropriate locations.

Dr. John Blackmore
Tokyo, November 1991

CHAPTER 1

SOME BIOGRAPHICAL DOCUMENTS

A.

JOHANN MACH 1840-1862

by P. Leo Fürst OSB*

The Catholic church in Untersiebenbrunn in Lower Austria [a small village which is about twenty kilometers east of Vienna] owned a large field which had been divided up among the individual farmers to be worked by them. The farmers in addition to paying a small fee for tilling the land also did tasks for the local priest including special field work for him. These were regulated by an 1811 contract between representatives of the community and the priest. The farmers, however, carried out these chores with growing reluctance.

Johann Mach, who was owner of house #22 and tutor of the children of Baron Breton and later Graf von Stadion, refused to work the soil of the church field, at least in the short run. The priest, Jakob Gruber, sent a rather sharp letter to Herr Mach, who requested from the local judge a look at the 1811 contract and turned to a legal friend in Vienna.[1]

The legal validity of the contract was thrown into doubt, because it was neither confirmed by the appropriate officials nor formally included in the basic book of written law. Outside of that it had been signed only by the community council and not the obligated farmers. Mach stressed his desire to be reasonable but wanted to make clear that all future help would have to be voluntary.[2]

The farmers of Untersiebenbrunn also refused to work for the priest. A seminary in the town of Melk appealed to a court of common law in the name of the priest. The court decided in 1850 [i.e. after the Revolution of

* Leo Fürst was Catholic priest for the village of Untersiebenbrunn during the early 1980's. He has since moved to the city of Melk. The editor is very grateful to him for his help in uncovering information about Ernst Mach's father. This document was included in a letter to John Blackmore, dated August 2, 1982. The numbered footnotes were added by Father Fürst. - Ed.

1 *Pfarrchronik Untersiebenbrunn,* I/1, p. 114 ff. [an unpublished village journal kept by successive village priests. - Ed.]

2 W. Kowarik, "Die Grundablöse 1848 am Beispiel des Stiftes Melk" in *Stift Melk - Geschichte und Gegenwart,* Bd. 2, 1981, p. 142 ff.

J. Blackmore (ed.), Ernst Mach – A Deeper Look, 1–28.

1848 had been suppressed] in favor of the priest. Alexander Bach, who was the lawyer for the Melk Catholic Seminary which had initiated the case, had meanwhile been promoted first to Justice Minister and then to Interior Minister of the Austrian Empire, replacing the conciliation-minded Graf Francis von Stadion, the very nobleman whose children Johann Mach had been tutoring.

That Mach had not been principally against the priest of his village came out in the fact that during 1848, the revolutionary year, after the elimination of the church fee [*Zehenpflicht*], he along with four other house-owners voluntarily gave their priest, Father Anton Fohringer (1845-1851), the usual poultry fee [*Geflügelzehent*].[3]

Only those who had paid at least one Gulden yearly in direct tax were eligible for election to the community council in Untersiebenbrunn. Johann Mach as owner of house #22 had to pay 90 Gulden and 17 2/4 Kreuzer... The small house #84 belonged to this complex [*Besitzeinheit*].[4] In the election of July 15, 1850, five farmers together with three gentlemen, Johann Mach, Johann Lausch, and Anton Kunz were chosen. These men then selected the village judge and the members of the two community councils. Johann Mach "rather unanimously" was elected to the first council. When the community government handed out offices Johann Mach was named secretary and chief of correspondence for the community as well as school inspector.[5]

Mach as [member of the] first community council attempted to eliminate abuses and settle grievances. He acted against waste and extravagance and protested against public drunkenness at community expense. At the end of 1850, Father Fohringer in his priest's chronicle would remark: "So much was saved in such a short time that significant debts could be paid off."[6]

But Mach's vigorous efforts at reform apparently stirred the envy of the village judge and members of the second community council who grumbled: "We can, like other communities, solve such problems very well without the help of gentlemen from Vienna."[7] The subsequent priests, Fathers Leo Deffenhart (1851-1854) and Friedrich Heilmann (1854-1862), do not mention Mach's name in the local chronicle.

The parish register, however, also contains two entries referring to the family Mach. Marie Mach [Maria Kunigunde Ursula Mach] was recorded as having been born on October 21, 1844. The father was listed as Herr

---

[3] *Pfchr.* U7 I/1, p. 157.

[4] Ibid. p. 179 & 161.

[5] Ibid., p. 182 ff.

[6] Ibid., p. 186.

[7] Ibid. p. 187.

Johann Mach, educator of the young Graf von Stadion in Vienna and son of Joseph Mach, farm owner in Liebenau, Bohemia and whose wife Johanna née Hürbe had come from Sastall, Bohemia. Marie Mach's mother and Johann Mach's wife was called Josepha and was the daughter of Wenzel Lanhaus, "*Rentmeister*" in Chirlitz, Moravia [where Ernst Mach was born in 1838]. She served as representative of the well-born Frau Gräfin Kunigunde Stadion.[8] The other entry in the priest's register was about Johann Mach's daughter Octavie, [Octavia Mach] who was born on May 19, 1839, married Josef Engelhardt on October 30, 1859.... [9]

It also seems that in 1859 Johann Mach was no longer the owner of the house #22 or its attached cottage for domestic servants #84. Octavia before her marriage that year and probably the whole Mach family apparently lived in #14, a rented house, which was a big farmer's dwelling on the western edge of the village. The parish register mentions a Herr von Jäger as owner of farm house #22 for the years 1860, 1862, and 1870.

B.

EARLY MEMORIES

by Marie Mach*

(1912)

My life which in itself is not remarkable has nevertheless encompassed a series of striking events which perhaps do deserve notice.

My childhood - I was born in 1844 - fell in a time when people and animals because they were still frightened by trains would run right and left when a locomotive puffed by. In my old age I have seen the development of airplane technology. In between, the telegraph with and without wires, telephone, gramophone, undersea cables, mountain lifts, electric trams, ocean voyages around the earth, expeditions to the North and South Poles,

---

8 *Tom. I* fol 137, Rz 13.

9 *Trauungsbuch Tom II* fol 5 Rz 2.

* This translation is from selected parts of the first chapter of *Errinerungen einer Erzieherin,* second edition, Vienna, 1913. It was written by Marie Mach but appeared in print as if by an anonymous author. The first edition had come out the previous year. To preserve anonymity or because the information was well-known to her readers, the author omitted names, dates, and other data. For the benefit of English-speaking readers of the late 20th century who prefer a smooth-flowing narrative and do not want to be troubled by too many footnotes such information to the extent it can be determined has been restored to the text. - Ed.

treks across Africa - the map of the world still had large blank spaces when I was young - and the building of the Suez and other canals would all occur. I have experienced the world celebration of Friedrich Schiller's hundredth birthday, his subsequent fall from fashion, his victorious return to life after the hundredth anniversary of his death, and by now have celebrated even his 150th birthday.

In political terms, my life has encompassed equally important events: absolutism, revolution, liberalism, clericalism, the "*Zurückschrauben*" of adult education, and the era of [Karl] Lueger, the growing brutalization of the masses by means of anti-Semitism. In France, I passed through the time of Louis Philippe, the Second Republic, the Second Empire, and its overthrow and the proclamation of the Third Republic. I also remember the epoch of Garibaldi, the unification of Italy under the monarchy, the unification of Germany, the unhappy national conflicts, and unfortunately the defeat of Austrian Germans, who were kicked out of Germany by Bismarck. - It is a strange irony of world history that the man who established German unity should have marred his work with an exclusion which will never be made good.

My memories from childhood reach back even to the tenderest years. Naturally, they are mere isolated pictures, which have imprinted themselves upon my faculty for recollection and which have only been explained later via questions. The first thing I recall is an altar with many lights which my brother [Ernst Mach] who was then very pious had built. I also see a second picture. I am sitting on a footstool near the oven. My father [Johann Mach] who has returned from a trip has brought me a toy, a pigeon loft, but no pigeon. Then my father shakes his traveling bag and to my great delight the pigeon falls onto the table. - In our garden - we lived at that time in the countryside near Vienna - the apricot trees are blooming. Fully enchanted, I go from tree to tree and put my finger in the blossoms, when suddenly - a burning pain! A bee has stung my finger. It was my first experience of suffering. I cannot remember whether I was able to infer something of moral value from this event, that one may not enjoy the fruit of every blossom.

Many strong impressions also remain from the Revolution of 1848-1849. I am sorry to have to say so - but as truth is my honor - at that time I was guilty of a major change of opinion, which also occurred to many other people, who were ten times or more older than I.

At first, I was enthusiastic for the students, whose felt hats [*Kalabreser*] overwhelmed me, but then came the billeting in the village of Imperial soldiers who had come to surround Vienna and put down the Revolution. A lieutenant took me by the arm and courted me. I played with the shining buttons on his uniform. It is always the shining buttons which play the main role in my memory; I cannot remember his face. It's even as if the man had no head. He solemnly promised to marry me when he became a general. I had the time to wait, fifteen or twenty years, but as I afterward heard he never rose above captain and married someone else. Fortunately, this jilting never caused me any heartbreak.

My admirer belonged to a battalion full of mutiny-minded Italians who wanted nothing more than to murder their officers. My father often told me later that the other officers had given up hope of controlling the battalion. Only the lieutenant still dared to command them. He put himself at their head and marched them away. The name of this brave man was Schuster. If the family members who came after him still live, they can be proud of him. Using all kinds of promises, he brought the mutineers to the railroad to Vienna and later shoved them into the front line of the assault where they splendidly made up for their disloyal attitude.

One day I sat among these rebels upon a big drum, feet-dangling, soul-happy, holding in each hand a large stolen pear eating first from one and then from the other, fruit which they had honored me with. My nursemaid when she saw the situation came flying, ripped me from the drum, and hurried me away. Torn from my wonderful perch, I bawled without end while being dragged home. Certainly, the mutineers had been very pleasant company, though I am not sure which language we had conversed in.

I remember another occasion [October 1848]. My parents stood in fearful expectation at the window, shaken by the last cannonshots from the Battle of Schwechat. A Hungarian army had come to relieve the encircled student regime in Vienna only to be defeated and driven away at the very edge of the city itself. I of course knew nothing at that time of what was going on; nevertheless, my feeling of terror was so great as to leave an impression which I have never been able to fully extinguish.

Another memory: The Augustinian Tower [in Vienna] burns. A blood-red glow flickers on the wall, over my bed. But this time the fire delights me. Still another picture: A victorious army [1849] was returning from Hungary and it camped on an open place right in front of our house. Many children, boys only twelve to fourteen years old, were being held as prisoners-of-war. Lifted up to the window to see them, I became angry, cried, and threatened the victors - with a plum switch.

My mother, who was a weak, nervous woman, spent the largest part of the revolutionary period alone with us children in the countryside, since my father was kept in Vienna by his profession where he was tutor [and also member of the National Guard]. How the gentle lady endured such trying times and yet knew how to be helpful under all kinds of circumstances remains puzzling. She once told me the following episode. She was sitting in the house alone when suddenly a ragamuffin vagabond entered her room. Thoroughly shocked, she sprang up. Still, as he came closer with a pleading demeanor, she recognized a Viennese professor, a friend of my father, who had fled from Hungary as a refugee from their revolution. She gave him a room, sent an old trusted servant to clean him and his things, and on the next morning he left the house furnished with fresh clothes and money. My parents never heard from him again and never knew whether he had saved himself or gone under. "I was only happy", concluded my mother, "that your father was absent and the responsibility fell on me alone."

....It all seems to me like a fairy tale those far off days. I remember my father telling us about the burial speeches for the many students who died in the March Revolution of 1848, touching speeches from priests and rabbis, and still unforgettable, after sixty-three years!

* *
*

I was a lonely, day-dreaming child. My sister Octavie who was significantly older could be no regular playmate for me. Contact with other children was always brief or fleeting. Therefore, I created my own fairyland and rarely lived in reality. Since I was always in the company of adults, I acquired adult reading habits quickly. By the age of nine or ten I had learned most of Schiller's poems by heart. Only a little later I read with great interest Alexander von Humboldt's *Kosmos*. I generally read everything which fell into my hands, and my good memory retained what I read. At this age I wrote my first drama, a strange thing, for there wasn't a single man in it. I believe that all of the main characters were Amazons.

Once my father took me with him on a wide sweep through the fields and since he wanted to walk further left me on the bank of a little stream, an old arm of the Danube. Pointing to the water, he said: "If you are good, the water-nymph who is your Godmother will come out." I then sat quietly and waited with great expectation, but she didn't come. I plucked flowers from the bank, bound a small bouquet, tossed it in, and called requesting my fairy Godmother to appear, but in vain! That was my first disappointment. For a long time I couldn't forget it. It helped spawn the distrust which some years later I would voice against fairy tales. We had an old servant in the house who was a wonderful storyteller. Even though I had been fully enchanted by fairy tales of houses made of marzipan, I soon developed considerable dislike for such stories, and the thought arose that it was all deception and make-believe. Perhaps, I just became too old to preserve such illusions any longer.

Our house stood in a large, wonderful garden, which my father tended with understanding and care. During the season when the fruit trees were blooming or when the roses blossomed it appeared like paradise. A small stream ran through the garden in which I caught fish, bathed, made boat trips on a washboard, and occasionally fell in. The whole garden was divided lengthwise by a path accompanied on both sides by magnificent cabbage roses [*Zentifolien,* i.e. having a hundred petals], which have unjustly gone out of fashion. These cabbage roses had been planted by Master von Moos, who once owned our house and who had been the fencing master of the Duke of Reichstadt, the Austrian son of the great Napoleon and his second wife, Louise. The tea rose from China which smells like tea is neither as beautiful in form, gentleness of leaf, nor equal in fragrance. But the tea rose blooms the whole year and the cabbage rose only once, therefore the former is in greater demand. If anywhere I find an old-fashioned garden with cabbage roses and bend my head over the fair blossoms and close my eyes, I am transported as if by a magical wand into the paradise of my childhood.

Was it really a paradise? We were rather poor and had to suffer many privations, but children seem better able to endure such things as these than grownups. Both my parents were guided by the highest ideals. They were unworldly people. My father, who in his time was a famous educator, would never take the opportunity to add to his assets. "As long as I lived, I have always kicked money away." My mother had an artistic nature; she only lacked training. She was gifted at music, painting, and poetry. The isolation of living on the land merely strengthened views which were so different from those of most people in the world.

Father instructed us three children himself, Ernst, Octavie, and Marie. I never attended school. As I later met girls who had been to institutes, I would envy them dreadfully. But the discovery of how comparatively little they had learned soon consoled me.

In the summertime, we always had a few guests from Vienna, relatives and many young people, who brought life and cheerfulness to the old house. Otherwise, I readily gave myself up to my daydreams. Among these guests was a cousin of the same age whom I attached myself to with boundless friendship. Did she feel the same enthusiasm? I find it hard to believe. My soul lay open before her, but I couldn't quite figure her out. She was always somewhat reserved toward my enthusiasm. She was cool, deliberate, and reflective [*altklug*], rare qualities for young girls of our age. But one summer morning she appeared to be swept up by my enthusiasm, and we promised that if there was a possibility to come back to this world after death, the one who died first would visit the other. She died at the age of twenty-one, but some words she spoke to her sister before she passed away "You were my only friend" broke my heart. This experience together with the pain of her death and my waiting for her to carry out her promise brought me close to the grave. In my fever, somewhat to the distress of my parents, I called out again and again for her.

But let us turn back once more to my childhood. In late autumn I sat in a small forest behind the barn and saw field mice enter winter quarters. My cat whose attention I had drawn to the event took note of their intrusion and where they would probably hide themselves, and in time cleverly caught many. Winter also had its agreeable hours. In the evening after our frugal supper, father would read to Ernst and Octavia. I was sent to bed "because I still couldn't understand anything". I pressed myself into the pillow and remained absolutely still and didn't lose even a single word from what had been spoken. My favorite was "Tales of a Grandfather" from the Scottish History of Sir Walter Scott.

The garden offered new pleasures with the coming of snow. The ice-covered stream and indeed every frozen puddle was used for sliding. Ice-skating was not yet widely done. The Empress Eugénie, the wife of Napoleon III, was apparently the first to bring it into fashion. Once I was found crouching in the snow under a tree with some salt in my stiff hand; I had listened to a birdwatcher who had suggested that I try to put the salt on the tail of one in order to grab it. Chickens, birds, pigeons, dogs, cats, horses, and sheep were my companions and remarkably tame. They also learned to negotiate with each other and avoid conflict.

I was fourteen years old [1859] when father sold the house at an unexpectedly good price. The same year saw my poor, beautiful sister Octavie marry and become an unhappy wife. We remained three years longer in Untersiebenbrunn in a rented house. I felt very unhappy at this time. I longed to do something with my life but still had to fritter the time away with daily chores. My brother brought me books and tried to fill the gaps in my education, but that wasn't enough to satisfy me. Then one day I gathered up my spirit and asked my father to let me be trained for the theater. The answer ran: "You can go, but please pick another name, because I don't want mine dragged across the footlights." - That naturally ended the matter.

My father longed to buy another farm. I think that if we had been more practical with the money paid to us, we could have laid the foundation for a very secure existence, but farming was my father's weakness and the very good deal he received for selling the old house would prove our undoing. Father looked everywhere until far away in a southern province [*Unterkrain,* in lower or trans-Slovenia, i.e. Croatia] he found to our ruin a beautiful piece of land, a paradise with hellish conditions.

Our last summer holidays in Untersiebenbrunn brought us agreeable social intercourse with a Viennese family which had settled in the village. The daughter was a talented, mentally-gifted girl who became very friendly with me even though she was significantly older. At the age of ten she had fallen into the hands of a clerical uncle who played on her highstrung nature with intense religious ideas from which she could never free herself. I tolerated her attitude, but in the argumentative manner of the pious she wanted to convert me, which caused me much discomfort. When her father died, she left her mother to live alone and became a lady of the Sacre-Coeur religious order. "God who can fill the place of everything can also fill my place by my mother" she wrote me. "I feel myself called to serve Him alone." - After many years, on the occasion of a trip to Vienna, I visited her in her convent. I had the feeling as if an iron partition had clanked down between us. We belonged to two different worlds and had nothing more to say to each other. I gritted my teeth, swallowed my tears, and left the convent, a place where so many wonderful human beings had been buried alive. Needless to say, she died young.

From the heights above where the road to Slatenegg our new acquisition led, between the oak-tree covered summits, I saw for the first time the little valley and the roof of our house, which the people of the neighborhood had pompously called a castle [*Schloss*], although it was only a modest, one-story thing. "Now I shall enter", I said to myself, "but how shall I ever leave?" It was eighteen years after [1880] when I left, unable to make out anything on a dark November morning, not even Slatenegg, that those first words returned once more. I had buried my parents, my youth, and my happiness. I had lost everything. Nothing remained....

* *
*

...Meanwhile I wrote small novellas which Johannes Nordmann took up in his newspaper and praised. He encouraged me to become a writer, to devote my life to literature. But during the last years at Slatenegg I could only devote here and there a winter's night. Had any of my acquaintances known about my short stories, I would have sunk into the earth with shame. I have never understood how authors could bare their soul, appear naked before the public, put their own signature below the title. - During my father's final year [1879] I lived in anguish before the unavoidable. I saw him grow weaker day by day without my being able to do anything about it. If my father was late to supper by only a few minutes I hurried to him, fearing the worst. Winter came and with it the catastrophe. Father died after a five day illness. I held myself upright until the last mourner had left the ceremony, lay down on a sofa, and that evening was carried to bed.

After having pulled myself together again, I made repeated attempts to sell the heavily mortgaged farm. Many kinds of unexpected things made that impossible. Many friends and acquaintances urged me to make a so-called "rational" i.e. forced marriage, but no one was interested. So I wrote to our creditor that he could take the property. He came, concluded the proceedings with a pseudo-sale, and I went my way with empty hands.

After a short rest in the house of my brother, who had just had a stormy year as Rector of the University of Prague, I looked for a job.

Above all I thought of becoming a writer. Although I had fabulous luck - almost always selling my essays - I quickly came to the conclusion that it would be very easy to starve to death. Above everything, I did not want to be a useless extra mouth. As it turned out, however, I had no chance to be choosy. I became in turn hospital nurse, magazine editor, lady's companion, governess, and lecturer....

C.

ERNST MACH

by Marie Mach*

The trace of his days on Earth
Cannot but endure for eons.

My nephew, Ludwig Mach, has requested me to write down what I can recall about the childhood and youth of his father, Ernst Mach. I'm afraid, however, that I can report very little of interest. My brother was almost seven years older than I, and from earliest youth had much more contact with our sister Octavie who was only a year younger than he and knew much more about him. As I was growing up he was a loving teacher who endeavored hard to fill the gaps in my knowledge. But at that time he was already studying at the University of Vienna and only came home for the summer or other holidays.

The difference between him and other boys of his age as far as I have been able to remember was that he never engaged in childish pranks or irrational tendencies. He seemed unaware of evil. Instead of playing, he would always show great interest in observing nature as a living process and would carry out experiments at his own cost of which some were rather dangerous. He made an attempt to discover if camphor would burn; the result was that he totally burned off his eyebrows and eyelashes and was lucky that worse didn't happen. My cousin Clementine Mauer, who was the same age as my sister and who had been educated together with her for two years by my father, therefore a reliable guest, said that the surest way of attracting my brother's attention was always to call: "Come and see what a remarkable animal is here!" When we children went walking, we girls

* The handwritten German original can be found in the Ernst-Mach-Institute, Eckerstrasse 4, Freiburg im Breisgau, Bundesrepublik Deutschland. Parts of it are very difficult to read. It was apparently written during the 1920's when Marie Mach was in her late seventies or early eighties. She died in 1928.

would help catch bugs and butterflies to add to his collection. A man frequently went in front of us who had hired my brother to lead the way. He also collected insects, but my brother didn't have to worry since he represented himself as only a harmless Viennese, a mere stockingmaker who would not even think of competing with him about the insect world.

Perhaps the most memorable event from his earliest years, at least according to my mother, was when Countess [Gräfin] von Stadion gave him a small steamboat as a gift. He placed it on the little stream which flowed through our backyard, fueled it with alcohol, and then had it go upstream against the current. This success made him so happy that he forgot everything which was going on around him.

When Ernst was only thirteen years old, at the wish of his father and understandably without stopping his education from books, he went for some days a week to the neighboring village of Lassee to a carpenter to learn how to make tables, chairs, and boxes, which he later said proved to be very valuable when setting up a physical collection [*Kabinett*]. As a little boy he had wanted to make himself Emperor of Lassee, but now he was a mere a carpenter's apprentice. Probably the unknown village had seemed rather immense to him at one time. In the evening when he came back, we three girls would meet him. The first sign of his coming would be a big torch emerging from the twilight.

I must not forget to mention that on the suggestion of the Countess von Stadion in whose house our father was tutor, Ernst was put in 1847-48 into the Benedictine monastery school of Seitenstetten, that is, when he was nine to ten years old. He never spoke gladly about this period. Probably his memories of the place left no agreeable feelings. He once said that he had caught the French teacher making mistakes about irregular verbs. In 1848 the inmates of the institution revolted and sent a stormy petition to the Catholic father in charge of cooking. They demanded the abolition of a particular type of food, so-called *Beindefleisch*. The chief cook thought it would be a good idea - foresight as we know is the mother of wisdom - to barricade himself in and only negotiate with the boys through closed doors. From what Ernst said, it is not clear what part he played.

In any case he was also stung by the revolutionary spirit of that year, for when he returned home to Untersiebenbrunn he made the proposal to us sisters to make catcalls [*Katzenmusik*] whenever an old lady from the house opposite ours angered us with criticism, as she often did. Though readily agreed to, the plan was never carried out. As a final stroke, the monks of Seitenstetten advised us that Ernst should be trained at something where he could use his hands, since he had no head for studying.

During 1848, our father left the employment of the Stadion family and took the education of his own son in hand himself. Since he, as already mentioned, was not satisfied with the progress of his pupil, he had him at the age of fourteen take the entrance examination of the Gymnasium of the Piarist Religious Order in Kremsier in Moravia. The result was that they proposed he enter the seventh level of instruction, but father wouldn't hear of it, so Ernst entered the sixth class. He stayed two years in Kremsier in the house of Uncle Mauer, the father of Cousin Clementine.

At that time, the age for pupils had not been legally decided, so Ernst could graduate at the age of sixteen. One of his teachers, a dear old man, Herr Peter Wessely, once told me: "Ernst graduated brilliantly and already spoke like a professor then." Peter Wessely died in 1909 while in his late eighties. Ernst corresponded with him until his death and sent him many of his works. He also did the same with *one* of his Seitenstetten teachers. In Kremsier, a teacher who was on rather bad terms with my father [*auf dem Kriegsfuss*] for some reason unknown to me, gave Ernst a low mark. Ernst took the note, went to the teacher, and said: "Sir, I didn't deserve this grade! Please give me a better one!" The unheard of boldness of the pupil raised a lot of dust, and Uncle Mauer as he wrote had to make a big effort to calm things down. Ernst had merely done the thing because his father had been so sensitive about the man. Then Ernst went to the University of Vienna and became a Doctor of Philosophy at the age of twenty-one.

When he was about twelve years old, Ernst was an eager hunter, but an incident seriously spoiled the pleasure he took in it. Our whole family was in the courtyard of our house when at sunset the gunsmith brought my brother's fowling piece which had been repaired. "Is the gun loaded?" asked my father. "No, I just shot at a swallow." - At that time one had guns with percussion caps. I don't know what demon impelled a little mischievous brat like me to ask my brother to click down a percussion cap in a dark room so I could see the sparks. He cooperated in the customary way. We went into a dark room, I sat down full of expectation on a footstool and I called, when he came up against me: "Let me hold it!" He answered: "No! No! Stop!" turned sideways and - the full shot exploded, ripping a big hole in the floor. The richocheted gun flew by the head of my father, who had just entered the room. In later years if someone looked at the twisted boards in the room and remembered the incident, my mother would say that since that moment Ernst no longer touched guns.

He inherited his outstanding musical gift from his mother, who had been able to play and sing on the piano all the operatic arias which she had heard without having been taught anything else than to strum a little on the guitar. She also composed short musical pieces. I am persuaded too that he could have also accomplished something significant in this field. But he once said: "Is all that lung effort justified merely to blow something into the air?" I answered: "Blowing air `a la Marseillaise can yet galvanize many people." He thought that was an exception. - Nothing gave me greater pleasure as a child than to sit still and listen as he sat and fantasized on the piano as the light faded at sunset. His playing always lifted me into a poetic fairy world.

On Sunday, brother and sister helped bring music to the village church. He played the organ and she [Octavie] sang. Later, as Ernst was no longer at home, she became the organist. Our music teacher when we started was Anton Schedling, the old schoolmaster of the village, who customarily slept during the hour-long lesson, but who woke at once when a wrong note was struck or the conception of the work seemed poor. My sister and cousin occasionally would play jokes on the old fellow, particularly when our parents were absent. I myself received some well-deserved slaps on the hand. He is said to have been a very good violinist, and still possessed as he once told us: "A genuine Cremona fiddle [*Geige*] from Stadelmann in Vienna". He often left this valuable instrument at our house when he didn't need it somewhere else. Naturally brown, in the course of time it had become black. This seemed to bother our old house servant who painted it pure white, to the subsequent horror of its owner.

While I was still growing up, my brother kindly filled a thin notebook for me with theory of musical harmony. It must have been a masterpiece for I, who was a vicious little rascal, could play the church music I read effortlessly by the age of twelve. I can only vigorously regret losing the notebook, for when I wanted to study music theory some years later the textbooks appeared so frustrating and hard to understand that I had to give up, and yet, I certainly had a much greater desire to learn at that time than when I was only twelve. I was never more than a merely average piano player, but when attempting four-handed classical music with my brother during his school holidays, an event which always took place between seven and ten in the evening, he went at it with such effort and I responded so strongly that one had to believe that I played well. Ernst once told me that his own daughter Carolina, who was distinctly a good player, was also encouraged to do better when playing with a more able partner.

At the age of fifteen to sixteen, Ernst wrote sentimental poetry which may well have influenced some of his later lectures. Even in afteryears, he occasionally wrote witty doggerel [*launige Knittelverse*]. I recollect once replying to his poem with something of my own when he wanted to give me a long, brown Dachshund as a gift: "But were she even longer and browner, we must let her go or impound her."

In Untersiebenbrunn at the end of the 1850's, Ernst wrote a one-act play for his sisters and our cousin Manon Merkl which we staged during the summer holidays. The hero was an old, retired officer by the name of Kunz, an original type, who owned a large house. He also had a nephew, Hugo Ordnung, who managed his estate. Both the uncle and nephew vainly courted a beautiful widow who lived nearby. Jealousy rose between the two relatives, and the nephew had to find a position elsewhere.

The play was entitled "Wenzel's Revenge", and Ernst's plot revolved around an intrigue by Wenzel, the village cripple who felt himself impelled with the help of the cook of Master Kunz to drive his nephew out of his position. Herr Kunz was masterfully played by my sister, Octavie. The cripple was acted in an outstanding way by Manon, Ernst took the role of the nephew, whose voice and way of speaking he imitated magnificantly, and I played the guilty cook. Our parents constituted the entire but appreciative audience for us. My brother also read the play to a blind gentleman who owned a house not far away, who shed tears because he believed that he could actually hear the people, who, he all knew, speak. He thought everyone had been characterized as he really was.

When Manon and I were thirteen, we had decided to act as if we were grown up damsels. We found that it was beneath our true worth to greet men before they greeted us. We began preparing ourselves for our new role by separating our clothes into neat piles, lengthening our short dresses, studying ourselves carefully in the mirror, putting on a dignified face, and testing the waters by venturing into those parts of the village where we were most likely to meet gentlemen from Vienna. And in fact to our great jubilation, we did draw appreciation from two gentlemen who took off their hats to us. Indeed, they were prominent people!

While we were still wondering whether our long dresses, serious faces, commanding glance, or the whole ensemble had been responsible for the desired effect, Herr Kunz surfaced in the distance. "Shouldn't we greet the old man?" I asked her somewhat doubtfully. "Stay resolute!" replied Manon energetically. Therefore, we first walked up to Herr Kunz and then threw a disapproving glance at his hat. And he actually did raise his hand to the tip of his cap. - But then he changed his demeanor and spread his forefinger and middlefinger wide, winked at us, and said in a pleasant voice: "May God be with you, dear children!" We were completely annihilated! But the story had an epilogue which I will now reveal. I don't know how or in which way our adventure came to the ears of my brother, but he wrote in my album: "Covet not to be what thou art not. And be no brat if thou wouldst raise the flag of virgin innocence." Great disappointment and tears, but finally better understanding won out, and I have treasured the page as a dear memory from a happy childhood.

In the Autumn of 1862 social intercourse became more agreeable as we became acquainted with the family of Counselor [*Rates*] Holzapfel, who had bought the house of the blind gentleman that we have already talked about. Almost every evening we came together for a poetry contest Everyone attempted to write down the bizarrest, most impossible end-rhymes, which the unlucky person who had drawn the piece of paper had to incorporate into a complete stanza or poem. Some of these verses I have kept and will now present in so far as they were composed by my brother...[1]

Whoever lost the poetry contest had to to try to attract both groups of parents into a game of cards. The rest of the children stayed in the Holzapfel parlor and kept the contest going. My brother never tired of playing classical music with Mieze Holzapfel, the daughter of the house, while I played chess with a cousin of my friend Mieze, who was a student and like my brother also spent his Summers in Untersiebenbrunn. Afterwards he acquired the title of Professor Ludwig Schiffner.

My Mieze was a talented, mentally gifted girl. She played the piano well, painted, drew, and wrote poetry. Without actually being beautiful, she was an interesting, pretty girl with wonderful black eyes and magnificant black hair. Unfortunately, as a young girl she had fallen into the hands of an uncle of religious disposition who encouraged her tendency to push piety to an excess. She sank herself into a convent in order to save her soul and died when barely forty as a Lady of the Order of Sacre Coeur after she had left her aged mother to fight alone against economic difficulties.

My father was once requested to extoll on the name "Holzapfel". He made a charade of the matter ending with the words: "Everyone came for a look at what they wanted, having already become enchanted." This direct homage to Mieze pleased her very much.

Although Mieze was a year older than my brother, Mama Holzapfel nevertheless suspected that a warm relationship was already budding. One evening when we came without our parents, she took the freedom - we did not stand in the odor of sanctity - to help guide my brother to a more pious attitude toward Jesus Christ. As we were walking home, I said in the words of Goethe's Mephisto: "Herr Doctor will learn his catechism here." He replied half in humor and half in anger: "I will shortly have a pair of horns made so that you can become a perfect devil."

In conclusion, I would like to rob oblivion of some sayings of my brother which have stuck in my memory:

---

[1] First he does a mighty hum
And drinks his flagon dry,
But if his tune is hoarse to some
He will not further try.

Zuerst tut er gewaltig brummen
Und trinken aus dem Krug
Dem hört man ein Liedchen sammen
Ja, dann - hat er genug

A journal new is *Sharpen Your Beak!*
My sister true will edit there.
With nosey fork to ply and tweek
She'll puncture lies, so beware!

Ein neues Journal heisst scharfer Schnabe
Mein Schwesterchen ist Redakteur
Die steht wie eine grosse Gabel
Ohn Rücksicht scharf auf Frau und Herr

(There is also a third poem, but it is difficult to decipher.)

"Nationalistic infighting in favor of one's 'culture' at least has the advantage of enriching our understanding of Civilization as a whole."

Grillparzer says: "From Humanity through Nationality to Bestiality." My brother thinks: "From Nationality through Bestiality to Humanity."

It was during the era of Karl Lueger, the anti-Semitic mayor of Vienna (1897-1907), as we were entering the City Hall that Ernst said: "This building is the bastion of anti-Semitism; only the torture chamber is gone." To which he added: "Who bests the beasts of his time, lives for all time."

While leafing through a botanical book, I remarked that the flowers had been touched up to look prettier than they really were. He replied: "No paint can match the beauty of real flowers or real women."

"Every man, even the most deeply educated, is potentially a coquette, and every woman, even the finest, is potentially evil."

"Those educated with painful strictness often think it best to be lax with their children, who in turn often lament lack of strictness and impose it in turn on their children. Thus we oscillate in educational method and belief."

During his youth, my brother had a colleague who fell for a beautiful but dumb girl, only to discover his mistake when linked by marriage to the ignoramus, so Ernst philosophized: "What makes the gosling glad, turns the old goose sad."

When the financial dealings of an acquaintance became a matter of general discussion, my brother said with goodnatured scorn: "Man basically deserves money only because he cannot live without it."

"The only dignified female profession is to keep out of the limelight."

"No fiction is as poetic or beautiful as the truth."

D.

ERNST MACH

by Ernst Mach*

(1913)

1. Ernst Mach was born in Turas**, Moravia on February 18, 1838. He was the son of Johann Mach, who was then tutor with the family of Baron Breton, and Josefine née Lanhaus. He was a weak miserable [*elendes*] child who developed very slowly. An old lady who was the stepgrandmother of the mother of the child was not satisfied with how it was being fed and gave it so much extra food as to bring about a serious situation. As the matter grew worse and hope for its life was abandoned, its father, a fanatic for the Priesnitzean water cure, dipped the child into the coldest possible water, thereby allegedly saving him.[1] Luck in slipping between the consequences of two opposed forms of fanaticism is what I actually owe my life to.

I still have many lively visual memories from before the end of my second year. They are naturally mere fragments, but whose accuracy has been largely confirmed with respect to individual details by the testimony of my grandfather, grandmother, and the youngest sister of my mother. But later on these memories become somewhat fewer in number even though I was put into a new environment when my father bought a farm in Untersiebenbrunn in the Marchfield plain in Lower Austria in 1840. Perhaps the slow-

* This manuscript was written for Wilhelm Ostwald in 1913, most likely June of that year, but was rejected for publication, presumably because the autobiography told little about Mach's life after 1870 and even less about his controversies with Boltzmann, Planck, and others. Mach's unnecessarily provocative claim to have anticipated Ostwald in the matter of Energeticism (section #10 of this paper) may also have discouraged the latter from wanting to publish it in his series of great men of science. See also Mach's agreement that it should not be published in his letter to Ostwald, dated July 23, 1913.

Friedrich Herneck discovered the manuscript in the Wilhelm Ostwald Archive which is now part of the Akademie der Wissenschaften Archive in East Berlin. He published large excerpts in his article "Über eine unveröffentlichte Selbstbiographie Machs", *Wissenschaftliche Zeitschrift der Humboldt-Universität*, **6** (1956/1957), 209-220. The full German version was published by John Blackmore in "Three Autobiographical Manuscripts by Ernst Mach" in *Annals of Science* , **35** (1978), 401-418. - Ed.

** According to Dr. Josef Sajner of Brno, Mach was actually born in Chirlitz/Chrlice a village about two kilometers away. He was christened in Turas/Turany. Both villages are close to Brünn/Brno the capital of Moravia in Czechoslovakia. - Ed.

1 According to the testimony of my mother.

ness of my bodily development at that time was responsible for the poverty and dullness of what I can remember. But from the earlier period I can recall seeing many men with heavy fur coats walking back and forth, which was probably a picture from the waiting room in Lundenberg [which was on the way from Brünn to Untersiebenbrunn]. But even from that period nothing remarkable comes to mind.

But somewhat later an event of intellectual importance occured and has remained unforgettable. It must have been approximately during my fifth year while in the company of my nursemaid that I accidentally won insight into how a windmill works. We had come to give some mail to the miller. As soon as we entered, the millwheels started to move round. The horrible noise frightened me but couldn't keep me from noticing how the teeth dovetailed during the grinding of the grain, how one tooth after the other would push forward. This insight helped make my understanding more mature and in my opinion lifted youthful thinking from the level of believing in mere wonders to that of causal understanding.

2. I no longer appealed to background marvels to interpret what I didn't understand, but looked for the lever or string in the broken toy to which one could ascribe the cause. This experience and others also accompanied me together with Kant's investigations into the concept of cause and also taught me to understand better than I would have otherwise done Levy Brühl's volume *Les fonctions mentales dans les sociétés inférieures.*

I received my first elementary instruction, which corresponded roughly to *Volkschulklassen,* from about the age of seven to nine from my father. But I must confess that grammar and foreign languages did not appeal to me. My fantasy was fully captured by windmills and other machines, which I had naturally not fully understood at first glance. Later it was even a problem why the arms of the windmill in a sense turned around the axis, while the wind was directed at the axis. Only later was the effect made clear to me by the principle of the inclined plane. There was at that time still another wind-mill remaining nearby in the Marchfield, one which had been built by a rather ingenious miller. It had a horizontal wind wheel [*Windrad*] with flaps which allowed the wind to come through but only in one direction but which nevertheless opposed enough to constitute significant resistance.; therefore the wind wheel itself always turned in the same direction while the mill itself did not have to be turned to face the wind. This mill was much more understandable to me and I thought about it a lot. In this manner, machines and their parts filled my youthful mind.

My father also carried out simple experiments which demonstrated the material composition of the air by using a garden tank full of water and then plunging a flowerpot into it. But my interest in natural science was already so strong that I scarcely needed more encouragement. Only later did I discover that my father's experiments were only slightly modified versions of what had been undertaken by Philo of Byzantium long ago. While in this frame of mind, my father put me for the year 1847-48 into the lowest class of the Benedictine monastery school in Seitenstetten.

3. If the boy was insufficiently drilled in Latin on the one hand, on the other he remained so deeply involved in his own fantasy world that he could not develop the slightest taste for sentences such as "Initium sapientiae est timor domini" or for declinations and conjugations. The only interesting hour which he remembered was the geography lesson. What seemed hard and strenuous for most pupils, however, appeared easy and enjoyable to him.[2] The Fathers found the boy lacking in ability [*sehr talentlos*], but while letting the boy pass offered the advice that he was unsuited for study and should be encouraged to do handwork or learn a trade. Given the circumstances, it was a correct judgment, for the boy would never have become a good office clerk [*Paragraphenkenner*] or legal factotum like most of his colleagues.

My father was deeply unhappy over his son's lack of success, took him home, and began teaching him Latin, Greek, History, and Elements of Algebra and Geometry. Once again the pupil showed little eagerness or understanding for languages. If he didn't want to concentrate on a grammatical rule, the pupil would often hear the impatient cries "Brain of a Northman" or "Head of a Greenlander"! But when he began reading the Classical authors, the old languages took on a kindlier face. The pupil soon became relatively adept at translation and understanding what was written. Instruction at home offered the advantage that much more could be read than was possible in public schools. Because of the pupil's greater interest in mathematics and physics he was soon left to teach himself in these subjects.

Since the mornings sufficed for instruction, the private pupil [*Zögling*] was able to carry out various types of work in the fields during the afternoons, experience which created well-deserved respect for handworkers.[3]

4. Something else should be mentioned. Ernst Mach's youth after the suppression of the Revolution of 1848 fell into a very reactionary-clerical period. Therefore, the boy who had grown up in a liberal family asked his father to let him learn cabinetmaking so that one day he would be able to emigrate to America. This wish was granted. He devoted two days a week

---

2 Without having worked at geography ever again, I still have the forms of the continents in my head now and can carry out imaginary trips without having to look at a map.

3 The difference in esteem often clearly came to expression in conversations with colleagues.

for over two years to this training under a skilled master in a neighboring village. This period remained an agreeable memory for the young man, and some of the experience in working with wood came in handy in his later profession. He also recalled with pleasure the comfortable feeling when tired in the evening he would sit on the fragrant woodpile and imagine future machines, winged vehicles, and the like. He also learned to rightly value the positive contribution of work.

Mach's father was particularly well-versed in Latin, Greek, and history, and at the same time was an excellent storyteller. Although he had never enjoyed a fundamental education in natural science, his love of observing nature and his stories about Archimedes and other ancient scientists, which he had learned from Vitruvius, Plutarch, and others, were quite stirring. The dazzling flora and especially the insect fauna strongly aroused the children and promoted the comparison of biological forms.

Guests rarely came to Untersiebenbrunn[4], our spot in the Marchfield. But on Sunday afternoons in Summer, my father collected the young people of the village, boys and girls, in order to instruct them in fruit culture and the tending of fruit trees while letting them eat a few. Otherwise, young Mach was restricted to the company of his mother, father, and two younger sisters. The need for greater social contact in order to avoid boredom could not be readily satisfied.

After a free and happy childhood, Ernst Mach, on passing an entrance examination, was admitted to the sixth level of the Public Piarist Gymnasium at Kremsier in Moravia. Initially, the youngster was not able to make a good impression, since he had not yet learned the school habits and slyness expected by teachers and fellowstudents.

5. In general, the Piarists were not outstanding scholars; nevertheless, Mach would remember some of them with pleasure, respect, and gratitude. F.X. Wessely, the teacher of natural history, was an especially kindly man, who not only was enthusiastic about his subject but also knew how to arouse his students to share his feelings. He told us about Lamarck's theory of evolution and the Kant-Laplace cosmogony without troubling himself about their incompatibility with the Bible. Mach would remain in written contact with him until he reached a very advanced age and just a few years before his death.

H. Philipp, the physics teacher, was a magnificent instructor [*Didaktiker*] who knew how to raise the interest of his students for his subject to the highest pitch; unfortunately, he suffered from unexampled impatience, so that in spite of his ability he could not be a successful teacher.

---

4 That very place remains today a wonderful memory.

Fabian Mathias, who was teacher of Greek, was an enthusiast for Greek culture who only wanted to show its better side. Mach remained especially thankful to him for his willingness to indulge those among his students who were not gifted at learning languages, an indulgence which made the subject matter endurable. Mach with his bad memory and lack of interest in dynasties and military history never stood particularly well with J. Dostal, the teacher of history. Nevertheless, while he tried to cram the students with dry chronology from Ploetz's book on world history, he also read out loud by the hour from comprehensive original sources, which let no one of us who heard him gain the impression that the worldly and spiritual leaders of mankind who had been entrusted in their role by "Divine Right" solely had the interests of their subjects in mind. In the higher classes the students could also learn about the oddities which Max Kemmerich presented in his *Kulturkuriosa.*

Although gymnasial education at that time was carried on entirely within a clerical-reactionary political context, there were still men who had not let all trace of the liberal era of the Emperor Joseph II pass them by and who did their best to retain high standards. The only thing which seemed disagreeable were the eternal religious exercises, which had the opposite effect than intended.

6. The Austrian gymnasium which Mach attended was no longer just a Medieval, narrow-minded, one-sided Latin cram school. It had already been touched by the refreshing Exner-Bonitz reform movement.+ In fact, I would like to maintain that the Austrian gymnasium of that time was better than it is today.[5] Not every specialist had yet reformed the teaching plan of his favorite field in order to fill it with as much as possible; there was still no overburdening the student with a mountain of information like today.

+ William M. Johnston writes in his book, *The Austrian Mind,* Berkeley, 1972, p. 67 about this movement as follows:

> During the 1850's the educational system was overhauled by the Bohemian-born Count Leo Thun (1811-1888), an exponent of Bohemian Catholicism who commissioned two Herbartians, Franz Exner and Hermann Bonitz, to draft a reform based on German and French models. Stress fell no longer on religion or obedience to the state, but on training students to engage in research at the university. In 1850 gymnasium was lengthened from six to eight years, and instead of teaching all subjects, instructors now presented their specialty. Natural science was introduced, and the younger livelier teachers taught the upper grades, where according to Josef Breuer, 'criticism and contradiction were encouraged rather than suppressed'.

5 This is well-known outside of Vienna.

One really acquired many new ideas and had the time to think about and clarify them. Mach had the same kind of luck as he had enjoyed under his father's tutoring while living out in the countryside. If he now has so few complaints about the Kremsier school and yet was never better than an average student, the reason was that his future profession had already been clear before he entered it.

He must mention, however, that perhaps the most important educational stimulus did not come from school. While ferreting among his father's books, Kant's *Prolegomena zu einer jeden Metaphysik, die als Wissenschaft wird auftreten können* fell into his hands. The fifteen year-old youth eagerly gobbled down this clear and relatively uncomplicated work. It made a powerful impression upon him, destroyed the young man's naive realism, stimulated an interest in epistemology, and with the help of Kant the metaphysician annihilated all tendency toward metaphysics within himself.* In equal fashion, the book stimulated my thinking about scientific and psychological matters. I soon detached myself from Kant's critical idealism. I recognized even as a boy that the thing-in-itself was a useless, metaphysical fabrication [*Erfindung*], a superfluous metaphysical illusion. I soon turned to the Berkeleyan ideas which had been latent in Kant and gradually came to a critical empiricism. If I could not remain loyal to Kant's thoughts, I remained thankful to him for the manner in which he had stimulated me. Kant also helped make my historical-critical treatment of mechanics pos-

---

* If we distinguish between the question of the reality of atoms or its "metaphysical" status and the utility of the atomic theory, then it seems clear that while Mach's attempt to reduce reality to sensations, which is clear from the passage above, was not compatible with accepting the reality of atoms as spatially extended causal powers, it was compatible with accepting a version of the atomic theory and even some non-standard conceptions of the reality of atoms. From about 1860 to 1863, perhaps under pressure from the persuasions of some of his teachers and colleagues, Mach attempted to employ a version of the atomic theory in his work and even find a non-standard way to accept the reality of atoms.(See Ernst Mach, "Über die Molecularwirkung der Flüssigkeiten" *Sitzungsberichte der kaiserlichen Akademie der Wissenschaften,* Vienna, **46** (1862), 125-134; *Compendium der Physik für Mediciner,* Vienna, 1863, and "Vorträge über Psychophysik" in *Oesterreichische Zeitschrift für Heilkunde,* **9** (1863), 145-366.) But as soon as he found a teaching position in Graz in 1864 away from his professors and colleagues he abandoned both the atomic theory and the reality of three-dimensional atoms. Did philosophy defeat science? Did Mach alone among his colleagues find scientific reasons to reject physical atomism? Was his theory of "elements" a kind of atomism?

Some writers during the 1970's supposed that Mach rejected the reality of atoms, the utility of the atomic theory, or both on scientific grounds, but this was before the above autobiographical manuscript became widely available, where Mach spells out how *early* he adopted phenomenalism and its incompatibility with "metaphysical" entities (such as physical atoms). Furthermore, the atomic theory which had utility in physics did not seem compatible with psychology, and Mach, as he often mentioned, was looking for a philosophy compatible with both physics and psychology. - Ed.

sible. From the same spring came the *Wärmelehre,* the initial stages of the *Optik,* and so forth.**

7. At the age of seventeen, Mach took and passed the final examination [*Maturitätsprüfung*] at the Kremsier gymnasium and entered the University of Vienna as a student of mathematics and physics. The education minister, Count Leo Thun, had already begun to reform the university and particularly in philology and history had been able to attract outstanding teachers, but the faculties of natural science and especially mathematics had remained for the time being untouched by these reforms. So it happened, for example, that beginners, because there was no instruction in integral and differential calculus, had to take private lessons to fill this gap in order to be able to follow the physics lectures which presupposed knowledge of calculus. Among these lectures, I would like to mention those of Professor Andreas von Ettinghausen which did not deserve in any way the overbearing, unbecoming criticism of Justus von Liebig, who was not at all well-informed in the field himself. It is correct that von Ettinghausen was no creative genius in this area, one which nevertheless he had mastered to a remarkable extent.

At that time, France still had the leadership in the field; Germany very rarely led. Von Ettinghausen was Director of the Physical Institute, the first establishment of its kind in either Austria *or* Germany. The physical lectures of Professor Grailich were also stimulating, but he died soon. Josef Petzval was a highly gifted teacher of mathematics, but was lazy and hardly ever accessible. He is known as the reformer of photographic optics.

In the Physical Institute, Mach succeeded in setting up a self-constructed and self-run apparatus to prove the existence of an acoustical phenomenon [the Doppler effect] whose reality had been denied by many mathematicians [i.e. by Professor Petzval] at that time. Much related work soon followed.

In January 1860, Mach became a doctor of philosophy after undergoing a medieval procedure which consisted of three two-hour examinations on

** The last sentence from Mach's paragraph above is also of current interest. It was apparently written a few days before Mach's anti-Einstein preface to his posthumously published *Optik,* which carries the date July 1913. What might surprise some people is that Mach refers to the unpublished work at all. *That* he refers to it, suggests it was on his mind during that time and that he intended to publish it shortly, and *what* he says is not only consistent with the liklihood that Ernst Mach wrote the preface himself and not his son, (one scholar, Gereon Wolters, has developed an argument that the son forged it), but to the possibility it refers to what the preface calls "Part II" of the *Optik,* something which was never published and which that same scholar has fancied was merely a figment of Ludwig Mach's imagination. See Gerald Holton's article in this anthology which answers many of the claims and allegations put forward by Wolters. From the German above, the last line is: "Derselben Quelle entsprang die Wärmelehre, die Ansätze zur Optik usw." - Ed.

the most diverse subjects, but above all nothing about what he had specialized in. Mach became through this demonstration of comprehensive ignorance "*magister liberalium artium*".

8. The wish to study under Franz Neumann at Königsberg had to be given up from lack of means. I had to support myself by giving private instruction. Under these discouraging conditions Mach won his *Habilitation* and became a *Privatdozent* in physics, but without knowing how he would keep himself alive. He kept himself going by giving private lectures on mathematics, Fechner's psychophysics, and Helmholtz's theories on the sensations of sound. Naturally, a lot of time was lost in this manner which could have been better used by concentrating on his own studies. Nevertheless, Mach did gradually acquire a circle of outstanding people who later became famous. Through contact with Ernst Brücke and Carl Ludwig, who were both leading physiologists, Mach gained insight into the scientific work taking place in Germany and entered the field of sensory physiology, where he could carry out experiments without costly equipment, and which also led him to critical examinations of epistemology. The teaching itself led him to the opinion that the historical presentation of material was the simplest and the most understandable. Such general conceptual connections revealed the economic motives of cognitive theory and the conception of science as part of a general phenomenon of life and development, a view which finally rounded out his biological and economic epistemology.

During that period, my work was primarily devoted to the needs of physicians and physiologists. This led me toward the theory of the kymography of pulsewave indicators [*Pulswellenzeichner*] and equipment which would register such phenomena, which in turn led to study of the auditory organ. The Vienna Academy allocated 500 Gulden, in part as acknowledgment for past published work and in part as encouragement for work still to be carried out. I accepted the money but with some hesitation, since I did not then know whether I would discover anything important enough to justify the gift. In fact, it was almost a decade before I contributed something important to the field. The crucial insight first came to mind during an accidental observation while taking a train trip, a discovery which threw light on the function of the labyrinth, a question previously raised by Goltz. But only a short time later, Breuer in Vienna and Crum Brown in Edinburgh reached the same conclusion as I, but in very different, rather singular ways.

9. In 1864, I was named Professor of Mathematics at the University of Graz, which was then a somewhat neglected institution. I read elements of integral and differential calculus, analytic geometry, and tried by reading the work of Möbius, Plücker, and other books which I had brought with

me from Vienna to strengthen my still threadbare mathematical background. I was able to quickly recuperate from the miserable life of a *Privatdozent,* rapidly went to work, and no longer worried about making money or about the fact that there were better positions and salaries than 1050 Gulden per year. But already in 1867, I received a professorship in experimental physics from the University of Prague, a more promising field which offered me the chance of being able to carry out experiments on a more extensive basis.

My work was indeed truly many-sided, in fact splintered. I had thought out [*konzipiert*] my epistemology early, but it ripened late; also, the order of presenting my lectures ripened slowly; The best manner of carrying out experiments, which I first had to learn myself, fell into place relatively quickly through working in common with the students, above all by means of acoustic investigations, which because of similarities to work carried out during my physiological period [in Vienna and Graz] were more easily and readily learned. Many experiments conducted in this way were published under the auspices of the Vienna Academy, but most of them would appear under the names of my assistants. One of my first and most gifted students, whom I remember with fondest thoughts, was Prof. Dr. Vincenz Dvorak in Zagreb.

Among my scientific publications in book form, only *Optische-akustische Versuche*, Prague, 1872, and some years later *Grundlinien der Lehre von den Bewegungsempfindungen*, Leipzig, 1875, appeared at that time under my own name. My stroboscopic investigations with Dr. J. Kessel also took place during this period. He later became professor at Jena. In the Summer of 1870, I caught infectious bronchitis, which compelled me to suspend experimental activity for considerable time. I used this period to pursue my historical investigations further, which I had already begun somewhat earlier. Since then I have remained as loyal as possible to the historical way of proceeding [*Gange*] in giving lectures.

10. My reading of the works of Galilei, Huygens, and Newton along with brushing up the ideas I had developed in Vienna and Graz so strengthened me, that I could write a sketch *Die Geschichte und die Wurzel des Satzes von der Erhaltung der Arbeit* [which was later titled in English: *History and Root of the Principle of the Conservation of Energy*] in which I dared to put forward my thoughts about scientific epistemology, above all with reference to the fields of mechanics and thermodynamics. The basic ideas about economy of thought, the task of science, and conceptually establishing what is factual, are expressed in the above work with the greatest possible brevity. Since 1899, this work which has been kept unchanged since it appeared in 1872, has been published in English, Russian, and again in German. It also helped prepare the way for my *Mechanik* which came

out in 1883 and *Wärmelehre* [*Principles of the Theory of heat*] which first appeared in 1896.

The essentials of a more general Energeticism were also very briefly contained in the 1872 work, and can scarcely be distinguished from Wilhelm Ostwald's version at all, except with respect to his particular terminology. My psychological and philosophical reflections about epistemology first appeared in more extended form in my *Analyse der Empfindungen* [*Analysis of Sensations*] of 1886. I thought the book would have more to offer philosophers than scientists, but I deeply deceived myself.

After I recognized Kant's "thing-in-itself" as nonsense, I also began to regard the "unchangeable ego" as an illusion. I can scarcely relate how lucky I felt to free myself from all tormenting, absurd [*albernen*] ideas about every kind of individual or personal immortality and to find myself introduced into an understanding of Buddhism, a good fortune which rarely falls to the lot of any European. The ego or self [*das Ich*] is not unalterable, but changes very slowly during life, and after this slow development it vanishes with death. The *continuity* of the self is the basis for the possibility of the existence of significant people *and* philistines; human beings can also with fewer or greater modifications *regenerate* themselves in every descendant. This does not invalidate the correct thoughts of Buddhism.

11. In *Die Analyse der Empfindungen* and in the later, riper book *Erkenntnis und Irrtum* [*Knowledge and Error*] which came out in 1905 I once again emphasized that the entire internal life of people breaks down into elements and that everything human beings experience or pass through represents the dependence of *two* groups of these elements, external experience, that is, the physical or perceptual life, and internal experience, that is, the psychical as conceptual life. The first kind of existence is no arbitrary creation of our fantasy as I believe that I have already sufficiently said. Therefore, there was no need for some physicists [Boltzmann and Planck] to have to misunderstand me, still less for certain philosophers [Stumpf and Külpe] to have helped guide them to that conception.*

I have also spoken about a monism of physical and psychical events. There are not two different worlds, and that is not what the issue is about. I am only concerned with examining the kind of dependence between the two. I am also required by this monism, however, to clarify the unity of human experience [*des Lebens*] in light of the distinction between my ego and those of other people. Professor Wilhelm Jerusalem [a longtime friend] also admits that one can arrive in this way at monism, but has told me more than once that he thinks that my position cannot be maintained, because

* See John Blackmore, "Mach vs. Boltzmann, Planck, Stumpf and Külpe" in *Ernst Mach-His Life, Work, and Influence*, Berkeley, 1972, pp. 204-231. - Ed.

without distinction between the world and the self no experience of any kind is possible. On the contrary, I think that, if, as it seems, experience without a distinction between self and world was not originally given, then one is obliged to show how such a thing could have genetically developed, since even Jerusalem himself, who wishes that philosophy could have an empirical connection, flatly *postulates* the development. If one seriously wishes to reach agreement on this point, then the matter must be thought about with discretion; It is not something suited for a hasty answer.[6] I think of progress toward knowledge, whether physical or psychical, developing by means of fitting thoughts with facts and in the discovery of logical relations via the accomodation of thoughts with each other.

As one can see, my epistemology is as much as possible simple and homemade; It originated partly from biological and partly from economic stimulation. I have never considered it something complete. For example, the fourth edition of my book *Populäre Vorlesungen* [1910], which contains the most diverse topics, has raised some quite new matters for discussion.

12. Let me very briefly summarize the rest of my life: I married Fräulein Luise Marussig in 1867, whom I had gotten to know as an orphan in Graz, although it took a little courage to take the leap on a salary of 1300 Gulden. My family gradually expanded until it reached four sons and a daughter. The simple questions and remarks of small children belong to the most stimulating which I have experienced. At the age of four, my eldest son [Ludwig] asked the question: "How would the light get in, if every candle were put out and there was no more light in the room?" I have commented on these questions in detail in my *Analyse der Empfindungen* to the extent that they seemed instructive and advanced investigation into the motives for them.

Epistemologists could learn a lot through interaction with children whose native wisdom [*Altklugheit*] had been awakened prior to entering school. My little daughter [Carolina] remarked at the age of two when she visited the country for the first time: "We are in a blue ball!" "The world is a blue ball!" The carpet design on the room wall would seem small and up close to her when she brought the binocular pictures into congruence by squinting. During a snowfall in the absence of a wind, she believed that the house with her in it were moving upward. Nor do I need to tell the experts

---

[6] See Wilhelm Jerusalem, *Einleitung in die philosophie* , fifth and sixth editions, pp. 190-195. See also Ernst Mach, *Analyse der Empfindungen* , [Jena, 1911] sixth edition in the index under "Physisches", "Verhältnis zum Psychischen". In addition, see Ernst Mach, *Erkenntnis und Irrtum* , [Leipzig, 1906] second edition in the index under "Physisches und Psychisches", also p. 461 & 462.

squinting. During a snowfall in the absence of a wind, she believed that the house with her in it were moving upward. Nor do I need to tell the experts how instructive it can be to notice the first signs when children have begun to speak.

13. Much later [1894], one of my sons [Heinrich], who was considered a gifted chemist, killed himself in Göttingen only a short time after having superbly passed his examination for the doctor's degree. Mindful of my own youth, I had not wanted him to suffer from want. If he had had to labor hard for his own bread, he would never have found the time for such a thing.

During my sixtieth year in 1898, I suffered a stroke, and when I had incompletely recovered I discovered to my horror that neither one of my youngest sons [Felix and Victor] had any material-economic foundation for life. Before that, I had considered education to be everything. I now saw that a father has other duties. This meant that in spite of my crippled condition that I now had to keep working! Nor would it be possible to do this without my wife's self-sacrificing care.

CHAPTER 2

ERNST MACH IN PRAGUE

First Study:

ERNST MACH AND THE CONFLICT OF NATIONS

by Dieter Hoffmann*

When reviewing Ernst Mach's life and work one will soon realize what an important part Prague played in his biography. The near three decades that Mach spent in this city on the Moldau river were full of highlights and turning points that mark the 78 years of his lifespan. This important period in the scholar's life also allows us to draw conclusions about his personality as well as his scientific and social activities. It was here in Prague where Mach contributed so much to modern ballistics and instantaneous photography, perhaps his most important experimental work in physics which later made his name famous under the expressions "Mach 1", "Mach 2", etc.. It was also in Prague that Mach did much of his work in psychology and physiology, completing a series of papers on an unusual inhibitory phenomenon of sensory perception called "Mach Bands" as well as discovering a principal function of the inner ear and partly explaining vertigo or dizziness in human beings in a psycho-physical way. In addition to all that, he began to deal intensely with problems concerning the theory of cognition and the history of physics, reviewing them in a philosophical manner that was marked by an empirical approach, which helped make him one of the co-founders of the movement subsequently called "Empiriocriticism". The majority of Mach's publications - some 150 - were written in Prague. In addition to that, many of his manuscripts for books - such as *Beiträge zur Dopplerschen Theorie, Beiträge zur Analyse der Empfindungen,* his textbooks of physics, and his most popular book *Die Mechanik* - went from his Prague desk to the typesetter, gradually aroused interest, and largely foun-

* This article is a revised version of a chapter which originally appeared in Dr. Hoffmann's doctoral dissertation at the Humboldt University in Berlin in 1989: *Studien zu Leben und Werk von Ernst Mach (1838-1916).* See also: Dieter Hoffmann: "Ernst Mach in Prag" in D. Hoffmann and H. Laitko (eds.), *Ernst Mach - Studien und Dokumente zu Leben und Werk,* Deutscher Verlag der Wissenschaften: Berlin, 1991, pp. 141-179.

J. Blackmore (ed.), Ernst Mach – A Deeper Look, 29–55.

ded the author's international reputation.[1] However, it was not only that Ernst Mach did work in Prague which made him more widely known and set the stage for his much greater reputation in the years which followed, but for a brief period he also became - against his will - a political figure in the nationalist controversies of his time. Owing to the fact that he was rector of the University of Prague on the eve of its division into separate Czech and German institutions, what he said and did had political repercussions.

To start from the beginning, Mach had considered moving from Graz to the venerable metropolis on the banks of the Moldau even before his appointment. But his application some months earlier for a chair in General and Technical Physics at the Prague Royal Bohemian Polytechnic, the oldest technological college in Central Europe, was not successful and Adalbert von Walthofen, a physicist from Innsbruck, was appointed instead.[2] In retrospect, Mach's rejection has to be considered a lucky turn for him, his career, and even science in Prague, since, had he been appointed to the Polytechnic, he would have spent only very few years in that post with so little to challenge a highly talented physicist like him that he would have almost certainly accepted a position from a more prominent school, very likely far from Prague. The fact that he did not abandon the city and its university during almost three decades, rejecting calls from Graz (1876 & 1890), Jena (1882), and Munich (1885) shows that the new post provided him with relatively good working conditions and opportunities to develop.

Ernst Mach took up office in Spring 1867 and was later confirmed "in the professorship by highest resolution [of the Emperor Franz Joseph] of August 4...after a (successful) provisional period".[3] The young professor, standing in the prime of life, finally had almost as much opportunity as he wished to carry out laboratory research. In both Vienna as a *Privatdozent* (1861-1864) and Graz originally as a professor of mathematics (1864-1867) he had been troubled by "complete lack of means" for carrying out other than the simplest of experiments. He now made use of the opportunities that had opened up for him as the director of his own institute with enthusiasm. The fact that the institute was housed in a building of the historic university complex in the city of Prague in which conditions for research were anything but ideal did not seriously impair his attitude. Actually "the decayed, unsteady ground on which the building was situated" constituted a serious obstacle to certain experiments, as Mach would complain about in

1 See J. Thiele, "Ernst Mach-Bibliographie", *Centaurus,* **8** (1963), 189-237.

2 J. Haubelt, "K Machove zadosti o profesoru na przske polytechnice", *Dejiny ved a techniky,* **5** (1972), 52-55.

3 Central State Archive Prague (SUA Praha, Fond CM 1856/1883, Box 1007 (Ministry of Culture to the district government [*Statthalterei*], August 9, 1867)

his short, technical book *Optisch-akustische Versuche*, (1873).[4] Nevertheless, he might have found consolation from the fact that hardly any university in a German-speaking region, not to speak of Bohemia, could provide better conditions, since the establishment of well-stocked institutes custom-built for the needs of physicists had only just begun.[5] And at the stage of development in physics prevailing then, it was still possible for bright and creative minds using relatively modest experimental apparatus to hit upon important findings.

The first decade of Mach's activities in Prague were marked not only by numerous experimental investigations into optics, acoustics, additional work confirming the Doppler theory, and soot and explosion experiments which helped prepare him for his later work on shock waves, but also fundamental investigations into the physiology of sensory perception[6] and above all an eventually successful effort to improve unsatisfactory working conditions by building a new, modern institute of physics for the university. As early as the summer of 1873, a note of optimism appeared, however, since the Emperor Franz Joseph gave his agreement in principle for the construction of the institute.[7] Nevertheless, it took more than five additional years before the plans could be realized. After lengthy debates about the actual site of the institute, it was decided during the mid-1870's to have the new building constructed to the south-east of Prague's New Town. The designs for the new building were such that it was to house in addition to the institute of physics, institutes of mineralogy and geology, one for zoology, and a section for mathematical physics. The sum allocated for the project amounted to some 300, 000 florin (Gulden).[8] In the Winter of 1879, the institute could be moved to its new location on Vinicna, which might be translated as "Vineyard Lane".

The Institute of Physics occupied the left wing of the building on the ground floor, but Mach also took over the basement beneath "in order to prevent any disturbance of his physical research".[9] Recalling this extremely complicated period, he wrote more than thirty years later in a letter to his colleague Otto Wiener in Leipzig (who at that time had been busy building the new Leipzig institute of physics): "I feel very sorry for you that you

4 E. Mach, *Optisch-akustische Versuche,* Calve: Prag, 1873, p. 1.

5 D. Cahan, "The Institutional Revolution in German physics 1865-1914", *Historical Studies in the Physical Sciences,* **15** (1985)2, 1-66.

6 D. Hoffmann, O. Poess, & J. Chrapan, "Ernst Mach as an Experimenter", a talk given at the conference: "Ernst Mach and the Development of Physics", Prague, June 19, 1988, publication forthcoming.

7 SUA Praha, CM 1856/83, Box 990 (Ministry of Culture to the district government, September 4, 1873.

8 Ibid., Box 989 (*Kolladierungsprotokoll*).

9 Ibid., Box 990 (Information of the Board of Works, November 23, 1877)

have so little time for work, though I do understand that, since I remember building the new institute in Prague and all the loathsome aspects of administration very well."[10]

Thus, Ernst Mach at the beginning of the 1880's now directed a modern though not very big institute of physics, and right from the start he knew how to make good use of the newly attained opportunities for carrying out experiments, as it was in those years that he began the fundamental ballistic work which would lead to his discovery and description of shock waves, experiments which would finally require the establishment of a special laboratory within his institute.[11] In addition to improvement in working conditions, there was also one concerning his family as well which moved into comfortable and spacious living quarters in the newly built institute, which had been especially prepared for professors. On the other hand, the modest budget of 1500 florin per year for upkeep of the institute scarcely allowed for more than bare maintenance of existing equipment, ongoing research, and giving lectures.

With the onset of the winter term 1867/68, Ernst Mach began offering courses in experimental physics, very largely for teaching candidates in disciplines such as pharmacy as well as for chemists and potential medical doctors. This was partly a continuation of what he had done earlier in Vienna and Graz. Since Mach's initial salary in Prague, 1365 florin,[12] was only insignificantly higher than in Graz and because he now had several mouths to feed, he still had to think of possible avenues to acquire extra income. But the fact that most of his students did not have physics as their primary field of interest shows how Prague played such a comparatively minor role in the ensemble of research and teaching institutes in physics at that time. Gradually, however, the number of students who wanted to study under Mach rose, though it is not certain how long this trend continued after the division of the university during the 1880's. Nor was the rise in students during the 1870's and early 1880's only because of increased need for physicists and subsequent changes in the training of physicists, but not least because Mach was extremely stimulating both as a teacher and a scientist. He had a radiant personality with great authority as memoirs of his

---

[10] University library of Leipzig, Department of Handwriting [*Handschriftenabteilung*]: Estate of Otto Wiener (E. Mach to O. Wiener, Vienna, February 9, 1912.)

[11] See D. Hoffmann, O. Poess, & J. Chrapan, op. cit.

[12] During the next few years, Mach's salary rose as follows: 2200 (1874), 2400 (1875), 2600 (1879), and 3000 (1882) Gulden; this amount was an average income for a Prague professor. In Vienna (1895) he got 5000 Gulden. Its interesting that the increase in his income correlates significantly with major events in his academic life. In 1874 he was appointed "*Regierungsrat* ", in 1879 he became rector of the university, and in 1882 he had declined a favorable call to the University of Jena. The move to Vienna in 1895 represented a switch in discipline as well as city.

colleagues and students prove.[13] Thus, for example, Cenek Strouhal - who belonged to the first generation of Mach's students - reports that he was fascinated by Mach's clear, deep-going lectures, not least because at that time it had not yet been general practice to conduct experiments during lectures. In addition to that, there was opportunity to carry out further experiments after the lectures with an assistant. The total number of Mach's students, however, was not too big; in the twenty-seven years of his activity in Prague there were just half-a-dozen doctor's theses written. And neither do the number of *Habilitationen* defended under his auspices[14] as well as his assistants increase the number of his pupils very much.

Though Mach was able to contribute significantly to shaping the progress of physics during the past 100 years and gave a host of impulses to its development, he was not able to found a school of experimental procedure in the real sense of the word - not in Prague and even less so earlier in Vienna and Graz. His range of experimental interests seems to have been too wide to allow the development of a tradition of research or laboratory programs within the small group of his immediate students, though the work of Vincenz Dvorak at Zagreb, Gustav Jaumann at Brno, and Anton Lampa in Vienna and later in Prague should not be overlooked nor the many readers of his books who may have patterned their experimental procedure after what they believed was Mach's approach. In philosophy of science on the other hand, starting with Mach's call to Vienna in 1895, he would indeed be able to found what might be called a "school", rather informal before the First World War, but increasingly organized and influential afterwards under such labels as the "Ernst Mach-Verein", "Wiener Kreis", and to some extent "Logical Positivism".

Ernst Mach did not confine himself to his professorship. He was elected member of the Society for the History of Nature, called "Lotos", in the spring of 1870. This society having been founded as early as 1848 obviously regarded itself as the Bohemian counterpart of the famous "Society of German Scientists and Physicians" [*Gesellschaft deutscher Naturforscher und Ärzte*] and wanted to play at least originally the same role for the whole of Bohemia, i.e. for both Germans and Czechs.[15] Through regular meetings Lotos did its best to contribute to the development of science and

---

13 See C. Strouhal: "Dr. A. Seydler...", *Casopis pro pestrovani mathematicky a fiziky,* **21** (1982), 196; W. Kienzl, "Erinnerungen an Prag", *Deutsche Arbeit,* **6** (1907-08), 62; and V. Sebasta, "Strouhal jako fysik, ucitel a clovek", *Technicke Listy,* 1950, s.l.

14 See the lists in part two of this chapter as well as D. Hoffmann, "Ernst Mach in Prag" in D. Hoffmann & H. Laitko (eds.), *Ernst Mach - Studien und Dokumente zu Leben und Werk,* Deutscher Verlag der Wissenschaften: Berlin, 1991, pp. 141-179.

15 Lotos invited members of both nationalities until the late 1870's, when differences between Germans and Czechs in Prague became more acute. It subsequently changed its name to "*Deutscher naturwissenschaftlich-medicinischer Verein für Böhmen* ".

medicine. To this end, some eight or ten lectures were read per year. They were held in the evenings and on scientific subjects which it was hoped would be of general interest. These lectures were then published in the Lotos magazine together with papers and reports on the society's activities. Ernst Mach was a very active member of the society. From 1872 he was a member of the executive board and was even vice-president from 1872 until 1875.[16] More importantly, however, he presented over twenty papers during his 25 years of membership of which many included aspects of ongoing research.[17] In addition to that, he published some other notes and papers in the society's journal of which a few parts or revised versions appeared in his books *Beiträge zur Analyse der Empfindungen* (1886) and *Populär-wissenschaftliche Vorlesungen* (1896). When Mach left Prague in 1895 and resigned from the society, he was made an honorary member for life in appreciation of his many services.[18] But important as Lotos was, the society and journal, during his years in the Bohemian capital, as vehicles for the initial introduction of some of his ideas, one ought not to forget that the more widely-known journals *Annalen der Physik* (Leipzig) and *Sitzungsberichte der kaiserlichen Akademie der Wissenschaften. Mathematisch-naturwissenschaftliche Classe.* (Austrian Academy of Sciences, Vienna), remained his principal publishing outlets for research articles in physics.

The "Royal Bohemian Society of Sciences" [*Königlich-böhmische Gesellschaft der Wissenschaften*] was another respected body which invited Mach to become a member. He was elected a full member in 1871. Mach considered membership in this society of scholars with its grand tradition as a great honor. In a letter to the society's president he wrote in this connection: "Please convey my deepest-felt thanks for having elected me to membership of the Royal Bohemian Society and tell them that I promise[19] to spare no effort to show that I deserve this honor."

During the first years of his membership in the society Ernst Mach contributed to the organization's activities by passing opinions on papers.[20] He also gave a major lecture before the society "On the History and Root of the Conservation Law of Work" which while contributing to history and physics became his first important book on philosophy of science: *Ge-*

---

16 SUA Praha, PP (Vereine), Lotos.

17 See the list of *Lotus* lectures in the second half of this chapter or D. Hoffmann, "Ernst Mach in Prag", op. cit., pp. 170-171.

18 *Lotos,* **44** (1896), 100.

19 Central Archive of the Academy of Sciences, Prag (UA CSAV), Fond KCSN, Files of the members (E. Mach to J. Jirasek, Prague April 4, 1871).

20 Ibid., Box 31.

schichte und die Wurzel des Satzes von der Erhaltung der Arbeit (1872).[21] His commitment to the society, however, lessened rather quickly and apparently ended abruptly about twelve years later during the development of increasingly tense relations between the Czech and German-speaking communities in Prague. On April 7, 1884 he wrote to the society's president: "After having been prevented by business and other nuisances from partaking in any activity of the Society ever since 1872, I would rather no longer stand in the way of other, more active members, and so I take the liberty to most obediently declare my membership of the Society terminated with the expression of the greatest thankfulness to all the gentlemen who elected me." This letter coupled with Mach's resignation from the rectorship of the University at that time was viewed by many observers as a nationalistic statement meaning that Mach was breaking all offical contact with the Czech community, but as we shall see, Mach was actually thinking of very different things. He was anti-nationalist, and both resignations, from the Society and the University, were very possibly connected with Mach's desire to continue his scientific, historical, and philosophical work untroubled by "trivial" matters such as administration and adjudicating between German and Czech interests.

The president of the Royal Bohemian Society of Sciences, J. Jirecek, the Czech philologist and scientist of literature, perhaps suspecting Mach's real reasons, noted that the tone of Mach's letter was neither brusque nor provocative, and in his response he simply stated that Mach's decision "deeply distressed him, since the Society has thus lost not only an outstanding member, but the ranks have also been thinned as well of men who have striven for harmonious joint work in science regardless of differences of opinion in other fields, which has been one of the goals pursued by the Society from its very outset."[22]

When Mach first came to Prague in 1867 only very sensitive observers of the local situation could have detected the social and national conflicts to come. Mach and most of his colleagues at the university surely considered that science and concern with it was an oasis safe from such remote possibilities. But a close look at his first official action might have planted seeds in his mind. It was one of the social duties of a newly appointed professor to pay an introductory visit to the rector of the university, the medical doctor Joseph Halla. Mach reported about this call later: "The reception by the rector would have made an extremely strange impression on an unprejudiced person. Owing to his name, I thought he was Czech, and he thought

21 The talk was held in the society on November 15, 1871 and was published as a book by the "Calve'schen Buchhandlung" in Prague.

22 UA CSAV, Fond KCSN, File of E. Mach (E. Mach to J. Jirecek, Prague April 7, 1884).

the same of me for the same reason. So, since no one would like to start an argument right on the occasion of an introductory meeting, both sides began probing cautiously..."[23] This description hints at the atmosphere which Mach would later experience, one less difficult to avoid given the increasing load of administrative responsibilities dumped upon him. He did not like national disputes, although, especially in later years, he could be quite sensitive to social injustice and political efforts to rectify it.[24] He regarded nationalism as an outbreak of irrational emotion. Ernst Mach's daughter, Carolina, once cited a draft letter from her father:"I hold that the conception of nationalities proves deplorable narrow-mindedness and is a terrible relapse. It was under the influence of this plus the pretext of religion that the worst brutalities of the 17th century were committed."[25]

Ernst Mach, who had been born in Moravia and never lost social ties to the land of his childhood years, had always felt a measure of patriotism about his *Heimat,* one inhabited by people of two different nations, rather than to any special nationality as such.[26] On the occasion of the celebrations commemorating Mach's 100th birthday held in 1938, the Czech engineer Frantisek Nachtikal explicitly appreciated him as a "Moravian" and added, "If we try, by the way, to put ourselves in the situation as it was in Moravia only a hundred years ago, we shall soon find that youths at that time regarded themselves as Moravians, thus expressing their affection for their native country. Those people had an equally good command of both the German and Czech languages, and the question of nationalities did not play as important a role for them as it does for us now."[27]

In other words, Mach's relationship to the people around him differed fundamentally from that of some colleagues who often were "celebrities of cultural life imported from Germany ... who miss no opportunity to provoke and sometimes even offend Bohemians."[28] His territorial patriotism which respected ethnic differences while not denying one's own nationality harks back to the thoughts of Bernard Bolzano, perhaps the best-known German-Bohemian philosopher, who had been very influential at the University of Prague during the 1820's, the period when Mach's father studied

---

23 K.D. Heller, *Ernst Mach,* Wien & New York, 1964, p. 17.

24 See Dieter Hoffmann, "Sozial- und Gesellschaftspolitisches bei Ernst Mach", *NTM,* **26.2** (1989), 61, and G. Wolters, *Mach I, Mach II, Einstein und die Relativitätstheorie,* Berlin, 1987, pp. 120-129.

25 C. Lederer [Mach's grandson] to F. Herneck, Essex Falls, March 2, 1959 (Private information from Professor Herneck, Berlin).

26 See J. Sajner, "Ernst Machs Beziehungen zu seinem Heimatort Chirlitz (Chrlice) und zu Mähren", *Bohemia,* **24.2** (1982), 358-368.

27 F. Nachtikal, "Arnost Mach", *Elektrotechnicky Obzor,* **27** (Feb. 18, 1938)7, 97.

28 *Politik,* Nr. 135, May 16, 1880, p. 2.

there while receiving such excellent marks in his studies.[29] But Bohemian industrialization during the 1860's and 1870's plus the ideological orientation of the Czech and German middle classes was soon accompanied by the growth of Pan-Slavism on the one hand along with greater German nationalism on the other, which soon spawned a number of romantic illusions. Thus, Mach found it increasingly difficult to ignore such "irrational" pressures.

Having been voted Dean of the Philosophical Faculty in 1872/73, it was thought that he was prepared for higher academic office, and was accordingly elected Rector of the university by his Prague colleagues in the early summer of 1878. In these elections he won the overwhelming majority of 13 out of 15 votes of the college of electors.[30] He thus held one of the most important political posts which a scientist could reasonably hope to obtain in the Dual-Monarchy. The rectorate usually lasted one year and was traditionally rotated between the four faculties at a festive ceremony at the beginning of the winter term in October of every new year. But when unfinished business loomed it was traditional for the rector to be appointed to a further period as "prorector" (an interesting verbal throwback to the Roman practice of making "consuls into "proconsuls".)

As rector, Mach was obliged to run the university both internally and externally, had to observe representative functions, was on several boards, committees, and governing bodies, and he was also a deputy to the Bohemian provincial parliament. Taking into account his one year as rector and an additional year as prorector, Mach was in a responsible political position from 1879 until 1881, a period which saw drastic changes in the organization of the university and much success in the Czech campaign to revive and extend their own cultural heritage, success manifest to the world in the music of Smetana and Dvorak and in their efforts to build and maintain a Czech opera house. But it was also a time when the conflict of nations reached its apex in Prague.

The summer of 1879 saw a change of government in Vienna; the Auersperg cabinet was out and Count von Taaffe's approach of reconciliation with the Slavic minorities, especially various Czech national groupings was in.[31] This change brought the question of equal rights in the University of Prague to the fore, a university which was considered to be the oldest in both Czech and German-speaking Europe and which both nationalities claimed as basically their own. Indeed, John Hus, the Czech religious reformer and political hero had been a professor there, and from there had begun his influential activities. German control via Habsburg power, how-

29 See K. Berka, "Bernard Bolzano" in *Bernard Bolzano 1781-1848 - Studien und Quellen,* Berlin, 1981, P. 27.

31 See W.A. Jenks, *Austria under the Iron Ring 1879/93,* Charlottesville, 1965.

ever, had been dominant since the Battle of White Mountain in 1620, (which had helped kick off the Thirty Years War) though using Latin as the main language of instruction had helped defuse tensions until the introduction of German following the "reforms" of Joseph II at the end of the 18th century.

By the 1860's, Czech moderates were already asking for an "utraquistic" solution of the problem of the university. "Utraquism", an old expression meaning "under both kinds", had earlier referred to an attempt by moderate Hussites to compromise with the Catholic Church "by drinking from both cups", but by the 19th century meant attempts to satisfy both Germans and Czechs in Bohemia through a reasonable compromise which would guarantee the basic rights of both groups.

At the start of Mach's first rectorate, the Czech Reichsrat Club in the parliament in Vienna revived these utraquistic proposals only to provoke a countermemorandum on the part of German professors which in December 1879 was also presented to the Education Ministry in Vienna . Holding fast to traditional privileges, they strictly rejected any utraquisation of the Prague University.[32] They feared that at an integrated university the Czech party might by degrees acquire superiority both in the numerical strength of students and teachers and in the qualitative terms represented by Czech culture and linguistic "Czechization". Such a prospect had to be opposed by all means. One such means might be to limit Czech influence by creating a special Czech university while leaving the "historic" one to the German-speaking minority in Bohemia.[33] Needless to say, most Czechs scorned that suggestion, alleging that the "historic" university had more Czech than German roots. How was Ernst Mach, uninterested and perhaps unequipped as he was in Bohemian history and nationalist controversy, to help solve the problem? As early as December 1879, he had suggested that the university be split into German and Czech halves, but this private recommendation was initially ignored. It was a compromise, and few people seemed ready to compromise.

Perhaps largely by accident, Mach was able to win a measure of backing from most of the German-speaking population in Prague, support which would enable him to make compromise seem more acceptable. In

32 For more details, see D. Hoffmann, "Ernst Mach und die Teilung der Prager Universität", *Jahrbuch für die Geschichte der sozialistischen Länder,* Berlin, 1991. [Should this work in fact fail to appear, presumably for political reasons, then the reader is requested to check Dr. Hoffmann's Humboldt University doctoral dissertation mentioned at the beginning of this article in an asterisked footnote which contains a version of the same article, pp. 79-123. - Ed.]

33 "Promemoria der deutschen Professoren und Dozenten der Philosophischen und Medizinischen Fakultät, Dezember 1879" in J. Goll, *Rozdeleni Prazske University,* Praha, 1908, pp. 83-84.

May 1880, the German *Burschenschaft* "Carolina", a student drinking, duelling, and nationalist society, celebrated the 20th anniversary of its founding. Numerous delegations of students and teaching staff as well as the Rector Magnificus accepted invitations to join in the merry-making. It was the rector's task to present the opening speech in which he tried to weaken the nationalist tenor and strike a clearly objective note.[34] Compared to this damper, the speech of the dean of the medical faculty, E. Klebs, was militant and provocative, alleging that "Just as the Carolina *Burschenschaft,* so the big Carolina - the University itself - will always remain German. The opponents of German spirit and ardent patriotism ("*Deutschtum* ") do not want a university of their own which has been proposed to them because they are unable to maintain it."[35] The nationalistic boasting of E. Klebs and his numerous political friends among the students tended to drown out E. Mach's moderate words so that "Mach's special note was soon forgotten in the flood of nationalistic rhetoric."[36] The Czech response was violent with demonstrations and rock throwing, which even broke several windows in Mach's own apartment. Mach reported these rows to the authorities in Vienna[37] and the local Prague newspapers commented on them extensively.[38]

The Czech newspapers which criticized the Carolina extravaganza as the instigating cause of the riots did not seem to clearly distinguish between the moderate position of Mach and the extreme one of Klebs,[39] but tended to treat Mach as rector of the university as the main German leader in the city, giving him an undeserved but subsequently useful reputation. His "defence" of the German position won admiring recognition in numerous newspapers in Austria and Germany as well as criticism from Czech publications, and years later this would be what most non-academics would appear to remember most vividly about him. Some of the labels applied to him were: "nationalist",[40] "opponent of Czechifying the university",[41] and even "ideologist of the German *Burschenschaften*".[42] At first, Mach was

34 *Bohemia,* Nr. 136, May 16, 1880, p. 6.

35 O. Hackel, "Die Geschichte der Burschenschaft 'Carolina'", *Deutsche Arbeit,* **9** (1909-10), 493.

36 J.T. Blackmore, *Ernst Mach - His Life, Work, and Influence,* Berkeley, 1972, p. 74.

37 SUA Praha, PM 1871/80, Box 1160 (Decree of the minister, May 18, 1880), and Box 1046 (several police reports).

38 See the issues of *Bohemia* and *Prager Tagblatt* for the German position and *Politik* for the Czech during the latter part of May 1880.

39 *Politik,* Nr. 135, May 16, 1880, p. 2.

40 K.D. Heller, op. cit., p. 19.

41 *Meyers Grosses Konservations-Lexikon,* Leipzig, 1906, Bd. 13, p. 20.

42 A satirical poem ("*Der importierte Professor* ") about the events in Prague in SUA Praha, PM 1871/80, Box 1046.

shocked by these reports and did his best to deny them, even publishing a leaflet in which he criticized the style of some Prague newspapers declaring that "politics concerned him primarily only when it touched on issues of teaching, especially questions concerning the university". He also warned young students about "deriving their judgments from the easily accessible but muddy manure of the party press."[43] But it no longer seemed to matter what he actually thought or said; he had become a symbol of German resistance to Czech demands. It is not certain if he was ever fully conscious of the fact, but the strong support behind him from the German community within Prague and even in Vienna and abroad, placed him in an unusually strong bargaining position should he follow his natural inclinations and seek a compromise or utraquist solution. On the other hand, his basic lack of interest in local politics naturally tended to make him for many months more a conduit for the thoughts of more concerned people. But this attitude also had its social advantages for him among fellow German-speaking colleagues.

The curious context where the unknowing among Germans and Czechs considered him a nationalist and the knowing among both bodies, his close friends, colleagues, and students, realized that he was not, permitted him to remain on good terms with many Czech colleagues and leaders. Another advantage was that as a student at the Kremsier Gymnasium during the 1850's he had studied the Czech language and presumably could converse in it,[44] something many of his German-speaking colleagues neither could nor apparently wanted to be able to do. Not only was he an honorary member of the Society of Czech Mathematicians and Physicists since 1870, though it had already assumed nationalist features, but he was the widely recognized and honored teacher of the first generation of Czech physicists (C. Strouhal, F. Kolacek, A. Seydler, and others.).[45] Indeed, the development of an independent Czech physics owed much to his efforts.[46]

When Mach handed over the honor and burden of the rectorship to his successor in October 1880, he dropped out of the public limelight but his new position as prorector gave him more opportunity to make a carefully-thought-through proposal to satisfy both Germans and Czechs. Since both parties claimed the "historic" university and neither would accept a new "unhistoric" school, then the only reasonable solution as he had privately suggested many months earlier would apparently be to model one's ap-

---

[43] SUA Praha, PP 1876/81, Box 761 (report of the printing office to the police, June 13, 1880).

[44] See Okresnis Archiv Kromeriz v Holesov and J. Haubelt. op. cit., p. 53.

[45] See V. Novak, *Vzpominky a pameti,* Brno, 1939, pp. 79-93 and J. Folia & E. Tesinska, "Cenek Strouhal a ceska fyzika jeho doby", *Dejiny ved a technicky,* **12** (1979), 1-8.

[46] See C. Strouhal, *Akustika,* Praha, 1902, p. 147 & L. Novy et al, *Dejiny exaktnich ved a ceskych zemich,* Praha, 1961, p. 284.

proach on that of Solomon, by literally cutting the "historic" university in half with walled passageways and separate entrances. In this way, both groups could wallow in tradition.

Ernst Mach gave his keynote report at a meeting of the Academic Senate in January 1881.[47] This document later became the basis for discussion between the university and the education ministry in Vienna.[48] In this document, Ernst Mach argued strongly and decisively that if the creation of a Czech university were unavoidable, then this should be done through division of the existing university. This proposal, while not satisfying extreme Czech nationalists, was much more conciliatory than that of the then Governor of Bohemia, Philipp Freiherr Weber von Ebenhof. The latter even formulated a counter-memorandum to the university's recommendation in which he explicitly used arguments about political power and rejected any concessions towards the Czechs as contrary to the survival of the Habsburg state. He not only wanted to preserve Prague University as German but thought that its German character should even be strengthened.[49]

Two months later in March 1881, the University and the Austrian Government began a series of meetings with Mach as an important participant and his plan as the main order of business. The result was that in the following month the Emperor made a "supreme decision" which declared that in Prague "there should be two universities, one German and one Bohemian-speaking".[50]

Though the decision to add a new university by splitting the old one did not fully please either the Czechs or the Germans in Prague, shortly after it was implemented the new, or if one will, renewed Czech institution developed very quickly into a national and scientific center of good reputation, and soon, within a few years, became the second largest university in the Empire. As early as the 1890's some 18% of all students in the Austrian half of the Dual Monarchy were enrolled in it compared with only 10% (about 1500 students) in the German University of Prague.[51]

Was the split a longterm success? Yes, to the extent that Bohemia remained a region with two major nationalities and two major languages (not to mention the large Jewish minority which tended to speak German or Yiddish) and as long as honorable leaders sought reconciliation.This is proven by the fact that even after the breakup of the Habsburg Empire and the

---

47 *Prager Tagblatt,* Nr. 10, January 10, 1881, p. 1.

48 *Petition des akademischen Senats der k.k. Karl-Ferdinands-Universität an das hohe Herrenhaus,* Prag, 1881.

49 Note des Statthalters an das Cultusministerium. In J. Havranek, *Ceska univerzita*...pp. 63-68.

50 Quote from: J. Vanacek, "V Prze budou ... dve univerzity...", *Historia Universitas Carolinae Pragensis",* **22** (1982)1, 13.

51 E. Pliwa, *Österreichische Universitäten 1863/64 - 1902/1903,* Wien, 1908.

establishment of the Republic of Czechoslovakia in 1918 both universities were allowed to continue their existence. Indeed, Mach's compromise lasted almost sixty years, that is, until Hitler marched into Bohemia in Spring 1939 and virtually all Germans were expelled from Czechoslovakia in 1945 with the end of the Second World War.

But the division during the early 1880's which resulted in shrunken enrollment in the German University also had considerable economic consequences for a number of German-speaking professors, since it resulted in reduced salaries, presumably even for Mach who wrote in 1910: "[I] tried to compensate for the loss through technical activity, and in fact managed to earn as much as [my] annual salary within some weeks, but at the same time recognized that it was impossible to reconcile this kind of life-style with scientific work".[52] Thus, E. Mach's successful excursion into lucrative jobs remained a mere episode.

But Mach's retirement from academic politics - something he much longed for - while it allowed him to publish his most famous book, *Die Mechanik in ihrer Entwicklung historisch-kritisch dargestellt* (1883), was of rather short duration. As early as the summer of 1883, he was called upon by his colleagues to become rector once again, but this time of course just of the German university.[53] Given the rotation of the rectorship between the four academic faculties (law, medicine, theology, and philosophy), this meant that his fellow professors were re-electing him at the first opportunity. As a scientist of rising reputation whose work on shock waves was about to enter a critical stage of development and whose book on mechanics would soon be considered a classic, the university clearly wanted to spotlight its best scientists and scholars. In fact, the number of those who bore the burden of the rectorship twice could be counted on one hand.[54] (One might add that Ewald Hering, the outstanding physiologist, was made prorector.)

Ernst Mach, who was strongly anticlerical in his beliefs, was quickly confronted with another Czech-German conflict, namely, what to do about the still undivided faculty of theology. The Catholic Primate of Bohemia, Cardinal Friedrich Fürst von Schwarzenberg (1808-1885), a member of the family which next to the Habsburgs themselves was probably the most influential in the Empire, had insisted on the ground of the supranational function of the Catholic Church that the theology faculty remain undivided

---

52 J.T. Blackmore, "Three Autobiographical Manuscripts by Ernst Mach", *Annals of Science*, **35** (1978), 410.

53 AUC, AS NU, Box 2 (academic elections).

54 M. Svatos, "Seznam rektora a dekinu byvale Nemecke univerzity v Praze", *Zpravy Archiva Univerzity Karlovy*, Praha, 1984.

and part of the German university.[55] Since most of the theology students and many members of the teaching staff were Czech and they naturally wanted to become part of the Czech university, tension perhaps unavoidably rose. Mach as rector was responsible for the religious oaths which theology students took at the beginning of each school year.

But in the autumn of 1883, shortly after Mach began his second rectorship, the Czech students refused to fill in the matriculation forms because they were written in German while also refusing to take an oath given to them by a German rector. Mach responded by suggesting to the education ministry in Vienna that the forms be written in the appropriately Catholic and "neutral" language of Latin, but he rejected the minister's proposal that the dean of the theology faculty accept student oaths by proxy, since this would weaken the traditional powers of the rectorate.[56] But given the uncompromising attitude of the Czechs, Germans, and Catholic Cardinal and in spite of Mach's objections, the education minister insisted during the week before Christmas that the matriculation ceremony be carried out *not* directly by Mach but under the authority of the dean of the theology faculty acting for him. Ernst Mach resigned two weeks later on the ground of poor health: "Due to my indisposition I shall not be able to manage the duties of rector[57] plus simultaneously fulfilling the tasks of my professorship".

Compared to the controversies that preceded and accompanied the division of the university, the conflict around the matriculation of the students of theology seems a bagatelle and not all that important, certainly much less than enough to justify resignation of the rectorship. But the issue was much larger than it appeared and there are levels underneath levels which have to be understood to comprehend the matter, not least being the question of Mach's underlying motivation.

The general public, both Czech and German, largely ignored Mach's "bad health" justification, and considered the resignations of Ernst Mach and the prorector Ewald Hering, which also occurred, as primarily demonstrative acts representing a German-nationalist attitude. This appeared to be supported by subsequent public events. After a delegation of leading professors of the university had expressed to Mach "deepest regrets of the academic senate about his resignation together with thanks for his manly stance in defence of the rights of the university", the German students decided not to fall short of their professors.[58] They organized a procession of

55 See K.A. Huber, "Die Prager Theologische Fakultäten von 1883/91 bis 1945", F. Seibt (Hrsg.), *Die Teilung der Prager Universität...* pp. 37-54.

56 AUS, AS NU, Box 1883.

57 AUC, AS NU, Box 2 (Letter of E. Mach, Prague, January 11, 1884).

58 *Bohemia*, Nr. 14, January 14, 1884, p. 4.

coaches to Mach's house on January 18, 1884. But the planned torchlight march that evening was forbidden by the Prague police.[59] Such behavior in support of "the nationalist" Ernst Mach naturally aroused strong protests in Czech newspapers. The pro-Czech but German language *Politik* headlined "rector strike" on its front page adding:

> The latent war, the skirmishing behind the stage props...has turned into open fighting in the Mach case. This is an act of haughty obstinacy, provocative rebellion. The minister of education is being pressured by the terror of a persistent rector strike and by German-nationalist demonstrations at the Prague university to revoke a decision that is totally in accordance with law and to do homage to the national chauvinism of the German professors... Thus the plan of war against the Ministry of Education is clear: The rectors and professors argue, protest, and resign, the German national press stirs up the music of war, and the students and their corps riot in the streets. The real thrust of the demonstration is that our local German university is to become a pure-blooded German National University, the Professors are to become German Nationalist politicians, and the students are to be misused as battering-rams to budge and break the Minister of Education.[60]

Pro-German newspapers replied in kind. *Bohemia* chirped in gleeful appreciation that Mach had become "very popular with both professors and students".[61] A lengthy leading article poured it on: "[What a shame] to spoil the joy of the young Czech university by painfully reminding them of the continued existence of the venerable German one. To be sure, not every Czech publication confesses this agony as openly as *Narodny Listy* or becomes as shaken up as soon as they see a German professor... Men such as Ernst Mach who are not moved by personal interest and are above mere provincial matters represent not pedantic disputatiousness but are guided by conscientious conviction that it is their responsibility not to yield to any massive system of mean attacks."[62]

After his resignation, Mach was able to concentrate more on his work as a professor, scientist, and scholar while showing his loyalty to his increasingly beleaguered colleagues by turning down good appointments in Munich (1885) and Graz (1890).[63] Concerning the theology faculty, after the death of Cardinal Schwarzenberg, it too was eventually divided.

---

59 SUA Praha, PM 1881/1900, Box 1677 (Letter of the Prague police chief, Prague, January 16, 1884.)

60 *Politik,* Nr. 15, January 17, 1884, p. 1.

61 *Bohemia,* Nr. 14, January 14, 1884, p. 4.

62 Ibid., Nr. 15, January 15, 1884. p. 1.

63 J. Thiele, "Briefe H.v. Seeligers an E. Mach", *Philosophia Naturalis,* **17** (1979)3, 391-399.

But while few people believed that Mach had resigned for reasons of bad health, there is also evidence that Mach's primary motive may have had little to do with what his "nationalist supporters" imagined either. During the last days of December 1883 when he was locked in controversy with the minister of education over whether the dean of the theology faculty or himself should administer oaths to theology students, he wrote a letter to his friend Josef Popper-Lynkeus:

> The rectorate embarrasses me a great deal. I cannot reflect in peace and must always defend myself from becoming occupied with mere trifles. I deeply want to write the *Analysis of Sensations*.[64]

In short, while Mach had threatened to resign as early as December 10 in a letter to the education minister if not allowed to handle the oath ceremonies such that his actual resignation was merely the carrying out of a promise, the above passage from the letter suggests that his deepest motivation for resigning was to get back to scholarly activity. Nor should one overlook that his laboratory work on ballistics had progressed far enough that articles would appear on it in 1884, the year of his resignation. Many things were on his mind.

But there is an even deeper consideration. In rejecting Mach's request to handle the oaths made by theology students, the minister of education had responded even earlier on November 7, that is, over a month before Mach's threat to resign and well before the letter to Popper-Lynkeus: "Theology students should not have to be in the position of taking sacred oaths from a non-Catholic rector."[65] Mach was not only anti-clerical but as a philosopher who reduced all reality to sensations he was clearly an atheist as well, which would become even more evident when his book on the analysis of sensations was finally published (1886). Mach's threat to resign if the dean administered the oaths, which the dean did, may have been intended by Mach as giving him the opportunity he wanted to get back to research and writing his book. On the other hand, the longer Mach stayed in office, the greater the possibility that both German and Czech theologians would be tempted to criticize their "atheist boss" and create a scandal.

But if there are simply too many possible reasons to be exactly certain why Mach wanted to resign, one can at least say that resignation probably killed a lot of birds with one stone: first, he would be fulfilling a promise; second, it would keep most of his colleagues and students happy; third, it demonstrated his strong support of academic freedom and the rights of the

---

[64] See John T. Blackmore, *Ernst Mach - His Life, Work, and Influence*, Berkeley, 1972, p. 79. For the original German see J. Blackmore & K. Hentschel (Hrsg.), *Ernst Mach als Aussenseiter*, Wien, 1985, P. 7.

[65] Ibid., Blackmore, p. 78.

rectorship; fourth, if his health was really weak it ought to benefit from the change; fifth, it would give him more time to work on shock waves and to finish his book on the analysis of sensations, and sixth, it would take the air out of potential scandal.

From 1884 until 1895, when he accepted a chair in the history and philosophy of the inductive sciences in Vienna, Mach continued both his experimental work and his publications in the history and philosophy of science. In addition, he authored several textbooks in physics and natural science and became co-editor of an educational journal. The call to Vienna was not only due to his scientific merits, but the press once again alleged that "Mach was considered to be a pillar of '*Deutschtum*' in Prague".[66] But even though such characteristics were entered into some contemporary reference books[67] and have continued traditionally to the present time,[68] it is clear from Mach's actual accomplishments, such as his compromise in helping to establish "historic" universities for both Czechs and Germans, that he was a moderate on national issues in the best sense of that word. While he was generally considered to belong to the "United Front" (*Einheitsfront*) of Germans in Prague, it seems beyond question that he actually held liberal views, and did his best to mediate in the delicate conflict of nations in the city of Prague, that is, in so far as the spirit of the age and his social allegiances allowed.

---

66 *Fremdenblatt,* May 11, 1895.

67 *Meyers Grosses Konservations-Lexikon,* Leipzig, 1906, Bd. 13, p. 20.

68 See the position of the late chief of the Prague university archive, F. Kavka (quoted in K.D. Heller, *Ernst Mach,* Wien, 1964, p. 19), or the very polemical standpoint of Czech writer N. Ivanov, that "E. Mach was an ideologist of the German *Burschenschaften* ". (quoted in: N. Ivanov, *Pozar narodniho divadlo aneb prilis mnoho nahod,* Praha, 1895).

Second Study

PRAGUE DATA ON ERNST MACH (1867 to 1895)

by Dieter Hoffmann & Ivo Tretera*

A. MACH'S *LOTOS* LECTURES[1]

1868	Some New Acoustical Research; New Research on Some Principles of Mathematics
1871	On the Beginnings of Mechanics; A New Acoustical Application of Spectral Apparatus
1872	On the Doublerefraction of Amorphous Substances
1873	On the Diffraction of Sound and the Interference of Light
1874	Some Experiments on Statics and the Perception of Motion
1875	On Some New Optical Research on Anomalous Dispersion
1876	On the Doublerefraction of Moved Liquids
1/27/77	On Explosion Waves
4/6/78	Some Stroboscopic Experiments
2/17/80	On the Theory of Radiometers and a Connected Experiment on the Arrangement of Radiometers
1/15/81	Demonstration and Explanation of Some Electrical Experiments
4/29/82	On Steady Electric Current
2/3/83	Experiment and Lecture on Electricity
3/3/83	On the Lightning Conductor System of Melsen
5/3/84	Experiment and Lecture on Optics
12/ 6/84	Experiments on Explosions and Connected Phenomena
10/30/86	Demonstrating the Speed of Sound Waves in an Apparatus
2/26/87	Demonstration of Polarization in Two Physical Apparatuses
10/15/87	Comments on Some Optical Experiments
10/13/88	Ballistic-Photographic Experiments Carried out in Meppen
12/13/90	Demonstration Lecture from the Field of Mechanics
1/23/92	Optical Experiments
11/4/93	On Optical Experiments

* These selections are taken from Dieter Hoffmann, *Studien zu Leben und Werk von Ernst Mach (1838-1916)*, Doctoral Dissertation at Humboldt-Universität zu Berlin, April 1989, pp. 64-71 and from Ivo Tretera, *E. Mach na Prazske Universite/E. Mach at the University of Prague*, (International Conference - Ernst Mach in the Development of Physics), Prague, 1988, pp. 5-12 & 25-30.

1 Hoffmann, ibid. pp. 70-71. See also *Lotos*, **17** (1868) - **44** (1896)

B. ERNST MACH'S ASSISTANTS IN PRAGUE[2]

Anton Karl Grünwald	Assistant	1865? 1871
Cenek (Vincenz) Dvorak	Assistant	1871 - 1875
Wenzel Rosicky (Rositzky)	Second Assistant	1874 - 1875
	Assistant	1875 - 1878
Jaromir Wosycka (Wosyka)	Second Assistant	1875 - 1876
Andreas Haberditzl	Second Assistant	1877 - 1878
	Assistant	1878 - 1880
Ottokar Tumlirz (Tumlir)	Second Assistant	1878 - 1880
	Assistant	1880 - 1890
Georg Dick	Second Assistant	1880 - 1883
Otto Biermann	Second Assistant	1883 - 1885
Gustav Jaumann	Second Assistant	1885 - 1890
	Assistant	1890 - 1893
Joseph Wanka	Second Assistant	1890 - 1893
Cölestin Krupka	Second Assistant	1894 - 1895
Joseph Ritter von Geitler-Armingen	Assistant	1893 - 1895

C. MACH'S SUCCESSFUL DOCTORAL STUDENTS

1. 1867-1877 (Prior to Surviving Official Records)

Cenek Dvorak	Gustav Gruss
Miromil Neumann	August Seydler
Jan Sommer	Jaroslav Simonides

2. 1877-1895 (Mentioned in Surviving Official Records)[3]

a) Received Doctor's Degree with Mach as First Advisor ("Erstgutachter"):

Franz Kolacek	On the Anomalous Dispersion of Light (Mach and Lippich, June 30, 1877)
Wenzel Rosicky	Four Physical Treatises Published in the

[2] Ibid., p. 66

[3] Ibid., pp. 64-65. See also Tretera, op. cit., pp. 25-26. The available original source was *AUC Praha. Rigorosa Filosofia - Protokoll über die Akte zur Erlangung der Doctorswürde an der philosophischen Facultät der Universität Prag. 24. III. 1877 - 18. XIII. 1913. Proto-kolarni zapisy z drivejsich let nejsou k dispozici.*

	Proceedings ("Sitzungsberichte") of the Academy of Science in Vienna (Mach and Lippich, July 18, 1877)
Ottokar Tumlirz	On the Propagation of Thin Airwaves Having a Finite Period of Oscillation with Special Attention Given to Graphic Representation (Mach and Lippich, June 23, 1879)
Stefan Doubrava	Examinations into Both of the [?] Electrical Conditions (Mach and Lippich, December 22, 1881)
Georg Fischer	Explanations of Spectroanalytic Appearances On the Chronic Aberrations of Telescopes (Mach and Weinek, March 18, 1885)
Hans Luggin	1. A Simple Method of Comparing Magnetic Fields 2. Experiments and Remarks on Galvanic Arcs (Mach and Lippich, April 16, 1888)
Anton Schlosser	Movement of a Fixed Circular Cylinder in an incompressable Liquid (Mach and Lippich, March 17, 1890)
Gustav Jaumann	Influence of Quick Deviations of Potential in the Process of Discharge and Three Other Published Treatises (Mach and Lippich, March 26, 1890)
Josef Wanka	On the Phenomenon of Condensation (Mach and Lippich, 1893/94)

b) Received Doctor's Degree with Mach as Second Advisor ("Zweitgutachter"):

Bohuslav Brauner	On the Transformation of Isobutyljodid in Trimethylcarbinol (Linnemann and Mach, July 26, 1880)

Heinrich Goldschmidt	On the Hypochlorite Saltpetre Acid (*Unterchloralpetersäure*) of Guy-Lussac (Linnemann and Mach, February 3, 1881)
Robert Leipen	On Some Compounds of Allylbignanids. On the Lactic Acid - Ester of Athelids (*Milchsäure-Athelidenester*) (Maly and Mach, December 19, 1887)

D. MACH's *HABILITANTEN* [4]

Doctors of Philosophy in Physics Who Received their *Habilitation* at the University of Prague (unpaid teaching tenure in that particular university) with Mach as a principal advisor:

Miromil Neumann (1871)	Cenek Dvorak (1874)
Otakar Hostinsky (1877)	Gustav Jaumann (1892)

E. CORRESPONDENCE AND NOTES

The following letters[5] are about some of Mach's students who became assistants, doctors, or *Habilitanten*. They range from the very friendly and rhapsodic (to Dvorak) to generally favorable (about Hostinsky), to rather unfriendly (about Doubrava) to notes left in his *Nachlass* which were severely critical (about Wosycka) to hostile (about Jaumann). But letters or notes written in particular moods may not fully reflect one's best judgment. For example, Mach had a high opinion of Gustav Jaumann's abilities and research and would recommend him for a professorship many years later in Prague (1910), which Jaumann for unclear reasons rejected.

---

[4] Hoffmann, op. cit., p. 65.

[5] For Mach's letters to Hostinsky and to Seydler (about Doubrava) see Tretera, op. cit., pp. 27-33. The letter and complete poem to Dvorak can be found in the physics library in the University of Zagreb. The notes about Wosycka and Jaumann are in the Ernst Mach Institute, Eckerstr. 4, Freiburg im Breisgau.

1. To Cenek Dvorak, Grosspriesen, August 23, 1873:

Dear Sir Doctor and Assistant:	Lieber Herr Doctor und Assistent!
To you out there so distant	Da Sie dort an der Welt End
In Dusajov town by your own choice	In Dusejov nach eig'nen Ermessen
Where books & pen have lost their voice.	Das Lesen and Schreiben gern vergessen
My letter I think tis best for you	So will ich nur zu Ihrem Besten
Though apt remark may rile or rue.	Mit einem Briefe Sie beläst'gen,
I would but ask how is your health	Und Sie fragen nach Ihrem Befinden,
And do advise but not in stealth	Und Sie bitten, sich nicht viel zu schinden
The fields of grain to partly shirk	Mit Feldarbeiten aller Art,
So one may better physics work.	Dass für die Physik noch was erspart
Thus men of mind to save their strength	Wird von Ihren besten Kräften,
Must keep their farming lust at length.	Nicht alles bleibt den Landgeschäften.
I'll dare disclose my aim some more	Ferner melde ich Ihnen weiter,
To be the boldest fighter for	Dass ich als ein tapferer Streiter,
Your promise as a mighty paladin,	Für Ihren Ruhm den hochgeschätzten,
A man of physics to glory in.	Die Correctur von Ihrer letzten
I read your treatise and marked it all	Abhandlung bestens ausgeführt,
With neater note than J.M.'s scrawl.	Statt Jochmann Math'sen hingeschmiert,
I saw in the work you did for me	Und an der Abhandlung Spitze
How a doctor's cap fits perfectly.	Notirt Ihre Doctorsmütze.[6]

. . . .

2. Ernst Mach about Otakar Hostinsky, Prague, May 16, 1877:

The written essay "Theory of Musical Sounds as Foundation of a Theory of Aesthetic Harmony" which was submitted by Dr. Hostinsky for his *Habilitation* consists of two parts. The first and larger part is a manuscript ready for publication; the second part is a summary, but which is detailed enough to give a clear representation of the standpoint and inclinations of the author.

This work stands entirely upon the foundations created by Helmholtz and v. Oettingen and seeks to widen and strengthen that foundation, partly by building on it an aesthetic theory of harmony. The writing therefore handles partly known and partly new questions. Concerning the known ones one can clearly see that the author has fully adopted the clarification and stimulation for which we owe so much to Helmholtz, and in the musical direction which is still more important, much has sprung from Oettingen. Concerning his handling of new questions, the author has showed great independence. He very rightly puts his finger on the weak points of Helmholtz and Oettingen, and in the opinion of this referee has sought the

[6] The poem continues with a second paragraph which contains thirty-two lines on diverse topics. Professor Dvorak also kept another poem by Mach of sixteen lines.

right means to overcome the difficulties. That the author is generally stronger and more fortunate at criticism than building something new, should not be held against him given the difficult nature of the problem.

This general judgment can easily be confirmed by going into details. Section #1 Pitch and Volume, #2 Timbre, and #3 Beat and the Combined Sounds are handled in physical terms. The author shows that overall he is well-oriented, and even concerning less popular questions has been well-instructed and knows the relevant physical literature. Sections #4 and #5 are physiological in character. In #6 the hypothetical "subjective undertones" are discussed and the author rightly, to his credit, leaves this provisionally unsolvable question open. In #7 the author in discussing sounds in series, shows the right scientific attitude. He assumes the same standpoint on this question as modern physiologists have had to assume.

Sections 9, 10, and 11 are about theory of harmony and consonance. The author rightly recognizes that because of the lack of beats or oscillations that consonance cannot be sufficiently defined, which means that the theory of Helmholtz could be merely negative. But the theory of Oettingen, which according to the detailed analysis of the author is much closer to that of Helmholtz than either of those scientists has been willing to admit, seems insufficient to him, which one must also agree with. But the way out proposed by Hostinsky himself also does not permit him to reach the desired goal and will probably be abandoned by him again.

But at the same time, Hostinsky's conception of dissonance is original and fruitful. While for v. Oettingen dissonance is sound with two or more ambiguous strains ("Zwei oder mehr deutiger Klang"), for Hostinsky it is a combination of *indirectly* linked sounds, and the hearer is the one who solves, resolves, or dissolves. On the basis of this conception one is left with, as Hostinsky shows, a series of rules for guiding the voice which are more logical and can be derived in a more consistent way than has previously been done.

If this work well articulates the fully scientific grasp of the author, still, here and there, one notices an unmistakable tendency to accomodate problems which belong to inductive science within the framework of a finished philosophical system. But this is the only thing which strikes this referee as unwelcome. Because of his love of truth, the author will no doubt accept this criticism and know how to find the right path.

3. Ernst Mach to A. Seydler*, Prague, January 12, 1882:

You would do [Dr. Stefan] Doubrova a service by helping him to put his head on straight. D. is remarkably gifted in terms of imagination. But he is utterly lacking in critical ability, and he is so sanguine and proud of every suddenly conceived idea that this makes him completely unable to *learn*. His character is not scientific.

For two trying years, during which daily he stubbornly spit his unchewed mouthfuls at *my* head, I gradually climbed beyond and freed myself from his hypothetical rubbish. Scarcely having reached a better standpoint through *my labor*, he once again considered himself to stand high above me and believed himself to have understood the fundamental nature of things, which it happens I had occasionally inquired into, but before I wrote anything down.

I have gone through two works with him in great detail; *more* I cannot endure, for I really cannot devote all of my time to D. If ever left alone, he would swish the air like a literary hussar.

That because of his naiveté and unclear understanding and because he is a student of mine he believes that he has to publicize my thoughts, or thinks that my ideas are his own, is not so bad. Much worse is that all the world will believe that I support what he has written and that the rest of my students who have like tendencies and who are willing to listen to him will be drawn after him.

Under such circumstances nothing remains left for me to do than in milder form and without naming him to circulate the basis of my views to the academy. I have withdrawn a very *clear* note, which was sent to Wiedemann [*Annalen der Physik*] during the first stages of my anger. I don't want to ruin anyone, but if possible cure them, since I myself am largely to blame. I would only request that you help clarify the situation in your circles and try to help the cure take effect. Lots of luck!

I apologize for not having returned your visit yet. Please accept that nothing else has been behind it but sluggishness and lack of time.

4. A note in Mach's *Nachlass* dated 1886:

In the year 1876, my assistant J. Wosyka was discharged at once from the institute for stealing books. I replaced them, but have fallen because of this behavior of Wosyka's, who was indebted to me for many things, into an emotional depression which has made me for years incapable of work.

---

* For more details about this letter please see Ivo Tretera, "Casus Stepan Doubrava (K neznamym dopism Ernesta Machs Augustu Seydleroviz roku 1882)", *Acta Universitis Carolinae* [Prague University], **XXV** (1985), 89-94.

First, after several years, I was so far composed that I could persuade myself that equipment was also missing. If I should die before everything has been replaced this will serve for explanation. I myself took care to lock up the Institute every evening. [But] naturally this does not suffice against a thief in the house, who in a clever way steals never used and therefore seldom contested things. These proceedings have robbed me of my joy in life, [but] one should not condemn him too hard.

5. Note dated 1894:

I have given respect to Wosyka. When he arrived again at a position after many years I held myself obliged to tell his superior of his past behavior. Also, Wosyka is aware that one knows and watches him. He is said to have been upright until now, and I don't know anything else detrimental about him. Since I have taken care of all of the damage which could be traced back to this source there is no longer any basis on which to prosecute him. But the facts shall not remain unknown.

6. Note also dated 1894:

I wish in the case of my death, that the Institute *not* be given to Jaumann, at least not immediately. I wish, that beforehand, *someone else* become acquainted with it. Jaumann has for the last ten years, so to say, worked as a guest, that is, for himself and for that drawn a salary. During the first few years I requested very little from him as an assistant, and then nothing at all. On the contrary, I was continually having to defend myself against his tendency toward extravagant behavior and disorder. The assistant's work: Preparing for the lecture, putting back the apparatus, ordering cost-free new equipment, organizing and administering material, and handling related correspondence has all been done during the last four years by my son Ludwig, solely out of concern to help me. He has with great effort by using the means available to Jaumann brought everything into tiptop order.

Probably because Jaumann thinks of this order as a reflection on himself, he has found it necessary for years to insult my son and to slander him in the most abject way.

How much I have tried to avoid being prejudiced or even seeming to be so, and how much I have preserved an objective judgment with respect to Jaumann's talent is known.

I may therefore allow myself to say, that I consider Jaumann to be capable of anything in the heat of the moment. One should keep this in mind when he is offered a position where conscientiousness toward everyone is required, toward people in front of him as well as those out of sight. How irresponsible Jaumann's behavior has been can best be judged by those who know how hard I have tried to promote his welfare.

F. FIFTY-SIX SEMESTERS[7]

Mach's Lectures at the University of Prague
from 1867 to 1895 as Listed by General Topic

Topic	Number of Semesters	Last Semester Given
Experimental Physics	51	SS 1895
Practical Exercises	36	SS 1895
Optics	27	SS 1895
Independent Research	26	WS 1894/95
Physics for the Public	14	SS 1895
Physics for Prospective Teachers	14	SS 1883
Electricity	7	WS 1883/84
Physics for Prospective Pharmacists	7	WS 1875/76
Heat Theory (Thermodynamics)	6	WS 1891/92
Physics for Prospective Doctors	6	WS 1887/88
Main Questions of Physics	6	WS 1881/82
Latest Developments in Physics	6	WS 1876/77
History of Physics	4	WS 1875/76
Theoretical Supplement to Experimental Physics	3	SS 1880
Mathematical Physics	2	WS 1875/76
Survey of Physics	1	WS 1869/70
Mechanics	1	WS 1870/71

---

[7] See *Ordnung der Vorlesungen an der k.k.Universitat zu Prag*, Prag, 1867 ff. See also Tretera (1988), op. cit., p. 5-12 for a semester by semester listing of courses.

CHAPTER 3

ERNST MACH AS A HISTORIAN OF SCIENCE AS SEEN IN BOOK REVIEWS AND NOTES (1896 - 1932)

A

ON MACH'S *PRINCIPLES OF THE THEORY OF HEAT*

E. Mach (1896)*

....Although a great many sources have been used, one should not look for the results of *archival investigation* in this book; The focus is much more on the *connection* and *growth of thought* than on interesting curiosities. Also, one will seldom find *biographical* details. Where the discussion is about people, they will only be viewed as *intellectual* and *ethical individualities,* and I think that the historical picture which will thereby emerge can only be the better for it....

J.E Trevor (1897)+

The object of Mach's book seems to be to present clearly the actual ideas which constitute the present theory of heat, to exhibit the mutual relations of its parts, and to show how it has come about that this theory has developed as it has. The undertaking is accordingly entirely similar to that of the author in his justly famous treatise on mechanics, - an exposition of fruitful ideas and their growth, and not a codification of archives nor a storehouse of interesting curiosities. This promising programme has been ably carried out, as must have been expected from the higher order of Mach's previous contributions to the philosophy of science. In the book are presented successively the topics of thermometry, heat-conduction, calorimetry, and thermodynamics. The treatment of these things is everywhere clear and well-balanced, and the development of the theoretical views

* Ernst Mach, "Vorwort", *Die Principien der Wärmelehre - historisch-kritisch entwickelt,* J.A. Barth: Leipzig, 1896, p. vii. The English version of this book is: *Principles of the Theory of Heat,* Ed. Brian McGuinness, trans. T.J. McCormack, with an introduction by Martin J. Klein (*Vienna Circle Collection,* vol. 17), Reidel/Kluwer: Dordrecht and Boston, 1986.

\+ This book review appeared in *Journal of Physical Chemistry,* **1** (1897), 430-431.

J. Blackmore (ed.), Ernst Mach – A Deeper Look, 57–70.

which come into question is in every case stated in a thoroughly original and appreciative way. The volume is concluded by over a dozen essays and sketches of philosophical import together constituting something like a psychological study of the methods of physical science. These articles are immensely interesting and important: special attention should here be directed to the one on the relation between physical and chemical processes, where it is shown that possible "chemical potentials" must be discontinuous, in contrast with the continuous potentials of physics, and the ground of contrast is sought in the definite mass-relations of the former case.

Mach's treatment of the early history of the theory, say of the period preceding Gibbs, arouses only admiration, but the remainder of the book has by far neither the same completeness nor the same finish. Horstmann's practical application of thermodynamic method to chemical action is neglected. Massieu's characteristic functions, Helmholtz's theory of free energy and its applications, v. Oettingen's antithetic developments and the magnificent work of Gibbs are little more than cited. This is indefensible, especially since the author brings his treatment so far down into the present as to touch upon the recent discussion of "energetics".++ Then again, the assemblage of both new and reprinted philosophical sketches at the close, instructive as they are, is very disconnected; the book as a whole is neither a collection of scientific papers nor a well-rounded critical treatise on its subject, - we get the impression that a splendid work, partially finished, has been dumped on the market in company with the materials for its completion.

A lack of care in revision is rather prominent; it surprises the younger of us to read (page 300) that Rankine has "recently" published a paper which appeared in 1853, and we note a consistent fault in the pretty uniform phonetic (from the German point of view) spelling of English words occurring in the text; examples are "commicatet", "ewer", and "electricyty". Other printer's errors are present also, and further evidence of haste in publication is the lack of subject and author indices, which could have greatly enhanced the value of the book. The dress is the usual satisfactory output of the publishers, but the accompanying portraits of Dalton, Black, Carnot, Joule, Clausius, and Kelvin are abominable and had better have been omitted.

These objections relate, however, more to form and finish than to actual content; the writer is impelled to recommend the book heartily, - It is an important book by an unusually able man, and no one interested in the phi-

++ It is possible that Mach decided to rush his book on heat theory into print before it was ready in an effort to counteract Boltzmann's apparent "victory" for atomism over Ostwald's "energeticism" in their famous debate at Lübeck in 1895. See John Blackmore, *Ernst Mach - His Life, Work, and Influence*, University of California Press: Berkeley, 1972, pp. 204-206.

losophy of science, whatever his special field may be, can afford to pass it by.

B.

ON MACH'S *SCIENCE OF MECHANICS*

E. Lampe (1901)*

Since the first edition of this book was reviewed by me in 1883, two further editions have appeared (1888 & 1897), and its reputation has risen so much that it is now considered a classic throughout the whole world. With satisfaction, the deserving author therefore says in the preface about the fourth edition: "The basic text from which later insertions clearly stand out can remain standing. I also do not wish, should another edition be necessary after my death, for the basic text to be changed even then." Concerning new insertions, 44 pages have been added to the existing third edition, an increase which is twice as much as from the second to the third edition. New historical research, new inquiry into the foundation of mechanics, but also the inclusion of critical objections against the author's presentation have caused the enlargement. Particularly significant is the review of Hertz's *Prinzipien der Mechanik* and the sections on the different conceptions of the author's thoughts which have been added to the first two chapters. Chapter two from page 269 to 292 has been increased by two additional sections.

For a new edition, one might recommend that some thought be given to the following details and peripheral matters: Abandoning the old spelling of "Huyghens" and "Leibnitz" to "Huygens" and "Leibniz", changing Sir William Thomsen into Lord Kelvin, shifting the date 1866 on page 195 to 1666, and respelling "Brachystochrone" as "Brachhistochrone".

In a work of as much weight as this, which is used by many as an authoritative source, it is definitely not advisable to abbreviate by changing the citation of titles. The treatise by Gauss on Capillarity has the title: *Principia generalia theoriae figurae* (not *de figura*) *fluidorum in statu aequilibrii* and was submitted in 1829 and not in 1828. - The principle of least force [*kleinsten Zwanges*] was published in an article by Gauss: " *Ueber ein neues allgemeines Grundgesetz der Mechanik*" but not as "*Neues Princip der Mechanik*" (p. 546). The treatise "*Intensitas vis magneticae*, etc." dates

* This review of *Die Mechanik in ihrer Entwicklung historisch-kritisch dargestellt,* Vierte verbesserte und vermehrte Auflage, F.A. Brockhaus: Leipzig, 1901, appeared in *Naturwissenschaftliche Rundschau*, **16** (1901), 488-489.

der Mechanik" (p. 546). The treatise "*Intensitas vis magneticae,* etc." dates from 1832 and not from 1833. The first volume of the complete edition of Gauss's works appeared in fact in 1863, but the fifth [volume] which includes mathematical physics, came out in 1867. - Euler's "*Methodus inveniendi lineas curvas*" was published in Lausanne in 1744 (not in 1741). H. Hertz's dates are missing (1857-1894), though other authors which are listed have them.

It is precisely the high respect which this reviewer holds for this book and its author which has caused him to expresses his wishes, which only concern peripheral details and do not touch basic content.

E. Lampe (1905)+

The new edition of this classical work which follows its predecessor by only three years has been augmented by eleven new pages but in a full-page manner which leaves the rest of the text typographically unchanged. The first addition (pp. 157-158) is devoted to a general consideration of the complete accomplishments of Galileo; the second concerns itself similarly with Isaac Newton (pp. 212-214). A good likeness of Newton has also been added at this point. The most significant interpolation (pp. 257-263), is concerned with the question of the relativity of motion; There is a short overview here of the most important writings which have appeared since the first edition of this work. A last added page (pp. 405 to 406) takes a position on the first beginnings of an energetic mechanics. Because the additions take up exactly one or more pages, the rest of the text has been presented in the same way as in the fourth edition.

Since the most worthy author has shown this reviewer the honor of taking his minor requests into account which had been made in my review of the fourth edition, I will allow myself to put forward a few more. The field of mechanics has been described in several volumes of the *Enzklopädie der mathematischen Wissenschaften* published by B.G. Teubner. The principles of mechanics have been magnificantly represented by Voss in Munich. Because positions on the most important questions have been enumerated here and a rich collection of citations from the literature have been added, I think that the article and the whole undertaking deserve mention. Perhaps one might also add the newer and newest writings of Hölder, C. Neumann, H. Poincaré, and other men.

+ Ernst Mach, *Die Mechanik in ihrer Entwicklung historisch-kritisch dargestellt,* Fünfte verbesserte und vermehrte Auflage, F.A. Brockhaus: Leipzig, 1904. The review appeared in *Naturwissenschaftliche Rundschau*, **20** (1905), p. 76.

The reviewer had already made another not-too-important criticism of the fourth edition, but it was not printed. On page 201 of the present edition it is mentioned that August Föppl in 1897 had put forward and used the hypothesis of negative mass. The same insight struck A. Schuster soon after, and without possessing knowledge of Föppl's thoughts he called it "potential matter - A Holiday Dream" (*Nature*, **58**, pages 367 & 518). In the second place, Schuster confessed that meanwhile Föppl's treatise had been brought to his attention concerning the hypothesis of "anti-matter", but that already before Föppl, Karl Pearson had employed a similar notion in 1891 in the work "Ether Squirts" (*American Journal of Mathematics*, volume 13, pp. 309-362).

In conclusion, this outstanding book, which has long been a dear old friend for all physicists and mathematicians, can be heartily recommended in its latest version for all lovers of deep analysis into nature.

W. Ostwald (1905)*

Since this work first appeared in 1883 as a volume of the *Internationalen Wissenschaftlichen Bibliothek*, which has meanwhile ceased operation, the appreciation of Mach's thoughts, which were in his representation from the beginning, have completed an extraordinarily strong turn. At that time the mechanistic world view or scientific materialism was unconditionally dominant, and neither room nor disposition was at hand for this critical-objective representation which treated mechanics as merely one field of physics among many others. Thus the second edition came in 1888 and the third only in 1897; but a fourth one was required after only four years and the fifth after only three. Meanwhile, as a sign of the times, mechanics has been forced from its position of absolute ruler, and nowadays the question has been discussed about how far it has to be conceived and represented as a branch of electrodynamics.

The content has meanwhile long become classic, which means that as a work acknowledged to be one of the most worthwhile components of world literature, nothing needs to be said further, since it may be assumed that every reader already possesses a copy of the book, even if only in an older edition. But an especially comforting pleasure should be mentioned by this reporter in this connection. In the preface to the fourth edition the honored author speaks of the wish that all possible future editions printed after his death should be done so in unchanged form.

* *Zeitschrift für physikalische Chemie*, **52** (1905), p. 251.

C.D. Broad (1916)+

This very useful little volume consists of two parts. The first one contains Mach's additions and alterations for the seventh German edition of the *Science of Mechanics*; the second contains a number of historical notes on the whole book by Mr. Jourdain. The first part brings the English translation up to the date of the latest German edition, while the second gives us additional matter of great value which has been approved by Prof. Mach himself.

Mach's alterations are mainly in consequence of Wohlwill's researches on Galileo and Duhem's on the history of statics. The result of the two men has been to exhibit a more steady and continuous development of mechanics from Greek scientists to modern ones. Thus Duhem discovered a manuscript of Jordanus Nomorarius, or rather a later elaboration called *Liber Jordani de ratione ponderis,* which anticipated in a large measure the theory of moments and of inclined planes. This author may be regarded as a forerunner of Leonardo da Vinci. We are now in possession of rough notes by Leonardo containing many sketches of mechanical principles, some correct and some incorrect. There is further reason to believe that Leonardo da Vinci's work was known to Cardan and Benedetti, and that through them it influenced Galileo, Stevinus, Roberval, and Descartes.

We are now acquainted with earlier works of Galileo than the *Discorsi* and can see the true notion of constant acceleration developing. Galileo at first thought that falling bodies were an instance of the law $dv/ds = K$, and that this agreed with all the facts. Later he rejected this in favour of the true law $dv/dt = K$; but his reasons both for his first acceptance and his final rejection of the erroneous law are not cogent. Mach also mentions Galileo's speculations about the mathematical infinite, which he compares with those of modern mathematicians, classing them both as 'mystical'.

Mach has in no way altered his views about absolute and relative motion. He says that probably there will soon be no reputable supporter of absolute motion; it seems curious that he makes no reference to the important chapters in Mr. Russell's *Principles of Mathematics*. They do not indeed appear to me to prove their point, and I believe that Mr. Russell has now altered his views; but at least they would have supplied Mach with a distinguished modern supporter of absolute space, time, and motion. Mach has also altered his former very obscure note about Lange's Inertial System. Unfortunately, it still remains obscure to me, and I could have wished Mr.

+ The book review is about the *Supplementary Volume* (1915) to the third English-language edition of Mach's *Science of Mechanics* (1907), published by Open Court Company. The book review appeared in *Mind,* **25** (1916), 118-119.

Jourdain had supplied a supplementary note on this subject. What is meant by one straight line being *warped* with respect to another?

There is also a far clearer statement than before of Mach's much quoted remark (in connexion with Newton's bucket) that 'the universe is not given to us twice, but only once'. It is now clear that Mach's meaning is that the Ptolemaic and the Copernican view are simply different ways of describing precisely the same set of facts, and that therefore there is no real difference between the bucket standing still with the fixed stars rotating and the bucket rotating with the fixed stars standing still. This is clearly a necessary result of the relative view, and it is one that is often overlooked.

Mr. Jourdain's notes have all his usual accuracy and wealth of historical knowledge. They consist partly of corrections and amplifications of some of Mach's references; partly of remarks on the Principle of Least Action, of which Mr' Jourdain has made a special study; and partly of remarks on the Calculus of Variations. On page 87 in the note about D'Alembert (line three of the note) surely *equalities* is a misprint for *inequalities* .

This volume is bound similarly to the translation of Mach's *Science of Mechanics* and is indispensable to anyone who has that work. But why did not the publisher make the supplement of the same height and width as the original? As it is, we cannot place the two side by side by side on our shelves without hideous irregularity.

V.F. Lenzen (1932)*

It is well-known that one of the landmarks in the history of mechanics is Archimedes' theory of the lever. In this theory he assumes as self-evident the proposition that equal weights attached at the ends of an equal-arm lever will maintain it in equilibrium. From this proposition, which is the

* V.F. Lenzen, "Archimedes' Theory of the Lever", *Isis,* **17** (1932), 288-289. An extensive analysis of Mach's theory of the lever can be found in Dora Reimann, "Historische Studie über Ernst Machs Darstellung des Hebelsatzes", *Quellen und Studien zur Geschichte der Naturwissenschaften und der Medizin,* **3** (1936), 554-592. Criticism of other aspects of Mach as a historian of mechanics, especially concerning his treatment of Galileo can be found in the writings of E.J. Dijksterhuis, I.B. Cohen, and Dudley Shapere. Unfortunately, most of Dijksterhuis's criticisms are in Dutch. (See the article by Henk Visser in this anthology for more particulars.) For I.B. Cohen's criticisms see: "History and the Philosopher of Science", in *The Structure of Scientific Theories,* ed. F. Suppe, 2nd edition, Univ. of Illinois Press: Urbana & Chicago, 1977, pp. 308-373. Dudley Shapere's criticisms can be found in his book *Galileo - A Philosophical Study,* Chicago, 1974. See especially the pages 4-6, 8-9, 44, 58, 63, 97, and 144. In larger terms, Mach's "empiricist" interpretation of Galileo was countered by Alexandre Koyré. Since then, Stillman Drake and others have attempted to revive it. - Ed.

principle of the lever for equal arms, he deduces the more general principle for unequal arms.

In his history of mechanics, Mach has given a classical discussion of Archimedes' theory. Mach argues that the principle of the lever for equal arms is not a self-evident principle, but an induction from experience. It is the purpose of the present note to indicate the elements of correctness and incorrectness in Mach's criticism.

I open the discussion with the remark that if the restricted principle of the lever is to be interpreted as an induction from experience, one would need an independent definition of the concept of equality in weight of two bodies. Now Mach's criticism is fundamentally inadequate because he does not base his analysis upon a prior definition of weight. One can not ask the question, will two equal weights maintain the equal-arm lever in equilibrium? Until one has some definition which enables one to decide whether or not two weights are equal.

In the opinion of the writer the standpoint of Archimedes requires the interpretation that the restricted principle of the lever constitutes a definition of equality in weight. By definition, two bodies are equal in weight if they maintain an equal-arm lever in equilibrium. Now the modern logic of our deductive systems has transformed self-evident axioms into postulates which thereby define the characters of the elements and relations referred to in the postulates. From the mathematical point of view Archimedes' theory of the lever is a deductive theory for a restricted field of mechanics. It is therefore in accordance with modern logic to interpret Archimedes' restricted principle of the lever as a postulate which defines the concept of equality in weight.

There is an element of correctness in Mach's discussion, however. The choice of the definition is guided by empirical considerations. If two bodies maintain a particular equal-arm lever in equilibrium, then the same bodies will maintain all levers in equilibrium, regardless of the time and position, length, material, color, temperature of the lever. Thus the relation between two bodies which maintain a lever in equilibrium is a universal character which exhibits itself in a multitude of particular instances. The relation is symmetrical and transitive and is therefore to be interpreted as manifesting the possession of a common property. The common property is a specific magnitude of weight. Thus the empirical generalization that equal-arm levers behave similarly under the action of two given bodies is a justification for the definition of equality of weight in terms of the restricted principle of the lever.

From the physical point of view the restricted principle is a definition of equality in weight which concept is useful because of its generality of application. From the standpoint of deductive theory, however, a postulate expresses certain properties of the fundamental concepts of the theory.

From this point of view the restricted principle of the lever is a postulate for a deductive theory of the lever. Thus on interpreting Archimedes' theory in the light of modern theory of deductive systems, his restricted principle is seen to be an essential element in the foundations of a deductive theory, an element which from the mathematical point of view is to be assumed at the outset, since it constitutes a definition of a fundamental concept of the theory.

C.

ON MACH'S *THE PRINCIPLES OF PHYSICAL OPTICS*

Max von Laue (1921)*

In this book, a man of undeniable significance speaks perhaps for the last time to his contemporaries ("perhaps" because a continuation is planned which hopefully will also come out). One will therefore certainly take it in hand with utmost respect. But in the reading of it one can scarcely avoid the impression that it is the work of an old man. Though the intent is to place it besides the historical-critical *Mechanik* by the same author, it has nothing of the originality which that 30 year old work had overall, and nothing as well of the scientific insight, which that very frequently shows. The "cognitive-psychological" aspects [which are mentioned in the title] are very faint, and what we actually have is only an admittedly good historical representation of geometrical and wave optics from Euclid to Arago. Later authors and their performances are on the contrary at least as I see it rather badly reduced; Thus, for example, Abbe's theory about optical illustration [*Abbildung*] which is certainly in principle an important theory is disposed of in a sentence, and there is nothing at all about recent apparatuses for the numerous kinds of interference [*Stufengitter & Lummer-Gehrcke-Platte*]. Nor do the suggestions in the foreword offer us prospect that what is lacking will be made up in part two. Indeed, the *electromagnetic* light theory is supposed to follow.

Nevertheless, even for current physicists who possess a good picture of the newer optics, the book remains interesting as a source for the older otherwise little accessible history about this branch of physical science. And the ten likenesses of significant investigators of optics raise the value of this work considerably.

* Ernst Mach, *Die Prinzipien der physikalischen Optik, historisch und erkenntnis-psychologisch entwickelt,* J.A. Barth: Lepzig, 1921. The review appeared in *Die Naturwissenschaften,* Jg. **9** (1921), 966.

But the preface is nothing short of "sensational". The author provisionally - the exact details will follow in the second part - takes a position with respect to relativity theory, and indeed, a rough, unfriendly one. [He sees that people] are gradually beginning to think of him as a "forerunner to relativity theory", and he believes that he must reject this [identification] with the same decisiveness with which he has denied "the atomistic dogma of the current school or church" for his own person. For the author, even relativity theory seems "dogmatic", and wherever he spots that word, he senses, as can be noticed in more than one place in the book which is before us, an utterly unhistorical, uncritical, horror.

If one were to characterize the situation, it cannot be correct to push scepticism so far that one denies one's own deepest thoughts the moment a person greater than oneself develops it into something positive. And yet the book has rendered a service to relativity theory: He never speaks of a palpable "Aether" and shows how well all of wave optics can be represented without it.

Ernst Bloch (1922)+

The announcement of Ernst Mach's death (February 19, 1916) visibly moved his friends and everyone familiar with his life's work. Obituaries reveal how unusually affected scholars from the most diverse scientific orientations were. He was able in an era of Darwinism and the first great triumphs of molecular theory to put forward the leading ideas of the next age, among them applications of the theory of evolution and criticism of molecular representation, and by his quiet, yet increasingly acknowledged scientific work he helped lend to contemporary positivism in Germany a good part of its special imprint....

Atoms and molecules, seen by his contemporaries as the most real of all realities, were already mere thought-symbols for the young Mach, which according to circumstances were either convenient or inconvenient for economic description of physical and chemical appearances....

[After his death] it soon became known that a posthumous work on the history of optics would come out. What lies before us is only the first part. In the foreword the author says that the book developed from his lectures when he taught in Prague [1867-1895]: "I hope not without success to have laid bare the concepts of optics and the historical threads of their development freed from metaphysical ballast. But this is no compilation of historical research, and details which have been exhaustively treated by others are [often] only touched on here."

+ This review appeared in *Isis,* **4** (1922), 560-562.

In the 1913 foreword, Mach distances himself from the theory of relativity, and his announcement will not fail to stir a sensation in the scientific world. He perceives that one wants to think of him as a forerunner of that theory. But personally he rejects it and will put forward his reasons in the second volume. The theory will retain a certain lasting value, but how long can its popularity last and in the history of science will it eventually signify more than a mere ingenious *apercu*? This is in some contrast with Einstein's obituary of Mach where he is treated as a forerunner of relativity doctrine [*Relativitätslehre*], and it is maintained that if at the time his mind had been young and fresh the significance of the absolute velocity of light had already aroused physicists, then Mach would have accepted the doctrine. Consequently, by sticking with older views when the younger generation came to the doctrine of relativity, who has seen the farthest? [Mach or Einstein?] One may now await the second volume with even greater suspense....

J. Stark (1922)*

Mach was a sharp, critical thinker, a relentless seeker and confessor of truth, a master of words - thus does he live in our memory. After a book from him, after this last book from his hand, let us seize it with interest. The title could raise false expectations; He treats the appearances of light diffusion in different media, not physical optics in today's sense, therefore, not the phenomena of the emission and absorption of light. The subtitle "historically and cognitively/psychologically developed" is thus to be understood only in this sense, for while geometric (radium-dynamic) optics has been developed up to a certain conclusion, this has not been the case with physical (atom-dynamic) optics in the narrower sense.

The titles of the most important chapters of the book are as follows: The Rechtilinear Propagation of light, Reflection and Refraction, The Early Knowledge of Vision, The Composition of Light, The Further Development of the Theory of Interference, and Polarization.

Mach's book is based on fundamental historical research. It therefore possesses the significance of a source tome which one cannot do without. The whole book is applied to physics of a bygone period. He does not treat problems of the present. Only in the foreword does Mach acknowledge contemporary physics. It is in personal and factual respects so characteristic and worthwhile, that it will be repeated here word for word:

* This book review appeared in *Deutsche Literaturzeitung*, June 10, 1922, Nr. 23, columns 503 and 504.

> But why and to what extent I reject for myself the doctrine of relativity, which is becoming ever more strange and dogmatic, and which [particular] sense-physiological considerations, epistemological reflections, and above all, experimentally won insights have caused me to adopt this position in detail shall be set forth in the continuation of this work. Certainly, the ever increasing amount of cognitive labor devoted to the study of relativity will not be lost; it has already been fruitful for mathematics and of lasting value, but can it be maintained that in the physical world picture of a distant age, which will have many times more insights to accomodate, that it will signify more than an ingenious *apercu* in the history of science?

In the interests of our science, it must be vigorously regretted that the death of this great scientist did not allow him to extend his promised discussion of relativity theory.

S. Brodetsky (1922)+

....As the title indicates, the book is an historical and epistemological account of the development of the principles of physical optics, the writer's aim being similar to that in his famous book on mechanics. The book is thus not really a history, but rather a statement of the progress of ideas and of the manner in which they arose in the minds of their originators. Nobody who has read other books by Mach will need to be told that this volume is interesting and instructive: the description of the way in which fundamental notions such as periodicity and asymmetry of the light ray emerged, is as fascinating as any romance. The present volume was intended as a first part of a larger work: subjects like theories of the ether and relativity were reserved for the second volume. But the author's foreboding that this second volume would not be written was unhappily justified.

Many points of interest emerge in the course of this history. Full tribute is paid to Newton's genius - it was Newton who first realized clearly the periodic nature of the light ray, later worked up into the wave theory, and its asymmetry, which became the theory of polarization and the notion of the transverse light-vector. When Fresnel first reached the conclusion that light oscillations are transverse, the revolution in ideas was too great for even Arago, who refused to be associated with Fresnel's idea, and in fact, although Fresnel and Arago collaborated in many pieces of work, the

+ This review appeared in *Nature,* **108**, June 3, 1922, pp. 706-707.

transverse vibration theory was put forward in a paper bearing only Fresnel's name. Young's experiments did not at once receive the recognition they deserved, owing to the mistaken and reactionary servitude to all that was supposed to have emanated from Newton. The author has not much good to say about Schopenhauer's attempt at a theory of colors, while faint praise is all that is allotted to Goethe....

L.C. Martin (1926)*

....It is not a book for the beginner. For such, nothing else can take the place of a careful systematic treatment adapted to develop the philosophical sense which is demanded of the reader of the present volume. From its very nature, an historical treatment must deal with matters with which a beginner need not concern himself, such as Newton's "theory of fits". Then, again, the standard of knowledge demanded of the reader of Chapter ii. (on the rectilinear propagation of light) is certainly higher than that of a novice; and the same applies to the several later parts of the book. Nevertheless, it is a valuable book for a teacher or for a student who wishes to get beneath the surface of the subject.

The translaters have performed their task well, although there are a few places in which the meaning is obscure. The text is full of interest, and the admirable range of portraits gives a human touch to all.

It seems to the writer that one or two references to recent work might well have been added by the translators. For example, the following passage occurs in the description of the Michelson interferometer: "It is desired to study an interposed substance which is non-homogenous, and upon which the eye must thus accomodate, parallel light cannot be used for this purpose, and the substance may not be traversed *twice* by the rays." Such a statement, although perhaps applicable to the case in question, should not have been passed without a reference to Twyman's modification of the instrument, in which the interposed substances are traversed twice by the "parallel" rays. There is still an unfortunate gap between the optics of the laboratory and the workshop, and there are several places where this is instanced in the present work.

Professor Mach occasionally manifests an amusing touch of dogmatism. His introductory letter speaks of "the only possible form of immortality". Has he, then, explored the possibilities of life so thoroughly? Dogmatism

* This account is about the English translation: Ernst Mach, *The Principles of Physical Optics - An Historical and Philosophical Treatment,* translated by Dr. John S. Anderson and Dr. A.F.A. Young, Methuen & Co.: London, 1926. The review appeared in *Nature,* **118**, September 25, 1926, pp. 437-438.

also emerges, too, in the discussion of color vision. The conclusive nature of the introspective analysis of a series of sensations is taken for granted, and Thomas Young is accused of introducing confusion into the subject! It is to be hoped that this statement may challenge some to go back to Young's own discussion. In another case where a psychological question is involved, namely, the question of light quantities, the book is vague and unsatisfactory. The coming of physical photometry is foreshadowed in terms which give no hint of the many and serious difficulties which have been encountered since 1913... Indeed, the statements on p. 20 would lead the unsuspecting to conclude that a physical measurement of the gross energy of "the *illuminating* rays" (whatever that may mean) is an adequate means of finding the "quantity of light emitted from a light source in unit time." We conclude with regret that the "philosophy" of the book is lamentably lacking where it might be most helpful.This is yet another instance of a clear call for the insertion of a translator's note directing attention to more recent developments. It is not forgotten that the book deals with physical optics, but every textbook ought to indicate the vital points of connexion between its own subject and allied branches of knowledge.

A feature of the book which deserves especial praise is its treatment of polarization and double refraction. For several years past, the present writer has encouraged optics students to begin the study of these subjects with the simplest apparatus and natural crystals, such as were available to the early workers. In this way they obtain a thorough acquaintance with the principal features of the subject; after a day or two of making retardation plates of mica, and using natural crystals of quartz, they are ready to appreciate the quartz wedge; the natural rhomb of spar affords a good introduction to the Nicol prism and its modifications. Thus by easy stages the properties of uniaxial crystals become familiar; and how many students are *really* ready to pass beyond this stage in the limited time of college courses? The book adopts a similar method of approach, and, in addition, describes many beautiful experiments which will be new to the majority of readers.

This book, then, may be recommended to all teachers who have occasion, in the midst of the multiplicity of subjects comprised in present-day "physics" to give some little attention to "light".

CHAPTER 4

MACH CORRESPONDENCE

A

ERNST MACH AND WILHELM OSTWALD*
(1889 - 1913)

#1

Mach to Ostwald — Prague, June 7, 1889

Highly Honored Sir!

Due to a long illness, I am today for the first time in a position to be able to thank you for your friendly communication.

About thirteen years ago, I put forward the proposal to B.G. Teubner [the publisher] to bring out significant mathematical and scientific works in editions similar to the Greek, Roman, and other ones which had already appeared. I received the answer that they would have no chance to be success-

* Wilhelm Ostwald (1853-1932) was Professor of Chemistry at the University of Leipzig from 1887 to 1909. He is famous as one of the co-founders of Physical Chemistry and for establishing a very productive laboratory which helped train many students. From about 1892 to 1908 he maintained the superiority of energeticism over atomism, but the work of Einstein and Perrin on Brownian motion finally helped change his mind. The fact that he received a Nobel prize in 1909 perhaps made his return to atomism easier. His debates with Ludwig Boltzmann from 1895 to 1904 over atomism were long remembered. The letters from Ostwald to Mach are located in the Ernst-Mach-Institut, Eckerstr. 4, Freiburg im Breisgau. The letters by Mach are in the Ostwald Archiv located in the German Academy of Sciences in East Berlin. Joachim Thiele published part or all of letters 3, 4, 6-9, 19, 25, 26, 27, & 29 in German with extensive footnotes in his book *Wissenschaftliche Kommunikation - Die Korrespondenz Ernst Machs,* A. Henn Verlag: Kastellaun, 1978, pp. 106-116. Four of the letters, however, which we have numbered as 25, 26, 27, & 29, were incomplete.[See Thiele, p. 306.]. Klaus Hentschel and Frau Joachim Thiele provided typescripts in German of letters 1, 2, 5, 10-18, 20-25, 27, & 29 for the editor of this chapter. With their help, full-length letters have finally been pieced together; they were then translated by the editor into English. Letter #19 has also been published in John Blackmore & Klaus Hentschel (eds.), *Ernst Mach als Aussenseiter,* Wilhelm Braumüller: Wien, 1985, p. 99-100. Letters from Ostwald's son Walter have not been included. The publication of *Ernst Mach - Studien und Dokumente zu Leben und Werk,* edited by D. Hoffmann & H. Laitko, Berlin, 1991, has added three letters which are numbered 3a, 5a, and 23a. Their book is the first German publication of the extant Mach-Ostwald correspondence as ours is the first in English. A few minor discrepancies between the two versions have been addressed in the relevant footnotes. -Ed.

J. Blackmore (ed.), Ernst Mach – A Deeper Look, 71–110.

ful from a book publisher's point of view. Even then I was persuaded that this opinion was not correct and today even more so, since many of the works which I then had in mind have since really appeared and actually been successful. I mainly had older not easily accessible works in mind. The idea which you seem to be thinking about does not appear to be exactly the same. The three issues, which I have seen, include what is rather easily accessible, which many physicists possess, even many times over.

I don't want to say that their accessibility can't and shouldn't be made even easier. I believe that small paperbacks will be bought by people who cannot purchase whole collected works. I wish you and your undertaking the best success. My very respectful greetings

Yours most devotedly

E. Mach

#2

Mach to Ostwald (postcard) Prague, September 26, 1891

H. H. C!

Please don't bother yourself about the textbook. I will send you a copy from the new edition. My very respectful greetings.

Yours most devotedly

E. Mach

#3

Ostwald to Mach Leipzig, October 5, 1891

Very Honored Herr Colleague!

Thank you very much for your friendliness in sending me a copy of your book, which I have looked at with lively interest. There is a very special need in physics for beginners to learn things from one who has not just copied information from other books but who has independently thought things through himself. How many problems, especially the "self-evident" ones, have again become clear to me by taking the opportunity to work them out on my own. With the most reciprocal greetings

Brüderstr. 34 Yours devotedly

W. Ostwald

#3a

Ostwald to Mach Leipzig, November 15, 1897

Confidential!

Highly Honored Herr Colleague!

Conditions which I cannot now set forth compel me to ask you, under the assumption that the current situation works out, whether you would be willing to accept a chair in experimental physics at a larger German university. I must emphasize that my feeler is completely private, and is only for my own clarification. Also, I must emphasize, that the particular eventuality about which I seek information, has not yet taken place, and possibly will only open up after several semesters or years. What I would like to know is whether in such a case one might look with hope toward you.

Naturally, any eventual information from you will be kept and handled with strict confidence. Let me confirm that the impression which you received at our last meeting remains true, that my feeler is made solely on factual grounds [because of your high qualifications}. I hope for a positive answer.

With the best greetings

Your wholly devoted

W. Ostwald

#4

Mach to Ostwald Vienna, November 18, 1897

Highly Honored Herr Colleague!

I heartily thank you for the friendly confidential inquiry and would reply to you in full frankness as follows:

In my youth it was my deepest wish to move to a German university, but at that time for obvious reasons no call came. When they finally did arrive, and some quite advantageous, it seemed too late to pursue them. I had already built [in Prague] an institute to my taste, and had become accustomed to my circle of students, and above all was engaged in unfashionable research far from the more conspicuous problems in my discipline. The isolation of us Austrians may also have its advantages, but it also clearly has its negative side, and I fear that this will become much too prominent were I to enter into new relationships and be faced with new tasks. I am self-taught in everything, in experiment no less than theory, for when I studied in Vienna one could learn in neither direction anything important. Please keep in mind also, that in a few weeks I will have finished my sixtieth year, so that perhaps "after a few semesters" you yourself will have no more interest in asking me again.

My present position [in Vienna] pleases me very much. I lecture on epistemology, history and theory of inductive science, can follow my own special interests, and I have a lot of time for myself. Also, everything has been done for me in material terms which is possible, given our modest conditions.

Our circumstances are indeed somewhat peculiar and unstable and no one can know how the future will develop. Nevertheless, should the question be raised again [about moving to another university], I would answer briefly and clearly in the same way without engaging in mock negotiations which would neither be honorable nor profitable. I currently have no ground to leave my position in Vienna.

I heartily thank you again for your friendly attitude.

with greetings full of high respect

Yours most devotedly

E. Mach

#5

Ostwald to Mach — Leipzig, August 2, 1898

Very Honored Herr Colleague!

In *Zeitschrift für physikalische Chemie,* volume 26, pages 181 and 568, you will find a polemic between Mr. Cantor* and myself on the significance of an experiment which was carried out by him (*Wiedemann's Annalen,* volume 62, page 482) concerning the kinetic hypothesis. From his reply it appears to follow that *after* the absorbtion of chlorine that copper can have no velocity, but not that *during* the absorption by the absorbing surface that it cannot be displaced, and that is also as far as I can see the meaning given in your *Mechanik* and in the case mentioned by him.

* Mathias Cantor (1861-1916) was a professor of chemistry at Strassburg, Würzburg, and Heidelberg. - Ed.

Because of the importance which could come from Cantor's experiment, if his interpretation were correct, I don't believe that I can rest fully satisfied just with my own opinion. I would like to request that you look at the matter and share with me in a friendly manner your judgment about it. If you do not wish to intervene in the controversy, I would still be thankful for your private opinion, which I would not use publicly. I would be even more appreciative should you decide to discuss the question definitively in the *Zeitschrift für physikalische Chemie,* even if you were to oppose my position, for in this matter I have a pressing need to understand the situation clearly.

If it would be awkward for you to get hold of this copy of the journal, I will be more than happy to send you one. (I will attach a copy of Cantor's answer with it.)

Let me also add in passing, that I have been able to put together a very light, movable reaction-turbine out of glass. On intake I always observed a small movement in front, but no acceleration, completely corresponding to what one should expect.

Physical-Chemical Institute Your wholly devoted
Linnéstrasse 2/3

W. Ostwald

#5a

Ludwig Mach to Ostwald Vienna, August 16, 1898

Most Highly Honored Herr Professor!

As you perhaps have heard, my father suffered a light stroke in the last month.** This is, therefore, the reason why he can no longer answer your worthy letter dated the second of this month. At the moment things are going somewhat better with him, the paralysis is retreating and his general health is relatively good. My father greets you heartily. As soon as his condition allows, he will answer your letter in detail, since as he has told me, he also finds the matter uncommmonly interesting. Please receive my expression of particularly high respect from yours

Sincerely

Dr. Ludwig Mach

** Ludwig Mach (1868-1951), Ernst Mach's eldest son had graduated from the University of Prague as a medical doctor three years earlier, but seemingly unable to become a *Privatdozent* he turned to working for the Zeiss Optical Works and later became wealthy due to an ability to sell some of his numerous mechanical and metallurgical inventions. Most of what was left of his fortune was apparently lost with the inflation after World War I. The right side of Ernst Mach's body remained paralysed until his death in 1916. - Ed.

#6

Ostwald to Mach Leipzig, May 31, 1901

Honored Herr Colleague!

The third edition of your *Mechanik* which has arrived and which even now is in my hands has spurred me to realize an intention, which has already been on my mind for some time.

This semester under the designation *Naturphilosophie* I am giving a series of lectures which are essentially on theory of knowledge, and whose foundations for the most part I owe to you. The large audience of three or four hundred people testifies to the growing interest of many young people in these questions which have been repudiated by a majority of your contemporaries and mine. I'm thinking about publishing my lectures and request your permission to dedicate the book to you.

At the same time, I have the plan starting approximately next year to start a journal which will focus on the same matters, but at first it will not appear regularly. My experience in physical chemistry has convinced me of the advisability of such a course for the activation and promotion of interest in the development [of philosophy of nature]. If you agree, it would be most desirable that the first issue include something from your pen, since it would establish the [sober] character of the project for all time, and would keep away the wild philosophizing of uncountable numbers of natural scientists. If you might even be willing to be *co-editor*, that is, guardian of the orientation and reliability of the journal without being troubled by responsibility for the technical side of editing, then that would be of decisive importance for the success of the journal.

Physical-Chemical Institute With the best greetings and wishes.
Linnéstr. 2/3

Your wholly devoted

W. Ostwald

#7

Ostwald to Mach Leipzig, June 5, 1901

Honored Colleague!

Many thanks for the pair of "yeses", which to some extent ease the pain for the "no" about the third matter; nevertheless, I still reserve the right to make additional proposals to carry out the plan to start a journal, which will ensure its success without putting you under undue pressure.

Concerning that, I had already heard that you were retired and hope that your decision works out well. For the influence which you have exercised through your writings is so great and fruitful, that every change in your external condition which gives you more opportunity and peace of mind to continue your scholarly activity would seem to me to be a plus.

Please forgive me for mistakenly having designated the fourth edition of your *Mechanik* as the third (in case you might have noticed the error); there is no point in clarifying why the mistake happened.

Physical-Chemical Institute
Linnéstr. 2/3

With the best Wishes

Your wholly devoted

W. Ostwald

#8

Ostwald to Mach [undated, about September 1901]

Honored Colleague!

If you could finish your prospective paper quickly, it would be able to appear in the first issue of *Annals*. In order to fix the character [and tone] of the new journal, this would be of the greatest value. Perhaps you might add a section from the new work which I hear you've been working on.

I am writing this from a week-end, country villa [Landhaus] at the edge of the forest, which I have bought and have christened "Energie" and where for the last five weeks my family has been happy and me too because of the amount of work done. I won't return to Leipzig until the very end of the vacation period.

Your wholly devoted

Grossbothen in Sachsen
Landhaus Energie

W. Ostwald

#9

Ostwald to Mach Leipzig, October 28, 1901

Honored Herr Colleague!

I come once again with a request. Every year for the Christmas Issue of *Zeitschrift fur physikalische Chemie* I like to include a picture of someone who has been important in the development of science. This time I would like to have your picture to get the chemists into studying your writings. You know yourself how bitterly necessary it is to do a bit of critical epistemology [*Erkenntniskritik*] on them. I therefore request from you a not too small photogram (not smaller than a cabinet) or still better the attached negative, which will be returned together with a published copy which you will allow me to send. There is no great hurry, but within 14 days would be welcome.

I believe I am right that you are no great admirer of such things, but it would give a lot of pleasure to my readers and me. Please think about it.

Naturphilosophie has just been finished and the printing will be over in 14 days. I'm a little worried about your judgment. The *Annalen* have experienced a small delay, since I still want to include a very respectable essay by Couturat* (Toulouse) on international language, whose composition is still to be carried out. Also, the necessary material has been brought together for the second issue. It will appear soon after the first.

Your wholly devoted

W. Ostwald

P.S. Thanks a lot for the Stallo [book]**. I advertised the English edition already in 1895 in the 18th volume of my *Zeitschrift für physikalische Chemie*.

#10

Mach to Ostwald Vienna, January 27, 1902

Highly Respected Herr Colleague!

* Louis Couturat (1868-1914) taught at Toulouse and Caen. He contributed to the development of symbolic logic and like Ostwald was in the forefront of the artificial language movement. In particular, he helped reform Esperanto by establishing the vocabulary of what was called "Ido". His death was somewhat ironic. As a pacifist he may not have been in sympathy with or not deeply interested in the war fever which struck Paris in early August 1914. While riding in a carriage he was struck and killed by a speeding vehicle which carried the French orders for national mobilization. He corresponded extensively with Mach, but Mach's letters to him have not yet been located. - Ed.

** John Bernard Stallo (1823-1900) emigrated from Germany to America where he became an educator and diplomat. He developed a philosophy of science which had close similarities to that of Mach. Mach only discovered Stallo's writings very late, but during the two or three years before Stallo's death in retirement in Italy and shortly afterwards Mach did all he could to interest his friends in the ideas of the neglected German-American philosopher. - Ed.

Overcoming all kinds of awkward handicaps, I was finally able a few days ago to read your beautiful book to the end. I cannot hold back my hearty appreciation for this brilliant and overwhelmingly clear representation of the significance of the energy principle in all fields. It will also certainly yield the finest fruit in a wide range of circles.

You may have already received the third edition of *Analyse der Empfindungen*. The work attempts to discover accessible ways from *given* perceptions into what is physical on the one side and into psychological regions on the other. The purpose is not to investigate physical details. In spite of that, if the form does not put you off from occasionally looking into it, you will find many points in common between us, and indeed, perhaps also some differences between our views. About discussing the latter, it would certainly be too soon for me so long as I am still in the process of sorting out my own thoughts, which I am now fully occupied with doing. *As a scientist, wherever we agree makes me happy; and let the philosophers live with the differences.*

With repeated hearty thanks for dedicating your outstanding book to me and with best wishes for further success in your work.

Full of highly respectful greetings

Your most devoted

Dr. Ernst Mach

#11

Mach to Ostwald — Vienna, May 14, 1902

Highly Respected Herr Colleague!

Since I am currently very busy and, apart from that, troubled by bad health, even with the best will I cannot write an article. Please be good enough to dispense [with my contribution].

Best wishes and with

Greeting full of high respect

Your most devoted

Dr. Ernst Mach

#12

Mach to Ostwald Seebenstein, August 16, 1902

Highly Honored Herr Colleague!

In Vienna, a newspaper will appear in grand style starting this Fall and it will have a scientific-technical supplement. You will be requested by the editorial staff [*Redaktion*] of this paper, which will carry the title "Die Zeit", to occasionally write for this supplement. I have been asked to recommend the paper to you, since several of the brightest lights of our university here will be contributing to it. I believe this can be done with peace of mind, under the presupposition that you are interested and have enough time to write for a newspaper. Best wishes for the holidays,

Full of high respectful greetings

Your most devoted

Dr. Ernst Mach

#13

Mach to Ostwald Vienna, December 25, 1902

Highly Honored Herr Colleague!

The movement for a so-called "Free Catholic University" [in Austria] is already sufficiently known to you. *Privatdocent* Dr. Ludo Hartmann*, who has done the great service of introducing a *countermovement,* will drop in on you, and in connection with this matter direct a request at you, which perhaps should not be hard to fulfill. I hope that these lines will allow Hartmann to come to you as less than a complete stranger, and would not thrust upon you the obligation of [suddenly having to find] an answer which would take up your precious time.

With the expression of utmost respect

Your most devoted

Dr. Ernst Mach

* Ludo Hartmann (1865-1924) was an Austrian historian, politician, diplomat, and adult educator. He was a social democrat and friendly to Mach's philosophy, even dedicating a publication to him. - Ed.

#14

Mach to Ostwald Vienna, November 20, 1904

Highly Honored Herr Colleague!

It has become known that you will soon be giving two lectures here. On the instructions of the council for the anticipated retirement home [*Volksheim*] which is now under construction, allow me to ask the question concerning their request whether you might be inclined to favor the home with a lecture on a theme of your choice. Your reputation and the great success which a generally accessible lecture would promise have attracted the interest of the council. I would naturally be very pleased if this request might be affirmatively answered, although I cannot conceal that an unwarranted expectation may be present in this matter.

Your most devoted

XVIII, Gersthopferstrasse, 144

Dr. Ernst Mach

#15

Mach to Ostwald Vienna, January 5, 1905

Highly Honored Herr Colleague!

I have just read your wonderful Faraday Lecture. In addition to the content, I was also very pleased with the acknowledgment given to Frantisek Wald.* I consider it one of my best experiences to have beheld the fall of the atomic theory in chemistry. In physics the thing plays too subordinate and variable a role to cause serious damage. The pressed physicist always says in the end: "Yes. Yes, but chemists can't do without the atomic theory". If the Faraday Lecture exists as a separate publication, please send me the details on a card like a friend.

With unmatched respect

Your most devoted

Dr. Ernst Mach

---

* Frantisek Wald (1861-1930) was a chemist and chief metallurgist in Kladno, Bohemia, but like Mach and Ostwald at this time, he opposed the reality of atoms. He corresponded extensively with both men. See *The Correspondence of the Czech Chemist Frantisek Wald with W. Ostwald, E. Mach, P. Duhem, J.W. Gibbs, and Other Scientists of that Time*, edited by J. Pinkava, Prag, 1987. - Ed.

#16

Mach to Ostwald (postcard) Vienna, February 9, 1905

I confirm with heartfelt thanks the arrival of the material which you very kindly sent me.

Respectfully

Your most devoted

Dr. Ernst Mach

#17

Ostwald to Mach Gross-Bothen, September 28, 1908

Very respected Herr Colleague!

Like you, I also consider J. Petzoldt* to be a very outstanding researcher; But your presupposition that I have something to say about selecting candidates for the position in Dresden is not correct. Nevertheless, I have sent your letter together with some accompanying words to the Minister of Culture [in Saxony] to bring it to his attention, so that he can't later apologize for not knowing anything about the matter.

Landhaus Energie

With the heartiest greetings

Your wholly devoted

W. Ostwald

#18

Mach to Ostwald (postcard) Vienna, February 8, 1912

Highly Honored Herr Colleague!

It will please me very much in any case, if you would visit me on your way back. I already possess the monistic preaching; please don't give any authorization [*Auftrag*] after you return home. I have for the most part already read it. It will please me a lot to speak with you.

---

* Joseph Petzoldt (1862-1929) was a philosophical follower of Richard Avenarius until his death in 1896 and then of Ernst Mach. During his later years he taught philosophy of science at the Technical University of Berlin. See Gerald Holton's article in this book. -Ed.

With best wishes and my heartiest and highest respect

Greetings

Dr. Ernst Mach

#19

Ostwald to Mach Gross-Bothen, February 24, 1912

Honored Herr Colleague:

With heartiest thanks for your full willingness to oblige, which you have proved to me and the *Monistenbund,* allow me once again to summarize the points we discussed:

1. You will have the goodness within three or at most four weeks to send an essay for our new journal: *Das monistische Jahrhundert.*

2. Connected with the honorary presidency of the local Vienna group of the *Monistenbund* you will assume no personal obligations.

3. You will agree to having your name placed on public announcements of this local group which will be printed at the appropriate time in Viennese (and other Austrian) newspapers, and whose content will be submitted to you before publication. The announcement [*Aufruf*] will presumably be issued in connection with a lecture which I will hold in Vienna at the end of April.

4. You will allow the *Brücke* [society] the honor of becoming a leading member of its council of honor. You will soon receive the appropriate printed matter. Also here the extent of the demands which will be placed upon you will depend entirely upon your [own] resolution.

Landhaus Energie With the heartiest and most devoted greetings

Your W. Ostwald

#20

Ostwald to Mach (postcard) (Postmark: March 10, 1912)

Very respected Herr Colleague!

May I expect the essay which you had the goodness to hold out in prospect by the 20th? I absolutely do not want to press you and will gladly wait. I only know how happy it would make me if I could finally get this first issue out. Faithfully yours

W. Ostwald

#21

Ostwald to Mach (postcard) (Postmark: May 13, 1912)*

Honored colleague!

I received your letter when it arrived and by return mail wrote what was necessary to my Vienna co-worker and adherent. I repeat, that pressure is utterly outside my intent. Your fundamental willingness to cooperate is already such an immense gift that we have every reason to be candidly thankful for it. With best wishes for a quick improvement of your health.

Yours devotedly

W. Ostwald

#22

Mach to Ostwald Vienna, November 26, 1912**

Most honored colleague!

I see from what you have written that you have taken with you a completely different impression of my atttitude than corresponds with reality. This may have been caused by my complacency in conversation, perhaps by my significant deafness as well. For many years my contact with other people has almost completely ceased. I no longer come to the Academy [of Science] or even the Upper House [of the Austrian Parliament], since the unpredictability of my bodily suffering could lead to incidents of a disagreeable kind at any moment.

You will perhaps believe me that under these circumstances I am completely happy if I use the time which still remains to me to correct the defects in my writings. Any kind of pressure to undertake new tasks or obligations, putting the least anxiety [*Sorgen*] on me which does not largely come from myself, does not interest me *at all*. If a young man or even an older one enters with a manuscript which I must read, I am very satisfied. All offices and functions which come amply to me? I have consistently re-

---

* D. Hoffman & H. Laito, editors of *Ernst Mach -Studien und Dokumente zu Leben und Werk,* Berlin, 1991, give the date as March 13, 1912 on page 371 of their book.

** Ibid. On page 369, Hoffmann & Laito give February 26, 1912. The problem is whether to identify the date with an arabic "11" or a Roman "II". A November date fits Mach's reluctance to write an article, but otherwise a February date seems superior. - Ed.

jected them for ten years, especially everything which touches on philosophy. Perhaps no one is more justified than I in saying: Don't waste energy!

If you succeed, most honored colleague, in bringing together a large number of people who have been guided exclusively by intellectual insight, this is a very praiseworthy cultural accomplishment, to which I heartily wish you luck. Concerning the majority of people, your group will always seem like a dwindling minority. If your society is named for monists, then it may attract people simply because of the name, but if we inquire about reasons for joining, it will turn out that there are approximately as many monisms as there are people "united" under the label.

Monism is a provisional goal toward which we all strive, but scarcely something finished or fully sufficient which could orient us about everything. In this connection, it seems to me as I briefly mentioned in my previous letter, downright comical, to establish a kind of religious sect which leaves the philosophical question about its nature [which is implied in its title] completely out of consideration. All of this doesn't matter much to the extent the movement is restricted to a small circle of intellectuals. But if [you succeed and] it takes on larger dimensions, then it will probably let loose a *counterreformation,* and for that *I* will definitely have *no sympathy*.

Already, the article for the *Monistische Jahrhundert* puts me in a predicament. 1. The honorary presidency I must gratefully decline. 2. The announcement, which to the extent it is very skillfully conceived, I would be able to sign. 3. Please don't name me to the "Brücke", since I have enough at home to keep me busy.

I thank you heartily for your letter of February 29,+ which has given me occasion for a clear expression [of my views].

With the same respect as of old

The typewriter doesn't work, unfortunately, at the moment I cannot do anything about it.++

Your most devoted

Dr. Ernst Mach

+ Ibid. Hoffmann & Laito give February 24 on page 370 of their book, which if true, would identify the letter referred to with #19 in our book.

++ Ibid. Hoffmann & Laito on the same page have included this sentence in their book, which is presumably a hand-written comment too faint to appear on most xerox copies.

#23

Mach to Ostwald Vienna, March 9, 1913

Most respected *Geheimrat* !

Since we last saw each other, I have been confined to bed lame and in pain for five months because of a fall due to the paralysis on my right side. I hope that you will understand that it was not mere indolence on my part that I was not able to cooperate in the grounding [of the *Monistenbund*], but correct familiarity with my own infirmity and frailness.

The book which I am sending you today is also not scientific but only one which I believe will have a beneficial impact to the extent it is promoted. If you will glance through it, you will see that it is quite simple and its style is effectively monistic. Your correspondent, Lakshmi Narasu, who has the name of the goddess of love, was also my correspondent over ten years ago. She was at that time, if I remember correctly, a teacher of natural science at the "Christian College" in Calcutta; being very poor she could not buy German books. I sent her several elementary books, among them some written by myself. I don't know if meanwhile she has made progress in German, since soon after that I became very sick.

Most honored colleague, please accept the expression of my especially high respect for your person and your inexhaustible cultural activity, as

18/II, Höhnegasse 25. Your devoted

Ernst Mach

#23a

Ostwald to Mach Grossbothen, April 12, 1913

Dear and respected Herr Colleague:

It is with great sympathy that I received your letter yesterday, and I wish to express to you my heartfelt commiseration concerning your past months of suffering, about which I had previously had news only in a more general form. But from the content of your letter I believe that I may be able to infer that things are going a little better now, at least my heartiest wishes to that effect.

The book which you have been so friendly as to send me I received yesterday and just as quickly read through. It is written in a fresh and lively manner and reveals a highly pleasing personality, so much so that one is sorry to have to end the book and leave her personal touch behind. I have in mind to publish a short advertisement of the work in the *Monistischen*

Jahrhundert, which might also help meet your wishes, since by that means one could probably open the way for a somewhat more detailed account.

Your report that my Buddhist correspondent is of the female sex has rather surprised me, since the earlier letter and the character of her book+ has given no indication that that inference could be drawn. Since then [the time you mentioned] she has certainly learned German very well, for I have been able to determine how carefully she has studied and used German books including mine, sometimes employing quotation marks and largely not. Concerning German Buddhists, my preaching in *Monism and Buddhism* [third series, Leipzig, 1913, pp. 113-120] has meanwhile resulted in my receiving other letters, and of us both, I have gathered that I am the better Buddhist.

Finally, I would like to mention that shortly my second son Walter Ostwald will presumably make a trip to Vienna, and because such a considerable part of his intellectual development has been due to you, he would like to take the occasion to have a personal chat. If you still have an interest in receiving strangers at all, it would please him and me a great deal if you might be willing to grant him a few minutes.

With heartiest greetings and wishes

W. Ostwald

#24

Mach to Ostwald — Vienna, April 13, 1913

Highly Respected Herr Colleague!

I confirm your friendly communication with heartfelt thanks. If your son still wants to visit me in Vienna, it will greatly please me. At the end of this month, I will resettle in the neighborhood of Munich. Since many books and items of furniture have already been sent, my home is already somewhat in disorder, but it shouldn't bother either your son or myself. I am very pleased by your reception of the book I sent you, and am very thankful for your promotion of it.

The authoress [of *Erinnerungen einer Erzieherin*] is my youngest sister; my father was very depressed during his last years. I also had to struggle against similar feelings when I was young. The same was true for a younger son of mine who was a promising chemist in Göttingen but voluntarily ended his life. It was the Sanskrit scholar Bühler, a colleague who later

+ P. Lakshmi-Narasu, *The Essence of Buddhism,* Madras, 1907, (second edition 1912). - Ed.

died in Lake Zürich, who conjectured to me on the basis of the name "Lakshmi" that Lakshmi Narasu was a girl.

I notice with happy acknowledgement that your *Monisten Jahrhundert* is winning visibly greater influence in practical life. Particularly hearty thanks for the book review of the work by Josef Popper [Lynkeus], *Nährpflicht als Pflicht*. This idea was first made public in 1878, for which I congratulated him as spadework for the millenium to come. But at that time it was generally laughed at.

Respectfully

18/2, Höhnegasse, 25

Your truly devoted

Dr. Ernst Mach

P.S. I hope shortly to be more capable of doing work again. Everything is still going badly at the moment, especially walking and standing, even writing.

#25

Mach to Ostwald München-Vaterstetten, May 29, 1913

Highly Respected Herr Colleague!

I do not know indeed whether you or anyone else would be very edified or draw benefit from my biography. But you may be able to use the manuscript belatedly or perhaps not at all. I leave the if and when entirely to you. I only directly know what I remember, namely just a few intellectual experiences which have made a lasting impression on me, the kind which probably everyone has encountered who spent the largest part of his youth alone on a farm.

I came here by train except for the last part painfully by ambulance from Munich. No one who has tried the latter will envy me.

I mustn't forget that I have already written an autobiographical sketch for Dr. Paul Carus in *The Monist* which I had thought of as my obituary. Dr Carus also added some philosophical comments for which he took responsibility. At that time, I merely had the purpose before I escaped into Nirvana of not allowing the clerical-reactionary period of Austrian history [which I experienced] from remaining unmentioned. The sketch can be found in *The Monist* for January 1910 or 1911. [The latter date is correct. - Ed.] This time I will present another side of what I can remember.

Wasserburger Strasse With respectful greetings
Kilometerstein 16

Station Vaterstetten

Your truly devoted

Postadresse:
München, Post Haar

Dr. Ernst Mach

#26

Ostwald to Mach

Grossbothen, June 3, 1913

Very Honored Herr Colleague!

Many and heartfelt thanks for your fundamental agreement. I am sure that everything which you have to relate will not only be filled with the most intensive personal interest, but will furnish our youth with the most worthwhile stimulation toward the acquisition of mental independence. Thus, I await your manuscript with the greatest longing and hope to receive it in order to be able to publish it as #1 in my planned series. I will look up your autobiographical sketch in the *Monist*.

With best wishes and greetings

Your wholly devoted

W. Ostwald

#27

Mach to Ostwald

München-Vaterstetten, June 18, 1913

Highly Honored Herr Colleague!

I am sending you the autobiographical sketch. I could not write more than a few pages without exaggerating matters and those few tell of rather insignificant but for me decisive experiences. But I would request that the manuscript be put by for a while. I feel in particular that with better health the thing could be formally improved at a later time.

I have recently read your book *Die Philosophie der Werte* with great interest. In the fourth edition of my *Populäre Vorlesungen* which appeared a few years ago one can find a short essay on "Time and Space"* in which I attempted to demonstrate by means of a simple picture that the second law of thermodynamics was not only physically valid but physiologically and

* E. Mach, "Eine Betrachtung über Zeit und Raum" in *Popular-wissenschaftliche Vorlesungen*, 4th edition, Leipzig, 1910, S. 492-508. [Since the English edition of this book stops with an earlier German edition, it does not include this article. - Ed.]

psychologically too. One can thus see the matter, so to say, directly. I have not sent you the fourth edition, because I feel embarrassed by the many printing errors, which Barth in the already thrice-printed text had allowed into it.

I request, therefore, that the autobiographical sketch not be published yet. With respectful greeting, your devoted

Dr. Ernst Mach

#28

Ostwald to Mach Gross-Bothen, June 21, 1913

Very Respected and Dear Herr Colleague!

Many thanks for your highly interesting manuscript, which indeed in this form is not quite suitable for publication. In particular, I miss, and other readers will also eventually miss, detailed information on your later activity in Vienna, as well as perhaps some further considerations on the gradual penetration of your views into the way of thinking in modern physics, in spite of the great resistance which you had to face at the start.

Your reference to the impossibility of reversing kinematographic pictures [in your "Time and Space" article] is well-known to me, and I have used this outstanding illustration very frequently in elucidating the general validity of the second law, and have not refrained from mentioning the author while doing so.

Perhaps it will interest you to know that Josef Popper [Lynkeus] has like you already declared his willingness to write an autobiography for our collection.

With all the best greetings and wishes

Landhaus Energie

Your wholly devoted

W. Ostwald

#29

Mach to Ostwald Haar near Munich, July 23, 1913

Most Honored and Dear Herr Colleague!

The day before yesterday I had the good fortune, though only for a moment, to see your son-in-law Dr. Eberhard Brauer and your daughter. My greeting was definitely not formal, since I wasn't able to rise from my seat, as I had hoped.

I consider your objections to my biographical sketch to be completely justified. I was simply not in condition, now, to deliver a better one given my miserable health, but I will endeavor under better circumstances to add a supplement. Our agreement remains the same, however, that the sketch will first be published when *you* are satisfied with it.

Both main mistakes, the formally deficient representation of the monistic point of view and slight progress from the side of official physics in acknowledging my standpoint, are well-known to me. Concerning the first mistake, I take it in the most cheerful spirit - though humor has not yet been enough to overcome it. To overcome the second mistake lies beyond my power, since at best I can merely voice my opinion, but can force neither the physicists nor the philosophers to accept it. It has scarcely been a year since independent young men like the mathematician Dingler, the philosopher Henning, Petzoldt, and gymnasial teachers like Dr. Jacob in Vienna have made known by their deeds that they have understood me. I will still remain alone for a long time or indeed always. With most respectful hearty greetings

Your loyally devoted

Dr. E. Mach

B

ERNST MACH AND HUGO DINGLER**
(1910 - 1915)

#1

Mach to the *Naturwissenschaftlichen Verein Aschaffenburg* Vienna, March 24, 1910

Highly Honored Sir!

I would be much indebted to you if you could tell me the address of Mr. Hugo Dingler. In addition, I placed an order with Gerold & Co. in Vienna to send me *Mitteilung* VI of the *Naturwissenschaftlichen Verein in Aschaffenburg,* which I hope in your goodness you will send directly to me, and merely make the cost known to Gerold [a Viennese bookstore and publishing company]. With highest respects

Prof Ernst Mach

18/2, Gersthoferstr., 144

#2

Mach to Dingler Vienna, April 4, 1910

Highly Honored *Herr Doctor* !

I was able to find your address through the publisher of your article "Grundlinien einer Kritik [und exakten Theorie der Wissenschaft, insbesondere der mathematischen]". I have possessed "Grundlinien" since 1907,

** Hugo Dingler (1881-1954) was a philosopher of science and mathematics and developed a very deep but unusual a priori point of view. He is often labeled as a conventionalist in the sense of Henri Poincaré but this does not do justice to his understanding. He was a teacher of Rudolf Carnap. Academically, he taught at Munich from 1920 to 1932 and Darmstadt from 1932 to 1934. With the rise of Hitler he lost his position. He was long an opponent of Einstein's general theory of relativity.

The letters to Mach are located in the Ernst-Mach-Institut, Eckerstr. 4, Freiburg im Breisgau. The letters by Mach were located in the Hugo Dingler Institut für methodologische Forschung, Eidos Verlag München 38, Postfach 23 during the 1960's when Frau Dingler graciously sent copies to this editor, but since then the originals are said to have been put in an archive in Aschaffenburg. Letters 2, 6, 11, 15, 17, 22, 31, & 32 were published in German in John Blackmore and Klaus Hentschel (eds.), *Ernst Mach als Aussenseiter,* Wilhelm Braumüller: Wien, 1985, pp. 78, 83-84, 93, 94-95, 106-107, 123, 149, & 156. The editor also had access to typescripts of the letters in German by Gereon Wolters, who has requested that his footnotes not be used. The editor is responsible for the English translations. - Ed.

but because of accidental factors I was only able to read it a short time ago. I would like to tell you that the differences interest me as much as where we agree, differences which come from diverse starting points: mathematics on the one side and physics on the other, which is quite understandable. Some time ago, I also wrote to the *Naturwissenschaftlichen Verein in Aschaffenburg* in order to learn your address and obtain *Mitteilung* VI from the *Verein*. But I received no answer. Perhaps you can help me get hold of a copy. At least I would be happy to learn how you stand with respect to Euclidean geometry.

With the expression of particularly high respect, Your devoted

18/2, Gersthoferstr., 144 Dr. Ernst Mach

#3

Dingler to Mach Munich, April 7, 1910

Highly Respected Herr Hofrat!

Since receiving your very friendly letter of the 4th, I have received a communication concerning the inquiry which you made with the *Naturwissenschaftlichen Verein in Aschaffenburg*. I'll send you what you wished as soon as possible. At the same time, I might perhaps mention that in the reasonable future a larger work of mine which has the purpose of examining the foundations of geometry in a deep manner will appear, in which I bring forward a series of conclusions [*Resultaten*] which stem from my investigations. I believed that such an approach in depth to this theme, by employing well-suited questions, could serve as a touchstone for how far one could penetrate into the theory of scientific cognition [*Erkenntnis*], an opinion in which I saw that I had not disappointed myself. I may perhaps add that the investigation has often followed paths where you have shown us the way, including many which helped find the goal. As soon as something comes from the press I will take it upon myself to send you a copy.

In conclusion, I would like to express to you what a great pleasure it has been for me to enter upon relations with the Nestor and Classicist of our research movement [*Forschungsrichtung*]. I sign my name most respected *Hofrat* with the expression of utmost respect and honor

wholly devoted

Dr. H. Dingler

#4

Mach to Dingler (a postcard) Vienna, April 11, 1910

Highly Honored *Herr Doctor!*

Let me provisionally confirm the reception of your friendly communication and the rich haul. I now possess *Grundlinien einer Kritik*... three times over, but will send extra copies to interested acquaintances.

With the highest respect, Yours truly

E. Mach

#5

Dingler to Mach Munich, April 27, 1910

Highly Respected *Herr Hofrat!*

My best thanks for the friendly act of sending me your beautiful essays. I enjoyed reading them very much.

With the expression of preeminent respect, Yours Truly

Dr. H. Dingler

#6

Mach to Dingler Vienna, September 4, 1910

Highly Honored *Herr Doctor!*

You have given me unusually great pleasure by your friendly act of having sent to me your work *Grenzen und Ziele der Wissenschaft* as well as through your kind dedication.

That you know how to judge preliminary work without prejudice and have succeeded with a happy stroke in bringing divergent conceptions under one unified perspective, will certainly be recognized as essential progress and will earn general acknowledgement for you.

As a mathematician, you stand completely with the apriorists, while I, as a physicist, stand closer to the empiricists. But when I reflect on more than forty years in my field, it was certainly less wilful aversion toward apriorism than the need to exhaust the prospects of my original standpoint, which determined my endeavors. Your performance heartily pleases me. Thank you very much for this [book] and I sincerely wish you the best.

With the fullest respectful greetings, Yours Truly

Dr. Ernst Mach

[P.S] Please do not think that the writings which I am sending you as a countergift impose a duty to read them; an occasional glance will perhaps keep you entertained.

#7

Dingler to Mach September 20, 1910

Highly Respected *Herr Geheimrat!*

On my vacation trip to England, I have now received the valuable letter which was sent after me, and the report of a package which you have kindly sent.

Let me express my best thanks for the friendly words which you, the master, have bestowed on the work of one so much younger, who works and seeks to earn his spurs in the same field of investigation in which you have worked so long and with such overwhelming success. I can only say that I could not wish for a more beautiful reward.

My larger work on the foundations of geometry is already being printed. It was very difficult to find a publisher for it, and I am indebted to the high-minded goodness of *Geheimrat* Wilhelm Ostwald for helping me to come to an arrangement with the *Akademische Verlagsgesellschaft m.b.H.* As soon as I receive a copy, I will take it upon myself to have one delivered to you.

Please allow me, highly respected *Geheimrat,* to convey my best wishes to you. With the expression of particular esteem I am Your devoted

Dr. H. D.

#8

Mach to Dingler (postcard) Vienna, February 27, 1911

Highly Honored *Herr Doctor!*

I thank you heartily for the *Grundlagen* etc., which I received today. I would like to know your Aschaffenburg address, since I do not know if everything will arrive safely to you at Arcisstr. 44? You also know well the address of Professor L. Burmester*, who as I notice you cite. Please, in this case if you could also send me his. Meanwhile, I am enjoying the study of your book. Most respectful greetings

E. Mach

* Ludwig Burmester (1840-1927) taught mathematics at the Technical University of Munich. Dingler was his assistant until 1912. - Ed.

#9

Mach to Dingler Vienna, March 16, 1911

Highly Honored *Herr Doctor!*

I have read through most of *Grundlagen der angewandten Geometrie*, but several disturbances have prevented me from carrying out a truly basic study, which is very necessary. The book is naturally more difficult to read than *Grenzen und Ziele der Wissenschaft.*

I find particularly interesting the investigation into fixed bodies. I believe, namely, that a geometry which starts from this one and which bases itself on it, which is similar to the one which Leibniz has already partly attempted, would be first a natural geometry, a physics of space.

Concerning the frequent, too frequent mentioning of my name - although I am very thankful for every acknowledgement, I am obliged to warn you in the interest of the matter and in your own interest. I believe that you will thereby create significant opposition. This has its natural grounds. I never had a teacher of importance, outside of the great, dead, Classical authors. For the time when I studied fell almost completely before the reform of the Austrian universities, which Kaiser Franz had allowed to rot completely. Since I didn't have enough money to attend a German university, I remained a stranger, an outsider with respect to the whole teaching community, and against whom one not only had mistrust, but visibly endeavored to arouse it. I'm not at all unhappy about it; perhaps the isolation is exactly what helped me discover some thoughts.

I thank you heartily once again for your work, from which I still hope to learn. I would gladly send you something in return, but have forgotten what I have already sent you. If you would be good enough to remind me about what you are lacking, I would be very grateful.

From the brochure of the Bologna Congress [for Philosophy] I see that you will make a contribution on the *a priori* and experience. I have had to confine myself on such occasions for over a decade [merely] to concern with the congress reports [since my health did not allow me to visit them]. I hope that you will have a good, refreshing discussion and will not let yourself be upset by national jealousies [*Eifersüchteleien*]. Should you find any traces of Franciscus Maria Grimaldi in Bologna, I would be grateful for any allusions.

I have received a contribution from Professor L. Burmester to be included in a new edition of *Analyse der Empfindungen.*

With most respectful greetings, Your sincerely devoted

Dr. Ernst Mach

#10

Mach to Dingler Vienna, July 20, 1911

Highly Honored *Herr Doctor!*

I thank you heartily for your friendly letter of July 24 and the information about the Congress in Bologna. I have also heard something through Kleinpeter* and von Engelmeyer** and will receive the Congress report. It is very understandable given the Italian way of doing things that you were unable to find anything about Grimaldi.+ Nor was Engelmeyer any more successful. He found the birthdate of Grimaldi in an archive and also an entry from his *father;* who most remarkably was eight years old when his son was born. Since this has not been the first attempt to discover something, I will probably have to give up on him. Also, the Jesuits don't know anything concerning who put the saying on his gravestone: "He lived among us without complaining ..."

Professor Burmester has given his permission for me to include his involution theory on some illusions of shape [*Gestalttäuschungen*] in the new edition of *Analyse der Empfindungen* and this has has encouraged me. The question now is on how to make the conception physiologically understandable.

Meanwhile, after numerous disturbances, I wish you fresher enthusiasm to work and renewed pleasure in doing work, Yours full of high respect, Yours truly

Dr. Ernst Mach

#11

Mach to Dingler (postcard) Vienna, July 25, 1911

Highly Honored Herr Colleague!

My heartiest thanks, especially for your Congress lecture. My previous attempts to understand Cantor's set theory have only experienced modest success. The problem is certainly in me. Nevertheless, even Poincaré ap-

* Hans Kleinpeter (1869-1916) was one of Mach's most dedicated followers. He taught in a Gymnasium in Gmunden, Austria and published a book appropriately titled *Der Phänomenalismus,* Leipzig, 1913. Mach's large correspondence with him has not yet been located. - Ed.

** Peter Klementitsch von Engelmeyer was a Russian engineer and philosopher who helped found a positivist society in Moscow in 1912 and who helped Mach publish several of his books in Russian. Mach's many letters to him have also not yet been found. - Ed.

+ Franciscus Maria Grimaldi (1618-1663) was a major source for Mach's posthumous book on geometrical optics. (He is mentioned on 22 pages.) Thus the suggestion is that he was still working on the book when the letter was written. - Ed.

pears to have little taste for "*le Cantorisme*". For me, it's as if I were listening to a Scholastic from the Middle Ages lecturing on mathematics. It would take more time [to master] at age 74 than I have left.

Greetings full of high respect

E. Mach

#12

Dingler to Mach Munich, September 1, 1911

Highly Respected *Herr Geheimrat!*

First, let me render because of many unfortunate circumstances my best belated thanks for your friendly package. I have again read with a feeling of particular pleasure some pages from your hand, and, as I may add along with a very special aesthetic feeling of what is hard than otherwise to call *gemütlichen* enjoyment, which I can scarcely remember having had from any other scientific author. Recently, I have re-read parts of your *Wärmelehre* and *Mechanik,* and am repeatedly astonished how deeply you as the first, [the pioneer], have thrown light on the actual rise and theory of science, where earlier, apart from occasional remarks of scientists and investigators themselves, many philosophers ... have only given absolute ignorance, idle rubbish [*Mist*].

I don't know what has given me the courage to write all of this to you. which risks the danger that you will consider me immodest, but I have the feeling that you will not misunderstand. - I work just now intensively on set theory, and I think I have clearly understood it. The paradoxes, which have really caused a bit of havoc [*Unheil*] and which the people who could have done better avoid, are completely obvious. (I don't know if the Afterword to Georg Frege's *Grundgesetze der Arithmetik* is known to you.) But in any case, what is more important, the so-called "continuum problem", whose two famous proofs by Zermelo I have been able to prove false in the brochure which I sent to you, has been strongly resisted [*verschleppt*].

There is still a certain charm in overcoming difficulties which for all their effort a considerable number of skillful scientists have not been able to do, and these problems appear to me to have real significance for science as a whole, [but] as their previous unsolvability has indicated, our abstract concepts are not yet compatible enough with the logical facts to be able to speak to you about them. These [issues] are quite fundamental and therefore rather important, even if in an abstract field of thought.

Finally, a certain mysticism has spread stepwise precisely in this field; nor can it be said that Georg Cantor the very significant and deserving creator of the discipline is innocent. Thus, it seems not entirely unsuitable to show that one can conceive concepts in a clear and sensible way, and by so doing, banish mysticism *ad diobol.* back to where it belongs.

One could bring forward a whole quantity of quite distressing examples to point out how narrow insight into the general theory of science still is and how important a possible comprehensive treatment by you of this first properly inaugurated discipline could well be. Indeed, a quite skillful mathematician has been tempted to speak out, for example, in several places precisely on these paradoxes that while mathematics must strictly defend the principle of contradiction, that it is not so settled, however, in logic (as if one could remove the paradoxes from mathematics and shove them into logic).

Nevertheless, I don't want to bore you any longer with these things and steal your valuable time, which to a large extent I have already done.

In lasting respect, most honored *Geheimrat,* I sign myself

Schraudolphstr. 14/11 Your devoted

Dr. H. Dingler

#13

Mach to Dingler Vienna, January 13, 1912

Highly Respected *Herr Doctor!*

I'm sending you a little piece of fun today, which is perhaps best suited to wake an interest in the growth of knowledge even in circles where the educational level is rather simple. Thank you very much for your friendly letter of September 1, 1911. I am very busily occupied with new editions. Most respectful and hearty greetings, Your devoted

E. Mach

[P.S] Still no Congress report from Bologna?

#14

Dingler to Mach Munich, January 22, 1912

Highly Respected *Herr Geheimrat!*

Hearty thanks for being so friendly as to send me your attractive essay, which I enjoyed [reading] very much. Nothing has come from Bologna yet. I am working just now on a kind of epistemology, since I have finally discovered some methods concerning basic questions (about the relation of the inner and outer world, etc.) which earlier I had not been able to treat in sufficiently exact form and where I had been essentially dealing with mere hypotheses, [but now] or so it seems, I am dealing with something objective and systematic and thereby capable of being handled in a basic but completely simple way. It gives quite remarkable results! I write these lines, most respected *Geheimrat,* with best wishes as Your most respectful devoted

Schraudolphstr. 14/1 Dr. Hugo Dingler

#15

Mach to Dingler Vienna, January 26, 1912

Highly Honored *Herr Doctor!*

For a long time I went without writing, because I was and am still the slave of my publisher.

I am very interested in your views on theory of knowledge and am pleased with your writings on that subject. You will reasonably soon receive the 7th edition of my *Mechanik;* it has become somewhat more homogeneous, but at 74 one no longer causes revolutions. Younger people must make them. The elderly console themselves by wishing them well, which in this case comes from the heart.

Though I esteem Poincaré with his conventions, still something is lacking in that approach in my opinion. Conventions are not at all arbitrary, but are pushed through with gruesome pressure. The bow, arrow, and sling were all discovered in prehistoric times without knowing how the principle of inertia was ordained. I also find a lot missing in Kant. I only see in the *a priori* a somewhat larger amount of circumspection among those who have studied a field for the second, third, or nth time. Also economy [Mach's own theory] can embody very different points of view.

With most respectful hearty greetings, Your devoted

Dr. Ernst Mach

#16

Dingler to Mach March 23, 1912

Highly Respected *Herr Geheimrat!*

Many heartfelt thanks for kindly sending me the 7th edition of your wonderful *Mechanik.*

The friendly remarks which you have so graciously bestowed on my modest works have somewhat shamed me, and caused me to reflect once again on how much remains to be done in order to justify your friendly opinion. Those remarks will spur me anew with you as my model in order to follow more diligently what I have set forth as my life's task.

About my present studies in epistemology, to which my free time is devoted to the extent currently possible, I would say that I am generally satisfied with my progress. But it is becoming apparent that it is impossible to carry out everything down to the same detail which would be desirable. Let me now say as I incorporate your general epistemological investigations and results [into my work] that it is self-evident, if I may express myself so, that the views which a scientist of your depth and lack of prejudice has acquired by your own efforts are unconditionally correct and unassailable; all that I can do given the extent to which my views are conditioned by yours might be to "augment in a certain respect" a few things.

I would be glad to go into more detail, but I fear that in this relatively small space that it is scarcely possible to make myself fully clear; but on the other side I hope that it will not be too long until I can lay my publication itself before you. Nevertheless, as it seems to me, I will have to face the fact that I am still going to have to spend a lot of time on it. The philosophical literature in this field is quite extensive and not always easily digested, but still, it does appear to be occasionally worth the effort, more so than I had sometimes earlier thought.

Once again my heartiest thanks for your kindness; I am, most respected Herr Geheimrat, [your] respectfully devoted

Dr. H. D.

[P.S.] I should add that I have already read some parts of your works again with great pleasure; I still remember buying the 4th edition of your *Mechanik* in Göttingen in November 1902, and with very great suspense together with the highest level of interest "devouring" it. That was indeed my first acquaintance with your book.

Already rather early at Gymnasium I had developed a naturally rather unconscious interest in science in general from my studies and reflections, and thus it was an unforgettable experience and the most essential ripening of that interest when one Christmas - I believe I was seventeen years old at the time (1898) - that my father, who knew what I was interested in, gave me as a gift your *Populär-wissenschaftliche-Vorlesungen*. I can freely say that this book and its manner of looking at things already made an extraordinary impression upon my memory at that time, an impact which has lasted ever since.

#17

Mach to Dingler Vienna, November 20, 1912

Dear *Herr Doctor,* Highly Respected Herr Colleague!

Your friendly letter of November 17 in which you informed me that you had obtained your *Habilitation* at the University of Munich for Methodology of the Mathematical Sciences has given me a great deal of pleasure, but not only for your sake but for the benefit of that field. Nevertheless, set theory lay rather far from my ken. That has its ground in the weaknesses of my mathematical education when I was young, which unfortunately I have not found an occasion to repair.

Things have gone rather badly with me recently. I fell and hurt my hip, which has kept me shackled to my bed immovable and helpless for a whole quarter-year. The result has been that the left-side of my body has been considerably damaged, while the right-side had been made almost completely unusable by a stroke I suffered almost fifteen years ago.

I don't know if you have joined the *Gesellschaft für positive Philosophie* which some young friends of mine have established? I am satisfied that there is not much more I can do in this direction. But I have experienced the satisfaction that my writings, which I have worked on with the greatest eagerness, have taken off on their own. They don't need me anymore.

Philip E.B. Jourdain, the set theorist, is currently harmonizing the 3rd English edition of the *Mechanik* with the 7th German one. Jourdain has also edited the 6th English edition of *Analyse der Empfindungen* and has prepared himself for the translation of *Erkenntnis und Irrtum.* All of this has somewhat lightened my misery. As a mutual friend has informed me, Jourdain, M.A. Cant., is completely lamed from what is presumably a progressive paralysis. In order to earn the 700 pounds which he needs to stay alive the poor man has to write and translate.

With hearty best wishes and deepest thanks for the letter and treatise

18/2, Höhnegasse, 25 Your loyal

Dr. Ernst Mach

#18

Mach to Dingler (a visiting card) January 10, 1913

In hearty reply to your friendly wishes [for the New Year]!

#19

Dingler to Mach undated

Best wishes and happiness for your most-honored Jubilee [February 18, 1913, Mach's 75th birthday]. Ad multos annos.

Dingler

#20

Mach to Dingler February 1913

(a pre-printed card thanking the recipient for offering best wishes for Mach's birthday.)

#21

Dingler to Mach Munich, June 24, 1913

Highly Respected *Herr Geheimrat!*

I have had the intention for so very long to write to you once again in more detail; I had even wanted to do it for your Jubilee early in the year, but again different things intervened [to prevent it]. The main reason it didn't happen was that at the end of winter I didn't feel up to par. There was clearly a relaxation of effort, which set in after the years of rather intense work during which I had fought for my *Habilitation* and after the first semesters of my new activity were passed, and with it my drive weakened. Nor could I rouse myself even over the Easter holidays. I also had difficulties finding a publisher for my book. Finally, *Hofrat* Ostwald was again so friendly as to have the book accepted by a publishing house with which he has close relations. I have called it *Die Grundlagen der Naturphilosophie* in order to give it a tag which can be identified with a particular group of people, since, as it seems to me, there is nothing worse for a book, than when people don't know "What they should make of it". The publication which is now almost finished has kept me busy the last month.

But I note that I am always talking only about myself. I don't know anything at all about the further consequences of your regrettable accident. I hope and wish from all my heart that you have fully recovered and are enjoying good health.

I am very pleased to be contributing to the *Zeitschrift für positivistische Philosophie.* I consider it a very promising thing. Previously there had been no central point for efforts to go further [even though] they always appear to be spreading more and more widely, [following] paths which for the most part had been opened up by you, and, though we may at present only think about it in hushed tones, in the future there will be calls to replace the current "metaphysical" philosopher-schools and systems - certainly at the very least in the fields of epistemology and "metaphysics". In the first issue, I don't rightly know if my little note "Übergreifende Begriffsbildung und Causalitat" was as generally understandable as I would have wished. Nevertheless, it includes - indeed in a very small space - what is essential. The two questions: What does so-called "logical dependence" really signify? and How do matters really stand with so-called "causality"? have kept me very occupied again and again over many years, and I have only made progress with the second problem after first passing through a series of provisional and inadequate formulations. How close these two things are related, only forcefully came to me with the writing of my new work [*Schrift*]. It has always given me great pleasure how closely these formulations follow in your sense, particularly in wiping out metaphysics. For if there is no immanent causality of things, then the last justification for every kind of metaphysical and eternally unexplainable powers, which are irresponsible and make things inconceivable in principle, falls. In other words, the last refuge of metaphysics collapses. That was already completely expressed in your acute remark on cause and effect in *Populär-wissenschaftliche Vorträge,* 4th edition, page 284.

Now, I am completely aware that when an experimentalist attempts to apply these logical formulations immediately to his daily scientific work he will have difficulty. There will be little progress, at least not immediately. This is because scientific theory must deal with the ideal case, and daily scientific practice for the most part is very far removed. It is therefore urgent to try to develop a methodology based on these principles which can be directly applied by working scientists, where it can be shown that the methods can be applied in all cases to investigations even about unclear and confused relations within an unexplored field, and finally, that these theoretical methods of science can only be grounded in this manner.

My larger work with which I am occupied now (and which unfortunately had to be delayed a long time because of proofs) is about the foundations of arithmetic and algebra. Until now, no one had approached the matter in your historical-critical way. The French appear to want to begin: Brunsvicg and in the new issue of *Revue de metaphysique et de morale* Boutroux. But they haven't gotten beyond some rather superficial historical tales and a pair of *apercus*. The core of their approach is therefore rather streaky [*gestreift*]. I find the topic wonderfully interesting, and have already found some results which have interest from a purely mathematical standpoint.Thus at least partly, wholly new mathematical methods of thinking are coming into existence. Still, they are mostly matters which over the course of years have only very gradually arisen for me from side concerns, and not from yesterday to today.

Highly respected *Hofrat*, it would be my above-mentioned wish that you should enjoy good health and recuperation, and I wish you the best.
I sign with expression of my respect

Giselastr. 22 Your devoted

H. Dingler

#22

Mach to Dingler Munich, June 29, 1913

Highly Honored Herr Colleague!

It pleases me very much to have received a sign of life from you. I am very curious about your new book. I read your article in the first issue of the *Zeitschrift für positivistische Philosophie*. Meanwhile I have received the second issue with "Paradoxien der Mengenlehre". Gradually, my library really is being put into order. Moving here has been accompanied with corresponding difficulties. I have been here [Vaterstetten] since May 15. I haven't yet made an attempt to visit anyone, since first I will have to learn how to walk, which I haven't yet mastered at all. I have a ledge running across my room which I use [for support], but without much success.
With most respectful hearty greetings, Your devoted

Short postal address:
München
Post Haar

Dr. Ernst Mach

Signpost:
München
Wasserburger-Strasse
Kilom. Stein 16
Station Vaterstetten

#23

Dingler to Mach Munich, July 1, 1913

Highly Respected *Herr Geheimrat!*

Many thanks for your considerate lines. I had no idea, that you have lived so long in the neighborhood of Munich, otherwise I would already have attempted to satisfy myself directly what your letter has led me to believe that everything is going much better with you. I am somewhat afraid that you will not find me to be the person you perhaps expect. It is hard to say something with lasting value about oneself; but at least one can say that an eye which is accustomed to unrelenting sobriety in science and the greatest possible absence of prejudice has certainly first exercised this capacity on his own mistakes and continues to do so.

I sign with expression of high respect and admiration

Giselastr. 22 [Your] devoted

H. Dingler

#24

Mach to Dingler (postcard) Munich-Haar, July 14, 1913

Highly Honored Herr Colleague!

I am very happy to accept your friendly request in order to receive the finished printed sheets [*Bogen*] of your [volume on] natural philosophy. The weather has taken on a more friendly shape and is more favorable to me. Full of high respect, Yours truly

E. Mach

#25

Mach to Dingler Haar near Munich, July 21, 1913

Highly Honored Herr Colleague!

Hearty thanks for the friendly act of sending me the proofs of your new book which I read with great pleasure in one gulp, but because of its rich content I will often read again. Most hearty thanks also for your friendly treatment of my writings and particularly on causality. As far as I know, you are the only reader who has not overlooked the citation on page 220, but have recognized that something about the structure of natural science was inserted into it, even if I did not allow myself to say exactly what it was.

In 1895, because I could suggest a lot more than I could prove I could not bring my investigation fully under logic, thus, I lectured on "the psychology and logic of investigation". I felt [the difficulty] still more when I wrote down the book, so I withdrew even more of my discussion on logic. But you are completely right, and I am grateful that you have reversed the process and corrected the situation. I hope and very much wish that your book [*Schrift*] will have the great success which it deserves.

With hearty greetings and full of high respect, Your devoted

Dr. Ernst Mach

[P.S] Hopefully, you at least had good weather toward the end of your excursion. Have you met [Joseph] Petzoldt yet, whom I gave your address to?

#26

Mach to Dingler Haar near Munich, September 5, 1913

Highly Respected Herr Colleague!

I am very pleased that you encountered better weather south of the Brenner [pass] and have recovered your spirits. I must conclude from your letter before your departure that you were not able to meet Petzoldt. I had somewhat erroneously thought you had met him.

Hearty greetings full of high respect, Your devoted

E. Mach

[P.S.] Could you give me the exact address of M.H. Baege? I believe it is time now to pay my annual subscription [to *Zeitschrift für positivistische Philosophie*]. Is Berlin-Friedrichshagen good enough?

#27

Mach to Dingler Post Haar near Munich, October 18, 1913

Highly Honored Herr Colleague!

I must tell you that I would be very interested were Professor Finsterwalder* to publish his lecture in the positivistic journal. If I have not deceived myself, his name popped up in a publication of von Seeliger** on the earth's shadow on the moon. It was my impression that I observed the facts in completely different form thirty years ago and also settled the matter. I am not acquainted with Finsterwalder's contributions.

Also, Burmester interests me a lot. He helped me in connection with the right expression for a pseudoscopic appearance of relief. But physiologically, something is still lacking in order gain full understanding. But please do

* Sebastian Finsterwalder (1862-1951) was professor of mathematics at the Technical University of Munich. - Ed.

** Hugo von Seeliger (1849-1924) was professor of astronomy at the University of Munich. He corresponded extensively with Mach. He is also known as a strong opponent of Einstein's theory of relativity. - Ed.

not bother Professor Burmester with the matter, since I am not in condition to make a visit to him.

Hearty greetings, Your devoted

E. Mach

#28

Mach to Dingler (postcard) Haar, January 14, 1914

Highly Honored Herr Colleague!

Before I could thank you for your card from Aschaffenburg, I received a letter from Munich once again, which brings me a lot of pleasant information, above all the report of your new work, which I anxiously look forward to; likewise with respect to the promised visit of Professor Finsterwalder, who I will receive gladly, even though at the moment I have some awkwardness in speaking, and therefore am little ready for lively conversation. Still, I hope to be able to make myself understandable. With hearty thanks and full of most respectful greetings, Yours truly

Dr. E. Mach

#29

Mach to Dingler (postcard) Haar, (undated)

Highly Honored Sir!

It would be nice if you might direct some lines to Dr. Joseph Petzoldt in Spandau, Wrohmännerstr. 6, about the coming publication by Professsor Finsterwalder: This might have the effect of making it better known to people who know nothing of this work, or like myself, almost nothing. I was not able to tell Petzoldt anything other than that the work was over a contrast in brightness and was connected with Seeliger's work on earth shadows on the moon. Full of high respect

Ernst Mach

#30

Dingler to Mach Munich, October 22, 1914

Highly Respected *Herr Geheimrat!*

I have wanted to write you for a long time, but from the beginning of vacation the military has once again called me in for an exercise*, and du-

* Mach's son Ludwig had asked Dingler not to mention the war, and since Mach had apparently ceased to read newspapers, it is not fully certain that he was aware that a world war had broken out. - Ed.

ring this time it has been difficult to write. I was already on the point of sending you a few lines when I received your nice card.

Concerning the matter which you spoke about, it seems that Mr. Finsterwalder first wants to undertake a more comprehensive investigation into his project before he steps before a larger public. In addition to that, he is extremely busy with taking over and giving new lectures on descriptive geometry and other things. So it will still be some time before something appears. At that time, I asked Mr. Petzoldt if he might not want to request a contribution from Professor Finsterwalder for the *Positivist. Zeitschrift.*

I am quite happy that this undertaking appears to bloom and is making good progress. It has also pleased me a lot to become personally acquained with Mr. Petzoldt, who is a very fine [*prächtiger*] man. I experienced some very interesting hours in his company.

Our university will start again on November third. This time I will try to begin with a "mathematical-philosophical colloquium" and of course on the newer literature on the foundations of arithmetic. If everything comes off, which is by no means certain given the novelty of the thing, you will occasionally have your ears burned.

With best wishes for your recovery, I sign with the expression of my warmest regard as Your devoted

Giselastr. 15 H. Dingler

#31

Mach to Dingler Haar near Munich, April 4, 1915

Highly Honored *Herr Doctor!*

Some days ago I received *Das Princip der logischen Unabhängigkeit in der Mathematik* from the publisher. Until now I have only browsed and not yet studied it, hence my thanks are not yet as complete as they will become. Reading it will make me happy to the extent that I generally understand it, for as I have told you, I have not been able to follow the development of modern mathematics for many years. Perhaps we can see each other again sometime. After this winter, my health leaves almost everything to be desired. I place my hopes as ever on the return of Spring.

Your old devoted

Dr. Ernst Mach

#32

Mach to Dingler October 8, 1915

Respected *Herr Doctor!*

Please do not hold it against me if with my bad physical condition and the disorder of things on my desk, that today for the first time I belatedly but just as heartily and sincerely offer my best wishes to you and your bride on your marriage.

May the coming years bring unbroken capacity to work as well as time and rest. For then it shall be certain that something very special [*Besonderes*] will take place. With this conviction, the old greets the young scientist.

Dr. Ernst Mach

CHAPTER 5

PHILOSOPHICAL DOCUMENTS

A

LECTURES ON PSYCHOPHYSICS - CONCLUSION

by Ernst Mach*

(1863)

"Thought is related to the brain like gall to the liver and urea to the kidney." Thus runs the famous materialistic line. *Lotze* has commented to the contrary in striking fashion that he would never have believed that thought could have originated in such a deeply poetical way had this saying not been so exactly proven. The above line rests indeed upon the entirely correct observation that psychical and physical processes stand in very close relation to each other. Nevertheless, it doesn't require deep thought in order to find the hitch in the figure of speech. Thinking is no secret. No one has yet discovered a thought-fistula like one finds urine-fistulas and gall-fistulas. One cannot say that thoughts are physical processes in the brain. If we examine physical processes very carefully, we indeed find molecular movement, but no thoughts. Electrical currents in the brain are definitely different than the green color which perhaps represents them when the subject is examined. They are *our* representations when we examine the brain, and not those of the examined brain.

Such crude ways of understanding owe their origin and influence for the most part to opposition to longstanding prejudices. But it is surely a very crooked way to the truth to fight existing prejudices and untruths by immediately replacing them with others.

* This is a translation of Mach's concluding remarks from a forty-two page article on Gustav Fechner's *Elemente der Psychophysik* (1860), apart from a short "Afterword". Mach's youthful commentary was originally published in twelve discontinuous segments in *Österreichische Zeitschrift für praktische Heilkunde*, Jahrgang **9**, 1863, ranging episodically from page 146 to 366. Joachim Thiele photoprinted the segments and put them together to form a privately published book in Hamburg in 1978 titled *Vorträge über Psychophysik*. A problem is that in this article - even in the conclusion - Mach did not always clearly distinguish his own views from those of Fechner. Nevertheless, one can detect more than one apparent shift in Mach's losing effort to convince himself that he believed in a generally acceptable form of the reality of physical atoms. - Ed.

J. Blackmore (ed.), Ernst Mach – A Deeper Look, 111–126.

Fechner has carried out his entire psychophysical investigations without starting from an hypothesis about the nature of the psychic and even without making an assumption about what might lay between stimulation and sensation. These things as we have seen are completely unnecessary if we are merely hunting for the facts of experience. Scientific progress even in this case is not to be rushed.

At the end of his book on psychophysics, *Fechner* comes back to questions about the connections between the physical and the psychical which he had only touched upon at the beginning and tries to answer how far these fit the current stage in the development of our experience-based science. Few would deny that there is a connection between the physical and the psychical, nevertheless, this is customarily only admitted for perception. *Fechner* considers it probable that this connection is also valid for representations and generally for all psychical processes, for the latter cross over so inconspicuously that one cannot fix the exact frontier where the well-grounded assumption about perception [which applies to them as well] could be justifiably given up. Sensations [*Empfindungen*] inconspicuously cross over into representations [*Vorstellungen*], and lively representations erupt into willing and muscular movement. Also, imagination [*Phantasiebilder*] can slowly assume the character of some objective occurences and can resemble blended images. Afterimages produced by dreams have already been frequently observed. The latter proves almost exactly that there is a nerve process which accompanies the representations.

But if the physical and psychical are so exactly connected as to be interpenetrable, what does this connection mean? Perhaps, thinks *Fechner*, physical and psychical processes are not really different at all. Perhaps they are the same merely considered from different standpoints. Perhaps what appears to the observer of the brain to be electrical current, seems to the possesser of the brain to be green color. The perspective varies with the standpoint. If one stands within a circle, then the convex outside is concealed by the concave inside. If one stands beyond the circle, however, one sees on the contrary only the convex outside. This way of viewing the matter serves very well at least as an hypothesis to account for the the appearances. It also suggests another option.

Certainly for most people, nothing seems clearer than matter, at least as long as they don't reflect carefully. In fact, many difficulties lay in the concept, even when looked at from a physical point of view. Certain physical appearances, such as the dispersion of light [*Farbenzerstreuung*] and transversal aether waves, which the undulation theory demands in order to fit the facts, can only be explained if one assumes that matter consists of discrete particles separated by empty spaces, that is, atoms.

Elsewhere I have shown (Schlomilch's *Zeitschrift für Mathematik,* **VII**, Heft 3) how one can infer the nature and constitution of gas molecules out of the results of the latest spectral investigations of light. That is, one can infer the nature of a system from the vibrations which shake it. Now, the vibrations which the heat-disturbed gas molecules radiate in the form of light teach us that the molecules consist of a finite number of atoms which vibrate against each other. Spectral analysis could therefore be called - to use an illustration familiar to physicians - fitting the percussion to the oscultation of atoms. If, however, we want to involve ourselves with as few hypotheses as possible, then we would have to say that physics always and everywhere leads to the assumption, that there are only a *finite number of substances and powers* which are active in sensory phenomena.

Now, how should we understand atoms? Are they colored? Luminous? Resonant? Hard?

These are sensory qualities which only atoms grouped together can share, for all sensory phenomena are the product of a variety of atoms. One atom in itself cannot possess such qualities. In fact, we may not [*dürfen nicht*] even think of atoms as spacially extended. For as we have seen, space is not an original condition [*nichts Urspringliches*] but probably results from an indirect cooperation of a variety of substances [*Wesen*]. - Even physicists have felt the difficulty of representing atoms *spatially*, and some, therefore, consider atoms to be mere centers of power, but even power centers are really nothing in themselves. Indeed, what could it mean to say that one power center influences another? - In short, let us make a confession. We cannot reasonably infer what the outside of an atom must be, but if we would think about atoms at all, we must attribute to them an inside, an inner character somewhat analogous to that of our own soul. In fact, where could the soul in a combination of atoms in an organism come from, if not present at least as a germ in individual atoms?

If we find in organisms that physical processes are the reverse side of what is mental, where is the seat of the soul to be found? *Fechner* keeps the previously known facts of physiology clearly in view and finds that noticable mental disturbance is not connected to any particular part of the nervous system. The whole nervous system is the showplace for psychic phenomena. Indeed, one could ask whether such an arrangement might not cause the unity of consciousness to be lost, but the facts speak too loudly [against this hypothesis] for such a contradiction to be endured. Also, the difficulties which lay in this situation for the process of thinking are not incapable of being overcome.

We can immediately know only *our* soul. By analogy we ascribe to other people a soul which is similar to ours on the ground that similar actions probably reflect similar thoughts and mental conditions. This type of inference is less certain when applied to animals, and indeed decreases the further we go down the size-scale. Where should these inferences cease altogether, asks *Fechner?* The mere question itself must be acknowledged as brilliant [*geniale*], and if we would ponder it well, we can only answer with *Fechner*: Nowhere! - An answer which has already become clear to us from the other [i.e. the physical] side by the considerations above.

[In short,] we are surely not yet authorized to deny a soul to plants and lower animals, even though, if we remind ourselves how essentially the organization [of animals], their capacity for movement with perception, hangs together with mental life, we must confess that the mental life of plants would have to be very different from that of animals.

With these somewhat idyllic considerations I conclude my lectures, and it will please me if I have been able to help my readers keep two things in mind above all which we currently feel more and more strongly able to strictly prove:

1. That physics, physiology, and psychology stand in indestructable connection [*unverwüsstlichen Zusammenhänge*] with each other, so much so, that for every one of these sciences their well-being is only to be found in conjunction with the others, indeed, that each one can be considered as an auxillary discipline for the others.

2. That it is not necessary to abandon exact research when one leaves the domain of the self-evident [*des Handgreiflichen*].

B

A DRAFT FOREWORD TO THE RUSSIAN TRANSLATION OF *DIE ANALYSE DER EMPFINDUNGEN*

by Ernst Mach*

(1906/7)

If the author, who is neither philosopher nor psychologist but only a physicist, indicates how he has arrived at his handling of the themes which are discussed, then the contents of this book will perhaps be a bit easier to understand and meet with more fair-minded judgment. In early youth, the author's naive realistic world view was shaken by Kant's *Prolegomena*, and all of his critical thinking stems from that. Very soon the author was led to recognize that the unreachable "thing-in-itself" was just illusion, and by that means he came much nearer to Berkeley and Hume, who had initially influenced Kant, without at first however being able to acquire from these authors more than the traces which had been left behind in Kant's writings.

Naturally, this frame of mind was maintained in his professional studies in physics as well. After graduation, he was short of laboratory equipment, even of a makeshift nature. Partly because of this and partly because of a tendency toward and interest in epistemology, the author hit upon experimentation with the sense organs [which did not require much equipment]. Thus originated his sense-physiological work which lasted from 1861 until 1875, which already contained everything essential which would appear later in more detailed and comprehensive terms in the first edition of this volume in 1886.

This concentration did not remain without feedback [*Rückwirkung*] on his understanding of physics. Adopting a critical standpoint and the historico-physical studies which he had developed during the period from 1863 to 1871 led his thoughts to the mature view that the goal of physical research consisted in the determination of the dependence of sensory experiences [*Erlebnisse*] on one-another, but that the concepts and theories of physics represent a mere thought-conserving [*denkökonomische*] medium

* This is a full translation of a much retyped manuscript located in the Ernst-Mach-Institute, Eckerstrasse 4, Freiburg im Breisgau. Only incomplete sentences and passages clearly intended to be superceded have been left out. The Russian book came out in 1907. The editor was G. Kotliar and the introduction was by A. Bogdanov. A few parts of the above manuscript also seem to have been used though employing slightly different wording in Mach's article "Sur le rapport de la physique avec la psychologie" in *L'Année Psychologique*, Paris, **12** année, 1906, pp. 303-318, which was years later included in a German-language version in *Populär-wissenschaftliche Vorlesungen,* 5th edition, 1923. - Ed.

for the achievement of this goal, in so far as they are not to be designated as physical analogues for illusory things-in-themselves which could even extend beyond possible experience.

This standpoint, which was reached in 1871, has merely been presented in more detailed form in his books on the science of mechanics (1883) and thermodynamics (1896). The author became acquainted with an ideal similar to that of physics as a student through Herbart's over-schematic attempt to form a mathematical psychology without any physiological foundation, and through that he came very close to Hume. At this point the rise of biology and the theory of evolution intervened to help correct his views, and already by 1865 had made plain that all of mental life, especially the activity of science, was to be conceived as a part of organic living, such that even economy of thought and the elimination of futile [*müssigen*] metaphysics found their deepest justification in biological needs. Thus the goal of science as a whole would be the most thought-sparing [*denkökonomische*] inclusion of reciprocal dependence of the sensory and conceptual experiences of mankind on one-another.

This goal would not be very different from that of Auguste Comte, except that for Comte the natural sciences are everything and psychology signifies almost nothing, a view which is contrary to that of the author, who considers both areas as equally important and would encourage the examination of what is only consciously presented to one person as strongly as what is consciously presented to everyone. If one abstains in each of the fields which have been named above from uncontrollable, futile, metaphysical assumptions, which disturb economy of thought, then one will recognize the simplest common *elements* of both fields, which in the psychological domain we call sensations and which in natural science are known as physical qualities, but which are actually identical and are only different in the different ways the relations are considered. This approach yields a further simplification and unification of *method.* But the author would remind the reader that in science one must keep to the next scientific step. This scientific analysis is not intended to disorient anyone for vulgar living. It is expected that the reader will not forget that the elements are connected parts of greater or lesser-known complexes (self, things, world). Scientific analysis can never have the goal of eliminating the vulgar concepts which are practical and worthwhile, but must critically refine and use them.

Although the author long had the impression that he was alone with his epistemological views, swimming against the tide, it was always far from his mind to believe that victory depended exclusively on his own efforts. Gustav Fechner's *Elemente der Psychophysik* [1860] stimulated him a great deal, and in the early seventies Ewald Hering and his orientation in the field of psycho-physiology had an effect on him. Gustav Kirchhoff's pronouncement on the significance of [science as factual] description (1874) was the first encouragement for him in the physical realm, but some twenty years later the first extensive agreement followed with the approval of Heinrich Hertz, support, however, which was not without reservations.

The author now knows today that Hermann Grassmann had already represented similar views in 1844, and can count men like Pierre Duhem and Henri Poincaré as fellow-seekers [*Strebensgenossen*] in epistemological matters, even if he is not able to follow the extreme nominalism of Poincaré in everything.

At the beginning of the 1880's, he became acquainted above all with the first writings of Richard Avenarius, and later, he persuaded himself to rest satisfied that also Wilhelm Schuppe, Theodor Ziehen, and a series of younger researchers, R. von Schubert-Solden, Frantisek Wald, Joseph Petzoldt, Hans Cornelius, and others while clearly not holding identical ideas, nevertheless, followed a similar path. One may therefore speak of a more general trend, which perhaps reaches back even further than has been customarily assumed, but which the author considers as a guarantee that his efforts satisfy a real need, and that he has not merely been chasing a personal phantom.

All that remains left to do now is to express my hearty thanks to the following gentlemen: S. Skimunt, Dr. J. Kotliar, and Alexander Bogdanov, who in the interests of the main text have concerned themselves with helping to expand its influence more broadly.

C

SENSORY ELEMENTS AND SCIENTIFIC CONCEPTS

by Ernst Mach*

(1910)

[The first half of this article was taken by Mach from a draft introduction to his *Die Prinzipien der physikalischen Optik*.]**

[Text of the First-Half]

We find ourselves experiencing, thinking, and acting in the world that is around us. The very simplest components of what we experience and live through which we do not know how to divide further are called *elements*. Experience shows us the elements as dependent on one another. I see a red piece of paper. The red is an element of experience which cannot be further divided. But if it is to enter my experience, then the sun, or a gas or oil lamp, must be present in my environment at the same time, therefore an exactly definable complex of other elements. With the disappearance of the sun and the like, red disappears as well. With changes in the conditioning complex of elements, say replacing the sun with a sodium lamp, another element replaces red, perhaps brown or yellow, that is, after the red has become darker or lighter. The same elements depend, however, also upon a special component of the environment, upon my body, particularly on my

* This article was published as a *Festschrift* for Ewald Hering's 75th Birthday. It originally appeared as "Sinnliche Elemente und naturwissenschaftliche Begriffe" in *Pflüger's Archiv für die gesamte Physiologie des Menschen und der Tiere*, **136** (1910), 263-274. It was republished first in 1919 in book form together with another article and again as an article in the fifth edition of *Populär-wissenschaftliche Vorlesungen*, Leipzig, 1923. The translation above is its first appearance in English. - Ed.

** Mach originally titled this manuscript "Entwurf einer erkenntniskritischen Einleitung". It was for his book on optics, a work which he had been preparing for many years, but first appeared five years after his death in 1921. This draft introduction was sent to Friedrich Adler in the year 1910 for his opinion about it. Adler had been trained as a physicist and was also a philosophical follower of Mach, hence it is possible that Mach in his half-crippled condition hoped that Adler might edit the work as a whole and prepare it for publication. Unfortunately for Mach, Adler became editor of a socialist newspaper that year and no longer had time to discuss epistemological issues or help Mach publish the *Optik*. Mach then turned to his son Ludwig. The draft introduction sent to Adler was apparently sent back to Mach who used it for "Sinnliche Elemente" and then presumably returned it to Adler. It can be found in the *Verein für Geschichte der Arbeiterbewegung* in Vienna. For a published version of the German-language original, including the second half of the introduction which was not used for""Sinnliche Elemente", see John Blackmore & Klaus Hentschel (eds.), *Ernst Mach als Aussenseiter*, Verlag Wilhelm Braumüller: Wien, 1985, pp. 163-168. - Ed.

eye, my retina. The chemist *John Dalton* or another red-blind person may call the paper black which I designate as red. In this special dependence of the elements upon a complex of elements which includes one's own body, one calls the former perceptions [*Empfindungen*]. This *second kind of dependence* teaches us to distinguish our body from the environment. The contrast consists of the variety of dependence.

Whether something happens on or in our body, whether something occurs around us, whether we or others perceive something, we always experience a change of elements or rather complexes of elements in the process. Only in so far as we establish the *dependence of the elements on one another* and *explore* those connections, whose stability is determined by the elements, can we orient ourselves in the world.

We are immediately certain of that which we perceive now, less certain of what having been carefully observed we have experienced and therefore only *remember*, and still less of what on the basis of analogy from experience we merely imagine, and not at all when imagination is based on what cannot be experienced. The non-experiencable has no conceivable sense and is absolutely undeserving of general (social) respect .

What we call spatial and temporal order is abstracted from the behavior of elements. If we want to spatially determine an element, we will say, by and between which other known *elements* is it to be found. For the temporal determination of the change of an element we say, with which and between which known *changes* of other known elements does this change take place. Every determination by means of spatial and temporal coordinates is only a convenient, convention-based circumscription of this procedure. If with reference to space and time it is also the case that not everything is clearly established,[1] indeed, not yet by far, nevertheless, let us still try to say something known and self-evident by maintaining that two elements in general stand in *looser* relation, the *further* they are separated from each other in spatial and temporal terms. On the other hand, however, we find with spatial and temporal *implosion* [*Zusammenfallen*], the closest possible interrelation of elements.

If I see a color in one place and then spatially bring this colored patch so as to cover the picture of my hand, then my hand, or the sight- or touch-sensitive hand of someone else, will experience a feeling of heat, cold, roughness, smoothness, pressure, and so forth. This close connection of elements in one space-time [*zeiträumlichen*] location, we call *matter*. Matter is therefore the *space-time connection* of different sense perceptions of *one* person and the sense perceptions of *different* people among themselves [*untereinander*]. Also, if we pay attention not to the dependence of the ele-

[1] On preparing the text for publication, Mach added this number to the text at this point to refer to the first of two long footnotes which we have placed at the end of the article. - Ed.

ments on the human body, but on the interrelations or *reactions* of the elements in general, then we can generally say that the space-time region where the *reactions of elements* are connected might be called matter. The earlier expression is physiological or psycho-physiological, and the later one is physical; it coincides with Wilhelm Ostwald's conception to the extent one conceives of all reactions as energetic.* A colored patch [*Fleck*] reacts during spatial *Coinzidenz* with another one by means of a stop in movement or sound. [For example], a red-hot wire not only can illuminate nearby paper, but can also heat or even help it catch fire. The *connection of perceptions* is only a special case of the *connection of reactions*. It would be utterly pointless [*müssig*] to leave aside factual connections between reactions which can still be explored further just in order to imagine something else which cannot be factually experienced. The material world *exists* exactly in the connection of the reactions of elements, of which the connection of the perceptions of people is only a special case.

If we think of such homogeneous, compact, interconnected elements ringed by a boundary surface [*Grenzfläche*] isolated from other kinds of places, then we have *bounded-matter*, a *body* before us from which space-time limited man may well derive his first experiences as a living person. We call such bounded-space the volume of a body.

The simplest and most accessible biological activity of man consists of the use of his *senses* and *organs of movement*. Already by means of touching and handling of a body new elements or combinations of elements enter into the situation. Grabbing a piece of iron already tells us its weight, hardness, volume, and capacity for movement; a piece of wax on the other hand shows itself to be pliant and yielding, rubber elastic, and so forth. All of this can be described as a manifestation of a combination of elements conditioned by another combination of elements, such as can link pressure with changes in the shapes of our fingertips. If let loose, a heavy body falls with visibly increasing velocity downwards; the farther down on its path that it is snatched, the more strongly it jolts our hand; and the harder the body it hits, the louder the sound. If one flings a body in a horizontal direction, it starts its movement horizontally and gradually turns ever more vertically. If the sun shines through a small opening in a window shutter and illuminates the smoke or dust particles of air in stuffy atmosphere, then the succession of tiny illuminated particles will resemble a stiff, tense string. If we bring a small, opaque shade [*Schirm*] into the path of the illuminated dust, the illumination between the shade and the open shutter is not disturbed, but the further succession of illuminated tiny particles is extinguished.

* This sentence was not present in the draft introduction to the *Optik*, but was added later for the published article mentioned above. - Ed.

Collecting such observations represents the rudimentary *qualitative* and (beginning) *quantitative physics* of primitive mankind. They only contain memories about his movements, which led to and accompanied characteristic appearances, and about perceptions of bodies untouched by his movements to which he himself stands in relation. The forms of movement of the kinds imagined such as grabbing, releasing, catching, throwing, and so forth already introduce something like classification and conception, perceived as directedness toward a goal. This also applies no less to observations of falling, opening, and tightening.

The simplest and most natural biological activity is enough to establish a foundation for the development of primitive physics. For primitive man under primitive conditions the reaction of objects within his grasp upon his own body is most important; and this is his most frequent experience.

To notice that bodies are heavy, somewhat heavier than another, that something feels hot, somewhat hotter than another, takes only modest intellectual development. But a *further look* is already necessary if we are to observe that a body can be lifted by a heavier one by means of a pulley and that a colder body can be warmed by a hotter one, for example, by putting fire under a pot of water. In advanced civilizations both handwork and technology can require the introduction of a whole sequence of bodily reactions in order to satisfy human needs. The acquired experience still exists in the recollection of all the precautions originally derived from sensory observation and in the sensory experiences and perceptions which have subsequently been associated with them. It makes no particular difference that the movements of our body no longer come exclusively into consideration or that the perceptions which become conscious no longer have personal interest for us. The *quality* of perceptions falls completely into the *background* before our *interest* in the *dependence of the elements on one another.*

[Text of the Second Part]

A prominent philosopher [Carl Stumpf][2] has objected that relations by means of scientific laws "absolutely never" exist between immediately given sensory appearances; what scientists mean by lawfulness is always completely different. The unprejudiced person, however, will already have recognized from the just-mentioned examples the expression of lawfulness in the appearances. But if one wants examples which demonstrate the existence of laws in sensory phenomena in striking fashion and if one really wants to avoid prejudice, then one can think of *Newton's* spectrum where one comprehends at a glance the dependence of the refraction of color on *Newton's* prism whose bands [*Ringe*] progressively draw together if one

2 Mach added this number to his text in order to refer to the second of two long footnotes, which we have placed at the end of this article. - Ed.

guides refrangible yellow, green, blue, and violet toward red spectral illumination. Also, should not the experimentally-inclined musician-philosopher [Stumpf] from whom the above-mentioned application stems remember that from two tuning forks which stand in the audible interval of an octave, the higher and the lower tracing waves on the same rapidly shifting plate of soot [*Russplatte*], that the higher draws exactly half as long waves at twice the number as the lower?*

Galileo had already noticed on the lathe, that his graving tool [*Stichel*] as soon as it skipped an octave at once left behind the impression of a half-interval on the rotated piece. Certainly one will not immediately behold a law in every sensory appearance which accidently throws its bright color into our path. But the scientist whose task it is to unravel what has been accidentally tossed together will nevertheless find the law. Even if sensory appearances merely had significance as "coupling points" for investigation into the "physical", "purely mathematical", "algebraic world", which we must lay down as existing "independently of consciousness", what right would we then have to read into these laws characteristics lacking even the slightest clear trace of sensations?

Let us now look more closely at *Stumpf's* purely mathematical world! Does he mean the slowly developing world of scientific concepts in contrast to the immediately given *Sinnlichkeit?* The general, conceptual progress which is concealed under these symbols seems to shroud what is living and sensory, disguise it, so that at first we believe ourselves to have scarcely anything graspable or gripable before us. Certainly it will seems so to those chiefly acquainted with the conceptual world through books; but it will be different for people who have gradually learned about concepts not at the study or writing table but from intercourse with nature.

Those who do not want to see the beginnings of concept-building in animals will also think that there is a deep chasm between human sensations and the world of concepts, but this opinion can be overcome if one follows the continuity of development. Perceptions, for example, the appearance of food or enemies, can let loose important biological reactions. *Charles Darwin* gave a lively depiction of an apes's horror while viewing a snake. But if an animal enters relations with the same object under different conditions, he will occasionally recognize the very manifold qualities of the objects. For example, the little mammal mongoose eats cobras and knows how to clutch the nape of its neck and avoid their dangerous bite as certainly as do Indian fakirs. A surely beheld goal belongs to this, an exact sensory-led movement, whereby the sensory expectation toward the conceived object is

* Stumpf and Mach were briefly colleagues at the University of Prague during the early 1880's and may have observed each other's experiments. Stumpf, a gifted musician and theorist about music, was later a professor at the University of Berlin where he became an influential philosopher and psychologist. - Ed.

movement, whereby the sensory expectation toward the conceived object is not disappointed by its behavior. One could almost say that the appearance of the snake wakes in the mongoose a practically acquired *concept*, reminding him about all the cobra's as well as his own ways of reacting.

Also, men acquire a number of practical concepts, for example, by means of *using* the somewhat hazy but partly very abstract features of everyday speech and through the application of fork, spoon, and tools in many-sided ways. It makes no particular difference whether for the purpose of movement the limbs of our body suffice or we are supported by instrumental means [such as crutches]. The reaction will always be aroused by the sensory goal, accompanied by sensory perceptions, and justified by the fulfilment of sensory expectation. Whether our expectation is satisfied by the behavior of a poisonous snake or a lever, or is disappointed, in both cases the value of the concept will be determined by testing. If prominent sensory memories are mislaid through a central-sensory malady such that we no longer know how to use words correctly, then we suffer from one or more forms of aphasia; but if it occurs that we can neither recognize the appearance of nor use a fork and spoon, then we have fallen from aphasia into apraxia.

Only gradually can we form more complete concepts. Raw biological and psychological reactions at first only distinguish what is crudest and most striking, perhaps the splitting of rays by refraction and reflection in contrast to their otherwise straight course, while other characteristics remain still unnoticed. Thus the spider, frog, stork, and cat hunt first only what moves or whizzes by. Gradually, living beings learn to better recognize what distinguishes the advantageous from the disadvantageous; they pay attention more carefully to similarities which could mislead. Concepts are clarified and sharpened through *unceasing psychic activity, through sorting sensory characteristics according to their importance and order of rank*; but this takes place only if suggested by *Sinnlichkeit.*

Sensory elements do not stand in an indifferent relationship to concepts, but on the contrary, are of fundamental importance. The slightest pressure can make the physicist conscious of the existence of *mass* and *acceleration*; the appearance of delicate *Haidinger fringes* proves that he has to do with polarized light. Indeed, the history of science teaches which concepts arise through the odor of *ozone*, which ones through the appearance of *crystal* shapes for the chemist, and which through a breed of *bacteria* for the physician; all this appears before our mental eye. Galileo proved that heaviness introduces acceleration. In Newton's *Principia* all forces are already treated as accelerating. How does one know that? Did he test them? Hardly! But every force could be perceived as pressure or tension, such that it could not differentiate weight from electrical or magnetic forces. It seems that here *homogeneous perception* suggested the thought, *ventured* to promote a homogeneous conception of all forces, a view both intellectual and practical considerations favored.

The conceptual connection of facts is certainly what first makes a compendious natural science possible, since without it only an endless, immense, scarcely usable number of individual facts would exist. But it does not follow from this that the conceptual system must include much more or something wholly different from the individual facts of sensation from which it has been drawn; it still includes them, only ordered such that they can be taken in at a glance [*übersichtlich geordnet*].

An experienced chemist knows many things well *immediately* by their sensory characteristics. But in order to put aside the slightest danger of error and so as not to waste time through superfluous testing, he appeals to the known tables of qualitative chemical analysis in which the sensory characteristics of different chemical compounds [*Stoffbegriffe*] can be clearly recognized at a glance. - I have myself sketched out a similar table for the qualitative analytic determination of the polarization of light. So long as the sensory aspects [*sinnlichen Merkmale*] of a concept are qualitative, we must hold it always to be possible that the concept can be corrected by a new experience or fall away altogether. But if the facts conceptually conceived together stand so near that their leading characteristics can only be distinguished by the number of like parts in which they can be split, then further conceptual classification can take place according to the number of these parts. Measuring, counting, and indirect measurement such as calculus now enter the situation, in short, mathematical treatment. The advantage lay in the fact that classification can be introduced without new discoveries arising every moment, forcing no end of additional qualification...

But sense perceptions are not thereby excluded, and it is pointless to talk about a world beyond *Sinnlichkeit*.... We remain, so it seems, with all our looking around in our customary sensory world; indeed, *Sinnlichkeit* itself drives us to enlarging concepts by letting us recognize similarities between processes, which from certain sides can again be represented as differences. Only intellectual rules by which we comprehend a sum of sense observations belong to a *freer world of thought*. But the value of them is determined solely by the exactness with which they represent sensory observations, which means that because of the absence of absolutely exact measurement always remain limited [in their application and value].

Though rules for the mathematical representation of *Sinnlichkeit* have also finally emerged, they still may only permit as much freedom of choice as *Sinnlichkeit* itself. For our familiar human activity of counting, calculating, constructing, in short, ordering, was first learned, applied, and used with respect to sensory objects, and is controllable by what is sensory.

In any case, *physicists* have *nothing* to seek "beyond the appearances". Whether *philosophers* [such as Carl Stumpf] will always find it necessary to affirm something real existing independently of consciousness, which is nothing in itself, and whose relations may only be recognized in the wholly abstract form of equations, may be left entirely for the philosophers to decide. Perhaps they will eventually ask whether such suspect relations are even meaningful? Perhaps they will even raise the question whether such an ordinance is necessary and what if anything is it good for?

Hopefully, physicists of the 20th century will not let their investigations be disturbed by such meddling! Was it the great Königsberger [Kant] in his attempt to cleanse science of metaphysics who left a little mildew of his own behind which has since sprouted into a forest of weeds? - When Ewald Hering honored *physiology* through his investigations into sensory elements, he added *much* to his field. It is my hope that through more exact attention to these elements that *physics* will add *something*.

[Mach's Subsequently Added Commentary]*

1. The original human conception of space was given by the sense organs. Geometry on the other hand has resulted from the socially idealized metrical experiences of mankind. The ancient astronomer, Ptolemy, probably on the basis of his experience with dioptera [optical units concerning the power of a lens], already gives the law of simple seeing through what are identical (corresponding) visual-rays, indeed somewhat inexactly but essentially in correct form.... His examination seems to have begun the clear distinction between visual space and geometrical space. But Ptolemy [in another context]... seems to have again confounded geometrical space with visual space, a confusion followed by Kepler, Descartes, and many modern scientists. Only in recent times have Panum, Johannes Müller, and especially Ewald Hering put the situation right.

Similarly, one will have to distinguish between metrical and physical space, with time included in the latter. I have already carried this out in my book *Erhaltung der Arbeit* (1872), p. 35, suggested on p. 56, and in *Erkenntnis und Irrtum* (1906), p.434 ff.; it is also a direction in which essential progress has been made by the work of A. Einstein and H. Minkowski.

2. At the opening of the International Congress for Psychology in Munich, a meeting held on August 4, 1896, Prof. Dr. Carl Stumpf gave an address in which he also undertook to criticize my epistemological views in psychology. I was indeed inscribed as a participant of the Congress, but because of its strongly hypnotic-telepathic program did not attend. Since I

* This commentary was not present in the extant draft manuscript, but was apparently added for publication in the form of very long footnotes. Some Latin quotations and references have been dropped. - Ed.

was then occupied with other things and just afterwards stricken by serious illness, Stumpf's speech came to my attention late. In a series of editions of *Analyse der Empfindungen,* I have indeed answered the objections of Carl Stumpf and others, which I view neither as personal nor malicious, but *typical,* but since Stumpf's talk has recently gone into a third edition, I want to add here a comparison between my representation and the main passages of his speech in so far as they relate to me. His words can only gain by being put in relief against mine. It is very convenient for my readers, and I am also fully satisfied with this kind of give and take. He writes:

> One could almost envy the adherents of this theory concerning the elevation of the epistemological and psychological standpoint which they believe that they have reached by such a short path. But both claims, which they support, have themselves no claims on the facts. Lawful relations, which are the object and goal of science, *never connect sensory appearances.* Between them, as each appearance is presented to one's own consciousness, there exists no regular [regelmässige] succession or coexistence, something which the scientist requires in what he means by laws.
>
> Laws exist exclusively within processes which we affirm and must affirm exist *beyond* sense appearances and *independently* of consciousness, that is, if one is to talk about lawfulness at all. If one seeks this reality, which is nothing [sensory] in itself, and can know its relations only in the totally abstract form of equations, then the appearance of space, within which we tend to treat relations as sensory, may itself be a superfluous symbol. Lawful relations and their contents form the "physical world" of science, while sensory appearances from which the physical world of common consciousness is built, merely have the significance of being starting points for the exploration of the purely mathematical, let us say the algebraic world. It is difficult to speak opposite someone as well-versed in the history of science as Mach, even when he misunderstands the real trend of physical investigation, indeed, turns it on its head. But the greatest personal and scientific respect cannot change convictions.
>
> That the second, the mental world, which we experience in thinking, feeling, and willing, is thoroughly reducible to sensory appearances, wins no warrant from the history of psychology, at least not up until now. On the contrary, all attempts since the days of Étienne Bonnet de Condillac to really carry through such reduction have failed. If this does not fully prove the unconditional impossibility for all the future, then one must still at least confess that dogmatic confidence in the reduction of everything to sensations as if it were a logical axiom needing no kind of prooof and should be set up as the highest goal of science seems to be even less justified.
>
> So, if I see the situation correctly, this *sensualistic monism* [of Mach] also dissolves into nothing. His contentions about the physical world certainly contradict the real course of science, and those about the mental world have not been confirmed in the least.

CHAPTER 6

THE MACH-PLANCK POLEMICS

A

THE UNITY OF THE PHYSICAL WORLD PICTURE - SECTION 4

by Max Planck*

(1908/9)

....What gives the future world picture in spite of all its disadvantages such decisive precedence? It is nothing other than *unity*, unity in reference to all individual parts of the picture, unity concerning all places and times, unity for all scientists, nations, and cultures.

If we look more carefully, the old physical system was not at all like a single picture, but rather a collection of paintings; each class of natural appearances had its own picture, and the different ones did not hang together well; we could remove each without impairing the others, something which will not be possible in the future. No single part can be left out as unessential. Each is much more the inalienable constituent of the whole and possesses as such a definite significance for observed nature, and the other way around, every observed physical appearance will and must find its exact corresponding place. In this lay an essential difference from normal pictures, which indeed, must correspond in the mind's eye, but definitely not literally in all respects with the original, - a difference, which, as I believe, has not up to now been sufficiently attended to in physical circles. One discovers precisely in the newer specialist literature remarks to the effect that

* Planck's original lecture "Die Einheit des physikalischen Weltbildes" was given at The University of Leyden before the Student Corps of the Faculty of Natural Sciences on December 9, 1908. It was first published in *Physikalische Zeitschrift*, **10** (1909), 62-75. One might add that the title of this article with its reference to a "world picture" (or *Weltbild* in German) was a manner of expression apparently introduced by Heinrich Hertz during the 1890's. Boltzmann, Planck, Einstein, and for a short while even Mach employed it around and shortly after the turn of the century. The only previous translation seen by this editor (*A Survey of Physical Theory,* translated by R. Jones and D.H. Williams, Dover: New York, 1960) had difficulty handling the notion of a "world picture", indeed, disregarding it altogether as if no one thought there was a difference between a world and pictures or models about it, even in principle or theory. This practice has forced us to attempt a new translation, which hopefully will reflect Planck's understanding more clearly while helping conscientious readers comprehend it.

J. Blackmore (ed.), Ernst Mach – A Deeper Look, 127–150.

one must always hold that recent applications of electron theory or kinetic theory of gases can only claim to give an approximate picture of reality. If this remark were so presented as if one could not or ought not to demand from all of the consequences of gas theory agreement with the facts of experience, then such rests on a serious misunderstanding.

When *Rudolf Clausius* about the middle of the previous century had inferred from the basic assumptions of kinetic theory that the velocities of gas molecules at customary temperatures measure in the hundreds of meters per second, the objection was brought against him that two gases only diffuse very slowly with each other and also that local temperature variations only equalize sluggishly in gases. Summoned to defend his hypothesis, *Clausius* did not do so by representing it as only an approximate picture of reality and that one ought not to demand too much from it; rather, he showed by calculation of the mean free path that the picture put forward by him really corresponds in both cases with physical observation. For he was very much conscious that firmly establishing a single definitive contradiction would irrevocably cause the new gas theory to lose its place in the physical world picture, a general point which is still just as valid today.

The persuasive power of modern science, which has finally compelled general acknowledgment, manifestly rests precisely on the legitimacy [*Berechtigung*] of these high demands which are now placed on the physical world picture. They are independent of the good will of individual scientists, independent of nationalities and centuries, even independent of the human species itself. To be sure, this last contention is indeed rather daring at first glance and may even seem absurd, but please let us remember, for example, our earlier opportune conclusion concerning the physics of [hypothetical] dwellers on the planet Mars. One must at least concede that the affirmed generalization is of the same type physicists form daily when making conclusions going beyond direct observation which can never be tested by human observation. This means that someone who in every case denies sense and cogency to such practice has removed himself from the physical way of thinking.

No physicist doubts the reliability of the claim that a creature gifted with physical intelligence who possess a specific organ for ultraviolet rays would acknowledge these as of the same general type as visual rays, although no one has seen either ultraviolet rays or such a creature. Similarly, no chemist doubts that the qualities of sodium attributed to the sun are like those found in earthly sodium, even if he can never hope to fill even a test-tube with sodium salt from the sun.

With these last observations we are ready to answer the question which I brought up at the end of my introductory remarks: Is the physical world picture merely a more or less arbitrary creation of our minds, or do we find ourselves driven to the contrary conception that it reflects real natural processes which exist in complete independence from us? Spoken more concretely: May we reasonably maintain that the principle of conservation of energy would still hold in nature even if there were no people left to think about it, or that the heavenly bodies would still move according to the law of gravitation, if our earth and all its inhabitants fell into rubble?

If I answer yes to this question with respect to everything previously mentioned above, I nevertheless remain indeed conscious that this answer stands in certain contrast to a direction of natural philosophy which right at this time under the leadership of *Ernst Mach* enjoys great popularity [*Beliebtheit*], particularly in circles of natural scientists. This view holds that there are no other realities than one's own perceptions and that all natural science in the last analysis is only an economic adjustment [*Anpassung*] of our thoughts to our perceptions by which we are driven by the struggle for existence. The boundary between the physical and the psychological is merely a practical and conventional one. The essential and only elements of the world are perceptions.[1]

If we attempt to bring this last sentence into conjunction with what we have gathered from our overview of the factual development of physics, then we will necessarily arrive at the peculiar conclusion that the characteristic feature of this development finds expression in the progressive elim-ination of the most basic aspects of the real world from the physical world picture. Every conscientious physicist should always carefully try to distinguish his own world picture, be it ever so idiosyncratic, from the full range of other views.

If one or two of his colleagues have carried out the self-same physical attempt and claim to have found contrary results, which occasionally occurs, then it would be a fundamental error were he to conclude that, therefore, at least one must be mistaken. For the contrast could also be conditioned by a difference in their respective world pictures. - [But] I do not believe that any real physicist would ever just chance upon such peculiar ways of thinking. Nevertheless, I will gladly concede that such an empirically enormous improbability is practically identical with what is impossible in principle; but what I would like to expressly raise at this point is that the attacks which have been raised against atomic hypotheses and the theory of electrons from that side are untenable and unjustified. Indeed, I would even put the very opposite contention to you - and I know that I do not stand alone - that atoms, little as we know about their more detailed properties, are no more and no less real than celestial bodies or the earthly entities around us; and I would even say: a hydrogen atom weighing 1.6×10^{-24} grams, is no less knowable than that the moon weighs 7×10^{25}. Ad-

[1] Ernst Mach, *Beitrage zur Analyse der Empfindungen*, Gustav Fischer: Jena, 1886, p. 23 & 142.

mittedly, I can neither put a hydrogen atom on a scale nor see it at all, but I cannot put the moon on a scale either, and about seeing, there are also invisible but knowable celestial bodies whose masses can be more or less exactly measured; the mass of the planet Neptune was measured even before any astronomer had turned his telescope toward it. In short, A physical method of measurement where every induction from perception [*Erkenntnis*] is excluded doesn't exist, which also holds for direct weighing. A single look at a precision laboratory shows us the sum of experiences and abstractions which is exactly included in such measurement which seems to be so simple.

It still remains left for us to ask, how has it come about that Mach's epistemology has become so widely spread among natural scientists. If I haven't deceived myself, it basically signifies a kind of reaction against the proud expectations of a generation ago which followed the discovery of the energy principle and was coupled with a mechanistic world view, as one for example can find laid down in the writings of *Emil du Bois-Reymond.** I will not deny that these expectations have led to some outstanding contributions of lasting merit - I name only the kinetic theory of gases - but when viewed in a larger context, it is clear that there was exaggerated promise. Indeed, through the introduction of statistics into physical considerations, hope of completely working out the [detailed] mechanics of the atom has been fundamentally abandoned.

Machian Positivism was a philosophical result of such unavoidable disillusionment. In the face of threatening scepticism, he deserves full credit for having rediscovered that the only legitimate starting point of all scientific research is sense perception. But he threw the baby out with the bath water, the physical world picture together with the mechanistic one, degrading them both to the same level.

I am firmly persuaded that the Machian system, when really carried through, contains no internal contradictions, but equally it seems certain to me that its basic significance is only formalistic. It does not really touch the essence of natural science at all; this is because the most important feature of all scientific research is the demand for a *constant* world picture independent of all evolutions in time and among human beings, a view quite foreign to Mach's perspective. Mach's principle of continuity offers no substitute; for continuity is not constancy.

As I have tried to show, however, a constant unified world picture is exactly the firm goal which real natural science comes ever closer to in all

* Actually, du Bois-Reymond's best-known contribution to philosophy was probably his notion that there were "seven world-puzzles" which he doubted that science would ever solve. A hundred years later they have still not been solved. See Emil du Bois-Reymond, "Die sieben Welträtzel" (1880) in *Vorträge über Philosophie und Gesellschaft*, Felix Meiner Verlag, Hamburg, 1974, pp. 159-187. - Ed.

of its shifts and evolutions, and we in physics may with justice maintain that already our current world picture, although it still glitters in the most diverse colors according to the individuality of the scientist, nevertheless, includes certain characteristics which no revolution in nature or human thinking will ever wipe away. This constancy which is independent of every human, especially every intellectual, individuality, is that which we now call *das Reale.* Or, to give an example, is there still any physicist who deserves to be taken seriously who still doubts the reality of the energy principle? Indeed, it is rather the reverse: One makes the acknowledgement of this reality a precondition for taking him seriously as a scientist.

Granted, that regardless of how confident one may be about having established the fundamentals of the world picture of the future, no general rules can be offered on how to do it. The greatest caution is in order. But a second way of approaching the problem may be appropriate. Regardless of how separately or uniquely it comes about, the acknowledgement of such an unshakeable goal, even if it is never fully achieved, is not adjustment of our thoughts to our perceptions *but the complete liberation* [*Loslösung*]*of the physical world picture from the individuality of the creative mind.* This constitutes a somewhat more exact rerendering of what I have already called the emancipation [of science] from anthropomorphic elements. [The emphasis on liberation from the *individuality* of the creative mind] is necessary in order to exclude the misunderstanding that the world picture should be liberated from the creative mind altogether, which would be a nonsensical beginning.

In conclusion, let us present still another argument for those who, in spite of their tendency to treat the human-economic viewpoint as if it were the only powerful one, will perhaps be more impressed by this than by all of the previous factual considerations. When the great masters of exact research contributed their ideas to science: when *Nicolaus Copernicus* tore the earth from the center of the universe, when *Johannes Kepler* formulated the laws named after him, when *Isaac Newton* discovered general gravitation, when your great countryman *Christian Huygens* put forward the wave theory of light, and when *Michael Faraday* created the foundations of electro-dynamics - we could continue such a series almost indefinitely - [Mach's] economical point of view was surely the very last thing [*allerletzten*] which steeled the resolve of these men in their battle against traditional views and towering authorities. *Nein!* - Whether it rested on an artistic or religious basis, it was their rock-solid belief in the reality of their world picture. In the face of this surely incontestable fact, let us not allow the suspicion [*Vermutung*] to be set aside that should Mach's principle of economy actually be shoved into the center of [scientific] epistemology, it might well disturb the thought processes of leading minds, lame the flight of their imagination, and by such means interfere with the progress of science in disastrous ways.

Would it not be more "economical" to allocate a more modest place to the principle of economy? By the way, you will already gather from how the question is presented that I am obviously far from wanting to deprecate or even ban concern for economy in a higher sense.

Indeed, we can even go a step further. These men did not speak about their world picture at all; they spoke about the world or nature itself. Is there any knowable [*erkennbarer*] difference between their world and our "world picture of the future"? Surely not. For there is not even a method to let us examine such a difference, [a thought] which through Immanuel Kant has become the common property of all thinkers. The compound expression "world picture" has only become somewhat customary as a precaution in order to exclude certain illusions from the start. We might be able, therefore, if we would only apply the required caution and if behind the word "world" looked for nothing beyond that ideal picture of the future, to again replace ["world picture"] by the simple word ["world"] thus insisting upon a more realistic form of expression, which precisely and by all appearances from an economical standpoint recommends itself much more than Machian Positivism, which under the surface is extremely complicated and quite hard to pin down in detail. Indeed, [the word "world"] is actually preferred by physicists when they speak in the language of their discipline.

I have spoken of illusions. Now it would most certainly be a dreadful illusion if I were to hope that I had been generally persuasive on the basis of the preceding remarks or even been broadly understandable; and I will take care not to deceive myself. Surely, much has already been thought and written on these basic questions, for theorists are many and paper is patient. Therefore, if we want to emphasize that which we can completely agree upon, which will be acknowledged by everyone, every time, without exception, and must be taken to heart: First, conscientious self-criticism, united with endurance in the struggle to genuinely know, second, honest respect for the personality of scientific opponents, which is also not to be shaken by misunderstandings, and last, serene trust in the power of the Word which for over nineteen hundred years has taught us the ultimate indubitable sign of how to distinguish true from false prophets: *By their fruits shall ye know them!*

B

THE LEADING THOUGHTS OF MY SCIENTIFIC EPISTEMOLOGY AND ITS ACCEPTANCE BY CONTEMPORARIES

by Ernst Mach*

(1910)

In order to briefly represent the epistemology which I have devoted a good part of my life to, I will begin with an account of the circumstances under which these thoughts have developed. While I was studying the achievements of the scientists I lectured on at the beginning of my teaching activity as a *Privatdozent* in physics in 1861, I spotted what most distinguished their approach by using a selection of the simplest, most economical, and most purpose-suited means. Through contact with the economist Emmanuel Hermann in 1864, who in his career had sought to trace the economical element in every profession [and who is known as the inventor of the postcard], I accustomed myself to designate the mental activity of scientists as economical. This is already obvious from the simplest cases. Every abstract conceptual, uniting expression of the behavior of facts, every substitution of numerical tabulation by a formula, rule of derivation, or of a law, every explanation of a new fact by a better-known one, can be conceived as an economical accomplishment. Even further, the more one goes into the methods of science, analyses systematizing, ordering, and simplifying logico-mathematical construction [*Aufbau*], the more one recognizes scientific activity as economical.

I had become acquainted with the teaching of Lamarck from my esteemed teacher F.X. Wessely already in 1854 while a *Gymnasium* pupil. This made it easy for me to accept Darwin's ideas on evolution which were published in 1859. These ideas were already present in my Graz lectures from 1864 to 1867, and were expressed through conceiving the competition of scientific thoughts as a life and death struggle resulting in the survival of the fittest. This view does not contradict my theory of economy, but helps enlarge it by merging it into a biological-economical representation of epistemology. Expressed in the briefest way, the task of scientific knowledge appears to be: *Fitting thoughts to facts and thoughts to each*

* See Ernst Mach, "Die Leitgedanken meiner naturwissenschaftlichen Erkenntnislehre und ihre Aufnahme durch die Zeitgenossen", *Physikalische Zeitschrift,* **11** (1910), 599-606. The article also appeared in *Scientia,* **7** (1910), 225-240 and was republished together with the article "Sinnliche Elemente" in 1919 to form a small book published by J.A. Barth in Leipzig. The editor was Friedrich Adler. To avoid repetitious description of Mach's philosophy, "Leading Thoughts" has been abridged. - Ed.

other. Every favorable biological process is one of self-preservation, and as such is at the same time an adjustment and more economical than a process which is disadvantageous to the individual. All beneficial knowing consists of special cases or parts of biologically helpful processes. This is because the physical [features] of biological behavior of the more highly organized species of being are determined and added to by an inner process of knowing, of thinking. In the knowing process one can otherwise observe the most diverse qualities; we characterize these as above all *biological* and *economical,* that is, as excluding pointless behavior.

I have presented these leading thoughts in different writings. They first appeared in *Die Geschichte und die Wurzel des Satzes von der Erhaltung der Arbeit* in 1872, which paid special attention to economy of thought. Both the biological and economical side were later considered in *Die Mechanik in ihrer Entwicklung* in 1883 and *Die Prinzipien der Wärmelehre* in 1896. *Die Analyse der Empfindungen* of 1886 considered the question particularly from the biological side. The most developed [*reifsten*] version of my epistemology can be found in *Erkenntnis und Irrtum* which was first published in 1905....

Although individual acknowledgement has not been lacking, yet it naturally occurred with respect to my first publications that physicists and philosophers as well reacted in an extremely cool and negative manner. Indeed, until the 80's of the century which has just run out I had the feeling of swimming against the current *alone*, although for a long time this had not been the case. Shortly before my *Mechanik* was published while looking for writings with a related content, I became acquainted with *Philosophie als Denken der Welt nach dem Prinzip des kleinsten Kraftmasses* which had been published by Richard Avenarius in 1876.+ I briefly mentioned it in the foreword to my book. Two years after my *Analyse* came out, the first volume of his *Kritik der reinen Erfahrung* appeared, that is, in 1888. I was also encouraged some years later by Hans Cornelius's *Psychologie als Erfahrungs-wissenschaft* (1897) and *Einleitung in die Philosophie* (1903) and by Joseph Petzoldt's *Einführung in die Philosophie der reinen Erfahrung* which was published in 1903. I thus saw that at least some philosophers were not as distant from my views as I had long thought, though admittedly, the long-dead Avenarius has found more public appreciation in Italy, France, and Russia than in his fatherland. Wilhelm Schuppe's work *Erkenntnistheoretischen Logik* (1878) first became known to me only a few years ago, when I saw that this author had already broken a path similar to my own but as early as 1870.

+ A letter has surfaced from Avenarius to Mach dated August 6, 1878, which suggests that Mach may have been familiar with some of Avenarius's ideas somewhat earlier than he later remembered. - Ed.

Physicists agreed with me much less often. Nevertheless, both my views and Ostwald's "energeticism" had a distinguished forerunner in W.J. Rankine who had already pointed out in his little treatise "Treatise of the Science of Energetics"[1] in 1855 the difference between *explanatory* (hypothetical) and *abstract* (descriptive) physics and recommended the latter as the essentially scientific, to which the first should only provide a preparatory stage. But the fact that Rankine's thoughts were and could still be unknown at the beginning of my scientific career sufficiently indicates how little spatial and temporal "action at a distance" they had. When I introduced the *economic representation of the factual* and the *dependence of appearances upon each other* in *Erhaltung der Arbeit,* which can now be seen as a renewal of Rankine's proposals, it remained equally and understandably unnoticed. The "general astonishment" of two years later [1874] which met Kirchhoff when he designated the task of mechanics as the "complete simplest description of movements" was also typical of the period. The isolated utterances which can be seen as agreement with the new conception have been mentioned in the foreword to the second edition of *Erhaltung der Arbeit* in 1909. How late is the word we hear from Heinrich Hertz that Maxwell's theory really consists of Maxwell's equations and even later the words of Helmholtz in the preface to Hertz's *Mechanik* (page xxi). It was first in 1906 with Pierre Duhem's *La Théorie physique* that a complete break was made with the old standpoint....

In order to make clear how my epistemological ideas have been received by eminent modern physicists, the reader is requested if possible to take in hand M. Planck's "The Unity of the Physical World Picture"* since I will be discussing this 38 page paper at some length. Without reacting to its form, much less imitating it - *le style c'est l'homme* - I would like to address the content in a purely factual manner.

On page 4, Planck distinguishes between two ways of doing physics, roughly in the sense of Rankine, that is, grasping individual appearances in terms of boldly generalized *explanation* and as modest *description*. For the first he mentions Thales' water theory, Ostwald's Energeticism, and Hertz's "straightest path"; for the second, he touches on Kirchhoff. Now I am glad to hear that he ascribes signifcant "impact" to Energeticism, while in Lübeck [in 1895 at the famous debate between Boltzmann and Ostwald over energy and atoms] he thought it had accomplished "nothing at all", but in my opinion both Energeticism and Hertz's thoughts about the straightest path would come under Rankine's second or descriptive method of doing

[1] *The Edinburgh New Philosophical Journal,* vol. II (New Series), 1855, p. 120.

* *Die Einheit des physikalischen Weltbildes,* Leipzig, S. Hirzel, 1909. The page numbers Mach refers to are from this separate publication. In most passages he also gives pagination from the journal version. See *Physikalische Zeitschrift,* **10** (1909), p. 62-75.

physics. Thus if Kirchhoff's "utterly simplest description" is exactly practised and not just description *per se,* then no room is left for explanation. For "if a fact is known from all sides, then by this procedure it is explained and the task of science is finished" (J.R. Mayer). Kirchhoff at least in his own case can have thought of no alternative. According to Planck, however, science cannot do without both methods; I indeed consider Kirchhoff's descriptive method to be correct, but without denying the historically proved usefulness of explanation.

On page 5, Planck asks how the physical world picture brought about by the application of these scientific methods came into existence? "Is the same merely a convenient but at bottom arbitrary creation of our mind, or do we find ourselves driven to the opposite conception that it mirrors natural processes which are entirely independent from us?" I can find *no irreconcilable alternative* here. Convenient [*zweckmässig*] it must be in order to guide us. How could we otherwise begin to do anything? But since it is also dependent on human individuality, it must also be at least in a certain sense arbitrary. This can be seen when we compare Newtonian with Huyghenian, Biotian, and Young-Fresnelian optics and Lagrangian with Poinsotian and Hertzian mechanics. Who can stop scientists from directing special attention to different sides of the facts? Perchance the decree of a sufficiently eminent physicist? But naturally the human world picture which is socially preserved by scientists succeeding or replacing each other will be more independent of individuality, progressively giving a purer expression of the facts. But in general, the observer comes into the picture in every observation and every opinion as does the environment as well.

Page 6 and 7 contain known historical facts which scarcely anything has to be said about.

Page 8 and what follows is about unifying systems in physics, which certainly no one has any objection against, least of all representatives of *Denkökonomie*, even if this unification be only a provisional, hypothetico-fictive[1] one. Only I believe in any case that electrodynamics, or should we say the Lorentzian theory, has more promise according to W. Wien's conception which assimilates mechanics as a special case than when the opposite is done.

Also, I basically agree with the comments which proceed from page 10 to about 29 relating to the first and second law [*Hauptsatz*] of thermodynamics, namely about the important distinction betwen reversible and irreversible processes....

But I cannot deny my aversion to hypothetico-fictive physics. Thus I have developed my own particular opinion about Boltzmann's probability investigations concerning the second law as based on the kinetic theory of

[1] The expression "fictive" is borrowed from Hans Driesch.

gases. If Boltzmann discovered that it was very probable that processes act according to the second law and very improbable that they don't, then I cannot accept that it has been proved that nature behaves according to this law [*Satz*]. Also, I don't think it is right for Planck to accept the first part without wanting to accept the second part, for both halves of the conclusion are inseparable from each other. Furthermore, how could an absolutely conservative system of elastic atoms be brought to that state even by the most skillful mathematics to which no objections could be raised as if the system were striving toward a particular endcondition?....

On page 29, Planck confirms that sense-perceptions may not be ignored as the source of our experience, but prefers the kinetic world picture of colorless particles because of its coherence and unity. In his opinion, this world picture is not only independent of individuals but also of all peoples and times. It is even valid for Mars dwellers who organize their lives differently. Whoever will not acknowledge this, he alleges has abandoned the physical way of thinking.

On page 34, we find the contention that atoms are no less real than heavenly bodies and that it is as certain that an atom of hydrogen weighs 1.6×10^{-24} g as it is that the moon weighs 7×10^{25} g. We find a similar utterance from H.A. Lorentz, the famous founder of the modern theory of electrons.

I also consider sense perceptions as the source of all experience, but I don't believe that they should be immediately forgotten after the establishment of physical concepts. I ascribe considerable worth to them, namely as a bond between physics and the other natural sciences. Elsewhere I have tried to show how one can gradually construct a unified physics without artificial [*künstliche*] hypotheses[2], though of course this cannot be done in a week.

Concern for a physics which is valid for all peoples and all times, even for Mars dwellers, while so many physical problems of the day press upon us, seems very premature to me, indeed almost comical. But I have already answered this question many years ago. All living creatures concerned with physics in the future will have to address themselves like us to questions of daily nourishment and survival, that is, above all to what is economically important...Yes, I also do not doubt that a being organized in an analogous way to ourselves, if it could observe the origin or end of the Earth somewhere in Space, would observe worldly happenings [*Weltlauf*] in a manner corresponding to how we would confirm them. Only in this *hypothetical* sense am I capable of rationally answering Planck's question which he rai-

2 "Über die Prinzip der Vergleichung in der Physik", *Populare Vorlesungen*, 3rd edition, 1903, p. 263.

ses on page 32. According to my conception, everything depends exactly on what is biological and economical, not at all on speculation [*in der Luft*] and definitely not at all on the *quality* of perceptions. - Regardless of the final answer about the "reality" of atoms whatever it is, I do not doubt at all that if the atomic theory quantitatively fits empirically given reality, then the deductions which are drawn from it stand *in some kind* of relation to the *facts*, only *which* kind remains in question. Concerning the first dark rings ["Newton's rings"] of reflected light, when a lens is put on a glass plate, the distance between the lens and the glass plate is half of what Newton calls a "periodic fit" [perhaps presupposing his particle theory of light] but a fourth part of the wavelength according to the Young-Fresnel wave theory. Similarly, results drawn from the atomic theory can experience numerous and *useful* changes in meaning and interpretation, also when one is not in such a great hurry to hold them as realities. Therefore, let us honor the convictions of physicists! But I cannot make this belief mine.

If Planck's lecture had ended with page 32 there would have been no cause for me to concern myself with it, but at this point he begins a polemic expressly directed against me, which also first made me conscious that in the preceding part there were noticeable cutting remarks which indeed had passed me by in unwounded fashion, but which had been intended with me in mind. I have therefore briefly mentioned the part before page 32. Concerning the polemic which makes up the final part, its unconventional form, complete ignorance of the matter at stake, and idiosyncratic conclusion have forced me to express a few words in reply. The rest of this treatise will be devoted to the putting right [*Richtigstellung*] of Planck's lecture in which he attributes to me a supposedly perverse understanding of perceptions.

As the reader has well noticed, the *biological-economical* conception of the knowing process is entirely sufficient to bring today's ongoing physics into a *consistent even friendly* relation to it without any difficulty. The actual difference which has been manifested up to now has been constituted by *belief in the reality of atoms*. That is also why Planck has scarcely been able to find sufficiently degrading words for such perversity [*Verkehrtheit*]. If one wants to enjoy psychological conjectures [about his motivations], then one must read his lecture himself, and I can only wish that such will occur. After Planck had already suggested it but still with Christian mildness and out of respect for his opponent, he finally stigmatized me with the well-known words from the Bible as a false prophet. One sees that physicists are already on the best way to becoming a church and have already grasped the readiest means at hand. Let me answer plain and simple: If belief in the reality of atoms is so essential to you [*Euch*], then I will have nothing more to do with physical ways of thinking (page 31); I will no longer be a genuine [*richtiger*] physicist (p. 33); I will give back my scientific reputation (p. 35). In short, thank you so much for the community of believers, but for me freedom of thought is more precious.

....Concerning page 34, I can only say that my "Positivism" has not been rightly judged if it is viewed as a reaction to the failures of atomistic speculation. Even if the kinetic physical world picture, which in any case I consider hypothetical without intending thereby to degrade it, could "explain" *all* physical appearances, I would still hold that the diversity of the world had not been exhausted, because for me *matter*, *space*, and *time* are also *problems*, which moreover, the physicists *(Lorentz, Einstein, Minkowski)* are also moving closer toward.

....I will not examine here what caused Planck's attack on my theory of knowledge and what goals he hoped it would help further. Others must judge whether he is correct and whether my views really stand in such screaming contrast to current physics. Planck thinks my position on economy of thought is presumptuous, but isn't it much more presumptuous of him to battle against it from the first disagreeable or alienating impression and on a matter which he knows nothing about and which lies completely outside the direction of his work and thought? I don't consider it a misfortune if thoughts attached to facts rattle differently in different heads, just the opposite. Nor do I consider disagreement tragic. It often brings illumination, like a torch to a tunnel, which can enrich one's thought-world. But a serious attempt to understand the views of one's opponent ought to come first.

(Received for publication on June 11, 1910)

C

ON MACH'S THEORY OF PHYSICAL KNOWLEDGE - A REPLY

by Max Planck*

(1910)

In a lecture held in Leyden two year's ago whose physical content[1] was rather general, I found myself with good reason turning against some points in Mach's theory of knowledge. While my comments found a certain interest and occasionally direct agreement even in circles which have stood rather far from physics, for example, from professional representatives of transcendental philosophy, they have as was to be expected experienced a sharp rejection from the side of those representing Mach's perspective.

Until then I had not cherished the intention of coming back again to these matters, since from my side I had nothing new to say, and outside of that had the feeling that I had presented the most important points in a sufficiently clear manner. But about the latter I have become less certain since the appearance of an article from Ernst Mach himself titled: "The Leading Thoughts of My Scientific Theory of Knowledge and its Acceptance by Contemporaries"[2] which has also been carried over into this journal.[3] Mach maintains not only that his theory of knowledge has not been refuted but he declines even to take my objections against it seriously, and even carries his conclusion so far as to deprive me of the capability of contributing to physical theory of knowledge at all.

But in light of this it does seem remarkable that [if] Mach has not been able to find any stronger opponent - and I am certainly not the only contemporary around - that I will be compelled by his idiosyncratic attitude to spell out my standpoint against Machist theory of knowledge even more clearly than I had previously thought it necessary to do. Concerning this, it may seem premature to emphasize the point, [but] my honored critics [*Herren Kritiker*] have indeed treated the matter somewhat too lightly.

---

* *Physikalische Zeitschrift,* **11** (1910), pp. 1180-1190 and *Vierteljahrschrift für wissenschaftliche Philosophie,* **34** (1910), pp. 497-507.

1 M. Planck, "Die Einheit des physikalischen Weltbildes", Leipzig, S. Hirzel 1909 & *Physikalische Zeitschrift,* **10** (1909), pp. 62-75.

2 E. Mach, "Die Leitgedanken meiner naturwissenschaftlichen Erkenntnislehre und ihre Aufnahme durch die Zeitgenossen", *Scientia - Internationale Zeitschrift für wissenschaftliche Synthese,* **14** (1910), pp. 225-240.

3 E. Mach, *Physikalische Zeitschrift,* **11** (1910), 599-606.

Because I have occupied myself intensely with Mach's theory of physical knowledge for many years, I believe that I am justified in uttering an opinion about it. When I was at Kiel (1885-1889), I considered myself to be one of the most committed followers of Mach's philosophy, which as I freely acknowledge exercized strong influence on my physical thinking.* But later, I turned away from it, mainly because I had begun to see that the glittering promise of Mach's philosophy of nature, which had attracted most of its supporters, the elimination of all metaphysical elements from physical theory of knowledge, could in no way be carried out. Proof of my claim is already implicitly present in the details of the lecture I gave at Leyden, but I will try to make myself even clearer in this article. For the most part, I will use Mach's own representation of his own theory as given in his own last publication.

In the first few pages Mach clarifies in familiar fashion the process of thought which led him to the conception that our knowledge of nature is fundamentally a biological and economic one. Once again he is fully clear and comprehensible. But whoever imagines that the essentials of Mach's theory have been expressed and that the criticism of my Leyden lecture simply follows in deductive fashion from his depicted standpoint would deceive himself. The main point comes later, and consists of a silently undertaken but very basic shift in [the meaning of] the introduced concept.

At the beginning of his discussion about my comments on thermodynamics he says: "The goals of economy of thought are not completely tied and limited to an examination of practical-economic needs". But that is something quite different from what he had previously said. If the goals of economy of thought are not tied to practical human needs, then what are they connected to?.... A few pages earlier, Mach had even stressed: "All favorable processes of knowing are special cases or parts of biologically favorable processes." Does economy of thought then serve any other aim than the promotion of human knowledge? - One vainly seeks an answer to this question - I will dare to insist, therefore, that his generalization of the concept of economy, even if in no more than this particular case, thereby loses its original signification and has become metaphysical in character.

....Mach's theory of knowledge, if consistently carried through is indeed without inner contradiction, but at bottom it only possesses formalistic significance. This is because the most important characteristic of every investigation in natural science is foreign to Mach: The demand for a *constant* world picture.

* Mach acknowledged Planck's interest in theory of knowledge as early as 1888 in the preface to the second edition of Mach's book on the science of mechanics where Planck is mentioned in the company of several of Mach's friends and allies. Planck also briefly corresponded with Mach in friendly fashion in 1893. - Ed.

....In reality, these two concepts [economy of thought and a constant world picture] are totally [*himmelweit*] different from each other. Economy is inseparable from purpose while the concept of stability has not even the slightest thing to do with purposefulness. One could just as easily make variability or capacity for evolution a demand of "economy". - Also, once again, one can clearly see that behind the officially-held slogan, that the concept is gradually being switched until it can be uttered to fit anything desired.

In truth, I am not inclined to continue this discussion any longer. I cannot hope to persuade my opponent; indeed on the contrary, I will have to be ready to defend myself again from the charge of having misunderstood all. I will therefore wait out the flood of objections [from Mach's myrmidons] in peace and quiet until something of a factual character comes up.

But in the last analysis who will render the decisive judgment? "By thy fruits shall ye be known." I am truly sorry, but I simply cannot think of any higher court of appeal. But Mach is strangely not satisfied with what I thought [would be a criterion of judgment] both of us could reasonably agree on. He smells concealed Christianity. But nothing will really help him. Mach and his theory of knowledge, like so many theories before him, will finally have to bow before that famous saying [*Spruch*], even if it does stand in the Bible. Let us now examine his "fruits" before us and study some from both sides.

Mach is able in general to agree with my remarks about the first two laws of thermodynmics. I regret, however, that he failed to make analogous admission in his *Prinzipien der Wärmelehre* [4], and it surprises me that he has not even noticed my detailed criticism. Thus, I am forced to point out more clearly that the study of the above work can only provide a superficial representation of the principles of heat theory.

He speaks very often in his book about perpetual motion, but he does not attach any clear-cut physical meaning to this expression. He continually confuses perpetual motion of the first kind (production of work from nothing) with perpetual motion of the second kind (uncompensated production of work from heat). When Mach says: "The principle [*Satz*] of forbidden perpetual motion can be detected most easily and clearly in the domain of pure mechanics, and in fact first struck its roots there" (p. 318), he is undoubtedly thinking of perpetual motion of the first kind. But on the other hand when he says: "A reversible circular process delivers the maximum amount of work, which can correspond to the transfer of a definite amount of heat from a higher to a lower temperature. This maximum is the same

[4] E. Mach, *Die Prinzipien der Wärmelehre, historisch-kritisch entwickelt,* Leipzig, Joh. Ambr. Barth, 1896, 2. Aufl. 1900. The following page references are valid for both editions.

for all matter [*Stoffe*], because otherwise a *perpetuum mobile* would be possible" (p. 302), then, if this process of thought has any sense at all, perpetuum mobile of the second kind is presupposed. Mach does not even devote a syllable to the fact that the two basic principles about the impossibility of perpetual motion are completely different from each other, that the first is reversible (the impossibility of abolishing work), but the second not (uncompensated production of heat from work being by no means impossible), that the energy principle (the first law) rests on the first and that the Carnot-Clausius principle (the second law) rests on the second, that the second is completely equivalent to the known Clausius principle of the transfer of heat from lower to higher temperature, and that this principle has as a presupposition that without the assumption of non-reversible processes proof of the Carnot-Clausius principle is not to be managed. All of this and much else besides of equal value is not even mentioned in Mach's book, although, as I must particularly stress, at the time Mach wrote his work (1896), the facts had already been made completely clear by the work of R. Clausius and W. Thomson, that is, forty years earlier.

Instead of this, Mach concentrates at length on the analogy between heat and electricity. In particular, it is maintained that our conception of heat instead of electricity as motion "merely has a historical and fully accidental formal and conventional foundation" for if Riess's electric air-thermometer would have been discovered earlier than Coulomb's torsion balance, then electricity would most probably be seen today as a condition of movement. As if Riess's air-thermometer and Coulomb's torsion balance were the only instruments by which we know something about electricity. As if Faraday, Feddersen, and Hertz had also not carried out experiments on electricity, which Mach's analogy, which he places so much in the foreground, reveals to be completely formal and external, because according to him electricity in contrast to heat possesses inertia.

Concerning absolute zero, he remarks: "In reality, one has assumed that it is unthinkable that anything can be cooled under this temperature, that bodies at -273° C. contain no heat energy, and so forth. I believe that these conclusions rest on an impermissable, all-too-bold extrapolation" (p. 341). Alas, belief in a physical law can be as little forced as forbidden. But that Mach thinks that this opinion is compatible with Carnot's principle, as the immediately following lines show, enlightens us once again about the quality [*Genüge*] of his understanding concerning the significance of Carnot's principle.

On a higher level than *Prinzipien der Wärmelehre* stands Mach's book *Die Mechanik*[5], which has proved useful, particularly by stimulating histo-

[5] E. Mach, *Die Mechanik in ihrer Entwicklung historisch-kritisch dargestellt,* Leipzig, F.A. Brockhaus.

rical interest and as a counterweight to a certain kind of dogmatism. This, however, I had already pointed out in my Leyden lecture. But I have not been able to locate anything of tangible physical consequence, either a physical principle or insight worth following up by physical investigation, which shows the characteristics of Mach's biologico-economical theory of knowledge. Exactly the opposite: Where Mach attempts to move forward by relying on his theory of knowledge right often he runs into error.

Here belongs Mach's strenuously fought for but physically entirely useless thought that a relativity of all translation movements also corresponds to a relativity of all rotary movements, that therefore, one cannot decide at all in principle whether the fixed stars rotate around the earth at rest or the Earth rotates around the fixed stars. The equally general and simple principle that in Nature the angular momentum of an infinitely distant body circling a finite, rotating axis cannot possibly possess finite value is therefore for Mach either false or not applicable. According to Mach's mechanics, one is just as bad as the other.

The conceptual errors about physical matters which this unallowable transfer of principles from the relativity of rotary movements, from kinematics into mechanics has already caused, if they were depicted more closely at this point, would lead us too far astray. It therefore naturally follows that Mach's theory cannot possibly account for the immense progress which is initimately associated with the introduction of the Copernican theory - a circumstance which should suffice by itself to put Mach's theory of knowledge into considerable doubt.

Thus, at present, his "fruits" scarcely amount to much. But perhaps one day in the future? I would be happy at any time to let the facts correct me. Mach doubts that the second law can be carried through on the basis of probability; he doesn't believe in the reality of atoms. Well, perhaps he or one of his followers will one day develop another theory which will be able to accomplish more. Let us wait.

By the way, one ought not to believe that Mach's [physical] views represent truly strict deductions frcm his theory of knowledge. Very far from it! So formalistic a theory lacks the capacity, as I already emphasized above, of bringing forth any determinate physical result, whether right or wrong. In fact, it would not surprise me were a member of Mach's school one day to come out with the great discovery that the probability hypothesis or the reality of atoms is precisely demanded by scientific economy. This would bring everything into such beautiful order. Atomism would be happily saved, and a special advantage would be that everyone could still interpret the word "economy" exactly as he likes.

In spite of this tempting picture, I still think that one should give the concept of economy real sense and therefore let it keep its human and practical significance. This would be important enough even for pure science.

Just as the original beginning of physics like that of every other natural science lay in the practical world, so scientific physics even at present receives not only its strongest stimulus but also its most effective support over and over again from the needs of practical life. Therefore, it is also in the most intrinsic interest of science to hold and extend ever further the closest possible feeling for the technical and applied arts.

But the principle of economy is useless in order to show the right path for physical science to follow, even when it is understood in the broadest sense. The simplest known grounds already show this, because right from the start no one can know which standpoint will preserve economy in the best and most lasting way. Therefore the physicist, if he wants to promote science, has to be a realist, not an economizer [of thoughts], which means in the flow of appearances he must search above all for that which is lasting, unchanging, independent of everything sensory. He must seek to uncover all [that lay beyond sensations]. Economy of thought can serve as means, but not as end-purpose. It has always been that way and will always remain so, inspite of Ernst Mach and his would be antimetaphysics.

Berlin, November 2, 1910.

(Received for publication on November 4, 1910)

D

MACH DEPARTS FROM THE AUSTRIAN ACADEMY OF SCIENCES*

(1913)

Ernst Mach to the General Secretary Vienna, April 28, 1913

Most Honored Herr Colleague:

Already last Autumn I had the intention to settle with my son near Munich. Because of an unlucky fall, which kept me chained to my bed for five months, my plan had to wait, so that now is the first time that I can dare to implement it.

At this time, concerning the Academy, I feel the need to take my departure [*verabschieden*], and to give special thanks to the society as a whole as well as to individual members to whom I have been greatly indebted. When I finished my studies at the University of Vienna, I took my doctor's degree according to the old system, after three, two-hour, so-called *Rigorosen* in different fields, where everything seemed to be touched upon except my main areas of study. Convinced of my comprehensive ignorance, I thus became qualified for *artium liberalium magister* at a Medieval university. Naturally, I now had the wish to visit a German university and learn my specialties in a thorough manner. But I soon had to abandon this hope because my means did not allow it. That meant I had to stay in Vienna and give [private] lectures and instruction in order to be able to live.

I was fortunate in the Physical Institute in 1860 to be able to show with the help of instruments which I made myself the existence of a phenomenon[1] which at that time had been placed in question from the mathematical

---

* This letter is from the archive of the Austrian Academy of Sciences and was previously published in German in John Blackmore & Klaus Hentschel (eds.), *Ernst Mach als Aussenseiter,* Braumüller: Vienna, 1985, pp. 113-116. It was apparently normal for a member of the Academy to resign when permanently moving to another country, but it was presumably abnormal not to ask to remain a corresponding member. The content of the letter suggests that Mach, who had retired from the University of Vienna many years before (1901), felt himself closer to biology and physiology at that time (1913) than he did to physics (See paragraph 5 of the letter), such that by shifting his primary interests, leaving the Academy, and not requesting to remain a corresponding member he had indeed carried out the threat mentioned in his 1910 reply to Max Planck "to leave the church of physics". That two months after this letter he apparently expressed reservations about Einstein's theory in his notorious preface to the *Optik* should not unduly surprise us. Most footnotes below come from the above mentioned book edited by Blackmore & Hentschel. - Ed.

1 The Doppler Effect.

side[2]. On the basis of this and some connected work I could now obtain my *Habilitation* [3]. But none of these publications had much worth. It was only after I first became acquainted with the physiologists Brücke[4] and Ludwig[5] and other outstanding medical doctors that I became familiar with questions of sense physiology, questions which were often discussed by these men and where fewer difficulties stood in the way of effectively putting to use my meagre experimental equipment.

Thus, I was less a pupil of the university than an apprentice to several members of the Academy. Moreover, it happened with unaccustomed liberality - for they were not as rich as now - that the Academy in 1864 awarded me 500 Floren [*Gulden*], partly as acknowledgement for work already published and partly for anticipated work on the human organ of hearing. I accepted this sum but not without concern, since I did not know If I could ever succeed in discovering anything further of value about this organ. The money proved most convenient for paying back small debts without which I could not have survived as a private lecturer [*Privatdozent*]. It was almost a decade later in Prague in 1873 that a lucky observation from a train relieved my concern. That together with modest insights and observations made in Graz led me into resolving an important question in the physiology of the human labyrinth[6], a discovery which Breuer[7] in Vienna and Brown[8] in Edinburgh also made very shortly afterwards using completely different approaches which reflected different personal touches.

The origin of my biological epistemological theory also owes much to this influence from the physiological side, a theory which has so alienated many physicists from me that I neither understand their speech nor they mine, which is why they have used means to proceed against me after the

---

2 Josef Petzval (1807-1891) is meant. See "Über die Unzulünglichkeiten gewisser populärer Anschauungen in der Undulationslehre und ihre Unfähigkeit, das Prinzip der Erhaltung der Schwingungsdauer zu ersetzen", *Sitzungsberichte der k.k. Akademie der Wissenschaften,* Wien, **8** (1852), P. 567-586. See also Ernst Mach, *Beiträge zur Doppler'schen Theorie...*, Prague, Calvé, 1874.

3 1861.

4 E.V. von Brücke (1819-1892).

5 Carl Ludwig (1816-1895).

6 See Ernst Mach, "Physikalische Versuche über den Gleichgewichtssinn des Menschen", *Sitzungsberichte der kaiserlichen Akademie der Wissenschaften*, Mathematisch-naturwissenschaftliche Classe, Vienna, **68** (1873), pp. 124-140.

7 Josef Breuer (1842-1925). He is perhaps best-known for his early collaboration with Freud. Concerning Mach, see J. Thiele, "...Briefe von Josef Breuer an Ernst Mach", *Centaurus,* **19** (1975), pp. 40-53.

8 Crum Brown (1838-1922).

fashion of Pius X.[9] Fortunately, this same theory in return brought me closer to biologists like Hering[10] and Haberlandt[11], so that in times to come I can again hope for a peaceful and more general reception [of my epistemological ideas].

Finally, I owe many varieties of stimulation to members of the academy since that time, that is, after I myself became a member [1880]. If I could be completely exact, scarcely anyone whose name has been included in the natural science division of the Academy could be left out. But if I gladly recall occasional conversations with Eduard Suess, von Tschermak[12], Lippich[13], v. Pfaundler[14], Grobben[15], Uhlig[16], v. Lang[17], F & S. Exner [18 & 19],

---

9 Pius X (1835-1914). He was Pope from 1903 until his death, but was much criticized by intellectuals for his condemnation of "modernism" in 1907, which seemed to many including Mach to resemble persecution. Mach often used the religious analogy as if Max Planck was a kind of inquisitor and was part of a Crusade against physicists who did not believe in the reality of atoms or who supported Mach's epistemology. Whether Mach's reference to "they" included Albert Einstein as well is not certain, but it is true that Einstein rejected the biological and psychological side of Mach's theory of *Denkökonomie*, and Mach, since Einstein's personal visit in 1910, was aware of that opposition. See Chapter Seven, Section B. of this book for Philipp Frank's account of the visit.

10 Ewald Hering (1834-1918) was a longtime colleague of Mach in Prague. He is perhaps best-known today for his Mach-influenced color theory, his notion of inherited memory, and his controversies with Hermann Helmholtz.

11 Gottlieb Haberlandt (1854-1945) was a botanist in Graz and Berlin.

12 Gustav Tschermak (1836-1927) was a minerologist, and after Stefan's death in 1893 wanted Mach rather than Boltzmann to be his successor, but Mach did not encourage Tschermak and Boltzmann was chosen. In fact, Boltzmann returned to Vienna in 1894 with a chair in physics and Mach the next year with one in philosophy.

13 Ferdinand Lippich (1838-1913) was Professor of Mathematical Physics in Prague. He was Mach's colleague for over two decades and the only other physicist among the faculty, But Mach rarely refers to him. There is something of a mystery here.

14 Leopold Pfaundler (1839-1920) was starting in 1906 Professor of Physics at the University of Munich.

15 Karl Grobben (1854-1845) had been Ordinarius for Zoology at the University of Vienna since 1893.

16 Viktor Uhlig (1857-1911) became Professor of Paleontology in Vienna in 1900 and in 1901 in Geology.

17 Victor von Lang (1838-1921) was a physicist and in 1908 Rector of the University of Vienna.

18 Franz Exner (1838-1921) was Professor for Experimental Physics at the University of Vienna.

19 Sigmund Exner (1846-1926) was Extraordinarius Professor of Physiology in Vienna in 1874 and became Ordinarius Professor in 1891.

von Escherich[20], Mertens[21], Wirtinger[22], Lieben[23], and Barth[24], especially from the 80's and 90's when still capable of lively conversation, I must not forget the enlightenment which members from the philosophical-historical division also transmitted to me from their store [of wisdom], men such as Gomperz[25], Heinzel[26], D. H. Müller[27], Reinisch[28], and many others.

For all of this, I owe the Academy and its members my heartiest gratitude, and I offer everyone my best and warmest wishes for their happiness and good fortune.

Should this letter be my last, please only assume that Charon, the old rogue, has merely abducted me to a station which does not yet belong to the World Postal Union.

Most respectfully your true servant

18 Höhnegasse

Dr. Ernst Mach

---

20 Theodor von Escherich (1857-1911) became Professor of Medicine in Vienna in 1902.

21 Franz Mertens (1840-1927) became Professor of Mathematics in Vienna in 1894.

22 Wilhelm Wirtinger (1865-1945) was a mathematician in Innsbruck at this time and later in Vienna.

23 Adolf Lieben (1836-1914) is well-known as a founder of organic chemistry in Austria.

24 Ludwig Barth zu Barthenau (1839-1890) was a chemist and together with Adolf Lieben founded the journal *Monatshefte für Chemie* (1865).

25 Theodor Gomperz (1832-1912) was a philosopher in Vienna and is perhaps best-known for his work on the history of Greek thought.

26 Richard Heinzel (1838-1905) was Professor for Germanistics in Graz.

27 David Heinrich Müller (1846-1912) was an Orientalist and since 1876 Professor at the University of Berlin.

28 Leo Reinisch (1832-1919) was an Africanist and Egyptologist.

CHAPTER 7

MACH AND EINSTEIN

A. MACH LOOKS THROUGH A SPINTHARISCOPE

by Stefan Meyer[1]
(about events of 1903, published in 1950)

A significant episode of this time has still not yet been mentioned: It became possible in the year 1903 through Elster and Geitel together with Sir William Crooke's apparatus to show alpha particles by means of flashes on *sidotblenden* screens.

Naturally, this still does not allow one simply to say that one can "see" atoms. One only observes the effects of individual particles. But no one raises objections if one says that he "sees" the sun, although actually one only observes a shining disk, and the inference that there is a big glowing ball with a mass of about 2×10^{30} kg. which circles the planets is very complex and far-reaching.

This has been mentioned beforehand in order to highlight the impression which was made on the great philosopher and physicist Ernst Mach, when he allowed me to demonstrate these beautiful appearances to him. At that time he was already sick and partly lame, but mentally fresh to the fullest extent. It was well-known that he had fanatically resisted the introduction of "pictures" as realities into physics. If one of the atomists who at that time had Ludwig Boltzmann as their leader spoke to him about atoms, he would most often intervene with the question: "Have you seen one?" [Habn S' eins gsehn?]. Until then, one had to keep silent. But this time it was different. It remains one of my most gripping memories that when the spinthariscope was demonstrated, instead of making stubborn objections, Mach

1 Stefan Meyer, "Die Vorgeschichte der Gründung und das erste Jahrzehnt des Institutes für Radiumforschung", *Sitzungsberichte der Osterreichischen Akademie der Wissenschaften,* mathem.-naturw. Klasse, **159** (1950), 5. About Meyer's assumption that the spinthariscope was introduced quickly into Vienna shortly after its invention in 1903, this view was supported explicitly by the testimony of his colleague Karl Przibram and implicitly by Meyer's own preoccupation with and important work on radiation at that time. - Ed.

J. Blackmore (ed.), Ernst Mach – A Deeper Look, 151–164.

simply said: "Now I believe in the existence of atoms".[2] A whole world picture had changed in a few minutes.

B. EINSTEIN'S VISIT TO MACH

by Philipp Frank[3]
(about events of 1910, published in 1947)

Einstein took this opportunity of his stay in Vienna [September 1910][4] to become personally acquainted with the physicist and philosopher Ernst Mach, who had had such a profound influence on Einstein's ideas. At the University of Vienna Mach had lectured on the history and theory of the "inductive" sciences - that is, sciences such as physics and chemistry that advance from individual observations to general laws. For more than twelve years, however, Mach had suffered from a severe paralysis and had retired from his position. He lived in his apartment in a suburb of Vienna, and occupied himself only with his studies and receiving occasional visitors. On entering his room one saw a man with a gray, unkempt beard and a partly good-natured, partly cunning expression on his face, who looked

---

2 This passage, which the reader will note seems inconsistent with most of the evidence about Mach's position on the reality of atoms, especially with his letters to Wald, Ostwald, and Höffding, can apparently be resolved when what Stefan Meyer seems to have meant by "atoms" is taken into account. As Ludwig Boltzmann's assistant, Meyer apparently accepted a common perspective of the period, at least in Vienna, namely, that the reality of atoms was just a "metaphysical" issue and that the difference between Mach and Boltzmann was merely about whether *Bilder*were indispensable in physics. In fact, Mach was not unalterably opposed to *Bilder* [pictures or models], hence, if Meyer wanted to identify atoms with them, then Mach had no difficulty accepting this variation of his own phenomenalism, since it seemed to reject the reality of atoms in any serious sense, i.e. as non-sensory or trans-sensory physical phenomena. According to the available evidence, Mach continued to express doubts about the existence of atoms in a serious sense well after 1903, even after the Radium Institute was built in 1910, and indeed, even as late as 1914, as is clear from his controversy with Planck, his notebook, his son's comments, the preface to the *Optik,* and his conversations with B. Brauner mentioned at the end of this chapter. - Ed.

3 Philipp Frank, *Einstein - His Life and Times,* translated from a German manuscript by George Rosen and edited and revised by Shuichi Kusaka, Alfred A. Knopf: New York, 1947, pp. 104 & 105. Frank also visited Mach at roughly that time, hence his account of Mach's appearance and mannerisms could well be accurate. The gist of the conversation over atomism is confirmed by two other sources who also interviewed Einstein and his recollections about the visit: Carlton B., Weinberg, *Mach's Empirio-Pragmatism in Physical Science,* New York, 1937, and I.B. Cohen, "An Interview with Einstein", *Scientific American,* **193** (1955), 71.

4 Different sources give different dates from 1910 to 1913 for the visit, but Gereon Wolters, who has studied the matter in some detail, has maintained that it was in September 1910.

like a Slavic peasant and said: "Please speak loudly to me. In addition to my other unpleasant characteristics I am also almost stone-deaf." Mach was very interested in meeting the originator of the new theory of relativity.

Even though Einstein greatly admired Mach's ideas on the logical structure of physics, there were many things he could not accept. According to Einstein's judgment, Mach did not give enough credit to the creative mind of the scientist who imagines general laws beyond a mere economic description of facts. Mach's opinion, that the general laws of science are only a means by which individual facts can be remembered more easily, did not appear satisfactory to Einstein. To him the phrase "remembered more easily" could in this connection apparently mean only "remembered with less effort". Mach's *economy* seemed to be *economy* in a psychological sense.

Hence, after conversing for awhile with Mach, Einstein raised the following question: "Let us suppose that by assuming the existence of atoms in a gas we were able to predict an observable property of this gas that one could not predict on the basis of a non-atomistic theory. Would you then accept such a hypothesis even if the calculations of its consequence required very complicated computations, comprehensible only with great difficulty? I mean, of course, that from this hypothesis one could infer the interrelation of several observable properties that without it would remain unrelated. Would it then be 'economical' to assume the existence of atoms?"

Mach answered: "If with the help of the atomic hypothesis one could actually establish a connection between several observable properties which without it would remain isolated, then I should say that this hypothesis was an 'economical' one; because with its aid relations between various observations could be derived from a single assumption. Nor should I have any objection even if the requisite computations were complicated and difficult."+

+ From the perspective of most historians, Philipp Frank's habit of putting speeches in the mouths of people on the basis of second-hand information is not a good one, even though it has the authority of Thucydides, perhaps the greatest historian of all time. The modern opinion is: If you can't quote literally, that is, accurately, then paraphrase and qualify. It is not at all certain that either Einstein or Mach said exactly what Frank has them saying. The supporting accounts from Weinberg and Cohen (mentioned in footnote 3) are much briefer and only hold for the general thrust of the conversation. Concerning valid content, Mach apparently did give a positive answer to a hypothetical question on the utility of the atomic theory, but Mach could have done that even if in fact he rejected the truth of the if-clause, therefore, it is not clear what one can validly deduce from his positive answer. If, nevertheless, Einstein (and Frank) were deceived, then it seems that this state was of short duration at least on Einstein's part, as one may infer from his remarks in Paris in 1922. - Ed.

C. ERNST MACH

by A. Einstein[5]
(1916)

In these days Ernst Mach departed from us. He was a man of rare independence of judgment, whose influence on the epistemological orientation of natural scientists has been most extensive. The immediate pleasure of seeing and conceiving, Spinoza's *amor dei intellectualis,* dominated him so strongly that until advanced age he saw the world around him with eyes of a child in order to enjoy unmarred by wish or desire clear insight into the connection of things. But how has it come about that such a respectable, gifted scientist should trouble himself at all with theory of knowledge? Aren't there worthwhile tasks in his own specialty? Such is the kind of thing I hear from some of my colleagues, or have traced in the feelings of many more. I do not share this attitude. When I think of the most able students who have been taught by me, those who have distinguished themselves by independence of judgment and not by mere quickness, then I must confirm that they were actively interested in theory of knowledge. They gladly raised discussions about the goals and methods of science and were dogged in fighting for their views; for them the issues at stake seemed to be important. Nor is this truly surprising.

If I turn to science, not from external grounds such as money or ambition, and not even, or at least not exclusively, because of the fun and pleasure of mental gymnastics, then, if devoted to my discipline, I must have a burning interest in questions such as: What goal can and should science attain which I can dedicate myself to? To what extent are the general results "true"? What is essential, and what rests on accidents of development?

In order to appreciate Mach's merit, one ought not to raise the question: What has Mach thought about these general questions which no one before him has done? In such matters, powerful natures will carve truth anew and recarve it again always corresponding to the needs of the time, because the sculptor works in its grasp; if not always made new, it will disappear forever. Thus it is difficult, and not at all so important to answer the question: "What has Mach taught which Bacon and Hume haven't?" "What distinguishes him essentially from John Stuart Mill, Kirchhoff, Hertz, and Helmholtz concerning the general epistemological standpoint over against that of the individual sciences?" The fact is that Mach through his historical and critical writings in which he followed the development of the individual sciences with so much love and traced historical details into

[5] This obituary first appeared in the April 1, 1916 edition of *Physikalische Zeitschrift,* **17** (1916), 101-104.

the inner sanctum of the brain [*Gehirnstübchens*] of pathbreaking scientists has had a great influence on our generation of natural scientists. I even believe that the people who consider themselves opponents of Mach, scarcely know how much of Mach's way of thinking they have absorbed, so to say, with their mother's milk.

According to Mach, science is nothing else than the comparing and ordering of factually given contents of consciousness by means of certain gradually and tentatively tried [*ertasteten*?] methods and viewpoints. Physics and psychology are thus to be distinguished from each other not by the objects they study but only by the manner of ordering and relating them. Mach seems to have thought that his most important task was to employ the same way of ordering details to the mastering of the individual sciences. The activity of ordering yields abstract concepts and laws (rules) as well as connections between them. Both are selected so that together they constitute a schema with ordering potential which allows sensory phenomena [*Gegebenheiten*] to fall into a certain and clear arrangement. Concepts only have sense according to Mach to the extent that things can be shown to be interrelated, as well as clearly arranged according to a particular point of view (analysis of concepts).

The significance of such minds as Mach does not depend in any way on having satisfied certain philosophical needs of the time, which the confirmed scientific specialist may designate as a luxury. Concepts which have proved useful in the ordering of things easily acquire authority over us such that we forget their contingent earthly [*irdischen*] origin and accept them as unalterable facts [*Gegebenheiten*]. Such errors often block the path of scientific progress for long periods of time. It is therefore absolutely not mere superficial amusement to get in the habit of analysing long-employed concepts in order to discover the conditions on which their justification and utility depend, how they have become individuated out of the given facts of experience. This analysis can break their all-too-massive authority. If they cannot be properly legitimized, they fall away; if the manner of arranging [empirically] given things is far too negligent, is replaced by another, or, for whatever reason we prefer, contributes to setting up a new system, then they are corrected.

This kind of analysis seems to the scientific specialist, whose gaze is directed toward details [*Einzelne*], largely superfluous, affected, and sometimes even ridiculous. But the situation changes when their science requires a traditional concept to be replaced by something sharper. Then people who have failed to analyse their own concepts will raise energetic protests and complain about the revolutionary threat to their holy property. Mixed with such cries will be the voices of those philosophers who do not believe themselves able to endure losing it because it had been preserved in their innermost treasure chest of "absolutes" of the *a priori* kind which they had proclaimed to be unalterable in principle.

The reader has already guessed that I am mainly referring to certain notions of space and time along with mechanics which have been somewhat modified by relativity theory. No one can assume [*nehmen*] that the epistemologists have made the way to this development easier; [nevertheless] concerning myself, I at least know that I have been directly or indirectly aided [*gefördert*] by Hume and Mach. I request the reader to turn to the considerations given in sections six and seven of Chapter Two in *Die Mechanik in ihrer Entwicklung*: "Newton's Views of Time, Space, and Movement" and "Synoptical Critique of the Newtonian Enunciations". There one will find ideas masterfully presented which in no way have become the common property of physicists. These parts are especially attractive because coupled with literal citations from Newton. Here are some highlights[6] [*Rosinen*]:

> Newton: Absolute, true and mathematical time, of itself, & by its own nature, flows uniformly on, without regard to anything external. It is also called duration. Relative, apparent, and common time is some sensible and external measure of absolute time (duration), estimated by the motions of bodies, whether accurate or inequable, and is commonly employed in place of true time, as an hour, a day, a month, a year...

> Mach: When we say a thing A changes with the time, we mean simply that the conditions which determine a thing A depend on the conditions that determine another thing B. The vibrations of a pendulum take place in time when its excursion depends on the position of the earth. Since, however, in the observation of the pendulum, we are not under the necessity of taking into account its dependence on the position of the earth, but may compare it with any other thing (the conditions of which of course also depend on the position of the earth), the illusory notion easily arise that all the things with which we compare it are unessential. Nay, we may, in attending to the motion of a pendulum, neglect entirely other external things, and find that for every position of it our thoughts and sensations are different. Time, accordingly, appears to be some particular and independent thing, on the progress of which the position of the pendulum depends, while the things which we resort to for comparison and choose at random appear to play a wholly collateral part.

> Newton: Absolute space in its own nature and without regard to anything external, always remains similar and immovable. Relative space is some movable dimension or measure of absolute space, which our senses determine by its positionwith respect to other bodies, and which is commonly taken for [absolute] space.

6 The English translation of Einstein's quotations from Ernst Mach is taken from the English-language, sixth edition of Mach's *Science of Mechanics,* La Salle, 1960, pp. 272-273, as translated by Thomas J. McCormack.

Then follows a corresponding definition of the concepts "absolute motion" and "relative motion":

> The effects by which absolute and relative motions are distinguished from one another, are centrifugal forces, or those forces in circular motion which produce a tendency of recession from the axis. For in a circular motion which is purely relative no such forces exist; but in a true and absolute circular motion they do exist, and are greater or less according to the quantity of the [absolute] motion.

The description of the well-known bucket experiment follows, which gives a visual [*anschaulich*] foundation for Newton's last claim. The criticism which Mach directs toward this standpoint is very interesting; let me cite some especially pregnant places.

> When we say that a body K alters its direction and velocity solely through the influence of another body K', we have asserted a conception that it is impossible to come at unless other bodies A, B, C, ...are present with reference to which the motion of the body K has been estimated. In reality, therefore, we are simply cognizant of a relation of the body K to A, B, C, ... If now we suddenly neglect A, B, C, ...and attempt to speak of the deportment of the body K in absolute space, we implicate ourselves in a two-fold error. In the first place, we cannot know how K would act in the absence of A, B, C,...; and in the second place, every means would be wanting of forming a judgment of the behavior of K and of putting to the test what we had predicated - which latter therefore would be bereft of all scientific significance....
>
> The motion of a body K can only be estimated by reference to other bodies A, B, C, ...But since we always have at our disposal a sufficient number of bodies, that are as respects each other relatively fixed, or only slowly change their positions, we are, in such reference, restricted to no one definite body and can alternately leave out of account now this one and now that one. In this way the conviction arose that these bodies are indifferent generally.
>
> Newton's experiment with the rotating vessel of water simply informs us, that the relative rotation of the water with respect to the sides of the vessel produces no noticable centrifugal forces, but that such forces are produced by its relative rotation with respect to the mass of the earth and the other celestial bodies. No one is competent to say how the experiment would turn out if the sides of the vessel increased in thickness and mass till they were ultimately several leagues thick.

The quoted lines show that Mach had clearly recognized the weak side of classical mechanics and was not far from calling for [*fordern*] a general theory of relativity, and this already a half-century ago! It is not improbable that when physicists were considering the significance of the constancy of the velocity of light that had Mach's mind been young and fresh at that time he would have come across the theory of relativity. [But] the lack of stimulation which ought to have come from the Maxwell-Lorentz electrodynamics also failed to prove sufficient for Mach's critical needs in order to awake the feeling of necessity for a definition about the simultaneity of spatially distant processes.

His comments about Newton's bucket experiment show how near to his mind the demands of relativity in the more general sense (relativity of acceleration) lay. Nevertheless, active awareness was lacking here about the insight that the equivalence of the inertial and gravitational mass of bodies required a relativity postulate in a broader sense, since we are not in a position to decide by experiment whether the fall of bodies relative to a coordinate system is to be traced back on the one hand to a gravitational field or a condition of acceleration of the coordinate system.

In terms of his mental development, Mach was not a philosopher who chose the natural sciences as the object of his speculation, but a many-sided, interested, diligent scientist who also took visible pleasure in detailed questions outside the burning issues of general interest. Among these, let us mention the almost uncountable individual investigations from the fields of physics and empirical psychology, which he published, partly alone and partly together with his students. The best-known among his physical investigations are those on shock waves which projectiles produce. While the basic thoughts which he applied were not new in principle, nevertheless, they illustrated an exceptional experimental gift. He succeeded in photographing the thickness of the air front in the neighborhood of a speeding projectile, which threw light on a form of acoustical processes which we had been ignorant of before. His popular lecture on the phenomenon will please everyone who can take pleasure in physical matters.

Mach's philosophical studies came entirely from the wish to develop a position which could help bring together the different directions to which his own scientific work had been devoted. He thought of all science as an effort to bring order to the elementary individual experiences, which he called "sensations" [*Empfindungen*]. His choice of words carried associations such that this level-headed and careful thinker was often considered to be a philosophical idealist and solipsist among those who were not deep students of his publications.

In the process of perusing Mach's works one feels the comfortable pleasure which the author must have felt when he effortlessly wrote down his pregnant, striking sentences. But not only intellectual enjoyment and appreciation of good style make the reading of his books over and over again so attractive, but also the humane, friendly, and hopeful demeanor which so often throws its lustre between the lines when he addresses the reader about general human things. This attitude also protects him from the sickness of our time, the national fanaticism which few can avoid today [World War I?

- Ed.]. In his popular essay "On Some Phenomena Attending the Flight of Projectiles" he has not been able to refrain in the last paragraph from giving expression to his hope for mutual understanding among people

D. F. ADLER AND L. MACH ON EINSTEIN
(written in 1917 and 1918)

(1)

Friedrich Adler to Ludwig Mach Vienna, September 6, 1917

Most Honored Herr Doctor:

I heard some time ago from Professor Jerusalem that you wanted to send me the proofs of your father's book on optics. That would please me very much . . . . Indeed, since last winter I have waited with great expectation. If it is not possible for me to receive the proofs, at least I hope that they will soon appear [in print] so that I can receive them as a whole. My father says that his bookstore tells him that it is supposed to come out "in nine months".[7] I know the introduction, at least that form of it which existed a few years ago [1910] which your father had sent me on his typewriter. But I am very interested in the whole thing, especially how far his representation deals with current matters.

If you could write me a line on the present state of the *Optik* I would be very grateful.

With hearty greetings, yours respectfully

Wien VIII Alsenstr. 7
or to my father Wien II, Blüme G. 1. Friedrich Adler

[7] Friedrich Adler (1879-1960) was a good friend of Albert Einstein, though not uncritical of some of his views. His father, Viktor Adler, was head of the Austrian Social Democratic Party until his death in late 1918. Friedrich Adler shot the premier of Austria in late 1916, presumably to help end Austria's participation in the First World War, but given his father's prominent position, he was not executed as one might have expected. At the end of the war, he was released from prison, became a leader of the party, and helped prevent Austria from becoming communist in 1919 as neighboring Hungary and Bavaria had briefly become. The two letters here by Adler to L. Mach are located in the Austrian National Library in Vienna. The letter by L. Mach to Adler as well as most extant correspondence between them is located in the Adler Archiv, which is part of the *Verein für Geschichte der Arbeiterbewegung* in Vienna. - Ed.

(2)

Friedrich Adler to Ludwig Mach February 13, 1918

Most Honored Herr Doctor:

Your friendly letter from January 28 has pleased me a great deal. I am very sorry that your letter from last September got lost. I had already been afraid that you had not received mine. From the *Lichtlehre*+ I have now received the first seven parts [of the proofs] and assume that more has not yet been set in type.

It would particularly interest me if therein a position is taken with respect to Einstein's theory of relativity. For some months I have been occupying myself exclusively with it and hope soon to have a publishable manuscript ready. I think that I have succeeded in pointing out Einstein's mistakes so clearly that other people will also have to be persuaded by it. In Vienna, I used the time to work so much on my older epistemological essays that they halfway approach a whole which hangs together, which will appear under the title of *Ernst Machs Ueberwindung des mechanischen Materialismus*. Indeed, it is already in type. Right now I am correcting the proofs. But given current conditions in the printing industry when it will actually come out is more difficult to calculate than the orbit of a comet.[8]

During this work, I became reimpressed with the value of the rather general final chapters of [Ernst Mach's] *Wärmelehre* (from page 365 on in the 2nd edition) which would also be important for non-physicists who customarily decide not to work their way through the purely thermodynamic parts. Perhaps Barth can be persuaded just this once to publish this part separately, since it has already been announced in the preface that these "epistemological sketches" are detachable.

The considerations which are printed at the top of this letter[9] fortunately do not apply to me, since I am being treated as a political prisoner and can receive letter and visit every week. Since all of my time is available for

---

+ This was Ernst Mach's pre-publication title for his book on optics which had been thought of as a companion to his *Wärmelehre*. - Ed.

8 It was published in 1918. Concerning other publishing ventures concerning Ernst Mach, in a letter to L. Mach dated October 4, 1918, Adler describes the need to republish two of Mach's polemical articles which with Adler's help appeared in book form in 1919. These two articles appear in English one for the first time in this anthology, though to avoid undue repetitiousness they have been somewhat shortened. See chapter five, section C, and chapter six, section B. - Ed.

9 Every convict may . . . *only once during every eight week period* receive a letter and send one . . . .

my own work, things are going essentially better for me than most people outside. Personally, I have no complaints to make; I only wish that the external world were more attractive than unfortunately remains the case.

Greetings to those around you. All the best. I remain yours respectfully

Friedrich Adler

K.K. Strafanstalt Stein an der Donau

(3)

Ludwig Mach to Friedrich Adler Haar, March 3, 1918

....You are entirely right if you think that the remarks from page 365 on [in the*Wärmelehre*] should appear separately. I will bring up the matter with Barth [the Leipzig publisher] and put a proposal to him . . . It is certainly correct that different groups of readers may prefer different sections of the book.

If you discover mistakes in the *Optik,* I would be much indebted if you would specify what and where they are. If no more [of the book] has been printed, that is because of a whole series of unfavorable circumstances, primarily, my lack of freedom, for I too have been drawn into the sufferings [of these times]. But I hope finally to get back to the remainder.

You will find little about relativity theory and nothing at all about radiation. - He repeatedly told me that these chapters are much too unclear to be brought into the work.

When the second volume of the *Optik* appears it will include a series of previously unpublished experiments which we had carried out together, and which will be exactly about such matters.

In his time, I brought his attention to certain insufficiencies in the Michelson experiment, and he acknowledged that he was uncertain about it - and I promised him that I would carry out a counterexperiment - The resources are there; only the time and rest are lacking. I am especially interested in your objections to Einstein, whose point of view I instinctively oppose. It would be too wonderful if we could experimentally and theoretically throw a bridge across this overgrowth of speculation. Up until his death, he voiced somewhat gentle irony for ions and the new views of the relativists - perhaps in a few years the whole thing will merely have been an interesting sidetrip [*Extraspaziergang*], but even then, it will still retain something of lasting value which no one will contest. I hope to be able to write you more about these matters soon...

E. EINSTEIN AND THE PHILOSOPHIES OF KANT AND MACH

(a note published by *Nature* in 1923)[10]

The *Bulletin de la Société Francaise de Philosophie* for July 1922 which has just been published (Armand Colin, Paris), contains the report of the reception of Professor Einstein in Paris on April 6, 1922. It is of exceptional interest, for Einstein did not make an original communication, but assisted at a discussion of the theory of relativity.

Professor Langevin introduced the discussion, and Messeurs Hadamard, Cartan, Painlevé, Perrin, Becquerel, Brunschvicg, Le Roy, Bergson, Meyerson, and Piéron took part. It elicited from Einstein two pronouncements of special significance in regard to the relation of his theory first to Kant and secondly to Mach. We quote them in full. The first was in reply to M. Brunschvicg, who had said that the Kantian philosophy in separating a *container,* space and time, from a *content,* matter and force, had ended in antinomies, while Einstein's conception, which is characterized by the fact that *container* and *content* are inseparable, had delivered us from them. To this Einstein replied: "I do not think that my theory accords with the thought of Kant, that is, with what that thought appears to me to be. What appears to me the most important thing in Kant's philosophy is that it speaks of *a priori* concepts for the construction of science. Now there are two opposite points of view: Kant's apriorism, according to which certain concepts pre-exist in our consciousness and Poincaré's conventionalism. Both agree on this point, that to construct science we need arbitrary concepts; but as to whether these concepts are given *a priori* or are arbitrary conventions, I am unable to say."

The second pronouncement was in reply to M. Meyerson, who had challenged him to declare how far he was in agreement with the theory of Mach. Einstein replied: "There does not appear to be a great relation from the logical point of view between the theory of relativity and Mach's theory. For Mach, there are two points to distinguish: On one hand there are the immediate data of experience, things we cannot touch; on the other side there are concepts which we can modify. Mach's system studies the existing relations between data of experience; for Mach, science is the totality of these relations. That point of view is mistaken, and, in fact, what Mach has done is to make a catalogue, not a system. To the extent that Mach was a good mechanician he was a deplorable philosopher. His view of science, that it deals with immediate data, led him to reject the existence of atoms. Probably, were he still with us, he would change his opinion. I would like

---

10 *Nature,* No. 2807, **112**, August 18, 1923, p. 253. (It is not clear who wrote the report, which is largely a translation of an account in French from the *Bulletin de la Société Francaise de Philosophie,* July 1922.)

to say, however, that concerning the other point, namely that concepts can change, I am in complete agreement with Mach."

F. THREE PARAGRAPHS FROM EINSTEIN'S KYOTO SPEECH as translated and published by Y.A. Ono[11] (Japanese 1922, English 1983)

Within five weeks, the special theory of relativity was completed. I did not doubt that the new theory was reasonable from a philosophical point of view. I also found that the new theory was in agreement with Mach's argument. Contrary to the case of the general theory of relativity in which his argument was incorporated in the theory, Mach's analysis had indirect implication in the special theory of relativity. [Paragraph Ten]

Mach was a person who insisted on the idea that systems which have acceleration with respect to each other are equivalent. This idea contradicts Euclidean geometry, since in the frame of reference with acceleration Euclidean geometry cannot be applied. Describing the physical laws without reference to geometry is similar to describing our thought without words. We need words in order to express ourselves. What should we look for to describe our problem? The problem was left unsolved until 1912, when I hit upon an idea that the surface theory of Gauss might be the key to solve this mystery. I found that Gauss's surface coordinates were very meaningful for understanding this problem. Until then I did not know that Riemann had discussed the foundation of geometry deeply. I happened to remember the lecture on geometry in my student years by Professor Geisser who discussed the Gauss theory. I found that the foundations of geometry had deep physical meaning in this problem. [Paragraph Seventeen]

About my work after 1915, I would like to mention only the problem of cosmology. This problem is related to the geometry of the universe and time. The foundation of this problem comes from the boundary of the general theory of relativity and the discussion of the problem of inertia by Mach. Although I did not exactly understand Mach's idea about inertia, his influence on my thought was enormous. [Paragraph Twenty]

---

[11] Y.A. Ono, "Einstein's Speech at Kyoto University, December 14, 1922" in *NTM. Schriftenreihe für Geschichte der Naturwissenschaften, Technik, und Medizin,* **20** (1983), pp. 26-27. First published in Japanese by Jun Ishiwara in 1922 and republished in 1971. See Bibliography.

G. EINSTEIN AND MACH

by Bohuslav Brauner[12]
(about events of 1914 and earlier, published in 1924)

In the issue of NATURE for August 18, 1923 on page 253, Professor Einstein is quoted as having said of Mach, at the reception given to him by the Société Francaise de Philosophie. "His view of science, that it deals with immediate data, led him to reject the existence of atoms. Probably were he still with us he would change his opinion." As one of the oldest [surviving] pupils of Mach, who worked with him on fluorescence in 1877 and enjoyed his warm friendship since 1876, I often had an opportunity of discussing atomism with him. Not long before the war, I was sitting with him in his garden in Vienna, and I remember that he said to me: "Atomism is a good working hypothesis for the study of chemistry; it must be used with great care on studying and working in science; but it is extremely dangerous as a noetic theory." It was after the memorable discovery of Rutherford and Geiger in 1908, who showed that the emission of alpha particles from radium-C is the same whether it be counted by the spinthariscope or by the electroscope, and the most incredulous chemists admitted it as proof of the existence of atoms, that Mach did not change his old view. At the beginning of the War [1914], I asked him in a letter whether he considered the results of all the observations in radioactivity as a proof of the existence of atoms, and he replied to me verbally:"I do not make myself a proselyte of my ideas - do not make yourself a proselyte on atomism." He never changed his ideas up to his death.

[12] *Nature,* **113**, June 28, 1924, p. 927.

CHAPTER 8

MACH'S RELATIVITY VS. EINSTEIN'S RELATIVITY

A

THE RELATIVITY CONCEPT IN PHYSICS

by Emil Wiechert*

(Submitted in the Sitting of December 3, 1920)

Among the conditions which have made discussion of relativity and ether difficult is that not enough attention has been given to the use of very different meanings of the words "relative" and "relativity" in physics. I intend in what follows to distinguish two concepts of relativity from each other by the designations "relativity of bodies" [*Körperrelativität*] and "relativity of representation" [*Darstellungsrelativität*]. - It was Ernst Mach who brought forth the idea of the relativity of bodies and who directed it against Newton's absolute theory of space and time. When one says with Newton that unattached bodies move uniformly in a straight line, Mach rejected the idea that therefore space and time must exist in themselves and become important as something given in the universe. According to Ernst Mach (*Die Mechanik in ihrer Entwicklung*, Leipzig, l. Auflage 1883, 7. 1912), absolute space is a "desperate" [*verzweifelter*] thought: "If we therefore say that a body maintains its direction and velocity in space, that is merely an abbreviated way of saying with respect to the whole world."

According to E. Mach, physical processes are determined by the relations of bodies with each other in the world. Since unmediated action at a distance is excluded by physicists today, then mediation must be assumed in order to carry this idea through. One way is to be seen in the effect of the "inertial field". We observe that matter is subject to it as is likewise a close-

* The editor wishes to thank Dr. Alois Fritsch of Bruch an der Mur in Austria who brought this article to his attention. Emil Wiechert was a Göttingen physicist who made his reputation in seismology, but his interests were much wider. He had apparently derived the Lorentz transformations kinematically without any reference to electrical concepts in 1911. (See C. Jungnickel & R. McCormmach, *Intellectual Mastery of Nature*, Vol. 2, Chicago: The University of Chicago Press, 1986, p. 322.) The document translated here is section #8 (minus the final paragraph which is on a different topic) from "Der Äther im Weltbild der Physik", *Königliche Gesellschaft d.Wissenschaftlichen Nachrichten* (Göttingen), Math.-phys. Klasse, 1921, Heft 1, pp. 29-70 [57-62]. - Ed.

J. Blackmore (ed.), Ernst Mach – A Deeper Look, 165–182.

ly connected field of light [*Lichtschreitungen*]. Thus, in the sense of Mach, one may consider the foundation of the world [*Weltuntergrund*], which is given by the inertial field in connection with the field of light, to be other than "space" and "time". We must represent it, rather, as somewhat corporeal [but] possessing spatial and temporal characteristics, and according to the mores of physics the fitting name would be "aether". - When Mach writes: "We must even allow room for the thought that the masses which we see and according to which we orient ourselves in a contingent way, are perhaps not what is ultimately decisive", it cannot remain in doubt that above all he was thinking about the aether.

Relativity of representation, that is, relativity of standpoint,** is something different from the relativity of bodies. The same man is considered by one much poorer to be rich and by one much richer to be poor. Among the same group, the minutes crawl by for one and race forward for another. The provisional principle [*Anscheinsatz*] connected with the diffusion of light [*Lichtschreitung*] teaches us that that as the diffusion changes for the observer other impressions [*Anschauungen*] arise about the mass of bodies, simultaneity, and the passage of time. All of these cases show the relativity of "representation". Let us recognize: The relativity of bodies concerns the connection of processes in the world; the relativity of representation is about its description. Nor is there even a contradiction between them. On the contrary, since they are about the same field of phenomena [*Erscheinungsgebiet*], the laws which control both, condition each other. The difference comes to light by means of diversely directed attentiveness. For relativity of representation, the most essential thing is about the reference systems of [different] standpoints, which are evaluated with respect to description. Thus, the impressions which we human beings retain from

** If anything, Professor Wiechert underestimated the ambiguity of the terms "relative", "relatively", and "relativity". For example, "relative" can mean comparative, approximate, probable, qualified, conditional, flexible, variable, dependent, connected, uncertain, partly certain, largely true, unreal, consisting of relations, having relations, being relatable, and being incapable of not having relations. Concerning different types of relations, one may speak of logical relations, causal relations, referential relations, observed relations, observable relations, conceptual relations, hypothetical relations, epistemological relations, and ontological relations. Further complicating matters, some things which are relative may also be absolute, may be "merely" relative i.e. unreal, may be relative in one sense but absolute in another, may be partly relative and partly absolute, and some relations may be relatively absolute or even absolutely absolute. Old jokes about philosophical relativism run: "Are you absolutely certain that everything is relative?"; "Is it a relative or absolute truth that everything is relative?" and "Is value merely relative? Absolutely!" How many of these jokes apply to relativism in science remains relatively uncertain. Concerning Newton, he appears to have meant by "relative" *merely* relative and by "absolute" having relations but not having the existence of "real" space, time, motion, place, and rest depend upon them. What Einstein meant by "relative" is often taken for granted by scientists, but is still not clear to most people. The question is a focus of this and the following short papers. - Ed.

processes are especially closely connected with the principles of the relativity of representation. [But] since we are concerned with physics here, the subjective element will not be pursued further. We will think of human perceptions as something which mathematical-physical procedure, it may be said, "eliminates". We can recognize at once how this is possible, if one keeps in mind, for example, that the different impressions which a building makes on us when considered from different standpoints is mirrored to a large extent in physically existing photographs. Only in so far as this occurs would the difference in phenomena [*Erscheinungen*] be kept in mind by us.

Einstein offers his principle of relativity in his "easily understandable" representation of 1920 in the following words, whereby by K a Galilean coordinate-system is to be understood. His "special" or "narrower" principle of relativity: "If with reference to K, K' is a uniformly moving, non-rotating coordinate-system, then natural processes moving relative to K' follow exactly the same general laws as in relation to K." For the "General" principle of relativity: "All Gaussian coordinate-systems are equivalent in principle for the formulation of general laws of nature."

It is clear that both principles have to do with the relativity of representation. - First of all, I want to remark that the general principle of relativity in the above formulation when taken literally asserts something obvious; it means that for the representation of world processes any coordinate-system whatsoever can be used. When E. Kretschmann pointed this out in 1917 (*Annalen der Physik*, **53**), A. Einstein answered in 1918 (Ibid., **55**) in a paper which designated the relativity principle as ***a)***:

> I believe that Mr. Kretchmann's argument is correct, but I cannot recommend the aforesaid novelty. Even if it is also correct that one could bring every empirical law into general covariant form, principle ***a)*** would still have significant heuristic power, which has already been proved by its brilliant handling of the gravitational problem and which rests on the following: When two theoretical systems are compatible with experience, preference is given to the alternative which is simpler and more penetrating from the standpoint of absolute differential calculus. One could even put Newton's gravitational mechanics into the form of absolute covariant equations (four dimensional) and one will certainly be persuaded that the principle excludes this theory, not theoretically but practically.

If the mark of correctness is to be sought in simplicity of representation, then with relativity theory one essentially toys with a point of view. What simply appears depends on the means of representation. Dictating and demanding simplicity of form merely means that one must make hypotheses. In order to gain insight into the significance of principles of relativity one must inquire into the hypotheses presupposed by Einstein's theory. The answer has already been given in the previous paragraph: The general formulation of the basic principles of the theory are opposed to the special instructions needed for the actual application of the theory. The key is the representation of four-dimensional space with its relations between masses which are given by Minkowski's real-imaginary element of length. Thus, the contrast between matter and mass-field and the stages or diffusion of light [*Lichtschreitungen*] - by means of the difference [*Grenze*] between real and inaginary lengths - will be fully accepted and serves the provisional claim [*Anscheinsatz*] which excludes light as a fundamental hypothesis. On top of that comes the hypothesis of the symmetry of field equations in all 4 coordinates of four-dimensional space and the hypothesis of geodetic lines in four-dimensional space as a path free of moving, material bodies.

Characteristic of Einstein's interpretation of relativity is always that certain coordinate systems for the description of world processes are seen as equivalent. Without doubt this gives a valuable extension to Lorentz's theory. But one should not overlook that this also introduces a contradiction with the basic conception of the relativity of bodies as soon as one assumes that equivalence exists in every possible physical respect. Every different system of coordinates signifies a different standpoint of judgment for the world as a whole. It is precisely when one thinks relativistically that we are led to the following conclusion: Since the coordinate systems of physics only apply to a limited range of phenomena [*Erscheinungen*], every different system represents a different physical standpoint. Therefore, in principle, no system of coordinates should be considered to allow different systems to be equivalent. - Epistemologically, the most essential results of the theory which we have won hold together, namely, the knowledge that with changes in the conditions of movement and of positions in the gravitational field, shifts of physical relations - shapes of bodies, bodily inertia, the acceleration of processes - are interrelated and unified, which would escape observation were the observer himself part of the changes and subject to the same shifts as the world around him. It therefore follows that only formal significance should be attributed to Einstein's principles: They are to be seen as hypothetical claims about the behavior of the relativity of bodies.

One may therefore confidently declare that Einstein's principles are not formally unlimited in their validity either: It holds for the special as well as the general theory that only those coordinate systems are allowed which belong to the field of phenomena moving at less than the speed of light [*Unterlichtschreitungen*]. On the other hand, a coordinate system at rest would bring this system of bodies into a disallowed condition. In short, among the peculiarities of the theory of relativity which is overlooked again and again is the limited validity of its principles. And yet, it has fundamental importance. It reveals the effect of the relativity of bodies: If one pursues its physical significance, then one will arrive by means of the idea of the relativity of bodies at a conception of the aether. - For the sake of trying to simplify matters, it is completely false to jump to conclusions as if "simultaneity, "earlier", and "later" had become completely relative concepts. The indeterminism [*Unbestimmtheit*] of simultaneity, which the provisional claim [*Anscheinsatz*] of Einstein's theory of relativity has left formally open is only valid within a narrow realm: For every distance of 150 kilometers, only 1/1000 of a second remains undetermined or indefinite. Outside of that, "earlier" and "later" are quite clear. According to section #6 of this paper, indeterminism is to be ascribed to human finitude and limitations, which the details of the diffusion of light have not allowed us to conceive. Whether perhaps the indeterminism of simultaneity in the occurrences of nature would still remain even if our powers of discernment were much sharper, is a question which must remain unanswered. If the means to send messages faster than light were available, then such a sharpening could be counted on. Perhaps gravitation will one day be the answer. The assumption of Einstein's theory of relativity that the speed of the diffusion of gravity is the same as that of light, is purely hypothetical, since it is merely a deduction from the hypothesis of the symmetry of field equations in four-dimensional space. - But even if there were no way to send messages faster than light in this world, we could still hope to sharpen our judgment about simultaneity if we could still manage to discover further physical details about gravitation or the speed of light.

Since Albert Einstein represents the foundations of the world and its qualities of mass as "space", his theory in sharpest contrast with that of Ernst Mach is brought back to a Newtonian representation of space. Only the form is changed when space and time are joined together into a unified manifold, one which contains variable characteristics from place to place. We find that what is most characteristic of the Newtonian conception is to be found again: Just as Newton left the material bodies to themselves in space and time, so Einstein keeps the same structure in four-dimensional space. This idiosyncratic standpoint was already suggested in the special theory of relativity, in which, by means of the principle of the constancy of light in "empty" space and in dependence on a determination of the inertial field, a speed of light emanations [*Lichtschreitungen*] was assumed which is independent of matter. Thus, it is quite understandable that Ernst Mach, in the 1912 edition of his *Mechanik,* could not include Albert Einstein among the "relativists". If we consider the relativity of bodies from Mach's point of view, Einstein's first theory in spite of its name appears to be an absolute theory. On the other hand, Lorentz's aether theory which was rejected by Einstein can be represented as a genuine theory of relativity. - Also, Einstein's new [general] theory is not a relativity theory in the sense of Mach, since it conceives of the foundation of the world as a space-time structure without bodies. Albert Einstein by means of the Lambda term which will be mentioned in the coming paragraphs has at least tried to fit his theory indirectly with Mach's thoughts by means of a special hypothesis ("Mach's Principle") in which matter and the foundation of the world are connected in a fixed way with each other. But if one thinks completely in terms of the relativity of bodies, then one may not ascribe bodily effects to space and time or to connections between them, but must assume that there is something really active in bodies in the foundation of the world. But such a change in Einstein's theory would make it an aether theory.

As the discussion before us shows, the path will be a long one which proceeds from Einstein's basic thoughts about relativity of representation to a [genuinely] definitive theory; he has brought in a whole series of hypotheses to help him whose physical character is very hard to unravel. Therefore, it cannot be excluded that friend and enemy will both misunderstand it numerous times over. Above all, the propaganda label "relativity" has caused mistakes....

B

ERNST MACH AND RELATIVITY

by Hugo Dingler*

[1924]

Many of Mach's writings have recently been employed to support the so-called Theory of Relativity. For example, Hans Reichenbach in his clear, informative, and meritorious but also one-sided paper "The Present State of the Discussion on Relativity"** writes: "A close relation exists between Mach's criticisms concerning the problem of movement and the general theory of relativity; indeed, Einstein's theory signifies the continuation of Mach's program."

....[But] how Mach understands his own "Principle of Relativity" can be discovered from his remarks in the second edition of his *Erhaltung der Arbeit* (1909), p. 60: "Space and Time are conceived not as independent beings but as forms of the dependence of phenomena on each other. I maintain the principle of relativity which I also do in my *Mechanik* and *Wärmelehre*. See also the chapter "Zeit und Raum physikalisch betrachtet" in

* Our short selection has been taken from the chapter titled "Ernst Mach und die Relativität" from Hugo Dingler, *Die Grundgedanken der Machschen Philosophie,* Leipzig, 1924.

** *Logos,* **10** (1922), 316-378. For an English version see: *Hans Reichenbach Selected Writings,* volume two, (*Vienna Circle Collection,* Volume 4), Kluwer: Dordrecht and Boston, 1978, pp. 3-47.

my *Erkenntnis und Irrtum* (1905) and in H. Minkowski *Raum und Zeit* (1909)."

Mach also says here that he even had the principle of relativity in mind in the [original] text which came out in 1872 which later appeared in his *Mechanik* and *Wärmelehre*. If one reads the places where the above utterance belongs as commentary, then one will discover that what Mach meant by "The Principle of Relativity" which he often mentions and refers to is that every natural appearance can be represented as a function of all other appearances, especially as a function of pendulum movement, i.e. time (See page 57). If Mach cites here the well-known treatise by Minkowski, the reason for it is that Minkowski also touches on the dependence of appearances on each other: "No one notices a place other than at a time or notices a time other than at a place."[1] At least Mach believed that he had discovered once again his own conception of space and time as forms of the dependence of phenomena. This agreement surely led to the citation, for it was Mach's mild manner to stress what was held in common more than what separated them. But he certainly did not accept Minkowski's merely formal-mathematical formulation *in extenso*; nor was he even inclined to share it. For the contributions of this brilliant mathematician, even where he attends to physical concepts and attracts physical experiments, naturally contain only the brilliant logical, mathematical-categorial transformations of certain groups of thoughts, while Mach as a physicist was much too immersed in [sensory] reality in order to see anything more than that. What attracted him to Minkowski's work was the general phenomenological tendency which stood in the background, and this was right up his alley ["*Geist von seinem Geist*"].

Thus, this citation in no way represents acceptance of Einstein's theory of relativity. In any case, If Mach had seen this theory as a necessary continuation of his own thoughts and been persuaded to acknowledge it as such, then he would not have let the opportunity escape for speaking out clearly on the matter on this occasion. But he didn't do so, just as he had failed to do so ever since 1905. But this citation is proof that he at least was acquainted with the theory of relativity by 1909, if we did not know that it could scarcely have escaped his attention right from the beginning.

Another place where Mach refers to research connected with relativity theory can be found in the treatise "Die Leitgedanken meiner naturwissenschaftlichen Erkenntnislehre und ihre Aufnahme durch die Zeitgenossen"

1 Lorentz-Einstein-Minkowski-Weyl, *Das Relativitatsprinzip,* mit Anmerkungen von A. Sommerfeld, 3. Auflage, Leipzig und Berlin, 1920, p. 55. [For an English version see: "Space and Time" in *The Principle of Relativity* by H.A. Lorentz, A. Einstein, H. Minkowski, and H. Weyl with notes by A. Sommerfeld, translated by W. Perrett and G.B. Jeffery (original English edition, London, 1923: Dover reprint N.Y. 1948 and later), pp. 73-91. - Ed.]

(first *Scientia,* **VII**, 1910, later in *Physikalische Zeitschrift,* **XI**, 1910, pp. 599 -606.) He says here:

> Even if the kinetic physical world picture, which in any case I consider hypothetical without intending thereby to degrade it, could 'explain' all the physical appearances, I would still hold the diversity of the world not to be exhausted, because for me matter, space, and time are also problems, which moreover, the physicists (Lorentz, Einstein, Minkowski) are also moving closer toward.

If one considers the real meaning here, even this cannot be regarded as acceptance of the theory of relativity. Mach is emphasizing here nothing other than the known but often overlooked fact that even if the kinetic world picture were able to be successfully carried through (which he does not at all *a priori* contest, though he is not certain about it), or some other theory, it would still remain a problem how this [or any other] world picture could come into existence. He is no longer naive toward "world pictures" but critical. According to his "theory of description" [*Beschreibungslehre*] he feels that such a world picture is not somehow "valid" (which would have to be empirically established for him), but that it raises additional questions. And he is acutely conscious that for him unlike most other physicists space, time, matter, and things still contain internal problems. But he also sees that that physics is gradually becoming conscious of them and that the investigations of the physicists he names have helped to bring out the problematic character of these concepts. Now, were one to hold the theory of relativity to be ill-conceived, then it would be a service one ought not to avoid to direct the attention of physical public opinion strongly toward these problems once again; which in fact the knowledgable have never let slip from mind. But more than this Mach will not in this place maintain. That he thinks the path followed by the physicists he names about space, time, and matter is not the only one comes out by means of his use of the word "gradually", and indeed, doubtless as well by his formulation that they have not found solutions but merely have come closer to "problems".

Even clearer proof of his views on the theory of relativity lies in the fact that in the 7th edition of his *Mechanik* [1912] he passes over the theory in complete silence. Since the theory had been known to him for some years, this would have been exactly the place to mention it. That it did not occur shows well that he did not perceive it as a useful [*brauchbare*] further development of his ideas. Given the mild mood which Mach normally observed in his utterances, this silence can be considered as a clear rejection of the theory of relativity.[2] But finally in the interest of scientific clarity

2 On the other hand, in the conclusion to the introduction Mach again emphasized his opposition to Newton's theory of absolute space.

which he had devoted his whole life to he roused himself to an expressly formulated rejection.

But such fine, distinction-based conclusions have not been drawn by the majority of people, and in the public realm, as we have already seen from various citations, Ernst Mach is valued as the author of several of the main thoughts in the theory of relativity - which taken in a purely formal and historical sense he was. For if one takes his utterances on the law of inertia and centrifugal forces literally and without the wide background of Mach's epistemological research, intentions, goals, and careful self-criticism, and above all ignores his basic criticisms of Newton's point of view and if one takes his utterances as an *a priori* foundation, then in fact [but] only in this sense can one say that Mach put forward [*aufgestellt*] the first theory of relativity.

Granted, however, in the broadest sense it may well be quite valid that the general epistemological background of the theory of relativity has indeed sprung entirely from Mach's phenomenological position. The following words reveal how strongly the founder of the theory of relativity himself feels the connection of Mach's utterances with his theory to be. In Einstein's obituary of Ernst Mach (in *Physikalische Zeitschrift,* **XVII**, 1916, pp. 101-104) he says:

> The quoted lines show that Mach had clearly recognized the weak side of classical mechanics and was not far from calling for a general theory of relativity, and this already a half-century ago! It is not improbable that when physicists were considering the significance of the constancy of the velocity of light that if Mach's mind had been young and fresh at that time he would have come across the theory of relativity. [But] the lack of stimulation which ought to have come from the Maxwell-Lorentz electrodynamics also failed to prove sufficient for Mach's critical needs in order to wake the feeling of necessity for a definition about the simultaneity of spatially distant processes.

Thus, for many, the appearance of Mach's posthumous work on optics probably signified a certain surprise, for in it he as clearly as energetically disavowed the theory of relativity and expressly rejected the role of having been a forerunner. In the foreword to *Die Prinzipien der physikalischen Optik* which came out in 1921, Ernst Mach says:

> ....thus I must now deny that it was I who paved the way for the relativists with the same decisiveness with which I have rejected for my own self the atomistic dogma of the current school or church.
>
> But the why and wherefore in rejecting for myself the doctrine of relativity, which is becoming ever more strange and dogmatic, and concerning the sensephysiological considerations, epistemological reflections, and above all, experimentally won insights which have caused me to adopt this position in detail, such shall be set forth in the continuation of this work....

Above all, the foreword includes the clearest rejection of all attempts to dogmatize his own thoughts on the questions of space and inertia. Mach's main objection to the theory of relativity consisted in treating it as a final solution of the problems which he himself had raised, and even as if these notions were an ultimate philosophy without submitting them to further critical examination, and to the extent such was done, without introducing a method of proving [or refuting] them. He is aware of and has always spoken out about how careful and cautious and especially how reserved one must be about the results of using inductive methodology in doing research in physics, and he must therefore have directly felt how tragic it was that he of all people should have been made the forerunner of a theory which in his sense was fully dogmatic, that is, which lacked sufficient examination of its [epistemological] foundations....

C

OBJECTIVE RELATIVISM

by Arthur O. Lovejoy*
(1930)

Introduction: The new variety [of dialectic stemming from the "objective relativism" of Dewey and Whitehead], somewhat paradoxically, arrives at much the same outcome as another philosophical tendency to which it at first seems antithetic - the type of positivism represented by Mach and Petzoldt. The affinity between the two may be illustrated by quoting Petzoldt's summing-up of the fundamental principles of positivism:

> We can think things only in the form in which we find them presenting themselves to us, never in a form in which they present themselves to nobody. We can apprehend the world only from the standpoint in which we actually stand, not from a standpoint in which we cannot think of ourselves as standing, or from no standpoint at all. There is no absolute standpoint, and there is no exemption from standpoints; there are only and always relative standpoints . . . . I can in reality think of

* Arthur O. Lovejoy (1873-1962) was a profound American philosopher perhaps best-remembered for his defence of epistemological and ontological dualism and for fathering and encouraging a branch of historical inquiry called: the history of ideas. He was as skilled in making distinctions as in his famous article about "the thirteen pragmatisms" as he was in discovering similarities between ideas prominent in different periods of history as in his book *The Great Chain of Being* (1936). Our selection here is a somewhat reorganized version of extracts taken from chapters III & IV of his most difficult and important book: *The Revolt Against Dualism,* Open Court Publishing Company: Chicago, 1930, (second edition, 1960), pp. 99-101, 105, 111-116, & 162-163. This article is on a philosophically somewhat more complex and sophisticated level than the previous two. - Ed.

no absolute whatever:[1] I [Petzoldt] always tacitly place myself upon the scene as the observer who is beholding things in their relation to himself.

The new relativism often sounds like an echo of the familiar strain. Yet its representatives still continue to assert a kind of objectivity in knowledge which the positivist cannot consistently admit - though he too has not always been able to refrain steadfastly from assuming it.

These older influences have of late been powerfully reinforced by the implications - or the supposed implications - of certain important new developments in natural science, especially in physics. A part of Einstein's special theory - not, in fact, the most significant or most characteristic part - has seemed to show that even the properties of natural objects and events which had most commonly been conceived, even by science, as inherent and absolute - their metrical properties, and especially their dates and duration - are "relative" (in a sense which we need not, just yet, attempt to define), and that certain relations of which the variations had been supposed to be dissociable from and irrelevant to certain other relations and properties, in fact "make a difference." This appeared to afford fresh and striking evidence of the general relativity of things; while at the same time, just this world in which things own their properties only in their diverse relations to other things is by some relativistic physicists conceived as the true objective order of nature. According to the physical theory of relativity, as it is frequently expounded, an identical body has many different lengths, in the direction of its rectilinear unaccelerated motion, depending upon the condition of relative motion of the reference-body with respect to which its motion is defined; it has therefore many different shapes; and these effects, the Lorentz-Fitzgerald contractions are often described by physicists, not as illusory ways in which the body merely "appears" to different observers, but as physical properties of the body, any of them as "real" as any other.[2] Thus a character which a material thing has only in a

1 "Wir können, die Dinge nur so denken, wie wir so vorfinden, nie so, wie sie niemand vorfindet. Wir können die Welt immer nur von dem Standpunkt aus denken, auf dem wir wirklich stehen, nicht von einem Standpunkt aus, auf dem wir überhaupt uns gar nicht stehend denken können, oder von gar keinem Standpunkt aus. Es gibt keinen absoluten Standpunkt und es gibt keine Standpunktlosigkeit, es gibt allein relative Standpunkte, diese aber auch stets . . . . Ich kann in Wirklichkeit gar nichts absolut denken: im Stillen stelle ich mich noch immer als der Beobachter vor, der die Dinge sich gegenüber sieht." Joseph Petzoldt, *Das Weltproblem von positivistischem Standpunkte aus,* Leipzig, date uncertain, p. 142. [This quotation cannot be found on this page in the 1906 edition. It is possible that Lovejoy meant to refer to the retitled version of 1912 or later which added the German equivalent of "relativistic" to the title. - Ed.]

2 It is perhaps unnecessary to insist that this is only half the story about the physical theory of relativity, and not an entirely correct version of that half; but it is the half, and the version of it, which helped promote the philosophical relativism with which we are concerned.

special context, only in its relation to another individual thing, it has none the less objectively and physically. This suggested that a character which exists only in relation to an individual sentient organism, or to an act of perception, may be in like manner objective and even "physical". . . .

Relativity, then, - and specifically relativity to a perceiver or thinker, or to the special conditions determining the content of his perception or thought - is so far from implying subjectivity that it is rather (objective relativists frequently insist) the very essence of that objectivity in the understanding of the nature of things which science has always sought to attain.[3] Error in science mainly arises, it is observed, through neglect of some relevant factor in a problem under investigation; a duly critical scientific procedure therefore "requires that all the conditions of a given event be taken into account." Now, the relation of the event to the perceiver, "what it is experienced as, must be accounted as one of the objective conditions." It was disregard of this type of condition by the older theoretical physics that Einstein noted; he pointed out certain contradictions which experiment had disclosed arose precisely from a failure to recognize that measurements of time, and of the linear magnitudes of bodies, are relative to the standpoint of the observer's particular reference-system. In calling attention to this oversight he was not lapsing into subjectivism, but on the contrary, was carrying forward "the high tradition of scientific objectivity".To him especially, then, "belongs the credit of having shown that the world is perceived as from any and all possible physical points of view constitutes the essential part, if not all, of its objective nature for physical science"[4]. . . .

[But in my opinion] in the scientific investigation of nature, men have been looking for views of things which have, at the minimum, *some* degree of generality, and are thus *not* relative to particular contexts or standpoints. You approach the truth about a thing, both common sense and science have usually supposed, only when, after changing your standpoint, you find some of the characters of the thing remaining the same; the more changes of standpoint you make, the more justification you have for identifying the characters found to be common in all with the sort of object knowledge aims at. The perspective appearances of physical objects, with their diversities of shape and size, have been treated as subjective, not because they did not truly appear, but because they varied with standpoints. The *cognoscendum,* in short, has always been regarded as a sort of constant

---

[3] See A.E. Murphy, "Objective Relativism in Dewey and Whitehead", *Philosophical Review,* **36** (1927), 122: "the ultimate incompatibility of objectivity and relativity" is destroyed; it becomes plain that "the objective facts of the world of nature and of reality are the very 'apparent' and relative happenings disclosed to us in perception" [as Ernst Mach had maintained - Ed.].

[4] M.C. Swabey, *The Monist,* 1927, pp. 310 & 313; the passage is a good illustration of the inversion of the older logic based upon the principle of the essentiality of relations.

- in the sense that it is not a function of variable relations to knowers or to conditions of knowledge external to its own spatio-temporal region of existence. And it is precisely in our time and in what is miscalled "relativity physics" that the craving of the scientific mind for such constants or absolutes has been most strongly manifested. It is true that the earlier steps in the reasoning of Einstein's special theory of relativity seemed to reduce the whole of the world of the older physics to a sort of relativity which looked very much like universal perspectivity. All of the characters previously supposed to be "primary" and truly physical were resolved into conflicting appearances presented from the standpoints of different reference-systems; no one reference-system could be determined only by an arbitrary setting up of coördinates, *i.e..*, by an act of mind. But so far from being content with this sort of world, physicists were profoundly dissatisfied with it; and the greater and more characteristic part of Einstein's effort, as has often been remarked, was a "quest for the absolute", which he found in the Minkowskian four-dimensional continuum with its absolute space-time magnitude, the "interval". The constructive part of Einstein's reasoning, in short, has throughout been determined by the principle that "the test of objectivity is invariance for all observers"; and it was through the application of this test that his reasoning led to the conclusions that, "unless we wish to continue to talk about such shadows as space and time, it is in the space-time world and there alone that we shall obtain universal objectivity."[5] It is precisely in the degree in which he denied that "what the world is perceived as from any or all" potentially differing "points of view constitutes the essential part of its objective nature for physical science" that Einstein really continued "the high tradition of scientific objectivity".

In so far as objective relativism in philosophy has been inspired by the theory of relativity in physics, it has been part of that theory which is regarded (though, undeniably, with much wavering and equivocation) by its author and by most physicists as proving that the world of actual appearance, at least in those aspects in which it varies with standpoints, is *not* "objective".

1. At least three distinct, though not necessarily inconsistent senses of the term "relativity" seem discernible in the objective relativist's reasonings. The first and most obvious we may call "conditionality". As applied to the case of perception, it consists in the dependence of the existence of a characteristic, or set of characters, actually given in experience, upon the existence or occurence of something else. The empirical datum which is declared to be so conditioned we shall, for brevity, simply call "the charac-

[5] These quotations are taken from A. D'Abro, *The Evolution of Scientific Thought,* 1927, pp. 222-223.

ter". If that by which it is declared to be conditioned is supposed to be itself a single existent, for example, a brain and nervous system, considered apart from *its* other relations, we have a simple asymmetrical two-term relation, in which the character would be one term and the brain the other, and in which the character would be conceived as dependent on the functioning of the brain, but not conversely. In fact, however, the objective relativist, in agreement with dualism and with common sense, regards perception as involving three factors; in other words, the relativity which he ascribes to the character is relativity to a relation between two other terms; this we may call the "conditioning relation". What, more precisely, the terms of the conditioning relation are conceived to be we have yet to inquire; so far as the general relativistic conception is concerned, they might be any one of several pairs, such as a body not necessarily organic and an organism, or a region of space and an organism. But in any case, one of them appears to be conceived as more closely related temporally and causally (though not spatially) to the character, and also is more closely connected (if not identical) with the percipient organism; it is the pole of the conditioning relation which takes the place, in objective relativism, of the subject-end of the traditional subject-object polarity, while the other term corresponds roughly with the object-end. We may, therefore, call the two terms of the conditioning relation the proximate and the ulterior relatum. Objective relativists usually appear to maintain, further, that the character does not "belong to" the proximate relatum, or is not situated where that term is, but is an attribute, or occupies the position, of the ulterior relatum.

2. With the relativity of the character that consists in its conditionality the relativist usually associates, for reasons, which have already been suggested, a second sort of relativity. The proximate relation, though commonly described as a percipient event, is also referred to as a "standpoint"; and it is asserted that the character "belongs to" the ulterior relation, or occurs at the same place with it, and can be perceived to do so, only *from* this standpoint. This apparently contains two distinguishable affirmations. Partly, I suppose, it expresses the fact of the numerical distinctness of the data of different percipients. My percept and your percept of the reading-desk are two things, not one. But the proposition seems also, and chiefly, to be intended to assert the necessary *qualitative* disparity of the content present to different percipients, even when they are said to be dealing with a common object. Because you and I have different points of view - not only in the literal, visual sense, but in its extended sense - and because the point of view determines the nature of the content present, we cannot experience the same kind of data. The hypothetical common object will exhibit different aspects of itself to each of us.

To assert this is manifestly to say more than that the character exists only by virtue of the conditioning relation and is to be allocated to the ulterior term of that relation. It might conceivably exist, and be so situated, "from" all possible standpoints; for there is nothing in the notion of the dependence of a character of an existent, A, upon the relation of the latter to another existent, B. which implies that the position of B constitutes the sole "point of view" from which the character of A can be observed. The development of objective relativism, however, has been much influenced by the notion of perspectives [perhaps from Nietzsche. - Ed.]; these tend to be taken as the best example, in familiar experience, of the type of logical situation in which the relativist believes perception, and knowledge in general, to consist. The perspective distortions of the visual images of objects - the convergence of the railway tracks, the elliptical shape of the penny - according to ordinary optical theory, are characters which I can perceive only by standing at certain distances and angles with respect to the "real" non-convergent tracks, or the "real" circular penny, and which I must perceive when I am so situated; and this may be, and by the objective relativist would be, expressed by saying that from those standpoints, and from them only, the tracks *are* convergent, or the penny *is* elliptical (though common sense prefers to say that, *e.g.*, the tracks are not convergent but only appear so). The kind of relativity** which involves, besides the general idea of conditionality, the further notions of the existence or appearance of the character exclusively from the standpoint of the percipient, and of the necessary diversity of the characters experienced by percipients having different standpoints, we may call "perspectivity".

3. The third type of relativity which it is needful to keep in mind in considering objective relativism may be termed "respectivity of the meaning of characters", or more briefly, "respectivity". A character is said to be "respective" when the term designating it has no meaning, as a possible predicate of a subject of discourse, unless, besides that term and the subject, some definite third term is implicitly or explicitly specified. A "respective"

** Two points. First, inspite of their many philosophical differences, there are some basic similarities between the method Mach used in writing his books on the history of science and the type of history of ideas which Lovejoy later recommended and practiced. And second, it is curious that while Lovejoy both helped launch the history of ideas as a recognized branch of history and warned against the dangers of "perspectivity" as a type of relativism that half-a-century later Michel Foucault extended Nietzsche's conception of perspectivity into a tool of historical research with which to *attack* the history of ideas, and by implication Mach's historical approach as well. It perhaps follows as a matter of course that Foucault should defend mentally unbalanced people who suffer from the disconnected type of thinking which perspectivism can lead to. On the other hand, in Mach's psychological studies - as contrasted with his historical writings - Mach himself often used the kind of relativism which Lovejoy later called "perspectivism". - Ed.

attribute, in short, always, by its logical essence, implies a triadic, not merely a dyadic, relation. Common sense and traditional logic have not been ignorant that certain characters are respective and others absolute. It has doubtless always been evident to the plain man that it is meaningless to say that one thing is "large", or at any rate to say that it is "larger", without mentioning something else with which the size of the first is compared; or to say that an object is "to the left" without indicating what it is to the left of. But it has not usually been supposed by the plain man that it is meaningless to say that a thing is blue or square without referring to another thing with respect to which it has those properties; on the contrary, such reference would probably seem to him meaningless. In short, the distinction between respective and absolute (which does not, in this context, necessarily mean unconditioned) predicates has commonly been identified with the difference between relations and qualities; that any terms except those designating relations are respective in meaning at least requires proof.

Between the supposedly absolute and the respective predicates of things, important differences in logical properties have always been recognized. The former have been supposed to be of a much more jealous and monopolistic temper; in other words, qualities, because of the assumed absoluteness of their inherence, have seemed more liable to fall into mutual contradiction. Even they, it is true, have been generally held to be readily classifiable into certain obvious genera between the members of which contradiction cannot subsist; but of no one subject or place could two positive qualities of the same genus be asserted without absurdity. A coin - it has seemed evident to the human species, until certain recent developments in philosophy - could not be at once circular and elliptical, though it could be at once circular and yellow, or elliptical and yellow; a stick could be objectively straight and objectively tapering, but not objectively straight and objectively bent; a liquid could be warm and sweet, but not warm and cold or sweet and sour. How such a relation of mutual repugnance can obtain between simple positive concepts has seemed to some philosophers a mystery. Leibniz, for example, in spite of the importance in one part of his system of the assumption of an inherent *a priori* incompatibility between many separately conceivable characters, and in spite of his faith in the principle of sufficient reason, admitted his inability to explain the reason of the incompatibility of different simple attributes or how it is that purely positive natures can be opposed to one another.[6] But however odd this peculiarity of certain groups of concepts may appear to some logicians, common sense, science, and most philosophy have unhesitatingly accepted it as a solid and indispensable basis for their inferences about things. Relations, however,

---

6 G.W. Leibniz, *Philosophische Schriften,* edited by C.J. Gerhardt, Band 7, 1890, p. 195.

have usually been regarded as less liable to these oppositions. They, too, it is true, so long as they were taken abstractly, as mere universals, fell into sets of which the members were antithetic to one another, though not to members of the other sets; but one thing or locus might without contradiction be simultaneously in any number of objective relations, even of the abstractly antithetic sort, with respect to different relata.

The relativity which Einstein's special theory emphasized, as usually interpreted, appeared to show that a great many more characters are essentially respective than had hitherto been supposed. Thus, for example (in the words of a typical exposition) the difference between the classical definition and the relativity definition of simultaneity may be expressed concisely in the following way. According to classical mechanics simultaneity is a dyadic relation, a relation between two terms. According to the relativity theory, simultaneity is a triadic relation . . . . According to the classical theory one would say, "A is simultaneous with B"; according to the relativity theory, "A is simultaneous with B with respect to K".[7]

Again, the theory of Einstein showed that "physical length, instead of being a dyadic relation between two intrinsic lengths, is a triadic relation between the intrinsic lengths and the reference-system used"; and so in general with the metrical characters of bodies. Thus the properties ascribable to the concrete entities in nature seemed to be very widely and deeply infected with respectivity of meaning; and it was because of this that it became possible to predicate of a single thing so great a variety of properties previously assumed to be mutually exclusive.

There is, consequently, in at least some forms of the theory now before us, a tendency to extend this notion of the respectivity of meaning, if not universally, at least so widely as to cover all the attributes of sensible appearances. Such attributes would thus come to enjoy that degree of comparative immunity from the constraints of the principle of contradiction which relations have always by general consent enjoyed. If shape or color are predictable of an object only with respect to something else, then the same coin may be both truly circular and truly elliptical, the stick thrust partly under water may be as objectively bent as it is objectively straight, the same surface of a material object may literally possess the color which is perceived by an observer having what is called normal vision and that which is perceived by a color-blind person - different relata, or loci of reference, being in each case presupposed.

The two previous senses of relativity. It is evident, do not include or imply respectivity. Conditionality, as defined may be merely an empirical relation of causal [or Machist] dependence of the existence of a particular having a certain character upon the existence of another particular having

[7] V.F. Lenzen, in *University of California Studies in Philosophy,* **5** (1924), 44.

another character, neither character being necessarily destitute of logical import apart from a reference to the other. And a perspective of a thing is not necessarily a character which, by its very meaning, contains a reference to the standpoint from which it appears. To say that a penny has an elliptical shape from a certain point of view does not obviously imply that "elliptical shape" in general cannot be so much as conceived unless one is thinking of a point of view and of the phenomena of perspective distortion. The *from*-ness, so to say, which is part of the generalized notion of perspectives is neither interchangeable nor reciprocal with the *to*-ness which is involved in the notion of respective attributes. It is therefore essential to be on the alert against illicit substitutions of one of the senses of relativity for one or both of the others, in the objective relativist's reasoning.

CHAPTER 9

ERNST MACH - THE SCIENTIST AS A BUDDHIST ?

by Ursula Baatz*
(University of Vienna)

When Heinrich Gomperz, an Austrian historian of philosophy living in the USA, entitled Ernst Mach as the "Buddha of Science" in his autobiography[1], he summarized a discussion which had begun at the turn of the century and ended in the 1920's. It was Mach's phenomenalistic approach to his *Analysis of Sensations,* his emphasis on the non-existence of the "Ego", which had prompted his contemporaries to compare his theory with Buddhist insights. The first translations of the Pali Canon were then available in German, and Viennese artists and intellectuals studied them for new insights. Mach did not oppose the interpretation of his adherents who claimed that he was akin to Buddhism. On the other hand, comparison between science (especially physics) and religion in whatever aspect were not a matter of inquiry among members of the scientific community or among scientific lay people in those days. The *Verein Ernst Mach,* which was the organizational platform of the *Wiener Kreis,* was basically socialist and was hardly interested in religions. This is not at all surprising, especially if one considers the strong ties between Austrian political conservatives and the Church in the 1920's and 1930's. Somewhat oddly, a recent volume of essays, *Ernst Mach - Studien und Dokumente zu Leben und Werk,* edited by Dieter Hoffmann & Hubert Laitko, Berlin, 1991 carried the pre-publication title *Ernst Mach - Buddha der Wissenschaft,* though it contained almost nothing about Mach's relationship to Buddhism.

Thus the question seems not to have gone off stage completely. Meanwhile, a new theoretical species has evolved - those who discuss possible similarities between "Eastern Religion" and "Western Science". If one tries to discover a basis for this comparison, however, one finds it a rather flimsy enterprise indeed. In terms of current scientific basics, one must concede that even the most esoteric physics today is based on knowledge apart from the knower, whereas the experience on which Eastern religions rely is entirely personal - in the Heideggerian sense of "In-der-Welt-sein". No kind of religious experience can be measured by scientific methods and equipment; it must be lived.

* This paper was especially prepared for this volume.

1 Heinrich Gomperz, "Autobiographical remarks". *The Personalist,* **24** (1943), p. 258.

J. Blackmore (ed.), Ernst Mach – A Deeper Look, 183–199.

If Mach's theories were really Buddhist at their core, they would provide an insight into the relationship between "Eastern Religions" and physics, as Mach was last but not least a physicist, although his credentials as an outstanding physicist seem slightly soiled because he rejected Einstein's theory of relativity (cf. for the controversy on the possible forgery of the preface to his *Die Prinzipien der physikalischen Optik,* see G. Wolters 1987 & J. Blackmore 1989).[2] Mach was a physicist by profession. His manner of arguing, his method and choice of topics was that of a scientist familiar with both physics and psychophysiology.

In addition, and this seems most important, this approach can illuminate problems connected with presentationalism and representationalism as well as with subjectivity and objectivity. These are standards of epistemology that are also basic to Buddhist "epistemology". So the inquiry into the possible correspondence or relation between Mach and Buddhism is not just a historiographic one, but a genuine philosophical investigation.

In this respect, we have first to scrutinize which features in Mach can be paralleled with aspects of Buddhism. The next step is to ask whether there were Buddhist influences on Mach, for since the publication of Schopenhauer's *Die Welt als Wille und Vorstellung* [1819] or rather since his ideas came into vogue during the 1850's Buddhism had become somewhat familiar to people of the period. In line with this, we must take into account that the development of science throughout the 19th century may have emerged with some similarities to Buddhist thought patterns. The philosophical impact of the comparison between Mach's thought and Buddhist doctrine can be analysed against this background not as a matter of cross-cultural references but more systematically. The significance of Ernst Mach's approach to epistemology lies in his philosophical interests and in his research in physics and psychophysiology. Both led him to an epistemological theory of science that begins with the living body. Neither physics nor physiology, which are both based on measurement, employ this point of view. Mach can therefore be seen as a scientist who moves beyond scientific method by connecting so-called "subjective" and "objective" accounts. The question at stake is not whether he was relevant as a physicist or as a psychophysiologist, but whether he can convey a new perspective, and it seems he can.

---

[2] Gereon Wolters, *Mach I, Mach II, Einstein und die Relativitätstheorie - Eine Fälschung und ihre Folgen,* Walter de Gruyter: Berlin, 1987, & John Blackmore, "Ernst Mach Leaves 'The Church of Physics'", *British Journal for the Philosophy of Science,* **40** (1989), 519-540. [See also chapters 14 and 15 in this volume for articles by Gerald Holton and Ryoichi Itagaki. - Ed.]

SIMILARITIES[3] AND DIFFERENCES:

The first to notice the seemingly Buddhist features of Mach's *Analyse der Empfindungen*, were not his Viennese contemporaries but Buddhist monks in what is now called Sri Lanka who came to know the English translation of 1897. They translated it into Singalese as "educational material"[4] for Buddhist purposes in a situation of upheaval against a Christianization forced by the British government. One can suppose that it was the vigorous remark in *Analyse* from "Antimetaphysische Vorbemerkungen": *Das Ich ist unrettbar* [the self cannot be saved][5] and the dissection of the "I" into elements which drew the attention of the Theravada monks to this theory. For it is a basic proposition in the teachings of the Buddha that there is nothing to be grasped as a separate entity called "I". And similar to Mach, the Buddhist tradition holds that human beings can be analysed in terms of five elements called *skandhas,* although Mach gave no number of elements. Furthermore, the Buddha and Mach rejected metaphysical speculations. These similarities were conspicuous for Theravadin monks as well as for Mach's contemporaries, e.g. Anton Lampa.[6] And finally, Mach considered his work himself to bear Buddhist features, especially toward the end of his life.

When Ernst Mach argued against a fictitious metaphysical "I", he turned against a Kantian theory of scientific knowledge which was based on a transcendental "I" rather than an empirical "I", uttered by individual human beings, as the condition for a theory of nature. The transcendental I is the "höchster Punkt der Einheit des Verstandesgebrauchs" [highest point of unity in the use of human understanding], as Kant states in his *Critique of Pure Reason* in B 134. The objects of cognition are appearances which can be perceived as constant and which are in accord with theoretical principles by the legislation of reason. The "Thing-in-itself" (*Ding an sich*) is inconceivable. When Mach criticizes the "I" and dissolves it into elements he also criticizes the world of constant appearances, which are dissolved in complexes of sensations or elements as well. Scientific knowledge is thus

[3] Parts of this discussion first appeared in Ursula Baatz, "Ernst Mach und der Buddhismus", *Conceptus*, Nr. 56, **22** (1988), 19-33.I am also indebted to Dr. Bettina Wahrig-Schmidt (Univ. Lübeck) & Dr. Rudiger Schmidt (Univ. Oldenburg) for the critical discussion of that paper.

[4] John T, Blackmore, *Ernst Mach - His Life, Work, and Influence,* Berkeley, 1972, pp. 288-289.

[5] Ernst Mach, *Analyse der Empfindungen,* 7. Auflage, Jena, 1918, p. 20.

[6] See Anton Lampa, *Ernst Mach,* Prag, 1918 and E. Schulzki, *Der Mensch als Elementen-komplex und als denkökonomische Einheit - Zur Anthropologie Ernst Machs,* Phil. Diss., Münster, 1979.

not knowledge that necessarily conforms to the principles of reason, but knowledge of logical regularities. Knowledge does not deal with the necessity for facts but with consitions and consequences of conditions, which can be interpreted as mathematical functions. Mach states: "Only the mediation of functional relations has value for us, since the question is merely one of learning how to distinguish the *dependent features within experience* from one another."[7]

The Buddha's teachings of *anatman,* the non-I, are critical of pre-Buddhist teachings of an unchanging self, the *atman*, which is fundamental to everything existing in time and space, but which is in itself an absolute and divine unity. Indian philosophy was not interested in gaining knowledge of contingent objects whose underlying aspects are change but in knowledge, which because it is not subject to time and space, liberates us from conditions of impermanence. The critique of the Buddha turned against rigid formalizations of this search for liberation, which interprets the *atman* as an unchangeable I; the Buddha also turned against the speculations of the brahmins, who had lost themselves in abstract subtleties instead of treading the path of liberation. This was quite different from Mach, whose critique remained in the realm of science and was rooted in the doctrine of "Denkökonomie". Nevertheless, Mach's relativism had a social background in its alliance of anti-metaphysicians, scientists, and social democrats, who were engaged in social improvements. The different contexts provided different arguments against metaphysics, too. Whereas Mach argued from the standpoint of sensory data, Buddha Shakyamuni compared metaphysical questioning with the curiosity of someone, who though hit by a poisoned arrow asks where the arrow has come from and what type of arrow it is instead of looking for an immediate cure. The Buddha considered such questions to be unhealthy. This notion of healthy and unhealthy knowledge, a notion intrinsic to Buddhism, is not found in Mach's thought, whose criterion of "Denkökonomie" aims to save time and labor by using concepts appropriate to facts and which will help in humanizing the world by realizing the "ideal of a moral world order with the help of our psychological and sociological insights."[8] When the Buddha speaks of healthy knowledge, he evaluates it according to its support for gaining ultimate liberation. Expressed in philosophical terms, this liberation is the insight that everything is conditioned and therefore empty, that is, that the "I" as well as everything else is empty. Buddhism refers to dependent co-arising, when dealing with the becoming and decay of contingent events of the the world. As world is relative to man, the twelve-fold net of conditions depicts the dependent co-arising of

[7] Mach, op. cit., p. 28.

[8] Ernst Mach, *Erkenntnis und Irrtum,* 4. Auflage, Leipzig, 1920, p. 463.

world and I; not as a linear sequence of conditions but as a cycle, which is interpreted in art as "the wheel of life".

The first condition is ignorance (*avidya*), which is unawareness of the relativity of all arising by means of conditions. Next are the creative impulses (*samskara*), the traces of misguided decisions and a perpetual craving for new appearances and forming consciousness (*vijnana*). This is differentiated in mind and form (*nama-rupa*); and this is in turn the condition of the six senses (*sadayatana*) seeing, hearing, smelling, tasting, touching, and as the sixth sense, thinking. If there are sense-faculties, there is contact (*sparsa*) with objects of these senses, and thus sensation is conditioned (*vedana*). Pleasant sensations are conditions for thirst for living (*trana*), and this longing conditions cleaving (*upadana*) to the wanted objects. Thus, new becoming (*bhava*) is conditioned, which leads to new birth (*jati*); being born conditions becoming old and eventually dying (jaramarana).

Simply stated, this formula of *pratitya samutpada* or dependent co-arising can be expressed as "If this, then that". In formula-language it could be expressed as $f(x) = y$. This is exactly the interpretation Mach gives to the relationship of sensations ("Abhängigkeit der Erlebnisse"). But Mach did not start his analysis with ignorance. Instead of this, he saw the world as a presupposition for any indispensable physical process, which can be set aside as a whole when doing research. So he started with the analysis of the elements of the "I" or the subject: "The subject builds itself [baut sich...auf] from perceptions which then react once again with perceptions"[9] So the contact (*sparsa*) of senses and objects of senses is the starting point of Mach's analysis, which remains exclusively in the realm of sensations. The thirst for life (*trana*) is taken for granted as a biological fact. Mach did not think of jeopardizing biological processes. On the contrary, he is interested in the reconstruction of the multiple dependencies which are basic for the development of certain organic structures. If there are no dependencies on the immediate environment, Mach hypothesized memories of past voluntary actions. Volition is:

> the totality of conditions of movement which are partly conscious and one with the prospects of success. If we analyse these conditions in so far as they are conscious, we find nothing other than memory traces from earlier experiences and their associations. It seems that the preservation of such traces and their connections is a *basic function of the elementary organism*, even if we can no longer speak of consciousness of a system of memories as something organized.[10]

This is more or less similar to the functioning of the creative impulses (*samskara*), which vivify the traces of former decisions. But whereas Mach

---

9 Mach 1918, op. cit., P. 18.

10 Ibid. p. 82f.

is giving an explanation for the development of organisms, the Buddhist teaching points to the conditions of the continuous turning of the wheel of life as a superpersonal process. Mach's reference to a "superpersonal life" of the contents of consciousness[11] again resembles Buddhist teachings. If the contents of consciousness, i.e. thoughts, are not bound to the limits of thinking persons, as Mach suggested in the *Analysis of Sensations,* then they have to be considered as the objects of thinking, as e.g. a tree is an object of seeing and so on. This is according to Buddhist understanding of perception, as thinking is the sixth faculty of the *sadayatana* or sense-faculties.

Whereas becoming is analysed in terms of *pratitya-sammutpada,* or dependent co-arising, the actual human being can be described as an aggregation of clusters (*skandna*), body, sensation, perception, volition and consciousness. These five *skandhas* again can be differentiated in themselves. Thus, the I is dissolved into elements and no longer seen as a compound entity. Similarly, each describes the I as an aggregation of changing elements:

> The I is not primary, but the elements. The elements form the self. I perceive green, which is to say that the element green occurs in a certain complex of other elements, but if I cease to perceive green, if I die, the elements no longer present themselves in their accustomed way. This is all that death means.[12]

According to Mach, what we call "inside" and "outside" consists both of sensations - our body as well as our perceptions or representations as well as colors, space, time, etc. Compared with the Buddhist view, Mach's is less differentiated. For him, the interest lay in epistemology and theory of science, so he aimed his analysis at research in physics and physiology.

This analysis that dissolves the tangible world into elements of sensation has its origin in the way we live in the world. Both a tree suddenly appearing within our field of vision and a hand before our eyes changes what is experienced as sensations. The relation of the seer and the seen is not a strict fixation, seen from a central point of view but changes continuously. Changes in perception alter the field of the seer and the field of the seen.

This analysis of elements of the perceptible leads to questions concerning similarities and dissimilarities in the realm of sensations. This research was continued later on in the field of Gestaltpsychology.

The Buddhist analysis of elements results from the practice of meditation. These elements called "*dharma*", are sometimes translated as "elements of consciousness".

> A *dharma* is an impersonal occurence which belongs to no person and hence to no individual, but acts itself out in its own objective way. It is valued as the most

---

[11] Ibid., p. 197.

[12] Ibid., p. 19.

praiseworthy deed of a Buddhist monk if he is able to give an account of the content of his thinking with the help of this impersonal *dharma* without introducing the nebulous and ruinous word "I". For this purpose, the monk has a precise list of *dharma* ready for use.[13]

But the number of dharmas varies according to the particular school of Buddhism. The *sarvastivadins* count 75 dharmas, the *yogacaras* 100. There are, besides sense-perceptions, thinking, wishes, ethical ("healthy") attitudes and their contraries, space and time and concepts, like holiness or differentiation, as well as dharmas, which belong to the practice of meditation. Especially in the Buddhism of Southeast Asia, the Theravada, there is a strong concentration on such meditative analysis. And the monks who translated Mach's *Analysis of Sensations* into Singalese were Theravadins.

According to orthodox Buddhist views, *dharmas* are non-substantial and respond to the law of dependent co-arising, except one *dharma*, namely *nirvana*, which is not conditioned.[14] Analogously, Mach refers to science, using similar references to permanence and change, and non-substantiality.

Unconditional invariability does not exist. We arrive at such a notion only by overlooking or undervaluing conditioning factors or by considering them as always given or by arbitrarily *abstracting* from the same. There is only one kind of invariability which comprehends all actual cases, the invariability of connection or relation. *Also substance, that is, matter, is not unconditionally invariable*. What we call 'matter' is a certain lawlike connection of elements (perceptions). The perceptions of the different senses of *one* man, like the perceptions of *different* people, are dependent upon one another in regular fashion. This applies to matter. The older generation - of physicists and chemists in particular - will be deeply disturbed by the assumption that matter is not be considered as absolutely unconditional but merely a *fixed interrelated pattern* of elements which in themselves seem much more *fleeting* than *invariable*. Just as it cost me a lot of effort to overcome [the materialist prejudice] in my time, so younger people will make the same effort to overcome it in theirs, and thus arrive at the unavoidable insight above.[15]

Mach, however, is only concerned with those parts of reality which can also be registered scientifically. It is a scientific preformed *Lebenswelt* that Mach dissolves into its elements. These exist in measurable dimensions of space and time, although Mach states that those dimensions are dependent on human beings in general and specifically on the person of the scientist. Thus those dimensions are "objective" in a narrow sense, but that does not mean Mach includes wish-projections, ethical attitudes, or meditative states (which are part of the dharmic analysis). Hence, the "scientific world view"

---

13 E. Conze, *Eine Kurze Geschichte des Buddhismus,* Frankfurt, 1984, p. 31.

14 J. Takakusu, *The Essentials of Buddhist Philosophy,* Delhi, 1978, pp. 92f.

15 Mach 1918, op. cit., p. 270.

becomes the background on which *Verbindungsgesetze* [connecting laws] crystallize as the only permanent point of investigation.

What Mach demanded was a completely radical way of thinking, but he did not completely give up the idea of permanence, since science demands landmarks for the construction and reconstruction of reality. Buddhism in turn claims that even dependent co-arising is not to be stated as a permanent law. The famous dialectic of Nagarjuna says: "We state that whatever is dependent arising, that is emptiness. That is dependent upon convention. That itself is the middle path."[16]

The "middle way" is neither assertive nor negative. To state that everything is dependently arisen or to negate it are both types of conventional truth (*samvrtisatya*). To *realize* emptiness is to realize that there is no difference between the circle of dependent co-arising and *nirvana*, which means the extinction of ignorance, greed, and hatred, therefore the conditions out of which the wheel of life keeps turning. That affects even the teaching of the Buddha: "The Buddha did not teach the appeasement of all objects, the appeasement of obsession, and the auspicuous as some thing to some one at some place."[17]

Here, even physics cannot pass through. Thus there is the final distinction between scientific reasoning and research, and realization of ultimate reality. Although it is realized nowhere else than in the concrete world, it cannot be stated in a law related to this world. On the other hand, it is just this concrete world to which the laws of physics, of chemistry, of physiology, etc., are applied. Insisting on the interdependence of related phenomena as a basic notion for scientific research, Mach seems to touch the realm of Buddhist teachings although he misses the depth of the tradition of the Buddha. He remains in the domain of conventional truth when he keeps to the frame of reference he is accustomed to as a physicist. Mach is not in search of ultimate truth, but he is interested in the evolution of scientific thought. He is a physicist, not a mystic[18], although he utilizes insights which are basically mystical. But he is a "half-hearted mystic" for he sticks to a description of phenomena that relies on a "Verbindungsgesetz von Elementen", which is asserted as enduring. His epistemology bears affinity with Buddhism, in the literal sense: *ad finem,* which means to be on or near the borderline.

Where does Mach's affinity to Buddhism come from? Usually in such a case one expects influences originating in books or persons representing

---

[16] Kalupahana, D., *Mulamadhyamakakarikas,* XXIV, 18. 1986, p. 339. (See also D. Kalupahana, *Nagarjuna - The Philosophy of the Middle Way,* New York, 1989)

[17] Ibid., XXV, 24, p. 369.

[18] Mystic means "knowledge of the absolute reality", which is apophatic. It is used here not in a strict sense restricted to monotheistic religions, but in a broader sense.

this tradition. This last assumption does not apply to the nineteenth century. Unlike today, there were no Buddhist teachers travelling to Europe. Nevertheless, the word "Buddhism" was not unknown in Europe, thanks to Schopenhauer, the first European philosopher to grant equal importance to Indian philosophy as to Kant or the Greeks. Schopenhauer read voluminously in the field of Oriental philosophy. Fundamentally, he adopted the views of the *Vedas* and the *Upanishads* in his concept of the world as a result of volition and imagination. Nonetheless, his assessment of Buddhism was imprinted by wrong interpretations, although his writings may be dyed by a "Buddhist mood".[19] His assumption that volition is the "Ding an sich" manifesting itself in all appearances, is contrary to Buddhist views; for in Buddhist teaching there is no ultimate principle to which appearances can be reduced. *Pratiya-sammutpada* or interdependent co-arising is no "Ding an sich". There is nothing else than what is manifesting. The crucial issues of Buddhism, the teaching of the inconsistency of the I and of interdependent co-arising were incorrectly interpreted until the publication in 1881 of Oldenberg's *Buddha: Sein Leben, seine Lehre, seine Gemeinde*. This popular, competent interpretation of the basic teachings of Buddhism is reprinted to the present day. Until Oldenberg's book appeared, Buddhism was assumed to be nihilistic, reducing the world to nothing, or it was regarded as pantheistic or pessimistic - the latter view favored by Schopenhauer. Moreover, what was mostly depicted were myths and esoteric teaching designed for the interest of the audience. In addition to Oldenburg, Karl Eugen Neumann's translations of the texts of the Pali Canon, published in 1892 and thereafter in small editions, expanded the understanding of Buddhism in the West. After his death in 1915, Neumann's work was spread, inspiring poets and literati to depict an image of Buddha and Buddhism which has imprinted the German-speaking world to this day. Hugo von Hofmannsthal and Fritz Mauthner are among those early discoverers of Neumann's translations.[20]

Mach published his *Analyse der Empfindungen* in 1886, so he may have been familiar with Oldenberg's book as well. The two catalogues of his library contain several books on Buddhism and numerous others on Chinese philosophy, though the oldest and strangest was probably Georg Michael Eisenbach's *Missiongeschäfte des Dalai Lama...* Nürnberg 1798 (*Bibliothek Ernst Mach,* Teil II, Nr. 105). As the catalogue mentions, it obviously belongs to the Protestant apologetic tradition, even depicting Martin Luther on the frontispiece. More informative were five books by Paul Carus: *Kar-*

[19] V. Zotz, "Einige Anmerkungen zum Buddhismus-Bild deutschsprachicher Denker", *Bodhi-Baum,* **11** (Nr. 2/3, 1986), 59.

[20] V. Zotz, *Zur Rezeption, Interpretation, und Kritik des Buddhismus im deutschen Sprachraum vom Fin-de-Siecle bis 1930,* Phil. Diss., Vienna, 1986.

ma. A Story of early Buddhism, 1894, *The Gospel of Buddha,* 1894, *Das Evangelium Buddhas,* 1895, *The Dhamma or the Religion of Enlightenment. An Exposition of Buddhism,* 5th ed., 1907, and *Der Buddha. Drama.,* 1913, but all these works are from the 1890's or later. Indeed, Mach did not openly confess his Buddhist affinities until 1902 in a footnote to the third edition of his *Analyse* (p. 273), and even then only to state that Buddhism came close to his own ideas merely from the practical side.*

He quotes no German author in this context, but instead Paul Carus and two of his works. He had become acquainted with Carus on the occasion of the translation into English of Mach's *Mechanik* (1893). Paul Carus though born in Germany, had emigrated to the USA in 1884, where he worked as an editor for the magazines "The Monist" and "The Open Court", which were supported by the German-born factory owner Alfred Hegeler. Both magazines were concerned with "scientific research into religion" and with "religious research into science".

Since their expansion in 1887 these magazines can be considered the leading publications of their time in the field of Eastern religions and philosophy and Western science. Among the contributors to these were Ernst Mach, Ernst Haeckel, Bertrand Russell, the Indologist Max Müller, Rabindranath Tagore, and a number of Mach's colleagues and followers including Ludwig Boltzmann and Hans Kleinpeter. Carus himself was not only a writer and publisher, but an active participant in the World Parliament of Religions in Chicago in 1893. He saw Buddhism as a "scientific religion", drawing parallels to the psychological theories of Fechner, Wundt, and Ribot.[21] Mach exchanged letters with Carus and read his books, and it is not at all surprising, that he appreciated this scientific reinterpretation of Buddhism. In a letter to his friend Popper-Lynkeus[22] he had previously expressed his hopes in psychophysiological research:

> I expect much greater enlightenment from this side than from earlier derivations from the physical. It will certainly not lead to a personal God, but most probably to a falling away of the apparent bounds of a person, between what is organic and inorganic.

---

* But Mach's interest in Hindu culture from which Buddhism arose was considerably older. Not only had Mach read Schopenhauer at an early date but he had read at least some Hindu drama, especially *Urvasi* by Kalidasa. Indeed, in 1876 he encouraged a musically-minded student of his, Wilhelm Kienzl, to compose an opera based on the Indian play, which Kienzl did (1884). See Wilhelm Kienzl, "Wilhelm Kienzl's Erinnerungen an Prag", *Deutsche Arbeit,* **6** (1907/1908), 62. - Ed.

21 C.T. Jackson, "The Meeting of East and West: The Case of Paul Carus", *Journal of the History of Ideas,* **29** (1968), p. 81.

22 J. Blackmore & K. Hentschel (Hrsg.), *Ernst Mach als Aussenseiter - Machs Briefwechsel über Philosophie und Relativitätstheorie mit Persönlichkeiten seiner Zeit,* Wilhelm Braumüller: Vienna, 1985, pp. 4 & 5.

Mach referred to animism, which was in his view "correct and capable of scientific development". Buddhism is, according to the footnote quoted above, a *practice* that meets with his theory of non-ego: "Buddhism has approached [our point of view] over a period of centuries, chiefly from the practical side."

But all this does not explain the origin of the affinity between Mach's epistemology and Buddhism. Mach himself denied that there was any direct or indirect influence. In a letter to Fritz Mauthner, a thorough critic of language as well as a poet and novelist (*Der Letzte Tod des Gautama Buddha*), Mach said:

> Ever since I first learned about it, I have had the greatest sympathy for Buddhism, even if my *Analyse der Empfindungen* was not stimulated by Buddhism but originated from the most naive reflections. Certainly, I can make no claims to originality vis `a vis Buddhism. Indeed, David Hume in his *Treatise on Human Nature* had already carried through an analysis of the self and destroyed the illusion of its perseverence long before me. Actually, I owe this insight neither to Buddhism nor Hume, but to familiarity with Lichtenberg's "It thinks", though he probably had had some acquaintance with Hume.[23]

Of what kind were these "most naive reflections"? According to Mach, they concerned Kant's epistemology. Mach studied *Prolegomena to any further Metaphysics* in his early years, and this marked a turning point in his development, and the end of his "naive realism". "What Kant did was merely to change Mach's particular form of epistemological presentationalism from a sense-physical-objects to a sense-sensation identification"[24]

The enthusiasm for Kant was only a transitory condition. The young Ernst Mach apparently suddenly realized that he did not need a "thing-in-itself". He was outdoors on a bright summer day in the fertile plain between Vienna and the Hungarian border when this insight struck him. "The superfluity of the role played by the 'thing-in-itself' abruptly dawned upon me." He writes:

> Out in the open on a bright summer day the world once appeared to me together with my self as one interrelated mass of perceptions, only more strongly connected with the self. Although the essential reflection [about that moment] was only added later, the original experience became decisive for my whole world view [*ganze Anschauung*]. Nevertheless, I still had to struggle long and hard to incorporate the insight which I had won into my special branch of science. One unavoidably accepts along with what is worthwhile in physical instruction a significant dose of false metaphysics, which can be very hard to detach [from valid ideas which must be retained], particularly when both are still current or fashion-

[23] Ibid. p. 105f.

[24] J. Blackmore, *Ernst Mach - His Life, Work, and Influence*, Berkeley, 1972, p. 11.

able. Also, traditional, instinctive, powerful conceptions may arise for a time which vigorously stand in the way. It was only by alternating my attention back and forth between physics and physiology of the senses together with historical investigation into physics which allowed me, after having vainly attempted to free myself from [mental] conflict by adopting a psycho-physical monadology, to attain greater stability in my views (approximately since 1863).[25]

And in his late autobiographical sketch of 1913 he described this experience once more:

After I recognized Kant's 'thing-in-itself' as nonsense, I also began to regard the 'unchangeable ego' as an illusion. I can scarcely relate how lucky I felt to free myself from all tormenting, absurd ideas about every kind of personal immortality and to find myself introduced into an understanding of Buddhism, a good fortune which rarely falls to the lot of any European.[26]

Aside from his own accounts, there are two further circumstantial proofs that the affinity to Buddhism was not induced by books. First, the only source coming into consideration could be Oldenberg's *Buddha,* as the book was edited in 1881, five years before Mach's *Analyse*. The wording Oldenberg gave there for *pratitya sammutpada* was not "interdependent co-arising", but "causal nexus", that is a concept fundamentally different from Mach's, as it points to causality and not to mutual dependence. Only mutual dependence can be interpreted as a function or, in Mach's words, as "Funktionalbeziehungen", as it is a logical description ("If - then") and not a description in terms of "action" and "reaction". As the notion of interdependence is crucial for both Mach and Buddhism, then one can assume that Mach did not use Oldenberg's book. (Besides, the interpretation of mutual dependency as "causal nexus" is not a genuine Buddhist view.) The other argument is, that Buddhism was always associated with pessimism. Thus Robert Bouvier stated e.g. that Mach was "un Bouddhiste sans le pessimisme"[27], trying to point out what seemed to be the most fundamental difference, besides the fact that Mach did not allow for "metempsychosis" and Nirvana in his theories. The opinion that Buddhism is a form of pessimism traces back to Schopenhauer, and it is known that Mach read his books. Obviously, he neither adopted this pessimistic approach nor the incorrect in-

---

25 Mach 1918, op. cit., p. 24.

26 F. Herneck, "Über eine unveröffentlichte Selbstbiographie Ernst Machs", *Wissenschaftliche Zeitschrift der Humboldt-Universität zu Berlin,* math. naturwiss. Reihe, Jg. **VI**, (1956/57), Nr. 3., p. 213.

27 R. Bouvier, *La Pensee d' Ernst Mach,* Paris, 1923, p. 245.

terpretation Schopenhauer gave.* And it is therefore not surprising at all that Mach did not identify his own views with Buddhist ideas, except subsequently, when he became familiar with them through Carus's activities. Hence, it seems probable that Mach's affinity to Buddhism resulted from his own experience as a young man of eighteen, from "naive reflections" and the later considerations which shaped the expression of his experience.

This raises the question whether such an experience is structurally analogous to religious or, more exactly, to mystic experience, and whether as such it can occur outside any religious framework. Mach's father after briefly teaching in the Piarist Order became anti-clerical and his son an atheist. According to Mach's colleague, William James, and to the psychological research of Abraham Maslow, such an experience of unity is due to the range of experience accessible to human beings: thus Mach need not be considered a Buddhist.[28] What most probably shaped his experience into an expression similar to Buddhism, were certain patterns available in the philosophy and science of those days. (If he had lived in the Middle Ages, he would have expressed his experience in a different way. But this is a *coniunctivus irrealis perfecti,* and therefore futile.)

Mach himself mentioned Lichtenberg, who in turn owed his phrase "It thinks" to Hume. But this philosophy was not among the customary reading of a young Austrian in mid-nineteenth century. The reform of the Austrian system of higher education in 1848 forced the philosophical system of Johann Friedrich Herbart to be taught as the obligatory philosophy not only in the University but at Gymnasiums as well.[29]

The impact of Herbart's psychology and philosophy on the development of the academic sphere in Austria in the second half of the nineteenth century cannot be underestimated. The textbooks of philosophy and psychology were rooted in the Herbatian system. One can assume that most proba-

* A little historical clarification may be appropriate at this point. It is possible that Schopenhauer did not distinguish between Hindu and Buddhist doctrines clearly enough. In much traditional Hinduism it is largely because the self cannot be eliminated that painful desire becomes such a persistent problem. But Buddhism claims to offer the "optimistic" possibility of eliminating the self as a shortcut to liberation from painful longing. On the other hand, readers need not be told how relative and controversial these terms are, and that most people in the West have traditionally tried to solve problems largely through "optimistic" action rather than mere "pessimistic" therapeutic elimination of distinctions, desires, and bad feelings, "pessimistic" because it implies that the problems themselves are unreal, unimportant, or incapable of solution. Schopenhauer talked about Buddhism but was actually closer to "pessimistic" Hinduism. Both normally seem "pessimistic" as understood in the West. - Ed.

28 U. Baatz, "Ernst Mach und der Buddhismus" in *Conceptus,* Nr. 56, **22** (1988), p. 31.

29 W. Swoboda, "Physik, Physiologie, und Psychophysik - Die Wurzeln von Ernst Machs Empiriokritizismus" in *Ernst Mach - Werk und Wirkung,* Hrsg. R. Haller & F. Stadler, Verlag Hölder-Pichler-Tempsky: Wien, 1988, pp. 356-403.

bly Mach was aware of these theories. Herbart described the psychical reality as consisting of *Vorstellungen*, which were acting on each other, partly retarding each other, partly stimulating each other. Like Hume, he started with the notion of simple perceptions, e.g. red or blue or sweet or sour, which originate in the senses at each moment. Reality consists in his terminology of such complex perceptions or *zusammengesetzte Vorstellungen*. Herbart stated his psychology against the idealistic German philosophy, claiming that the interacting of *Vorstellungen* (perceptions, mental images, memories, and so on) can be put into mathematical formulas analysing the psychical process in terms of statics and mechanics, the notion of an "absolute subject". In consequence, he stated:

> Fichte considered the self to be an identity of subject and object and thereby leaned toward the grammatical conept of the "I", for the first person singular can be a subject who is grammatically allowed to talk about himself as an object. But can we find in actual self-consciousness of ourselves merely and exclusively self-knowing? Definitely not. Some kind of additional character always slips in; one finds thinking, willing, feeling, suffering, and acting with definite, particular relations to what is thought, willed, felt, suffered, done, and so forth.[30]

All of these modifications of consciousness of oneself are kinds of *Vorstellungen*. Herbart and Fechner together with Wundt are considered to be the founders of the new science of the human soul and its perception. Especially Herbart and Fechner were important for Mach's development.

This new psychology leaned towards the natural sciences. Philosophical descriptions are replaced by experiments and mathematical methods. Psychology separates from philosophy forming its own experimental scientific methodology and, in the course of the 19th century, constitutes a new scientific discipline among other academic fields. Mach takes his terminology and mental constructs from this new psychology which defines itself as a mathematical natural science. In this new paradigm he formulated his theories.

It has to be added that throughout the nineteenth century there is a more or less tacit undercurrent of change in fundamental philosophical perspectives. The old model of substance and accident is replaced by a model favoring functionality. There were many different reasons for this; one was the increasing mathematization of the sciences, including such formerly "qualitative" fields as psychology, biology, and medicine.

Another reason for this change lies in the increasing importance of mechanical calculators. The basics for such machines were formulated by Leibniz; however, only in industrial mass-production offered the opportu-

---

30 J.F. Herbart, "Psychologie als Wissenschaft, neu gegründet auf Erfahrung, Metaphysik, und Mathematik", in *Sämtliche Werke*, hrsg. von G. Hartenstein, 5. Band, I, Leipzig, 1850, p. 267.

nity to construct practically applicable calculators. In the1820's, Charles Babbage constructed the first of these "computer prototypes".[31] The processes with which data are processed rely on logical sequences. Thus pure linguistic expressions were equated with formula-language. These deepgoing changes in worldview are reflected in a new branch of literature, the detective novel. From its very inception by E.A. Poe to our own era the characteristic thought pattern of these novels are logical sequences. Computer and detective novels are key features of a *Lebenswelt* based solely on functionality. Thus Plato's and Aristotle's view leaning towards forms and their qualities are superceded by a new view of *Lebenswelt*.[32]

On the other hand, there is an increasing interest in the bodily existence of human beings, not only in psychophysiology or medicine, but in philosophy as well. Schopenhauer was the first to include the physical, sensory human body into a philosophical epistemology, thus overthrowing the framework of the *Bewusstseinsphilosophie*.[33] He points out: "When thinkers consider the live body, they are led to an ultimate theme: intuition into the living organic Form, the Physis itself." The history of the influence of Schopenhauer's philosophy is congruent with much history of *Lebensphilosophie*,[34] that places itself against the dominating mathematical and technical models of the world and of human beings by insisting on the quality of the factual being which cannot be interpreted correctly in general terms. Factual existence is becoming a philosophical topic.

Mach was a scientist and as a scientist he used a model favoring functionality. But Mach was interested in the processes of actual living beings (unlike his contemporaries, he did not countenance vivisection), and his *Analyse der Empfindungen* gives an account of the description of phenomena as research into the qualities of actual life. This questioning of the process of life from a viewpoint characterized by functionality and therefore by non-substantiality constitutes Mach's affinity to Buddhist epistemology, especially the analysis of *dharmas* exercised in Theravada. But the sources of Mach's philosophy are rooted in the intellectual situation of the nineteenth century.

Does Mach convey a new perspective for his epistemology? Surprisingly enough in the very first lines of the *Analyse* Mach cites Goethe, Schopenhauer, and in particular Johannes Müller, the German physiologist, a

---

31 See H.P. Babbage, *Babbage's Calculating Engines*, 1889.

32 This was the content of an interview with Dr. Wolfgang Pircher (Univ. Wien) and Univ. Prof. Dr. Friedrich Kittler (Univ. Essen). The interviews were made for a transmission by the Austrian Broadcast Corporation (ORF): Ursula Baatz, "Kinder der Vergangenheit - das 19. Jahrhundert und wir", March 26-29, 1990.

33 A. Gehlen, "Die Resultate Schopenhauers", in A. Gehlen, *Gesamtausgabe*, Bd. 4, Frankfurt/M, 1983.

34 H. Schnädelbach, *Philosophie in Deutschland 1831-1933*, Frankfurt/M, 1983, p. 177.

teacher of Helmholtz, Du Bois-Reymond, and Brücke, among others. Although these scientists turned to a strictly physical interpretation of the sciences of life, Müller himself was greatly indebted to Goethe and the tradition of *Naturphilosophie*. Müller's breakthrough in physiology is characterized by exact observation and qualitative description of living processes. He understood the motions of living beings according to traditional European philosophy as teleological formative movements, as manifestations of the whole in nature. Although Müller abandoned this idealistic - in modern terms holistic - perspective, he did remain a consistent phenomenalist. The influence of Müller on Mach will be analysed in another paper. For the moment, it is enough to seize the fact that Mach strongly emphasized just those people who were controversial with respect to the new program of scientific research, men who concentrated not on mathematics but on intuition, particularly with respect to sense-perception. At the same time, Mach subscribed to the current program of psycho-physics, empirical psychology, and physics. What he really did in his *Analyse* in an attempt to show the psychological interpretation of sensations which can be interpreted equally well in terms of physics, was a phenomenalistic analysis. He did not develop a formal model of how the interaction of elements could be reconstructed; he merely analysed perceptions.

Such an analysis necessarily starts with subjective sensations which present the world. However, the analysis itself cannot accurately reflect this "presence", but has to turn the "presentation" into a representation. But before interpreting these as either coherent elements (physically) or as dependent on the brain (psychologically) it must be noted that sensations are neither psychical nor physical.[35] What then are sensations? Mach did not answer this question. But in his notebook he wrote at the end of his life, in 1910: "1. no objects beyond sensory limits /2. not independent of consciousness /3. frontier between objective and subjective runs right through the middle of sensation and consciousness."[36] And in the same notebook, he recorded three curious but very abbreviated sentences which are not easy to understand or translate: "Die Rose sehen und drinnen sein. Das Auge sehen und Netzhautreizungen desselben haben. Reaktions-Ordnung eine andere."[37] They seem to constitute a Chinese puzzle with intuition changing into scientific observation and vice versa.

Intuition is directly "into it", allowing for no distance, sensing the qualities of life immediately. Scientific observation needs distance, needs therefore, a frame of reference for comparison - reactions, that are more or less different or the same as last time etc.. Mach saw the anthropological pri-

35 Mach 1918, op. cit., p. 51.

36 Haller/Stadler 1988, op. cit., p. 208.

37 Ibid., p. 209.

ority of qualitative experience; and he tried to give it significance in a situation where quantification had overlapped into the world of evidence. This is an ambiguous approach in Mach's work itself, as he does not separate the "immediately given" world from the world of "facts", i.e. scientifically shaped features of our environment. Only in the changing perspectives of his experience can we - trained by Heidegger and others - recognize the difference in approaches. Nowadays ecological questions are not the least of those which pose epistemological investigations as formal scientific distinctions about the nature of reality. Emphasizing the qualitative aspects of life, one could trace in Mach's thoughts a way to a new understanding of the living process of this world and of ourselves living in it.

On one hand, Mach is an ardent proponent of science, progress, and enlightenment, on the other, he keeps an experimental interest connected with a qualitatively-based perspective. Therefore, he can refer to the tradition of *Naturphilosophie* without giving up his status as a representative of mathematical natural science. It is exactly this connection between attention to direct sense-experience and scientific, analytical thought-processes that reveals similarities to Buddhist traditions.

The "I" criticized by Mach, relying on Herbart and his own experience, is the "ego cogito" of modern philosophy and the bourgeois consciousness of any kind. It is the "I" that strives to control everything. Conversely, the traditional world-view regards the "I" as part of the whole. In such a view the whole is necessarily expressed in a non-scientific cosmology.

The development of natural science and technology in bourgeois society led to the disappearance of a cosmology, which was primarily derived from Classical sources. The quest for parallels between Eastern Religion and Western science attempts to create a fundamental identity based on superficial structural similarities. The goal lies in an understanding of the whole. Instead of merely creating a cosmological appendix to science, Mach's approach offers a realistic alternative for a continuous discussion of immediate experience and subjectivity and objectivity in the realm of natural science.

CHAPTER 10

MACH'S PHENOMENALISM AS A LINK BETWEEN PHYSICS AND PSYCHOLOGY

by Miodrag Cekic*
(University of Nis, Belgrade)

The essential aim of Mach's scientific and philosophical endeavours was not to build a coherent theory of knowing, but to overcome the gap between the science of inorganic matter and the science of psychic activities, and to create a conceptual system which could equally well be applied to physics and psychology. The intermediary role between these two groups of sciences should be played by biology. With the help of this discipline, unique scientific knowledge could be achieved, i.e. unity in subject-matter and methodology in physics, biology, and psychology could be reached. Mach believed that biology could intertwine with physics and psychology in such a way that they could help each other in describing the corresponding phenomena of their separate fields. The unity of science in methodology and in subject-matter as well as the unity of philosophy with the sciences could thus be attained by means of biology, especially the physiology of the sense organs, that is to say through the analysis of sensations.[1] Mach himself carried out a large amount of laboratory work on the physiology of the sense organs, psychophysics, and psychology of the sensations. He studied auditory and kinesthetic sensations as well as the perception of space, time, and movement. As a physicist, Mach successfully dealt with important aspects of mechanics, thermodynamics, optics, acoustics, and wave mechanics. Many of his discoveries were of special interest for ballistics and later hydrodynamics, and aerodynamics. Out of his work photographing and measuring the shock waves of bullets and shells a number of expressions utilizing Mach's name have become widely known such as Mach number (M) which signifies the relation between the speed of movement of a body and the speed of the sound within the medium through which that body has been

---

* This paper is an extensively revised and augmented version of "Mach's Phenomenalism and Its Consequences in Physics", which appeared in *International Philosophical Quarterly,* **21** (1981), 249-259.

[1] E. Mach, *Die Analyse der Empfindungen,* Jena, 1922, pp. vi, 77, 81-82, 83, & 278; *Erkenntnis und Irrtum,* Leipzig, 1920, pp. 305; *Populär-wissenschaftliche Vorlesungen,* Leipzig, 1923, p. 292. (These works shall henceforth be abbreviated as *A.d.E., E.u.I.,* & *Pop-wiss. Vor.*)

J. Blackmore (ed.), Ernst Mach – A Deeper Look, 201–214.

moving. Other Mach labels include "Mach angle", "Mach stem", "Mach shock wave", and of course "Mach 1", "Mach 2", etc..

But what Mach constantly had in mind throughout his longlasting scientific and philosophical work was the methodological and subject-matter unity of all the sciences; and it was out of this aspiration plus his theory of economy that he found scientific justification for his phenomenalistic theory of knowledge. Mach's laboratory work encompassed not only the problems of theory of knowledge in the narrower sense but also what might be called the biology of knowing plus the problems of scientific methodology. He also contributed an important point of view to the history of scientific knowledge through his generally empirical interpretations of the work of famous scientists such as Galileo and Newton. Mach did not deal with the history of science as a professional research historian, that is to say he was not willing to take the trouble to travel to archives and check original material, rather he studied certain puzzles from the history of physics in connection with contemporary problems in the methodology of physical science. It should be said that Mach did not want to create a new system of physical or some other science, nor to create a new system within the theory of knowledge. In short, in this treatise, I do not want to reconstruct a systematic theory of knowledge which in fact Mach denied that he had, but instead out of his various contributions to science and philosophy try to explain what remains relevant in his theory of knowledge in the narrower sense, and finally to put forward a historical and critical evaluation.

According to Mach's opinion there are many concepts, such as subject, object, matter, the self, etc., which are created instinctively and which satisfy certain practical needs, such as the notion of matter as solid substance which through motion changing its place in space could be of use in science if it were limited to a specific field (for instance motion in inorganic nature). But this same notion of matter in respect to some other fields, for example, psychology, seems totally useless.[2] If one desires, as did Mach, to join the sciences into a unity, then one has to reach out for some other concept capable of being equally useful in all the different fields of scientific research, first of all in the three basic sciences - physics, biology, and psychology.

The common notion for all the sciences Mach called the "element"; sensation was to be understood as an element, that is to say as something which was equally physical and psychical. If the old notions of hard matter

[2] As a physicist who liked to consider the problems of mechanics (Mach's work *The Science of Mechanics* (1883) has long been considered one of the best of its kind), Mach naturally was impressed by its breadth of application within physics, but he resisted all attempts to reduce other branches of physics such as heat, light, and electricity to mechanics, and he also saw such reductionism as harmful to psychology. Unity of method did not mean reduction of content. He was opposed to "mechanistic imperialism".

and of the indestructability of mind (self) were retained, then the unity of physics and psychology would not be insured. Mach's laboratory work in both areas and especially in interdisciplines such as psychophysics made the philosophical problem very real to him. But if notions such as matter and mind were analysed into elements (colors, tones, spaces, times, etc.) which were in one respect psychic and in another physical, then the unity of science might very well be assured.[3] According to Mach, at least in many of his writings, space and time were just as much elements as colors, tones, and sounds, and thus it is clear that he rejected Locke's distinction between primary and secondary qualities, whether they are understood as if the former were physical and the latter mental (as some direct realists imagine) or as if there were two sets of both (as indirect realists often allege), one set mental and the other physical with secondary qualities in the object merely being surface characteristics which reflected light or sound waves in such a way as to stimulate particular sensory secondary qualities in the mind of the perceiving observer.

Mach began with the presupposition that there were no essential differences between the physical and the psychic. If it were possible to find exact bonds between certain physical phenomena, then there would be no reason to doubt the possibility of finding similar bonds between physical and psychic events. If colour were observed in relation to the source of light, then it would be a physical object; if it were viewed in relation to the retina, then it would become a physiological, that is, a biological phenomenon; and if it were considered with respect to other colors *as colors currently being experienced* it would be a psychological phenomenon. Thus the physical, physiological, and psychic did not differ in content but through what they were related to within conscious experience. In this manner, gaps between physical, physiological, and psychological or psychic research, indeed, all "ontological" questions of outside and inside, bodies and sensations, and the material and spiritual world, would be eliminated.[4] Thus the basic constituent parts of the physical and psychic areas were the same for him, only the order of their connection differed. Accordingly, the total internal and ex-

3 E. Mach, *A.d.E.*, pp. 254-255. See also *E.u.I.*, pp. 12-15 where Mach mentions that the thing and the self are only temporary fictions, since the aim of science is to study the versatile dependence of elements. For Mach, the thing is an instinctive concept created for practical purposes; the terms "thing" and "matter" were used alternatively.

4 F. Nietzsche like Mach rejected Fechner's theory that mental and physical appearances were "two faces of the same substance". Both men tried to reject the notion of substance itself. See F. Nietzsche, *Volja za moc*, Beograd, 1937, p. 327, fr. 523. [in English: *The Will to Power*, edited by W. Kaufmann and translated by W. Kaufmann & R.J. Hollingdale, Random House: New York, 1967, Section 458, pp. 269-270.]

ternal world was composed of a few elements of the same kind that find themselves connected by more or less rigid bonds.[5]

Bodies do not produce sensations, as was regarded in traditional materialism and some forms of mind-matter dualism, but rather groups of elements (groups of sensations) make up bodies. Bodies are only cognitive symbols for the group of elements. Colours, tones, odors, times, spaces, and so forth, are for the time-being the ultimate elements whose mutual connections are to be investigated by science. When scientists study them, they should not be confused by temporary connections and concepts which are abstracted from them for practical purposes (such as everday notions like body, matter, self, mind, etc.). In place of casual and instinctive understanding, the serious scientist must seek a freer, keener type of comprehension which helps sharpen his powers of observation and experience.[6]

In using Mach's concept of element correctly and to maximum advantage, the researcher must not just *talk* about freeing himself from narrow practical ways of thinking, he *has to do it* and get in the *habit* of doing it. He must develop a constant *practice* of focusing on elements. To the extent he succeeds he has found an *economical* method of doing science, a method he can use in physics, physiology, and psychology. But *finding* a right method still needs wise *use* if it is to advance science. Naturally, there have been opponents, but few of them have really mastered what they criticize.

For example, proceeding dogmatically from his Neo-Kantian viewpoint, Herbert Buzello argued that in Mach's theory of knowledge there was no basis for establishing a criterion of truth, since the elements simply existed and their connections could not be considered as true or false.[7] In my opinion, this criticism was not justified, since the element as Mach defined it was not the sensation which was given in natural perceptual totality, hence, while indeed it was not capable of being either true or false to something transendent, it could be accurately described, which means that there was both a clear criterion of truth and, to the extent that the description was properly carried out, there was actual truth. Hence, Buzello was mistaken.

Furthermore, since Mach wanted to reduce the physical and psychic to that which was most simple in experience, to the sensation or element,[8] if the goal of science was the simplest and most economical understanding of

---

5 E. Mach, *A.d.E.,* pp. x, 12-14 & 17-18. According to Max Scheler, Mach's elements are in fact physical things and not neutral elements, and thus Machism was, if taken formally, materialism. See Max Scheler, *Der Formalismus in der Ethik und die materiale Wertethik, dritte Auflage,* Halle a.d.S., 1927, pp. 419-431.

6 E. Mach, *A.d.E.,* pp. 23-25.

7 H. Buzello, *Kritische Untersuchungen von Ernst Mach's Erkenntnistheorie,* Berlin, 1911, p. 12.

8 E. Mach, *E.u.I.,* p. 44.

what could be experienced, then Mach had provided an ideal method and goal not only for psychology and biology but especially for physics. What has made this method and goal possible? His underlying philosophy of *phenomenalism,*[9] his epistemological rehabilitation of sensibility as the only certain reality.

His epistemological phenomenalism has managed to overcome the old dualism of the internal and external world, the opposition of subject and object, or, we could say, all animism in theory of knowing. Mach had two main goals: the unification of science and the elimination of metaphysics. Are there scientists, perhaps even a majority, who share these hopes? If the answer is yes, then the heart of Mach's Philosophy not just should be, but *is* part of science as a conceptual activity and will remain part of it until those aspirations cease to be honored.

Nor would this development have displeased him. Mach often pointed out that he did not consider himself to be a philosopher. He was and wanted to remain a naturalist, that is, a lover of nature, but a naturalist who was interested in the methodology of research and the psychology of knowing and who wanted to find a way to protect his colleagues from wasting time and effort. How could that be done? By unifying the methods of science and avoiding the futility of metaphysical conflict.

If some people still considered him a philosopher, then that was their judgment, not his. He learned quickly that scientists were his best audience.

---

[9] It has often been stated in secondary literature (for example, Wilhelm Scherer & Oskar Walzel, *Geschichte der deutschen Literatur,* Leipzig-Berlin, 1928, p. 651.) that impressionistic music, literature, and art fitted well with Mach's phenomenalism, and indeed, their growth in the last decades of the 19th century paralleled Mach's own development and influence, but while the causal relations have not yet been definiively pinned down between his epistemological phenomenalism and these artistic movements, I would like to present my own non-causal theory about the resemblances.

(1) The paintings of impressionism (Monet, Renoir, and others) and especially Pointilism (Seurat, Signac) originated, in addition to everything else, on the one hand from the artistic proceedings of breaking the starting colours into sets of isolated points of primary colours which merge only in the eye of the beholder and giving the colour sensation from which the painter started from nature, and on the other hand, under the influence of non-Euclidean geometries of Lobatchevsky, Bolyai, and Riemann from which a new conception of space based on a new conception of geometry had arisen, (2) The discovery of photography (1838) considerably changed the visual culture of contemporary man, since photography enabled us to become better acquainted with movement and perspective. It speeded up and improved human communication, and as a scientific method it has become irreplaceable in some scientific disciplines. The later inventions of the film (1895) and television (1935) not only emphasized sensory phenomena, but in particular, seemed to reduce color to pointilist or Machian elements which could be related together in moving fashion to constitute identifiable pictures, as if the world had indeed become phenomenalistic in character. Thus, one may speak of Mach, who was himself very interested in photography and whose mother and one son were artists, as a forerunner in the arts as well as selected aspects of science.

He insisted that Machian philosophy did not exist.[10] His "philosophical" goals were part of science. In Hugo Dingler's opinion, Mach as a naturalist looked upon his results as more or less temporary attempts to solve problems;[11] Mach was thus anti-dogmatic and anti-systematic. But for the few scientists and many philosophers who do not accept his goals, then I think he has to be taken seriously not just as a scientist and historian but also as a theoretician of knowing *so that they can learn from him what those goals should be*.

But at the same time, one must somehow avoid the danger of comprehending Mach's theory of knowing with the spirit, terminology, and categories of traditional philosophy. His aim was to aid science so that it would no longer need philosophy, or at least conscious philosophizing. Philipp Frank rightly pointed out that Mach was bound to be misinterpreted by all of those who thought that his main focus was to answer the old ontological question of "whether the basis of the world is spirit or matter".[12]

But in thinking of Mach as a scientist's scientist and not as a philosopher, one must also appreciate his breadth of outlook. Mach did not wish to use his epistemological phenomenalism to remove the concepts which arise in practical life, nor the common notions within particular sciences. He honored each and every moral approach which had heuristic value both within and outside of science. Rather, he thought that through temporary neglect of external practical needs, a standpoint could be reached which would benefit the most intensive and detailed work in the most rigorous aspects of science, which in turn would ultimately have far greater practical value.[13]

As is widely recognized, Mach was correct in this, particularly with respect to the unity of science, since his relativization of the distinction between the physical and the psychic and especially the relativization of the difference between so-called sensory qualities on the one hand and space and time on the other, so freed and liberalized the scope and conditions of scientific inquiry as to help suggest and encourage the brilliant new ideas which helped give birth to modern physics in the hands of Max Planck and especially Albert Einstein. Without Mach's notion put forward in his book on heat theory in 1896 that chemistry unlike physics favored discontinuous change, would Planck four years later have ever have thought of discontin-

---

[10] E. Mach, *A.d.E.*, pp. 24, 39, & 300; *E.u.I.*, pp. vii-ix.

[11] H. Dingler, *Die Grundgedanken der Machschen Philosophie*, Leipzig, 1924, pp. 54 & 84.

[12] See Ph. Frank, *Modern Science and Its Philosophy*, Harvard University Press: Cambridge, Mass., 1950, pp. 81-89.

[13] E. Mach, *A.d.E.*, pp. 253-254 & 298; *E.u.I..*, p. 13.

uous energy emission or "quanta"?[14] And without Mach's criticisms of Newtonian absolute space, time, and motion would Einstein have ever developed, persevered in, or published his theory of special relativity? And could Mach himself have ever made these suggestions without his relativism and identification of physical phenomena with sensory elements? Only historians can fully answer these questions, but who can deny the influence of his methodology and goals *within* the development of science? Let philosophers criticize Mach as much as they like. My answer is to look at the concrete fruit of his ideas, the changes they have helped make in physics, changes which are never likely to be reversed. As long as quantum and relativity theory have a place in science, so surely will Mach's highest methods and goals.

Since the time of Galileo, through Newton and Laplace until the end of the XIXth century classical mechanics was the basic branch of physics. Classical mechanics was considered to be the privileged field. Other branches as mentioned before - such as optics, acoustics, and electricity - were considered largely as adjuncts. Mach was one of the first among reputable 19th century physicists who began to doubt major aspects of Newtonian mechanics. On the basis of a detailed analysis of this mechanics he showed that if physical phenomena are identified with sensory elements that absolute space, time, and motion as something non-sensory or non-relatable to what is sensory cease to have physical meaning. Newton's concepts of absolute space and absolute time have a "hyperphysical" character; they are empty metaphysical concepts, and therefore, do not have any strict scientific value. Furthermore, it seems that Mach often understood geometry as a physical science,[15] since he believed that geometrical figures were inferred from sensory experience and could be wisely read back into it under some scientific conditions. This occasional understanding of geometry as a division of physics could conceivably have been of major use to Einstein.

Though it is not certain whether Einstein shared Mach's phenomenalism when he read Mach's book on mechanics, presumably in 1897, Mach's challenge to Newton's absolutes during the next few years seems without doubt to have encouraged Einstein to accomodate the light constancy axiom into his special theory of relativity by attempting to relativize space and especially time.

Joseph Petzoldt mentioned that Mach already in 1866 and independently of Riemann had postulated a physical space which contained time and would represent relations between sensations. This mathematical construct, which Mach himself tended not to apply to physics, perhaps because it did

---

14 Heinrich Löwy, "Historisches zur Quantentheorie", *Naturwissenschaften,* **21** (1933), 302-303.

15 E. Mach, *A.d.E.,* pp. 87 & 282-283; *E.u.I.,* pp. 344, 350-351, 391-392 & 446.

not seem very economical, was in a larger sense taken up again by Einstein in his general theory of relativity. Thus, Mach may also be considered as a forerunner of that theory as well.[16]

Mach's ideas on mass and inertia also appear to have been influential. Rejecting Newton's definition of mass as "quantity of matter", he attempted to redefine it in terms of interacting particles as if mass were a purely relational notion, a view not fully unrelated to Einstein's equivalence principle in general relativity. It was Einstein himself[17] who gave the name "Mach principle" to his understanding of Mach's relativistic solution of Newton's bucket experiment as if inertia reduced to a functional relation between the rotating system and the rest of the universe. Thus Mach relativized Newton's law of inertia, reducing it to a relation to the "fixed stars". The assumption apparently was that the inertia of an isolated body was non-existent, slight, or meaningless.

Mach's principle postulated that that which is measurable in space and time could be defined by distribution of matter and energy and that movement of a body arises under the effect of surrounding bodies (including the rest of the universe), and not as Newton supposed that the movement of a body was the result of the "effort" of that body to maintain its direction in an absolute space.[18] For these reasons, I can, together with Philipp Frank, assert that Mach was Einstein's immediate precursor and that Einstein was the veritable successor of Mach's ideas in the field of physics.[19]

---

16 J. Petzoldt, "Das Verhaltnis der Machschen Gedankenwelt zur Relativitatstheorie" in E. Mach, *Die Mechanik in ihrer Entwicklung,* achte Auflage, Leipzig, 1921, p. 499.

17 J. Bradley thinks that the so-called Mach principle, as it was formulated by Einstein and other authors, cannot be found in Mach's works. Therefore, according to Bradley, it is necessary to make a distinction between Mach's principle (comprehended in Mach's manner) from Mach's principle (comprehended in Einstein's manner). See J. Bradley, *Mach's Philosophy of Science,* London, 1971, pp. 157-158 & 167.

18 A. Einstein, *Geometrija i iskustvo,* Osijek, 1923, p. 10 ["Geometry and Experience" in *Sidelights on Relativity,* trans. G.B. Jeffery & W. Perrett, Methuen: London, 1922] ; Ph. Frank, *Einstein, Njegov zivot i njegovo doba,* Naprijed: Zagreb, 1959, pp. 71-72. [*Einstein - His Life and Times,* trans. George Rosen, ed. and rev. by Shuichi Kusaka, Knopf: New York, 1947, p. 29.]

19 Ibid., Frank (1959), pp. 70, 86, & 246. This agreement with Frank does not mean in any way that I approve of some other of Frank's interpretations of Mach's philosophy. Mach was in any case the intellectual initiator of European and Anglo-American logical empiricism and logical positivism. But this does not necessarily mean that what Frank wanted to show the reader was most important in Mach's philosophy. Frank, in general, interprets Mach only in the spirit of the neopositivism of the Vienna Circle. For instance, Frank especially pointed out that Mach proposed the so-called "positivistic criterion of science" , which stipulates the use of only those presuppositions from which the statements of perceptible phenomena can be derived. (Ibid., p. 71.)

Hans Reichenbach in 1921 wrote that Einstein's theory of relativity was the fulfilment of Mach's program.[20] Hugo Dingler at first negated the inner connection between Mach's conceptions and the theory of relativity[21] but later admitted that Mach's criticism of Newtonian space and time had been used as the starting point for Einstein's theory. And according to Joseph Petzoldt, while Einstein's science was not the rectilinear continuation of Mach's science, still, there was no contradiction between Mach's views and theoretico-physical modifications made by Einstein in his theory of relativity, since both Mach and Einstein had done their best to overcome the mechanistic view of nature.[22]

In my opinion, Mach was in fact the originator of the theory of relativity, because in constructing his theory Einstein started from Mach's critique of the classical physical concepts of space, time, force, mass, and inertia; nor is this contradicted by Mach's apparent reluctance to accept Einstein's theory, nor by Einstein's reluctance to adopt a number of aspects of Mach's theory of knowledge.[23]

Now don't misunderstand me, I am not claiming either that Einstein's theories of relativity were the same as Mach's or that Einstein was able to eliminate all of the "absolutes" from his own approach. I merely maintain that it was Ernst Mach under the influence of his own version of phenomenalism who began the elimination of absolutes in physics, that Einstein eliminated several largely under the influence of Mach, and that physics will probably eliminate even more absolutes in the future. In short, Ernst Mach, the phenomenalist and relativist, was the author of an incomplete program and continuing trend *which few if any physicists think will ever be reversed.*

Mach's phenomenalism is so frequently misunderstood that more clarification is needed, especially with reference to how it affects the notion of physical objects. In a positive sense, Mach identified physical objects with groups of interelated elements or sensations. In a critical sense, he tended to reject as "metaphysical" all alleged objects which were not sensory or which could not be related to what was sensory. Indeed, his focus was not

---

20 G.J. Holton, "Mach, Einstein and the Search for reality", in *Ernst Mach - Physicist and Philosopher,* (*Boston Studies in the Philosophy of Science,* Vol. 6), Dordrecht, 1970, p. 195.

21 Dingler (1924), op. cit., pp. 65-71 & 77-78.

22 Petzoldt (1921), op. cit., pp. 501-502.

23 I freely acknowledge that Einstein did not agree with Mach that all physical concepts could be derived from experience. Einstein felt sure that the basic notions and laws could not be taken or borrowed from experience. According to Einstein, the essential constituent parts of the physical picture of the world were free inventions of the human spirit. These free creations of mathematical thought represented the key for understanding nature. (For more details about Einstein's epistemological conceptions, please look at my book: *Saznanje i rad* [Knowledge and Work], Beograd, 1976, pp. 104, 110-111, 114, 118, & 146.)

primarily ontological at all. He was not interested in the ontic base of either sensation or knowing. Most of his critics have not taken this into account. Mach was indeed overwhelmingly interested in the scientific study of natural phenomena, and he built a theory of knowledge inspired solely for the needs of exact natural-scientific research. In that theory, the notion of matter as a solid substance outside of the knower had no place. If someone, such as Lenin, were to approach his theory from another standpoint and request from it an answer to traditional metaphysical questions, such as whether the basis of the world is matter or spirit, and in addition, proceed from the assumption that knowing can be understood only as a reflection of external objects in consciousness, then Mach's cognitive-theoretical conceptions will inevitably seem subjective-idealistic and agnostic.[24]

If matter is only a definite order of connection of elements, then the task of scientific knowledge is to express conceptually the relatively stable connections of elements, that is, to find out the laws of nature. That which physics up to now has called "force" is only the permanence of a connection of elements, and not something substantial in a body as the bearer of that substantial something. Like matter, force was also for Mach just one cognitive symbol, an auxillary concept, an abbreviated form of expression.[25] Thus Mach destroyed the notion of matter and the notion of force as substantial concepts. He considered usage of the notions of matter and force in natural science to be no more than a new mythology, which replaced the old animistic-demonic mythology of nature. This modern mythology has operated with corpuscles of light, atoms of matter, electrons within atoms, and electro-magnetic waves, and for him all of these were a fantastic conceptual residue hampering serious research into the demonstrable elements of nature.[26]

Of course believers in theoretical entities still seem to exist, but the development of physics has made "realism" about alleged substances increasingly difficult. If substances are both particles and waves, what are they? How are fields different from spaces? If particle/waves only "exist" when being measured, and when their position is determinate their velocity is not and vice versa, then how can coherent "realism" still have determinate substances to mentally reify? If rest mass and relativistic mass are different, what could "real" mass possibly be "in itself"?

---

[24] It is to be noted that the Austrian Marxist, Karl Kautsky, a man who held a sober and world-historical attitude, wisely said that Marx himself would never have let controversies in natural science be turned into political questions, much less be "solved" by political decree. (See Karl Kautsky, "Ein Brief über Marx und Mach", *Der Kampf*, **2** (1909), 451-452.)

[25] E. Mach, *A.d.E.*, pp. 271-272.

[26] E. Mach, *E.u.I.*, pp. 106-107.

All of these ontic questions arising from interpretations of quantum and relativity theory strongly suggest that while models and pictures may continue to have heuristic value for those who crave the impression of something concrete or determinate, that a majority of up-to-date scientists are or will soon be incapable of any longer taking seriously the impossible "realities" which such models or pictures "represent". Ontology seems to have refuted itself, and Mach's clear theory of elements, his phenomenalism, beckons us like a bright talisman toward the future. He was far ahead of his time, but the world is beginning to catch up.

In conclusion, let me answer a few criticisms which have been raised against Mach's "philosophy" over the years:

(1) Ernst Bloch, the Marxist philosopher of hope, used a passage from *Erkenntnis und Irrtum*, in which Mach did not fully explain his view about knowing, for an attack on him, which I believe was unfounded.[27] According to his epistemological phenomenalism, Mach simply refused to admit the existence of an unchangeable substantial particle which could be used to explain natural phenomena. He did not wish to see the goal of natural science reduced to a metaphysical search for an unchangeable substance or Platonic abstraction which could be used to "explain" particular physical phenomena. Surely, Mach's caution has been confirmed in the recent development of physical science, which has "progressed" through numerous Platonic mirages from indivisible atoms, to indivisible nuclei, to indivisible protons, and perhaps the next candidate will be indivisible quarks.

(2) Mach's phenomenalism was criticized as naive realism* by Johannes Volkelt, Ernst Cassirer, and Branislav Petronijevic, who reproached Mach for not comprehending the connection between sensations and reflection. They argued that Mach did not see that pure sensation was not given, since

---

[27] E. Bloch, *Subjekt-Objekt*, Naprijed: Zagreb, 1959, pp. 101-102.

* The term naive realism or rather its German form "naive(r) Realismus" often appears to mean something very different in Central Europe than the former seems to in Anglo-American philosophy. Mach's phenomenalism is sometimes considered to be a form of idealism in the West but never a form of realism, let alone direct i.e. naive realism. Part of the problem is that in England and America philosophical idealism is idea-lism with Berkeley as the arch-example, whereas in Germany "Idealismus" means "Ideal-ismus" with Plato as the arch-example. One big result is different definitions of what realism means. Realism in the West traditionally means belief that there is a real physical world and it exists independently of all ideas, sensations, and consciousness, but in Germany and lands where German philosophers have been influential such is often called "Idealismus", while their word "Realismus" often tends to be understood as direct sensory awareness in epistemology and as phenomenalism in ontology. Put another way, their "Realismus" emphasizes sensory experience, "Idealismus" trans-sensory experience, and many claims about trans-experience are reduced to ideas about it. This wide divergence, which is almost a full reversal of definitions and may have its source in Leibniz or Kant, is by no means a feature of all German or Germanized philosophy, but it does pop up from time to time. - Ed.

the knowing subject always brought into it an element of thought.[28] However, this criticism was misplaced; Mach knew that without thought there was no sensation. He did not eliminate the moment of reflection, but rather pointed out the complete living experience, out of which sensations appear as elements of knowledge only through analysis. Similarly, Mach's principle of the economy of thought was not derived from sensations, and if the cause for everything were to be sought in sensations (as should follow, according to critics, from Mach's conception), then this principle would have to be rejected. Instead, since Mach accepted phenomenalism as a methodological principle, he indirectly acknowledged reflection as a source of meaning and understanding. In this respect, he differed from the logical positivists. Or put in another way, he held that elements or sensations gave very high certainty, but that while thought could give less certainty, nevertheless, it was still useful and indeed indispensable.

Concerning "Denkökonomie", that is, his theory of economy, it was a very useful conceptual means to help further valid ends. It was not always useful. Other means sometimes helped further valid ends better, but on the whole the use of economical, simplified means was expeditious and more practical than other means, especially in many areas and aspects of exact science.

Sometimes Mach did express himself as if sensations were everything,* but since he did not always clearly distinguish between questions of reality and certainty, what he appears to have meant was that sensations could give the highest degree of certainty,** not that there was nothing outside of sensations. Indeed, a major task of science was to find observational and experimental procedures to help make what was previously unconscious or merely an object of thought into something sensory *as in fact he did when he photographed what we now call Mach shock waves, thereby helping to make their reality more certain.*

(3) Let us now turn to still another attack on Mach's theory of sensations. Herbert Buzello, who we have briefly mentioned before, pointed out

[28] J. Volkert, *Erfahrung und Denken*, Leipzig, 1924, pp. 53-134; E. Cassirer, *Philosophie der symbolischen Formen*, III Teil, Berlin, 1929, pp. 29-36; B. Petronijevic, *Osnovi teorije saznanje* [Foundations of the Theory of Knowledge], Beograd, 1923, pp. 19-20. However, Leszek Kolakowski correctly understood that Mach's elements are still a result of analytic work. L Kolakowski, *Filosofija pozitivizma*, Beograd, 1972, p. 165.

* E. Mach, *Analysis of Sensations,* New York, 1959, p. 12: "The assertion, then is correct that the world consists only of our sensations. In which case we have knowledge only of sensations." - Ed.

** E. Mach, *Sinnliche Elemente*..., Leipzig, 1919, p. 19-20: "We are immediately certain of that which we are perceiving, less of that which we attentively observed and now remember, still less of that which we have experienced by analogy and construe as possible, and not at all the unexperienced which has no conceivable sense . . ." - Ed.

that Mach made a mistake in dogmatically treating sensations, the abstract notions of one science, such as psychology, as the ultimate datum of all knowledge in all sciences. A theory of knowledge should examine the cognitive value of concepts as well as theoretical suppositions of particular sciences and not take the concepts of one science as determinant for all. According to Buzello, there are no ultimate elements, no"simple sensations"; sensations are something qualitative and cannot be treated merely as mathematical quantities. Mach came to the concept of "simple sensation", according to Buzello, because he had to introduce hypotheses for the explanation of physical experiments. Therefore, Mach's sensations were really abstract notions and his doctrine of elements could not legitimately be called phenomenalism, because it was really metaphysics. Buzello seriously objected to this. On the other hand, Buzello reproached Mach for his empiricism, not admitting that Mach had arrived at a notion of knowledge or a notion of a knowing subject.[29] This criticism was only partly justified. Mach's conception was to be sure empirical, but not in the sense of Locke's empiricism. It was in fact epistemological phenomenalism. Mach himself admitted that his elements were abstractions, since they were reached only after a long period of scientific experience and reflection.[30] Also, Mach's elements were not the ultimate unchangeable bricks either of the world or knowledge, but simply for the time being the last known elements. Nor does every scientist start from the same elements.[31] When Buzello in the spirit of Neo-Kantianism reminded us that the object of knowing is that which is required rather than that which is given, then that cannot be seen as a criticism of Mach's conception, since Mach himself did not consider that he had found at the beginning of the process of knowing what was to be known by means of the process. *The shock waves caused by small projectiles such as bullets were normally not given to the senses; they had to be discovered.*

While the objects of knowing for Fichte and the Neo-Kantians were largely formed through the activity of thought, Mach claimed that the objects of knowing were formed at the level of the sensations. In place of the Neo-Kantian synthetic activity of thought, we have in Mach (perceptive) apperception as a factor in building the object of knowing. Thus Mach also attributed to the knower a great role in the process of knowledge, but of course

---

29 H. Buzello, *Kritische Untersuchungen von Ernst Mach's Erkenntnistheorie,* Berlin, 1911, pp. 7-13.

30 This is also why Max Scheler was wrong to state that Mach's sensations were directly given by pure reception and experience and that he should have understood them as indirect. Mach had already practised Scheler's solution. See M. Scheler, "Erkenntnis und Arbeit" in *Die Wissensformen und die Gesellschaft,* Francke Verlag: Bern & Munich, 1960, p. 259.

31 E. Mach, *E.u.I.,* p. 12

he meant the knower within the field of sensibility, since that was where the process of knowing occured.

(4) To recapitulate, he believed and readily emphasized that his epistemological phenomenalism was without metaphysical foundation and that it only served for the purposes of scientific research, but not just in physics but at least potentially in all sciences. Thus, Mach's phenomenalism, ungrounded in "absolute reality" and tied to scientific utility, was he thought an abstraction from wider experience, but still fell short of what Einstein would later call a "free creation of the human mind". But it is hard to accept Mach's position without qualification. Theory of knowledge, including that of Mach, has a metaphysical foundation, as every science does. Why? Because, as Heidegger showed,[32] every age has its own metaphysics i.e. it makes decisions about what is being and and truth. The age of modern science (and mathematical physics as well) has its own metaphysical foundation upon which it comprehends the essence of being, views nature and questions it. Modern science (and Mach's conception of it as well) is based upon research, which has been sketching natural events. This sketch consists of the following: Nature is a concluded bond of particles with mass connected in space and time; all natural events are treated in advance as having spatial & temporal values of motion. The discovery of natural laws by means of experiments has given an explanation of nature. Nature has become a subject of research within this explanatory setting. In short, the world is now comprehended as a picture. The metaphysics of contemporary science is the being which is found in the study of the presentability of what exists. By emphasizing his status as an observer and treating the world as an observable picture, man has acquired a new position in the world. He has become an existing-in-the-world which gives measure and sets the rules for all existing things. This is also the true metaphysical foundation of Mach's philosophy of science which he did not see through. But Mach's negation of metaphysics was not in accord with his own presuppositions. He apparently thought he could eliminate all metaphysics, but at the same time, as remarked by Erwin Hiebert,[33] he did not conceive that the rejection of metaphysical foundation of some epistemological theory still implied some metaphysical foundation. This was unwillingly admitted by Mach himself, when during the later years of his life (especially the controversy with Max Planck, circa 1908-1911) he felt compelled to employ philosophy in order to help justify his phenomenalistic clarification of science.

32 M. Heidegger, "Die Zeit des Weltbildes" in *Holzwege*, (*Gesamtausgabe*, Band 5,) Frankfurt am Main, 1977, pp. 75-96.

33 E.N. Hiebert, "Mach's Philosophical Use of the History of Science" in *Historical and Philosophical Perspectives of Sciences*, Ed. R. Stuewer (*Minn. Stud. Philos. Sci.* **6**) University of Minnesota Press: Minneapolis, 1970, p. 203.

CHAPTER 11

POETIC IMAGINATION AND ECONOMY: ERNST MACH AS THEORIST OF SCIENCE

by Rudolf Haller*
(University of Graz)

A

Mach's epistemological phenomenalism, the magnificent idea of constructing the world from those elements that are supposed to be nothing but sense impressions seems to have left the scene of philosophical controversy as a loser.

Since the attempts by Carnap and later Goodman[1] to erect a constitutional system as a rational reconstruction of the structure of the world based on subjective psychology revealed the unforseeable difficulty of reducing physical to psychological phenomena, many people immediately and from the beginning have sought to abandon the foundation of empiricism together with the epistemological foundation of the subjective psyche.

It is therefore not surprising that Mach's philosophy, always a philosophy of empirical science, lost that attraction and even fascination, which it possessed during the life-time of the great scientist and philosopher as well as within the Vienna Circle. With the verdict of the end of positivism and the historical judgment that somebody just killed logical positivism,[2] the relevance of Mach's philosophy seemed to have receded into the misty and

* This English translation originally appeared in *Scientific Philosophy Today - Essays in Honor of Mario Bunge,* edited by J. Agassi & R.S. Cohen, (*Boston Studies in the Philosophy of Science,* vol. 67), D. Reidel Publishing Company: Dordrecht - Boston - London, 1982, pp. 71-84. For the article in German see: "Poetische Phantasie und Sparsamkeit - Ernst Mach als Wissenschaftstheoretiker", *Ernst Mach - Werk und Wirkung,* Verlag Hölder-Pichler-Tempsky: Vienna, 1988, pp. 342-355. Footnote references for chapter eleven have been expanded in a number of places to include both German and English language titles. -Ed.

1 R. Carnap, *Der logische Aufbau der Welt,* Weltkreis: Berlin, 1928 (*The Logical Structure of the World,* University of California Press: Berkeley, 1967); N. Goodman, *The Structure of Appearance,* 3rd edition, D. Reidel: Dordrecht, 1977; Philipp Frank, "Ernst Mach - The Centenary of His Birth", *Erkenntnis,* **7** (1938), 247f.

2 K.R. Popper, "Der logische Positivismus ist tot. Wer ist der Täter?" in K. Popper, *Ausgangspunkte,* Hamburg, 1979, pp. 120f. ("Who Killed Logical Positivism?" in K. Popper, "Autobiography", *The Philosophy of Karl Popper,* ed. P.A. Schilpp, Open Court: La Salle Ill., 1974)

J. Blackmore (ed.), Ernst Mach – A Deeper Look, 215–228.

ancient realm of the classics. Only in this way can the judgment of the last occupant of Mach's Vienna chair for the history and philosophy of the inductive sciences, Victor Kraft, be understood. In the postscript to the beautiful volume, *Ernst Mach: Wegbereiter der modernen Physik* by K.D. Heller,[3] Kraft wrote in 1964 that Mach surely was "a historical figure of far-reaching importance", but that his attitude towards epistemology has to be viewed as "outdated" at this time. This Judgment is a misjudgment, a mistake, although it comes from a logical positivist. That is not to say that a point of view that is now outdated was wrong originally. On the contrary, it is typical for all scientific research and systematizing that it gets improved, perfected, outdated by further and new research. In that way the researcher contributes to decreasing the number of open scientific questions of his time in his specialty or to raising new questions, research into which will broaden or at least promote knowledge. But even here the scientist has no guarantee. On the contrary, it is not uncommon that, with hindsight, his assumptions and convictions did not turn out to be fruitful, but rather unfruitful and even misleading.

If we listen to the judgment of Mach by contemporary philosophers and theorists of science, the assertion of mere outdatedness comes close to a euphemism. Rather, according to this one-sided judgment, it would have to be counted among the philosophically misleading and unfruitful teachings. There are three assertions about Mach's philosophy of science designed to support the accusation of obsoleteness: First, it is phenomenalistic, second, it is narrowly instrumentalist-inductivist, and third, it is not adequate for science and its theories.

Before we form a judgment in this vague area of philosophical rumors and second-hand hearsay, it may be appropriate to call to mind Mach's basic ideas about the construction and change of scientific ideas and examine them in light of the controversy about philosophical theories in our day.

B

"Science emanates from life"[4]

In contemplating scientific activity there is an advantage in proceeding at first from our life-world, the environment in which we live, in order to examine continuity and differences of familiar behavior in a familiar environment from this vantage point. In this life-world of ours, the experiential

---

[3] K.D. Heller, *Ernst Mach. Wegbereiter der modernen Physik,* Springer: New York & Vienna, 1964, p. 168.

[4] "Die Wissenschaft geht aus dem Leben hervor" from Ernst Mach: "Notizbuch April 1882", given in H. Dingler, *Die Grundgedanken der Machschen Philosophie,* J.A. Barth: Leipzig, 1924.

sphere in which we grow up, we learn just like other beings. Yet we are privileged, because of the possibility of receiving, so to speak, the experiences of other individuals, indeed the experiences of past generations, all at once, and of adopting them to our needs. This process is analogous to the process that transfers the "experiences" gathered in the course of millions of years to living beings so that these experiences are available to future generations as inherited dispositions. Some people have viewed and characterized this very process as "knowledge producing" in which one generation genetically passes on to the next all information that is vitally necessary and promotes selection.

Man, however, belongs to two worlds insofar as he is given not only the "information" of nature, but also that of the mind, or more precisely, language. "We are the same sort of thing as those in our physical surroundings and it is through ourselves that we become acquainted with them."[5] The process of this acquisition of the conceptual apparatus is highly complicated and takes place initially without reflection and consciousness. If we become conscious at this stage of our learning process, which moreover is always practical, then what we find "within us is already a fairly complete world picture"[6] But this conception of the world is, of course, not a priori, as might be suggested by our "instinctive knowledge", which inspires a great deal of confidence.[7] This world picture is nothing other than the treasure of experiences which just as with material worth, concentrates the work of many individuals in one individual. "Not magic therefore," says Mach "but economy." In science, too, the situation is basically the same. Only here the work of many is collected and unified without the necessity of exploiting anybody's energy, or, as Mach himself says: "without robbing others of anything"[8]

Mach's inexhaustible and often praised gift for reducing complex or opaque relationships to simple concepts has found a fruitful beginning in the concept of economy expressed here (the conception of a cunning of reason, as it would have been called elsewhere) for a theory of the history of the progress of knowledge. The value and extent of this conception are only now fully appreciated.[9]

---

[5] Ernst Mach, *Erkenntnis und Irrtum,* J.A. Barth: Leipzig, 1905, p. 135. (*Knowledge and Error* [*Vienna Circle Collection,* vol. 3.], D. Reidel: Dordrecht & Boston, 1976, p. 99.)

[6] Ibid (1905), p. 135 & (1976), p. 120.

[7] See Mach's notebook from January 26, 1881, p. 15, as published in H. Dingler, op. cit.

[8] Ibid., February 1882, p. 86.

[9] See P.K. Feyerabend, "Philosophy of Science: a subject with a Great Past" in: *Historical and Philosophical Perspectives of Science,* ed. R.H. Stuewer, (*Minnesota Studies in the Philosophy of Science* V), University of Minnesota Press: Minneapolis, 1970, p. 174f. See also A. Gargani, *Wittgenstein tra Austria e Inghilterra,* Stampatori: Turino, 1979.

Mach does not tire of emphasizing the principle of economy, the economical nature of science as the principle characteristic of all investigation and achievement of knowledge of man. Again and again he points to the Austrian political economist Emanuel Herrmann, whom he met during his time at Graz as the person who inspired this idea.[10] While Mach wants to solve the problem of the dynamics of theories, i.e., the question of which principle determines change and progress of hypotheses and theories, with the aid of Herrmann's central concept of *economic theory,* namely the concept of economy, Herrmann himself strives to examine economic problems resulting from the principle of economy by means of the methodology of the *natural sciences*. Both scientists were convinced that a separation of man and nature can only be achieved artificially:

> The petty lines of division drawn between man and the rest of creation by superstition and false beliefs must dissolve in the light of science like fog in the sun and disappear; we know only *one* nature in which man finds a small and certainly quite insignificant place as a participating and cooperatively creative being. It is the very economy of *nature* (in the old and rotten sense of extra-human existence) which constitutes the most beautiful, most magnificent part of economic doctrine. What is the saving in human labor, our buildings, our machines, compared to the economy according to which the vital organs of plants, and the same organs and mental organs of animals, operate and create.[11]

Out of the idea of the unity of the physical and the psychological, however, grew Mach's theory of elements, which was so often and readily dismissed as a sensualism that would lead to solipsism. Such an interpretation, however obvious, 0nly too easily obscures the fact that Mach himself wanted to and did take a much more subtle view. Mach argues that a coin appears large and round or large and elliptical if viewed close up, and small and round or small and oval if viewed from a distane. This is to say that the 'effects' an object has are imparted in perception, and thus an object does not seem to be anything but a bundle of perceptions, a complex of perceptions. Taken at face value, Mach's position would seem nothing but a new version of Bishop Berkeley's epistemological idealism or idealistic empiricism. But only a superficial cliché of a specter of positivism can overlook the fundamental differences.[12] While according to Berkeley the fact of being perceived constitutes an entity and the *esse est percipi* repre-

---

10 E. Mach, *Popular-wissenschaftliche Vorlesungen,* J.A. Barth: Leipzig, 1896, p. 209 (*Popular Scientific Lectures,* 3rd ed., Open Court: Chicago, 1898.); E. Mach, "Die Leitgedanken meiner naturwissenschaftlichen Erkenntnislehre und ihre Aufnahme durch die Zeitgenossen", *Scientia,* **7** (1910), 225.

11 E. Herrmann, *Allgemeine Wirtschaftslehre,* vol. 1, part 1, J. Pock: Graz, 1868, p. 211.

12 See K. Popper, "A Note on Berkeley as Precursor of Mach", *Conjectures and Refutations,* London, 1965, p. 172 f.

sents a metaphysical determination of what *is*, the question *whether* something is an object depends, in Mach's opinion, only on the *mode of observation* of the elements, which, however, are ontologically neutral: "It depends on the definite purpose of the observation whether we view *A* [i.e. an element] as sensation or object"[13] Here Feyerabend, who is not always dependable because of his hedonistic tendency to reinterpret, is quite right when he states: Mach's "elements are not sensations"[14] It is rather the very concept of an element that was introduced *not only* as the irreducible part of the complex of sensations, *but also* as the irreducible part of a physical complex which is Mach's proper and independent safeguard against metaphysics.

In his confrontation with Max Planck - one of the few polemical writings of Mach - Mach clearly formulates his point of view which follows the natural world and is beyond idealism and realism. Here he asserts that it is experience which taught us that what at first sight appears to us to be immediately given is in fact not immediately and unconditionally given. Even ordinary people are aware of the influence of external circumstances as well as of subjective psychological and physiological conditions on our impressions of the world. This is even more true of scientific observations which, if they appear to be hallucinatory, must be corrected by the other senses or other persons if they "are to have scientific, and therefore *social* value. . . Overestimating this exception [i.e. through hallucination] easily leads to monstrous idealistic or even solipsistic systems."

And Mach adds a consequence which his interpretation seeks to avoid: "It would be strange if the experience of the world, because of its refinement, were to abolish [aufheben] itself and leave nothing of the world itself but unattainable phantoms."[15]

Here Mach opposes to the hypothesis of a shared world as the sum of the objects of experience the hypothesis that this world itself cannot be experienced independently of sense perception and the physiology of the living being. These, he believes, are the facts from which any theory of knowledge has to start. Of course, this is not yet the solution to the fundamental problem of epistemology. It is, nevertheless, a beginning the clarity of which is convincing and which is not surpassed by the so-called "realists".

Similarly to Hertz and Boltzmann, Mach views the models which we create from our experiences, namely thoughts, as building blocks of a world which is to be perfected in its epistemological value by means of

---

13 E. Mach, "Tagebuch-Notiz", 1880, in Dingler, op. cit., p. 16.

14 P.K. Feyerabend, op. cit., p. 176; See also R.S. Cohen, "Ernst Mach: Physics, Perception and the Philosophy of Science", in *Synthese*, **18** (1968), 132f.

15 Mach (1910), op. cit., p. 235.

continuous and increasing adjustment. "We seek pictures of facts", Wittgenstein writes in his *Tractatus Logico-Philosophicus,* and the picture is composed of elements which represent the objects. It is a picture that itself is a fact and to which belongs the "representing picturing relation" so to speak as a feeler toward reality.[16]

Due to this similarity to Mach's conception and the identification of the elements with sense impressions a phenomenalist program in the Vienna Circle and especially in the early Carnap was bound to appear as a continuation of Mach, and the *Tractatus* was also interpreted in this vein. Also according to Mach the goal at first is: "The representation in thought of facts."[17] However, while ordinary cognitive interest - or, as Mach puts it in the manner of his time, "vulgar thinking" - pursues the conceptual completion of observations for the sake of practical cognitive purposes and is, therefore, wanting, scientific investigation is free from such considerations and therefore not only moves away from ordinary views, but constitutes one of them.

The representation of facts in thoughts and the adjustment of thoughts to facts by no means guarantees that the thoughts *agree with each other.* To put it differently: in constructing a scientific theory it is not only a necessary condition to use theoretical concepts; the concepts have to "fit" with each other and be adjusted to each other. It is in this working out of the logical connections, which can be represented in their purest form in axiomatics, that Mach sees the essential difference between everyday thought and scientific investigation.

It is, however, a fact that "the facts are not compelled to conform to our thoughts"; "it is of course not possible to achieve perfect adaptation to every individual and incalculable future fact".[18] *How* is it then that we adapt our thoughts?

The answer to this question shows anew the fruitfulness of a principle of economy if properly understood. If we did not have the means to connect the objects of the observable and the *un*observable world with the help of simplification and analogy, to reduce the pluralities to unities, we would senselessly stumble about in a disconnected realm of set pieces belonging to

---

16 L. Wittgenstein, *Tractatus-logico-philosophicus,* Harcourt Brace: New York, 1922, 2.1 to 2.174; cf. Wittgenstein's early diary entry of October 15, 1914: "In the sentence we combine things, *on a trial basis,* so to speak; however, they do not have to be like that in reality." in L. Wittgenstein, *Schriften* 1, Frankfurt, 1960, p. 101.

17 E. Mach, *Erkenntnis und Irrtum,* op. cit., p. 3 (*Knowledge and Error,* op. cit., p. 2.) See also Robert S. Cohen, "Ernst Mach: Physics, Perception, and the Philosophy of Science", in *Ernst Mach Physicist and Philosopher*, edited by Robert S. Cohen and Raymond J. Seeger, (*Boston Studies in the Philosophy of Science,* vol. VI), D. Reidel: Dordrecht, 1970, pp. 126-164.

18 Ibid., p. 447 (also *Knowledge and Error,* pp. 355-356); See also E. Mach, *Populärwissenschaftliche Vorlesungen,* op. cit., p. 211.

all kinds of possible worlds. If, on the other hand, we divide the complexes of facts into theoretical elements, or elements of thoughts, then we simplify, schematize, idealize. "The aim of research is the discovery of the equations which subsist between the elements of phenomena."[19] We must of course not expect that Mach is in a position to present a semantically clarified definition of these concepts. Brentano pointed out in a sharp and penetrating critique of Mach's *Erkenntnis und Irrtum* [*Knowledge and Error*] that both Mach's concept of similarity as partial identity as well as his concept of analogy were not sufficiently and satisfactorily analysed.[20] And the physical equations express here "indeed a mathematically conceivable relation", but the logical scope of these relations is limited by experience. Those who so readily advocate and propagate the total theory-dependency of empirical concepts, such as observation, all too easily forget one of the pillars of their critical argumentation, namely, that "there is no inductive necessity other than a logical necessity, such as a physical necessity"[21] Even if we assume that the simplest law corresponding to our experience holds, there is nevertheless no ground to assume that the simplest case will occur. Wittgenstein's formulation in the *Tractatus* is a clear echo of the Humean-Machian position:

> A Necessity for one thing to happen because another has happened does not exist. There is only *logical* necessity. At the basis of the whole modern view of the world lies the illusion that the so-called laws of nature are the explanation of natural phenomena.[22]

Therefore, there is no explanation. The logic of inquiry shows us that the adaptation of the thoughts to the facts is nothing else than their description. It was a credo of Mach's philosophy that to describe meant the concentration, the condensation of all experience, and that theories are nothing but the systematic summary [Zusammenfassung] and ordering of these experiences. But Mach did not characterize these descriptions as assertive, much less apodictic, judgments, as propositions which are either true or

---

19 Ibid., *Erkenntnis und ...*, p. 221; *Populär....;* 264f.

20 F. Brentano, *Bemerkungen zu Ernst Machs 'Erkenntnis und Irrtum'*, edited by R.M. Chisholm & J. Ch. Marek, (volume 6 of *Studien zur Österreichischen Philosophie*), Rodopi: Amsterdam, 1988, pp. 94-95.

21 E. Mach, *Die Principien der Wärmelehre*, J.A. Barth: Leipzig, 1896, p. 435. (For a laer English version see: E. Mach, *Principles of the Theory of Heat*, edited by Brian McGuinness with an introduction by Martin Klein, (*Vienna Circle Collection* #17), D. Reidel: Dordrecht & Boston, 1986, p. 393.)

22 L. Wittgenstein, op. cit., 6.37 & 6.371.

false. Rather he maintains[23] that *all* empirical statements can only be hypothetical:

> A proposition of natural science always has a merely hypothetical sense: If a fact A corresponds precisely to concepts M, then the consequence B corresponds precisely to concepts N; the two correspondences have the same degree of accuracy. Absolutely exact and perfectly precise and unambiguous determination of consequences from a presupposition in natural science, as in geometry, does not exist in sensible reality, but only in theory.

Mach is of the opinion that mainly two reasons compel us to view descriptions not as assertive or apodictic, but as hypothetical judgments or statements: The first and deepest reason is that science provides an adequate picture of the facts and their conceptual real connections only as a *goal.* As inquiry or, as Mach says, as "*becoming science*", it moves in the realm of *conjectures*. Each hypothetical or theoretical value, therefore, is at first related to an area of particular circumstances with the hope of extending it to corresponding, similar and analogous cases. This "intended area of application" of a basic theoretical idea arises, so to speak, with the birth of the hypothesis or theory itself. From here the new problems grow which lead us to new hypotheses and so on. Viewed from the goal or retrospectively, hypotheses have a "self-destructive function" as Mach characterizes their tendency to change into the "conceptual expression of the facts," that is to say into successful adaptation to facts.

The second reason consists in the changeability of events and the changeability of opinions or conceptions of events and facts. Mach says in the introduction to the lecture "The History and Root of the Principle of the Conservation of Work" (1872): "History has made all; history can alter all".[24] If a change of conception did not have to be expected, it would be sufficient to store the accomplishments of the past in one's memory, or, even better, to relinquish this work to automatons. But just as there is no necessary sequence of events in nature itself, so is there none in the history of nature or human ideas. Mach reminds us that there are hypotheses that last only moments, namely when we dismiss an idea at once. There are those, however, which persist for hundreds or thousands of years, such as

[23] E. Mach, *Erkenntnis und Irrtum,* op. cit., pp. 303 & 448. (For the English version see: *Knowledge and Error,* op. cit., pp. 228 & 356.)

[24] E. Mach, *Die Geschichte und die Wurzel des Satzes der Erhaltung der Arbeit,* 2nd ed., Leipzig, 1909. (For the English version see: *History and Root of the Principle of the Conservation of Energy,* Open Court: Chicago, 1911.) See also Alois Höfler, "Zur Geschichte und Wurzel der Machschen Philosophie" in *Zeitschrift für physikalischen und chemischen Unterricht,* **23** (1910), 1f., and E.N. Hiebert, "Mach's Philosophical Use of the History of Science" in *Historical and Philosophical Perspectives of Science,* edited by R.H. Stuewer, (*Minnesota Studies in the Philosophy of Science,* vol. V), pp. 184-213.

Ptolemaic cosmology. "Absolute forecasts, consequently, have no significance in science"[25] But despite this realization - "History made everything; history can change everything" - Mach did not become a proponent of an anarchic epistemology, as is common today among so-called anti-positivists. And he is not ailing from the semi-scepticism of Critical Rationalism, the clichés of which he definitely opposes. Whoever reads Mach will easily discover how much even those who are so eager to attack him owe him directly, even take from him.

I already mentioned at the outset that Mach emphasizes the principle of economy as the characteristic principle for distinguishing scientific from non-scientific ways of knowing. Most importantly and primarily, history, methodological conjecture, investigation and representation of an area of facts is to be preferred as scientifically justified to a merely accidental gain of knowledge. Thus the method of finding as well as the systematics of representation are subjected to the principle of economy. Or, as Pierre Duhem states in full agreement with and in the spirit of Mach: "The economy achieved by the substitution of the law for the concrete facts is redoubled by the mind when it condenses experimental laws into theories".[26]

I am surprised that something in this controversial principle which is fundamental to it was evidently forgotten and could be rediscovered. The interpreters and opponents of the principle of economy overlooked the fact that in the rational theory of decision-making we proceed, of course, analogously to Mach. Someone who does not agree with the rules and norms of a rational theory of a rational theory of action, or of the realization of desires and purposes, strictly cannot be refuted any more than in the history of science a single theory could be falsified by exceptions. We would have to say of such a person that he is acting unreasonably, just as an anomaly disturbs the rational order of concepts in a system. However, the concept of rationality itself has by no means been clarified even to this day.[27] Furthermore, it is a relative concept which considers and refers to the kind of knowledge, or basis of information, somebody has at his disposal and the goals he pursues. As everyone knows, there are well-developed theories in which the expected utility, the worst or the best result possible among

---

25 E. Mach, *Populär-wissenschaftliche Vorlesungen,* op. cit., p. 223.

26 P. Duhem, *Ziel und Struktur der physikalischen Theorien,* translated by F. Adler and with a foreword by Ernst Mach, J.A. Barth: Leipzig, 1908, p. 24. (For an English version see: *The Aim and Structure of Physical Science,* translated by Philip Wiener, Princeton University Press: Princeton, 1954. The original French version was: *La théorie physique: Son object et sa structure,* Paris, 1906.)

27 R. Haller, "Wie vernünftig ist der Common Sense?", in *Wandel des Vernunftbegriffs,* edited by H. Poser, Alber: Freiburg/Br., 1981.

possible alternatives, can be calculated.[28] These models and theories, as incomplete and irregular as they may be in the present state of investigation, seem to me to be a splendid confirmation of Mach's initial attempt [Ansatz] to investigate how the most economical way can be found which will be most useful in attaining a given goal, namely economy, is nothing else than the basic hypothesis of the rationality of science, a rationality which is what makes understanding and comprehension possible.

C

At last at this point the question arises what role is to be given to imagination* - mentioned in the title - in such a process which after all has its own genesis. Is not poetry above the facts, the comprehensive knowledge of which is the aim of investigation:

To be sure, the conjunction "and" is patient, and with its help even the greater incomsistency can be connected. What is the reason, then, that justifies placing poetic imagination next to the basic principle of the rationality of scientific investigation as if the structure of science rested on both of these pillars?

The view expressed here is precisely that - following Mach - we need only these two principles in order to understand genesis, dynamics and the goal of scientific investigation. To whit, if the veil of is pulled away from the birth of scientific hypotheses, a veil created especially by relegating heuristic to the realm of the incomprehensible, to that which cannot be rationally reconstructed,[29] then we could be unimaginative enough to attempt to discern behind it again the discredited induction. And we are disappointed if we do not find it. Thus we read in *Die Prinzipien der Wärmelehre* :

> Those who have dealt with research or the history of research will have difficulty believing that discoveries are made according to the Aristotelian or Baconian

---

28 See B.R.D. Luce and H. Raiffa, *Games and Decisions,* New York, 1957; W. Stegmüller, *Normative Theorie des induktiven Räsonierens,* Berlin, 1973; C.G. Hempel, *Aspekte wissenschaftlicher Erklärung,* Berlin, 1977. (For an earlier English version see: *Aspects of Scientific Explanation and Other Essays in the Philosophy of Science,* Free Press: New York, 1965.); W. Sporn, Grundlagen der Entscheidungstheorie, Scriptor: Kronberg, 1978.

* Translator's note: Throughout this essay the German word "Phantasie" has been translated as "imagination" rather than "fantasy".

29 K. Popper, *Logik der Forschung,* Vienna, 1935. [For an English version see: *The Logic of Scientific Discovery,* Science Editions: New York, 1961.]

scheme of *induction* (by enumerating identical cases). In that case discovery would be quite a cosy business.[30]

What misconceptions in the estimation of positivism had to occur that *such* passages in Mach could be overlooked, missed or not read? And how much propaganda is necessary to give such mistaken judgments a measure of credibility! At the end, I intend to return to this point briefly, a point closely connected with the latest developments.

If, as we have heard, the activity of discovery cannot be a very cosy business because it is not possible to rely on past experiences in order to imagine new things, what then is it that enables us to see or to learn to see facts in a new light?

It is the work and role of the imagination which, according to Mach, is alone in a position to lead us away from what we habitually expect again and again. He who begins to let his imagination roam, leaves the shackles of everyday life behind him, plunges into the infinite realm of possibility and works on any worlds possible, no matter whether he enjoys it or not. Ultimately, he even leaves these worlds behind and crosses the line of what is *really* possible to the objects that are *logically* impossible. If the concept of the *perpetuum mobile* is an example of the former, every indirect *mathematical* and *logical* proof which shows the assumption of the opposite of what is proved to be absurd is an example of the latter. This is also the place for associations which lead us without any effort on our part to places which we would not reach without their seductive, capricious charm.

Without going into detail it maybe appropriate to distinguish between two forms of imagination. One form continues the possibilities of actual events and thus makes it possible for us - afterwards, as with dreams - to remain in the everyday world without wants and anxieties, even if the stream of its events has long flowed past us. The other rises, so to speak, *above* the events and creates a separate structure of relations where the intellect, unconscious of its actions, is quite capable of intervening and, out of the overlapping nets, to spin its own web, if I may use this image. Expressed less picturesquely, while the first mode follows the usual patterns of life and experience and substitutes only alternative courses of events for

30 E. Mach, *Die Principien der Wärmelehre,* op. cit., p. 443. [The English translation of this book appeared after Professor Haller's article was originally published and is titled: *Principles of the Theory of Heat*, edited by Brian McGuinness with an introduction by Martin J. Klein, (*Vienna Circle Collection,* vol. 17), D. Reidel: Dordrecht & Boston, 1986, p. 402. - Ed.]

the real ones, the latter mode constructs things, stories and structures, whose nature was hitherto unknown; this really is poetic imagination.[31]

Mach himself does not distinguish these two types, but simply assigns to imagination the role of *differentia specifica* for the animal organism of the human being. For him poetry is "the first step up from the everyday burden of life",[32] and poetic dreaming not only the very beginning of all mental development, but the source of completion of hypotheses and theories. What, after all, to remind ourselves, are the constituents of those facts that we subsume under laws? "Simplification, schematization, idealization", every theoretical concept also is a product of this disposition:

> Every alteration appears as upsetting stability, as a dissolution of what until then existed together, abolishing the accustomed condition and so disturbing us and posing a problem, which drives us to look for a new connection and enquire into the cause.[33]

> Abstractions and the activity of phantasy does the main work in the finding of new knowledge. The fact, emphasized by Whewell, that method can do little here, makes clear the mysterious feature that he thinks attaches to so-called "inductive" findings.[34]

First, the result helps us to understand what always seems to have been difficult for other interpreters, namely, why Mach assumes that the *knowledge* of facts as well as *error* with respect to them have the *same* sources. From a phenomenological point of view, our imagination is *active* in both instances, either following the event and alternatively, or constructively and poetically. In this way it is possible for us in the empirical sciences to add to the experimental methods the important method of the thought-experiment, which the scientist shares with the builders of castles in the air, namely the novelists and science fiction writers [*technischen Utopisten*]. While it is hard to say how to achieve thought-experiments, Mach indicates a way: by means of paradoxes:

> These [paradoxical situations - tr.] not only give us the best feeling for the nature of a problem since paradox is what makes a problem, but contradictory elements will not allow thought to rest.[35]

---

31 E. Mach, *Erkenntnis und Irrtum,* op. cit., p. 154. "Man kann auch in Begriffen phantasieren." (For the English see: *Knowledge and Error,* op. cit., p. 112-113: "One can exercise phantasy in concepts too.")

32 Ibid., p. 97 (Ibid., p. 72).

33 Ibid., p. 272 ff (Ibid., p. 205).

34 Ibid., p. 313 ff (Ibid., p. 236)

35 Ibid., p. 193 (Ibid., p. 143).

If knowledge and error therefore cannot be distinguished by means of their genesis, how then can truth be separated from falsehood, if everything is not to sink into the unknowable realm of mere conjecture?

Mach's answer to this question once more connects the two principles with his biological-evolutionary basic outlook[36] and with the realistic attitude of a great physicist and natural scientist who himself frequently erred. His answer is simple: Only experience provides this distinction. He talks about an "intellectual risk", that one takes, and he asserts that the intellect of a genius differs from the normal intellect precisely because of "the certain expectation of the success of an intellectual move". In other words and expressed in theses, the following picture of the structure of scientific theories emerges:

(1) "The essential function of a hypothesis is that it leads to new observations and experiments, which confirm, refute or modify our surmise and so widen experience."[37]

(2) The negative result of an experiment, that is, the falsification of a hypothesis must, however, "never be regarded as decisive".[38]

(3) An "unambiguous determination of consequences from a presupposition in natural science ... does not exist in sensible reality, but only in theory".[39]

(4) The result obtained through any quasi-inductive procedure has to be checked by deduction.[40]

(5) Science, like any other creation of man, is a product of history. History easily leads to prejudice, to clinging to old ways.

(6) Not only is there an evolution of life, but also of knowledge; knowledge itself promotes life.

(7) Only poetic imagination makes it possible to discover new methods and problems, the value of which always remains dependent

---

36 See M. Capek, "Ernst Mach's Biological Theory of Knowledge", in "Symposium on Ernst Mach" in *Synthese,* **18** (1968), 171f.

37 E. Mach, *Erkenntnis & Irrtum,* op. cit., p. 237 (*Knowledge & Error,* op. cit., p. 176).

38 Ibid., p. 210 (Ibid., p. 157).

39 Ibid., p. 448 (Ibid., p. 356).

40 Ibid. (Ibid.).

on success either by adjusting to the existing theoretical structure or, sometimes, by breaking it.

Contemplating these seven theses, which could easily be supplemented with many other compatible ones from Mach's work, is like opening a work in the theory of science of our century: All these thoughts can be found again almost without modification as the living accepted treasure of knowledge. And this thought can also be found in *those,* who elsewhere in their writings list Mach among those who contributed to a false understanding of science. Can there be greater, more vivid success for a concept than that it is used and recognized after a hundred years even by its alleged opponents?

CHAPTER 12

MACH'S RELATIVISM VERSUS APRIORISM AND THE MECHANISTIC WORLD VIEW

by Keiichi Noé*
(Tohoku University)

MACH'S RADICAL CRITICISM OF APRIORISM

First of all, for Ernst Mach our scientific activities are nothing but a form of biological adaptation among human beings, and thus for him: "The adaptation of thoughts to facts, accordingly, is the aim of all scientific investigation."[1] He explained this point of view as follows:

> In our view the laws of nature are a product of our mental need to find our way about nature, so that we do not stand estranged and baffled in front of natural processes. This manifests itself clearly in the motive behind these laws, which always correspond to this need as well as to the current state of culture.[2]

Since laws of nature are none other than a product of our mental needs, the distinction between truth and falsity must become relative. Mach excluded the transcendent character of truth and admitted an origin and growth in the concept of truth. He tried to grasp human knowledge under the aspect of a dynamic development instead of static structure. In short, he introduced the "evolutionary point of view" into his epistemological studies, which saw human knowledge as constantly growing and developing in the stream of time. From his standpoint, non-evolving notions such as "truth-in-itself", an "eternal truth" which was the goal of scientific research as well as the "necessary truth" of logicians and mathematicians, were mere idle fancies. Thus, Mach rejected the absolute dichotomy of true and false and adopted a more pragmatic distinction between "knowledge and error".

> A judgment, our own or conveyed to us, that we find appropriate to the physical or mental finding to which it relates, we call correct and see knowledge in it, especially if it is new and important. Knowledge is invariably a mental experience directly or indirectly beneficial to us. If, however, the

---

* This paper was specially prepared for this anthology.

1 Ernst Mach, *The Analysis of Sensations*, New York, 1959, p. 316.

2 Ernst Mach, *Knowledge and Error*, Dordrecht-Holland, 1976, p. 354.

J. Blackmore (ed.), Ernst Mach – A Deeper Look, 229–241.

> judgment does not stand up, we call it an error, or, if we have been deliberately misled, a lie. . . . Knowledge and error flow from the same mental sources; only success can tell the one from the other. A clearly recognized error, by way of correlative can benefit knowledge just as a positive piece of knowledge can.[3]

Certainly, it would be easy to reject such a pragmatic view of truth as a version of "psychologism" or "biologism". In fact, Edmund Husserl criticized Mach's claims in *Logical investigations* because he naturalized logical law and abandoned the ultimate grounding of logical validity.[4] But we have to note that Mach's theory of "knowledge and error" was very closely related to his criticism of apriorism in mechanics, and, as we will see, the mechanistic view of nature. Since Mach's world was ultimately composed of sensory elements, all scientific concepts were based on sensory experiences with laws of nature being descriptions of the mutual dependence of these elements with regard to biological and psychological benefits. Surprisingly, Mach extended this line of thought to the formal sciences. For him, even logical laws and mathematical theorems were based on sensory experiences. Thus, there was no room left at all for accepting apriorism in his sense-based epistemology.

A suitable example for understanding the nucleus of Mach's anti-apriorism would be his interpretation of Euclidean and non-Euclidean geometry. But we need to remember some of the details of Mach's epistemology. Even space and time were not special forms of intuition in Kant's sense. Mach clearly said that "If we investigate the matter accurately, then we must certainly admit that our sensations of time and space are just as much sensations as are our sensations of colors, sounds, and odors".[5] He called such a space "physiological space" and clearly distinguished it from "metrical space" which plays an important role in physics and mathematics.

To be more concrete, physiological space included visual space, tactile space, and so forth. In these spaces, location and distance were recognized not in quantitative terms but qualitatively as in notions such as front-back and right-left. Therefore, physiological space was not isotropic and homogeneous, nor unbounded and infinite. From a structural point of view, it should be compared with non-Euclidean rather than Euclidean space. On the other hand metrical or geometrical space had the same properties everywhere and in all directions, and was nothing but a concept acquired through experience. Generally speaking, the formation of geometry needed not only sensory experiences but also abstract concepts. Nevertheless, geometry was not merely a formal and coherent system of abstract concepts,

---

[3] Ibid., p. 85f.

[4] Edmund Husserl, *Logische Untersuchungen*, Bd. I, Tübingen, 1968, ch. 9.

[5] Ernst Mach, *The Science of Mechanics*, La Salle Illinois, 1974, p. 610f.

but was based on sensations of space. "Just as without sensations of heat there would be no theory of heat, so also there would be have been no geometry without sensations of space."[6] Mach then continued as follows:

> But both the theory of heat and geometry stand in additional need of experiences concerning bodies; that is to say, they must both go out beyond the narrow boundaries of the domain of sense that constitutes their peculiar foundation.[7]

Then, what did experience concerning bodies mean? In the opinion of Mach, it was an experience about the mutual spatial behavior of bodies. In other words, it was experience of measurement. "Measurement was experience involving a physical reaction, a superposition-experiment,"[8] and such an experience was indispensable to the formation of geometrical concepts. He explained the inseparable relation between the experience of bodies and the fundamental assumption of geometry as follows:

> Geometry consists of the comparison of rigid bodies with one another and rests on the simple assumption, that bodies, which exactly coincide with each other according to their extension, can be placed on one another (are congruent), and show the same behavior elsewhere too.[9]

Here, Mach pointed out the relation of being congruent in geometry depended upon the possibility of spatial displacement and comparison of rigid bodies. From recognizing the congruence of rigid bodies, we could proceed to understanding the universal comparability of geometrical objects with each other. This kind of comparison became more exact by way of introducing physical measurement. Thus, metrical or geometrical concepts had to be formed operationally by means of the act of measurement. "In the place of an individual and non-transmittable intuition of space are substituted the universal concepts of geometry, which hold good for all people."[10] Geometrical concepts were not Platonic ideas which transcended our experiences, but intersubjective ideas which stemmed from our special experiences concerning physical bodies. Therefore, geometry should be regarded as a kind of empirical science. Its task was none other than to investigate the mutual dependence of physical measurements. This was be-

---

6 Mach, *Knowledge,* op. cit., p. 264.

7 Ibid.

8 Ibid., p. 277.

9 Ernst Mach, *Populär-wissenschaftliche Vorlesungen,* Vienna, 1987, p. 494.

10 Mach, *Knowledge,* op. cit., p 300.

cause, to use Mach's phrase: "The fundamental assumption of geometry reposes on experience, although of the idealized kind."[11]

As a natural result of the above arguments, Mach rejected the a priori validity of Euclidean geometry. The universal validity of the Pythagorean theorem was thus not different in kind from Newton's law of gravitation. For him, the distinction between geometry and physics was not in kind but only in degree.

> The knowledge that the angle-sum of the plane triangle is equal to a determinate quantity, namely, to two right angles, has thus been reached by experience, not otherwise than the law of the lever and Boyle's and Mariotte's law of gases.[12]

> The difference between geometry and physics, taken by Duhem (*La Theorie physique,* p. 290) as basic and qualitative, I regard as only a difference of degree.[13]

Such a view of geometry seems very strange for us who are accustomed to the contemporary axiomatic refinement of geometry according to Hilbert. It might be possible to blame Mach for confusing "pure geometry" with "applied geometry" or "physical geometry", but from his epistemological perspective, our pure geometry which is completely detached from experience, would be unintelligible. As far as one is able to understand the meaning of any abstract geometrical concepts, there had to be, at least for Mach, connections with sensory experience as a basis for their intelligibility. This was the most important point in Mach's version of positivism. Nor can we avoid the impression that his radical empiricist interpretation of geometry opened up Pandora's box for a positive estimation of non-Euclidean geometry, not just in mathematics which he approved, but even in physics where he seems to have been more sceptical.[14]

When Lobachevsky and Riemann proposed the possibility of non-Euclidean geometry in the middle of the nineteenth century, most of the mathematicians at that time generally thought that non-Euclidean geometry was nothing but a kind of fiction or an artificial hypothesis, while Euclidean geometry was the only genuine geometry which coresponded to the real physical space. Surprisingly, even Gottlob Frege said that "No man can serve two masters. One cannot serve both truth and untruth. If Euclidean

[11] Ibid., p. 266.

[12] Ibid., p. 275.

[13] Ibid., p. 329, note. 23.

[14] See E. Mach "Eine Betrachtung über Zeit und Raum" in *Populär wissenschaftliche Vorlesungen,* op. cit., pp. 492-508.

geometry is true, then non-Euclidean geometry is false, and if non-Euclidean geometry is true, then Euclidean geometry is false."[15] And furthermore, of all things, he classified non-Euclidean geometry among the pseudo-sciences like astrology, alchemy, and so forth.

Against such a conservative argument, Mach sternly rejected the traditional apriorism concerning geometry and the true-false dichotomy. As we have mentioned, for him, the fundamental task of geometry was to investigate relations of mutual dependence among measurements. Of course geometrical concepts and relations though obtained from our experience of measurements had to be not only coincident with our empirical observations, but also logically consistent with one another. But to borrow Mach's words: "These two requirements can be fulfilled in more than one manner, and hence [we have] the different systems of geometry." [16] Thus, he suggested that there could be a plurality of different geometries each equally correct, but he still felt that "physicists have as yet no reason for departing from the assumtion of $k = \infty$ of Euclidean geometry".[17] Hence, it would be natural for him to derive the following conclusion from these arguments:

> If we lived on the surface of a planet with a turbid, opaque atmosphere, and, on the supposition that the surface of the earth was a plane and our only instruments were square and chain, we undertook measurements, the increase in the excess pf the angle-sum of large triangles would soon compel us to substitute a spherometry for our planimetry. The possibility of analogous experiences in three-dimensional space the physicist cannot as a matter of principle reject, although the phenomena that would compel the acceptance of a Lobachevskian or a Riemannian geometry would present so odd a contrast with that to which we have been hitherto accustomed that no one will regard their actual occurence as probable.[18]

But we need to remember that these passages were written in 1903 for the American journal, *The Monist,*[19] that is, two years before the appearance of Einstein's special theory of relativity and twelve years before the general theory. Needless to say, Mach's supposition of physical possibility about non-Euclidean geometry was later realized in the general theory of relativity, though in 1903 few people apart from Mach may even have allowed for its coherent conceivability.

It was Mach's epistemological considerations against apriorism, particularly his relativism, which made it possible for him to imagine physical

---

15 Gottlob Frege, *Posthumous Writings,* Chicago, 1979, p. 169.

16 Mach, *Knowledge,* op. cit., p. 321.

17 Ibid., p. 322.

18 Ibid. p. 323.

19 Ernst Mach, "Space and Geometry from the Point of View of Physical Inquiry", *The Monist,* **14.1** (1903), 1-32.

possibility where few other people did. Following his thinking, "the fundamental truths of geometry have thus, unquestionably, been derived from physical experience."[20] Therefore, geometry was not a logically consistent formal system but an instrument to describe the mutual dependence of sensory elements by observing the principle of economy of thought. From this point of view, the distinction between geometry and physics was only in the difference of ways of description. Hence, even geometrical truths might be refuted by the physical experience of measurement. In this sense, we could also say that Mach's radical characterization of geometry anticipated a widely accepted aspect of recent philosophy, in particular, W.V. Quine's famous thesis of "gradualism", namely, his abandonment of the analytic-synthetic distinction. Mach's own "gradualism", which was based on his critical attitude toward apriorism, led him to his far-sighted understanding of the physical possibilities of non-Euclidean geometry. Whether Quine was aware of Mach's position or whether it influenced his thinking would be a matter for further research, but could have interesting results.

THE MECHANISTIC WORLD-VIEW AND MACH'S CRITICISMS OF IT

In April 1900, Lord Kelvin, the famous physicist, who was 75 years old at the time, gave a lecture titled "Nineteenth Century Clouds over the Dynamical Theory of Light and Heat".[21] He was proud of the great fruits of modern physics and pointed out that there were two dark clouds which seemed to hang ominously over the beautiful theories of mechanics. One was the problem of the relative movement of a body with respect to the ether, and the other was the problem of the equipartition of energy concerning thermal phenomena. Lord Kelvin expressed the hope that these clouds could be blown away during the next century. But as we know, contrary to his hopes, it was the clouds which have blown many of the beautiful theories of mechanics away, or at least seriously qualified and restricted their application. Two of the better-known clouds, which were hardly visible at the time, have subsequently been called "quantum theory" and "relativity theory".

The system of Newtonian mechanics seems to have been for the most part formally completed by D'Alembert and Lagrange before the end of the 18th century, and has since held a special position in the physical sciences. In terms of this mechanics, given an initial condition, the movement of a body was causally determined by physical laws described in terms of

---

20 Mach, *Knowledge,* op. cit., p. 291.

21 Lord Kelvin (William Thomson), "Nineteenth Century Clouds over the Dynamical Theory of Heat and Light", *Philosophical Magazine,* July 1901.

differential equations. As a result, all natural phenomena appeared to become at least potentially understandable, provided they were reduced to a "mechanical model", namely to the movement of mass points in space and time. These assumptions became incorporated into what has subsequently been called "the mechanistic world view" which is commonly thought to have dominated nineteenth century physics and which appears to have influenced efforts to reduce electrical and thermal phenomena to a mechanical or even atomistic approach.

Hermann Helmholtz, whose nominal philosophy was based largely on Immanuel Kant, became at least in part a spokesman for the mechanistic world view. This seems to have happened quite early, namely with his formulation of the first law of thermodynamics in 1847:

> If we imagine the universe divided into elements with fixed qualities, the only changes which are still possible in such a system are spatial changes, i.e. movements, and the outer relations through which the action of forces is modified can only be spatial. Hence, there are no other forces except motive forces and these depend on spatial relations in their action. Therefore, to put the matter more precisely, natural phenomena have become reduced to movements of matter with invariant motive forces which depend only on spatial relations.[22]

Helmholtz then added that the task of natural science was nothing less than "to reduce natural phenomena to invariant attractive and repulsive forces whose intensity depends on distance."[23] In other words, the final goal of scientific inquiry was understood entirely within the context of the mechanistic world view. This ideal of knowledge adopted the principle of sufficient reason which stated that for every fact there was a reason why it was so and not otherwise. This was treated as an *a priori* axiom of physical inquiry.

The deterministic world view, perhaps most frequently identified with Pierre Simon Laplace, was an indispensable torchlight for much scientific research during the 19th century. It even seemed to go so far as to treat mechanical laws as necessary truths which could be derived from *a priori* reason or reasoning. According to the dedicated Kantian philosopher, Ernst Cassirer, both Helmholtz and Wilhelm Wundt, an acknowledged founder of experimental psychology, "endeavored simply to derive the axioms of mechanics from the general principle of causality".[24]

---

22 Hermann Helmholtz, *Die Erhaltung der Kraft - eine physikalische Abhandlung,* Berlin, 1847, pp. 4-5.

23 Ibid., p. 6.

24 Ernst Cassirer, *The Problem of Knowledge,* New Haven, 1970, p. 91.

But even if there is a measure of exaggeration in these charges, it is clear that *a priori* reasoning was widespread in mathematical physics and that *a priori* assumptions were present in the widely held conception that the study of mechanics required the existence of objective nature, a world fully independent of epistemological considerations. Physical reality existed in itself and had its own structure unrelated to subjective activities. Truth was understood as correspondence with reality and there was a sharp contrast between truth and falsity.

It was this view of nature and truth which Mach wanted to challenge. He focused much of his criticism on the "imperialism of physics", the notion that biology and even psychology would eventually be reduced solely to physical terms. Similarly, he opposed the "imperialism of mechanics" as if acoustics, optics, heat theory, and electricity could or would all eventually be reduced to mechanical terms. Philosophically speaking, the targets of his criticism were metaphysical realism, the correspondence theory of truth, and the reality of "substances" such as atoms.

First of all, Mach opposed all *a priori* science, all attempts to build science on other than empirical, testable foundations. Second of all, in his epistemological works, Mach totally rejected any world picture restricted to the movement of physical bodies in a system of absolute space and time. According to Ernst Mach, the ultimate components of the world were not physical bodies (particles having mass), but sensory elements such as colors, sounds, pressures, spaces, times, etc.. These elements were functionally interrelated with each other and formed numerous different kinds of "complexes". The "body" was none other than the special name for a complex which exhibited relatively greater permanence. Mach chided what he considered to be some prejudices of physicists in the following terms:

> Bodies do not produce sensations, but complexes of elements (complexes of sensations) make up bodies. If, to the physicist, bodies appear to be real, abiding existences, whilst the 'elements' are regarded as evanescent, transitory appearance, the physicist forgets: In the assumptions of such a view, that all bodies are but thought-symbols for complexes of elements.[25]

> Purely mechanical phenomena do not exist. . . .[They] are abstractions, made either intentionally or from necessity, for facilitating our comprehension of things.[26]

Therefore, we have no grounds to give priority to mechanical phenomena over others. In a word, the mechanistic view of nature should be called "an artificial hypothesis" or "a kind of prejudice". On the other hand, it

[25] Ernst Mach, *The Analysis of Sensations,* New York, 1959, p. 29.

[26] Ernst Mach, *The Science of Mechanics,* La Salle, Illinois, 1974, p. 596.

must be admitted that Mach enlarged what we have come to mean by the mechanistic view by attributing epistemological and ontological assumptions which all practitioners of Newtonian science have neither accepted nor employed. Nor is it the case that post-Newtonian physics necessarily imples the rejection of those attributed philosophical assumptions. Mach himself was not always as conscientious as he should have been to distinguish between Newtonian mechanics, the philosophical assumptions held by Newton, and the philosophical target of his attacks, which was the enlarged amalgam which we have called "the mechanistic world view".

Mach's attempt at what he perceived as reform did not end with mere criticism of a methodology associated with Newtonian science as seen from the perspective of the mechanistic world view. He positively proposed an alternative methodology of science linked to what he called "phenomenological physics". The contrast between it and "mechanical physics" may well have looked as if it stemmed from "the old opposition between Hooke and Newton."[27] Phenomenological physics was one which removed all artificial hypotheses and metaphysical elements (e.g. atomism, absolute space and time, causality, ultimate substance, and so forth) from physical inquiry and described physical properties in pure terms directly or indirectly linked to observation. In this sense, Mach's proposed methodology incorporated Newton's famous but often violated maxim *Hypothesis non fingo,* which Mach seems to have taken more seriously than Newton himself.

What is perhaps not widely known is that Mach's specific use of the expression "phenomenology" influenced the formation of "transcendental phenomenology" by Edmund Husserl, as Husserl himself recollected in the "Amsterdam Lecture" (1928).[28] Both phenomenologies of Mach and Husserl regarded the phenomenal field as the starting point of philosophical considerations and rejected the concept of "things-in-themselves". Furthermore, they both adopted the method of pure description without metaphysical presuppositions. But Mach's phenomenology lacked the transcendental moment completely in comparison with Husserl. Contrary to him, Husserl required a change of attitude, from the "natural attitude" to the "transcendental attitude". In this sense, Mach's position still remained within the natural attitude, and his physical "phenomenology" was very near to the empiricist's position of "phenomenalism". Therefore, we can say, though Husserl was stimulated by Ernst Mach's attempt, his own phenomenology was

[27] Ernst Mach, *Principles of the Theory of Heat,* Ed. Brian McGuinness (*Vienna Circle Collection,* vol. 17), Reidel: Dordrecht and Boston, 1986, p. 333.

[28] Edmund Husserl, "Amsterdam Vorträge", *Husserliana*, Bd. IX, Den Haag, 1968, 302. [For Evert W. Beth's charge that Husserl "stole" basic ideas from Mach, see Henk Visser's article in this anthology (chapter #20) and especially footnote 99. It is not yet clear whether Beth heard the Amsterdam lectures, but it is hard to exclude the possibility that they somehow influenced his attack. - Ed.]

formed by the "transcendental turn" from Mach's mundane phenomenology.

If seen concretely, Mach's main focus of attack during the last decades of the 19th century and the first few years of the new century, but especially during the 1890's, was Ludwig Boltzmann's kinetic theory of gases.[29] Mach objected to Boltzmann's atomism and his attempt to protect the primacy of Newtonian mechanics by interpreting the second law of thermodynamics in statistical terms.[30] Mach thought that thermodynamics understood in a phenomenological way without reference to atoms was logically and physically prior to both Newtonian mechanics and the statistical mechanics of Maxwell and Boltzmann.

It is commonly imagined that Boltzmann won this confrontation because of the apparent indispensability of atomism, the success of the statistical interpretation of the second law, and the resulting apparent priority of statistical mechanics over phenomenological thermodynamics,[31] but if we look at the larger picture we can see that Mach's point of view was actually more sound and successful than has generally been acknowledged. First, Newtonian mechanics no longer dominates physics, second, phenomenological thermodynamics perseveres, third, many aspects of the "statistical" second law still seem to be speculative (i.e. the reversibility of time), fourth, many aspects of Boltzmann's work on gas theory are not accepted or are incompatible with quantum theory, and fifth, even Boltzmann himself appears to have gradually abandoned belief in metaphysical atoms in favor of identifying them with mental or sensory "pictures",[32] a view since refined by calling atoms "theoretical entities". Mach won enough battles to make sure that the last word has not been rendered about phenomenological physics. The revolution still smoulders.

It is perhaps natural that historians, physicists, and philosophers should look at the relation between Newtonian mechanics and the Newtonian world-view differently, and tend to be in disagreement about Mach's role in helping to overcome both. Many historians see Mach's attempt to treat thermodynamics as if it were logically prior to atomism and statistical mechanics as mere mistakes on Mach's part, even though Boltzmann's approach did have many technical problems and phenomenological thermodynamics still has unfazed advocates.

---

[29] Mach, *Principles of the Theory of Heat,* op. cit., pp. 333-335.

[30] Ibid., p. 335.

[31] Max Planck, "Zur Machschen Theorie der physikalischen Erkenntnis. Eine Erwiderung", *Physikalische Zeitschrift,* **11** (1910), 1186-1190.

[32] Stephen Brush, "Ludwig Boltzmann and the Foundations of Natural Science", in *Ludwig Boltzmann - 'Principien der Naturfilosofi',* edited by Ilse M. Fasol-Boltzmann, Springer Verlag: Berlin, 1990, p. 54.

As for physicists, most appear to treat Newtonian mechanics as still valid as an approximate, special, or limiting case and think that it was experimental discoveries about radiation and electrodynamics plus incompatibility with quantum and relativity theory which reduced Newtonian mechanics to its present marginal status. But many recent philosophers insist that Newtonian mechanics was doomed from the start because of the mechanistic world view which Newton and especially mathematical scientists of the 18th and 19th centuries appeared to take for granted. The real point is that if even if the historians are right that Mach in his time failed to prove the physical priority of "pure" thermodynamics over atomism and statistical mechanics, it still remains the case that Mach did anticipate the demise of Newtonian mechanics, did arm Einstein with some of the conceptual insights needed to amend or replace it, did contribute to undermining the mechanistic world view, at least to the extent it has been toppled, and opened an option for phenomenological physics to eventually succeed. These were important accomplishments which no amount of historical nitpicking or polemics by Lenin,[33] Planck,[34] or Popper[35] can argue away.

MACH AND THE NEW PHILOSOPHY OF SCIENCE

As is well-known, Logical Positivism emerged from the *Wiener Kreis* and *Verein Ernst Mach* at the end of the 1920's. Moritz Schlick led the first and was titular head of the second group, though Otto Neurath and Rudolf Carnap seem to have been the most dynamic contributors to the latter. Mach was so to speak the mother of both Viennese phenomena, if we allow that Ludwig Wittgenstein was the father. Logical Positivists severely criticized traditional philosophy under the slogan of "eliminating metaphysics" and did their best to eliminate all metaphysical statements as meaningless by the principle of verification. The end-goal was to establish a kind of genuine scientific philosophy without any metaphysical presuppositions, but in practice they soon encountered many difficulties such that the concept of verifiability had to be weakened. On the other hand, with the development and refinement of logical analysis and the emphasis on "the linguistic turn", as if all questions of epistemology and ontology were "merely linguistic", they left Mach's epistemological and methodological concerns far behind. In the middle 1930's members of the Vienna Circle and *Verein Ernst Mach* were compelled to defect to Anglo-American countries first by local fasc-

---

33 V.I. Lenin, *Materialism and Empirio-Criticism,* (1908), New York, 1927.

34 Planck, op, cit.

35 Karl R. Popper, "A Note on Berkeley as a Presursor of Mach", *British Journal for Philosophy of Science,* **4** (1953), 26-36.

ism and then by Naziism. They generally found a warm welcome in America, but by the early 1960's they reached a turning point. The more the method of logical analysis appeared to become accurate, the more philosophy of science distanced itself from the problems of concrete science. Logical Positivists increasingly restricted their role to the logical justification of scientific theories, and many had become content just to flunkey as "maidservants of science". In other words, philosophy of science at that time was mostly engaged in the static explanation of scientific development by hindsight and had lost the power of foresight or critical function which Mach had always maintained.

Against such a dull coals-to-Newcastle rut, would rise The New Philosophy of Science of the 1960's. A number of acute, critically-minded theorists appeared including N.R. Hanson, S. Toulmin, T. Kuhn, M. Hesse, and P. Feyerabend. They largely tried to introduce a historical or genetic point of view into the methodology of science. It was Thomas Kuhn's *The Structure of Scientific Revolutions* which had the most telling influence, not only in philosophy and history of science but also the methodology of the social sciences and humanities. The wide acceptance of Kuhn's notion of "paradigm" and Hanson's concept of "theory-ladenness" provided a start for altering the direction of philosophy of science. The transformation could be characterized as "from static to dynamic philosophy of science" or "from logical analysis to conceptual history" (Toulmin). This novel movement required history of science to have an epistemological viewpoint as well as for philosophy of science to have a historical viewpoint. As a result, the fusion of history of science and philosophy of science showed signs of becoming real with the two branches becoming closely interrelated. But if we look back into past history, we find that such a fusion had already occured in Mach's historical writings a century before. In this respect then, The New Philosophy of Science, which was legitimately new in a number of ways, was just catching up with its even more critically-minded but often overlooked predecessor.

In other words, Mach's philosophy should not be pigeonholed into the "empiricism" or "positivism" of his immediate successors. Logical positivists did talk about and try to realize his plan of "eliminating metaphysics", but they also departed from Mach by adopting a "physical science is everything" approach, neglecting the biological and economical motivation for his work as well as his anti-apriorism and intent to make physics and psychology more compatible without reducing one to the other. In their hands, philosophy of science lost its critical mind and became a worshipper of formal success.

According to Paul Feyerabend: "The unity of science [between science and philosophy] was achieved *and at once lost again* by the empirical successors of Mach",[36] and he continues as follows:

> The first and most noticable change is the transition from a critical *philosophy* to *sense-data dogmatism.* Elements are replaced by sensations not just temporally, and as a matter of hypothesis, but once and for all . . . . Secondly, criticism of science is replaced by *logical reconstruction* which in plain English, is nothing but a highly sophisticated brand of conformism. As a result the idea of logical reconstruction became conformist. The task was now to *correctly present* rather than *change* science.[37]

In the context of the logical reconstruction of science by Logical Positivists and Empiricists, Mach's fundamental motivation, namely, his challenge to apriorism and to the mechanistic world view were almost completely forgotten. Indeed, there has been an increasing return to apriorism in philosophy, particularly on the wings of uncritical acceptance of formal logic and mathematical physics. A Japanese historian of science, Tetsu Hiroshige, has pointed out where Mach stood rather exactly: "Mach's purpose in *The Science of Mechanics* was to break down apriorism in mechanics, and as a result of it, to strike a blow at the mechanistic view of nature."[38]

As we have seen, Mach's attempt to introduce phenomenological physics as a substitute for mechanical physics aimed at bringing about a "paradigm-shift" in our traditional view of nature. This literal *re-vision*, at the same time, would imply a change in self-understanding of scientific activity itself. In his critical attitude toward established science,[39] Mach's philosophy of science could be characterized as a forerunner of the New Philosophy of Science, rather than Logical Positivism. On the other hand, his thesis of anti-apriorism - that the difference between geometry and physics was only a matter of degree - while it anticipated Quine's "holism" or holistic view of science, has still not been as widely or *deeply* appreciated as the idea deserves. What all this may mean is that in order to properly evaluate Mach's achievements from a contemporary perspective, we must locate him in the context of new developments in The New Philosophy of Science, that is, as it has developed *since the 1960's.*

---

36 Paul K. Feyerabend, "Philosophy of Science: A Subject with a great past", in *Minnesota Studies in the Philosophy of Science,* Vol. V (1970), p. 174.

37 Ibid., pp. 179-181.

38 Tetsu Hirosige, "The Ether Problem, the Mechanistic World View, and the Origin of the Theory of Relativity", in *Einstein Studies,* edited by S. Nisio, Tokyo, 1977, p. 229.

39 Ernst Mach to the Austrian Academy of Sciences, Vienna, April 28, 1913, published in E. Mach, *Ernst Mach als Aussenseiter,* edited by J. Blackmore & K. Hentschel, Wilhelm Braumüller: Wien, 1985, p. 114. [See also Ernst Mach to Wilhelm Ostwald, Haar, July 23, 1913 in this anthology. - Ed.]

CHAPTER 13

MACH'S CRITIQUE OF NEWTONIAN MECHANICS

by Mario Bunge*
(McGill University)

INTRODUCTION

A distinguished experimental physicist and psychophysiologist, Ernst Mach (1838-1916) was an influential worker in the foundations of physics. His main contributions in the latter were critical rather than constructive, yet important because they were concerned with certain fundamental ideas. Many of Mach's criticisms, particularly those of atomistics and Newtonian mechanics have had a profound influence. While the former was entirely negative, Mach's work on the foundations of mechanics was ambivalent.

On the one hand, Mach made a correct criticism of the concepts of absolute space and absolute time. On the other, he tried to minimize the role of mechanical theories and, indeed, of all theory. These two aspects of his work were as [different as?] many consequences of the empiricist theory of knowledge he adopted and popularized. In fact, for a radical empiricist, every inscrutable idea is damnable and, moreover, no idea is to be accepted unless it concerns experience. By the first token, untestable assumptions are discarded from factual science, and rightly so. By the second, every theory is distrusted, or even condemned, because theories proper overreach experience and do not refer to it; and this was wrong.

Faithful to his empiricist philosophy, Mach attempted to cleanse physics from untestable hypotheses by the simple technique of eliminating hypotheses altogether - in fact, by keeping only "experimental propositions" and definitions. Thus his criticism of Newtonian mechanics was more a criticism of theoretical physics than a criticism of classical physics - so much so that he attempted to replace that theory by a single empirical statement and definition, and that he sanguinely opposed every attempt to go beyond classical physics, particularly relativity and atomic theories.[1] By criticizing Newtonian mechanics, Mach undermined the prevailing dogmatism which

* This article first appeared in *The American Journal of Physics,* **34** (1966), 585-596.

[1] See, e.g., E. Mach, *The Principles of Physical Optics* (Dover Publications: New York, 1953), p. viii. For the opposite opinion, that Mach prepared the way to modern physics, see K.D. Heller, *Ernst Mach, Wegbereiter der Modernen Physik,* (Springer Verlag: Wien, 1964).

J. Blackmore (ed.), Ernst Mach – A Deeper Look, 243–261.

regarded that theory as final, and thereby helped to create the freer climate of opinion necessary to invent new theories. But he rejected these new ideas because they were incompatible with his philosophy. Furthermore, it is a conceivable but untestable conjecture that the strictures imposed by Mach's narrow and dogmatic empiricism have delayed scientific progress, which does not consist in sticking to what has been experienced, but in introducing daring new ideas enabling us to understand reality and to plan experiments unthinkable without those ideas.

Let us now examine the main features of Mach's criticism of Newtonian particle mechanics, the negative aspect of which has received more publicity than the positive aspect, to the extent that most "modern" textbooks on mechanics include Mach's blunders concerning mass and force.

I. THE RELATIVITY OF TIME

Mach rejects the Newtonian idea of absolute time which "of itself, and from its own nature, flows equably without relation to anything external".[2] He adopts instead what is properly called the relational and, improperly, the causal theory of time, which had been expounded by Plato, Aristotle, Lucretius, Leibniz, and others whom Mach does not name in this connection. (Mach acknowledged his ignorance of the history of philosophy. That he was a poor historian of mechanics is shown by the fact that he ignored the whole Middle Ages and - as shown by Truesdell[3] - that he drew from unreliable sources.)

According to the relational theory of time, time does not exist in and of itself but is something like the pace of events: there are changing things and "time is an abstraction at whch we arrive by means of the many changes of things".[4] The concept of absolute time, of a time independent of change, is "an idle metaphysical conception".[5] Like a good empiricist, Mach arrives at this conclusion by examining the way time is measured rather than the way the time variable occurs in theoretical physics. (This was also the approach of H. Reichenbach in his popular *Rise of Scientific Philosophy*.) From this examination he concluded that, "For the natural inquirer, determinations of time are merely abbreviated statements of the dependence of one event

---

2 I. Newton in *Mathematical Principles of Natural Philosophy,* F. Cajori, (ed.), University of California Press: Berkeley, 1947, p. 6.

3 C. Truesdell, *Archive for History of Exact Sciences*, **1** (1960), 3 and the Introduction to L. Euler, *Opere omnia,* **11** (*Soc. Sci. Natur. Helv.*, Zürich, 1960), especially page 409 on Lagrange as Mach's historical misguide.

4 E. Mach, *The Science of Mechanics,* 9th edition., Open Court Publishing Company: La Salle, Ill., 1942, p. 273.

5 Ibid., p. 274.

upon another, and nothing more".[6] In time measurements, one of the two processes being compared is "an arbitrarily chosen motion"[7] that functions as a time standard or clock.

Time, then, is relative; moreover it is local and even dispensable. In fact, although a time sequence can be set up for every part of the universe by means of some standard sequence of events, "For the universe [as a whole] there is no time."[8] In brief, there is local time but no cosmic time. (Consequently certain questions, such as the one concerning the age of the universe, are meaningless.) Moreover local time does not exist *per se* but is a relation among phenomena and can, so believes Mach, be eliminated. He thinks in fact that we could eliminate the time variable from any law statement by putting in its place a phenomenon dependent, say, on the angle of rotation of the Earth. (But he does not suggest how to implement this idea mathematically.) As can be seen, Mach's ideas on time were radical although they were neither too precise nor too new. Let us examine them briefly.

The idea that time is a relation between events rather than self-existent seems reasonable but is half-baked. Mach had in mind pairs of simultaneous events and was content with remarking that timing an events consists in pairing it to another event - its match in the standard sequence. This is true but insufficient; not only in science but also in modern everyday life we need a more complex concept of time, one with one foot on concrete events (whether actual or possible) and another on pure number. We need in short a quantitative concept of duration if we are to have a physical theory in addition to the empirical recognition of equal durations.

A more sophisticated relational concept of time can be built in the following way. The lapse between two events x and y can be assigned a real number t . In general letting E designate the set of point events and the Cartesian product $E \times E$ the set of ordered pairs $< x,y >$ of events, the duration is a function D from $E \times E$ to T, the range T being included in the real line. In short, $D : E \times E \rightarrow T$ with $R \supseteq T$. An arbitrary member of t of$T$ can be called an instant of time. In order to determine the duration function D we must subject it to certain requirements (postulates). A possible set of conditions on D is this: (a) for every x in E, the duration of $<x,x>$ is zero; (b) for any x and y in E , the duration of $< x,y >$ equals, with changed sign, the duration of $< y,x >$; (c) for any x,y, and z in E, the duration of $< x,z >$, This theory of (nonrelativistic) time can be refined by

[6] E. Mach, *Popular Scientific Lectures*, 4th edition, Open Court Publishing Company: Chicago, 1910, p. 204.

[7] Mach (1942), op. cit., p. 275.

[8] E. Mach, *History and Root of the Principle of Conservation of Energy*, Open Court Publishing Company: Chicago, 1911, p. 63.

introducing a third undefined physical concept in addition to those of event and duration, namely the one of reference frame. In any case it makes the relational or causal doctrine of time, adopted (though not invented) by Mach, somewhat more precise.

II. TIME STILL TIMELY

Yet, it is likely that Mach would have rejected the foregoing elucidation (not definition) of the time concept. First, because this analysis, far from eliminating concepts in favor of phenomena, shows that time is a certain relation (function) between pairs of events - whether phenomenal or imperceptible - and real numbers. Second because it makes no reference to clock readings and moreover does not apply exactly to such empirical data. For these reasons the preceding analysis is not a formalization of Mach's ideas on time but rather a synthesis of the relational view of Leibniz and the Newtonian concept of cosmic time.

Moreover, the above theory of time does not enable us to eliminate the time concept, but rather reinforces it with the help of mathematics (more precisely, set theory). To see what would get lost if the time concept were eliminated, consider these two statements:

a happened on 1 January 1966, (1)

b happened on 1 January 1966, (2)

where *a* and *b* name certain events. From (1) and (2) and the definition of simultaneity we can eliminate the time datum, obtaining

a happened simultaneously with *b*, (3)

b happened simultaneously with *a* (4)

which, by virtue of the symmetry of the simultaneity relation are equivalent. Yet, from the second pair of propositions we cannot reconstruct the first: the date has been lost and cannot be retrieved. And the date may be of importance; it will be indispensable if we expand the initial set of events. In conclusion, it would be foolish to eliminate the time concept: if we did it we would lose valuable information and we would be unable to build theories of change. Time must be analysed, not eliminated. The same is true of the other abstractions condemned by Mach for being abstractions rather than for being wrong abstractions.

Let us finally comment on the alleged arbitrariness of the time standard. According to Mach "we select as our measure of time an arbitrarily chosen motion (the angle of the Earth's rotation, or [the] path of a free body) which proceeds in almost parallel correspondence with our sensation of time".[9] This is false: (a) time standards are supposed to be regular rather than arbitrary and are corrected or replaced as soon as they are suspected of some irregularity; (b) since the beginning of scientific time reckoning the perception of duration has had little say in the choice of standards; it has none in the adoption of atomic (actually molecular) clocks; (c) what is decisive in the adoption of time standards is not our unreliable time sensation, which is rarely parallel to physical duration, but some theory which can justify the assumed regularity of the chosen process. Once again, the progress of science does not consist in sticking to sensations but in enriching our body of ideas and subjecting sensations to their control as well as conversely.

In brief, Mach's criticism of the Newtonian concept of time did contain a grain of truth - namely the old idea that time is relative. But this idea could not be developed along the line of Mach's philosophy, which was diffident to mathematics: it was made precise with the help of mathematical concepts. And even this refinement proved eventually to be insufficient: Einstein (1905) showed that time is not only anchored in events but is also relative to the reference frame. But this further relativization of time was disowned by Mach.

III. THE RELATIVITY OF SPACE

Mach opposes Newton's idea of an absolute space existing "without relation to anything external."[2] He regards absolute space and absolute motion as "pure things of thought, pure mental constructs, that cannot be produced in experience."[10] Being transempirical they are to be condemned.

Mach was certainly right in rejecting the concepts of absolute space and absolute motion. But the philosophical reason he gave was wrong, as it boiled down to the requirement that physical theories be limited to human experience. He rightly remarked that the changes in position and velocity b under the influence of another body b' can only be ascertained relative to a third material system. We may rephrase his criticism in the following way: free space has no intrinsic coordinate lines; if we wish to use the concept of coordinate system in physics we must at least imagine that it is the conceptual reconstruction of a material reference frame - a physical system, not a mathematical construct. Consequently, as far as physics is concerned all

---

9 Mach (1942), op. cit., pp. 275-276.

10 Ibid., p. 280.

geometrical properties are determined with reference to physical systems, not to absolute space. These were not Mach's own words: his criticism, though correct, was not as clear as Leibniz's. Moreover, it was mixed with certain mistakes.

A first mistake was the failure to realize that C. Neumann (1870) had already supplied a substitute for Newton's absolute space - namely his Alpha body, a forerunner of our inertial or Galilean frame of reference. Mach regarded the Alpha body as a fiction[11] and instead advocated referring location and motion to the universe as a whole: "When . . . we say that a body preserves unchanged its direction and velocity *in space*, our assertion is nothing more or less than an abbreviated reference to *the entire universe*."[12] This was a strange thing to say: (a) the universe, being known only in part, is not a suitable reference frame and is never used as such; (b) a good empiricist does not commit himself to assuming the objective existence of the universe - a realistic hypothesis. In any case, we can now reformulate Newtonian mechanics without either Newton's absolute space or Neumann's Alpha body or Mach's universe: one simply uses the set of inertial frames.

A second mistake of Mach's in relation with his criticism of the ideas of absolute space and absolute motion was that he thought position and motion to be *determined by* the reference frame rather than being *referred to* it. He failed to realize that a material frame of reference interacting significantly with the object of study is not a suitable frame, if only because it is not inertial. But he could not escape this confusion without leaping beyond his empiricist philosophy: in fact for an empiricist there is no determination apart from the operation whereby some human subject determines (perceives, measures, makes) properties of objects in his field of experience. In other words, an empiricist is bound to mix the ontological concept of determination (as exemplified by causation) with the epistemological concept of determination (as exemplified by measurement and prediction).

This confusion is at the root of the so-called Mach principle, according to which the motion and consequently (for Mach) the mass of every single body is determined (bestimmt=caused=produced) by remaining bodies in the universe.[13] This "principle", the child of a verbal confusion,* has been restated in a number of ways, not all of them faithful to Mach. (E.g., the hypothesis that the geometry of physical space is determined by matter.) Yet, though it has not been stated unambiguously, it passes for having been

---

11 Ibid., p. 291.

12 Ibid., p. 286.

13 Ibid., p. 283.

* For Einstein's own admission that he was not clear about the principle see chapter 7, section F in this anthology. - Ed.

confirmed by experiment, and it is often invoked in gravitation theory, cosmology, and in elementary particle physics for no better reason that that it enjoys the alleged authority of Mach. Its spurious origin is not realized, nor is it fully realized that, being a causal hypothesis, it is inconsistent with Mach's proposal of replacing causation by functional interdependence.[14] Furthermore, if the Mach *principle* were true, the mass of the universe as a whole should be zero, for it interacts with nothing.

This is not all: Mach stretched his criticism of absolute space and absolute motion to the point of claiming that the heliocentric (Copernican) and the geocentric (Ptolemaic) "systems of the world" are equivalent: the motions of the universe are the same whether we adopt the Ptolemaic or the Copernican mode of view. Both views are, indeed, equally *correct;* only the latter is more simple and more *practical.*"[15] This thought has been repeated *ad nauseum* ever since. It is a favorite with the positivists in their criticism of realism; with the conventionalists and pragmatists in their advocacy of simplicity as a substitute for truth; and last but not least with those who wished to rehabilitate the Inquisition and blame Galileo. Yet, notwithstanding the authority of Mach, Duhem, Pearson, Le Roy, Frank, Reichenbach, Goodman, and hundreds of textbook authors, a relativistic view of space and motion does not warrant the grotesque claim that the heliocentric and the geocentric "systems of the world" are equivalent except geometrically, in the sense that any trajectory in one of the "systems" can be transformed to the other.

Among the many reasons for the physical inequivalence between the two doctrines, the following can be mentioned. (a) Planets are not inertial systems, whereas the sun is nearly inertial; equivalently: the laws of mechanics do not hold exactly if referred to the earth, but they hold approximately if referred to the Sun - hence the heliocentric system is the one recommended by mechanics. (b) The solar gravitational field, which codetermines the planatary motions, cannot be transformed away - except locally - either in classical physics or in general relativity; equivalently: the equivalence principle is a local not an integral law. (c) The Copernicus-Kepler model of the planetary system, unlike the Ptolemaic model, can be embedded in a number of physical theories, and in this way numerous phenomena can be explained - e.g., the stellar parallax, aberration, the Doppler shift, the irregularities in the planetary motions - which remain separate and mysterious on the geocentric doctrine. For some of these and other reasons Poincaré, who had initially embraced the view that the two

---

[14] M. Bunge, *Causality: The Place of the Causal Principle in Modern Science,* second edition, The World Publishing Company: New York, 1963.

[15] E. Mach, *The Science of Mechanics,* ninth edition, Open Court Publishing Company: La Salle, Ill., 1942, p. 284.

"systems of the world" are equivalent, on reexamining it concluded that "the truth for which Galileo had suffered is still the truth."[16]

In short, Mach's critique of absolute space and absolute motion was basically correct but it was mingled with several important mistakes, among them the confusion of "relation to a frame" with "determination by a frame", and a purely kinematical conception of motion, oblivious of the force fields which can be involved in it. This latter mistake enabled Mach to regard force and mass as dispensable. To this point we now turn.

IV. THE ELIMINATION OF MASS AND FORCE

Mach disliked the concepts of mass and force, which were at the heart of the popular materialist philosophy he opposed. He objected, in particular, to the prevailing idea that mass was a measure of the quantity of matter - a concept he could not accept. Indeed, since matter is by definition supposed to exist by itself, it should not be characterized by reference to sense impressions alone, and is therefore unfit to enter an empiricist schema, in which existence must be identified with capacity to have or produce sense impressions - Berkeley's famous "To be is to perceive or to be perceived". Science according to Mach, must endeavor to record phenomena (observable facts) and "to establish the interdependence of phenomena". Since the phenomena of mechanics are motions of perceptible bodies, we must limit our attention to the kinematics of bodies. This had also been d'Alembert's program in his influential *Traité de dynamique* (1743).

To implement the kinetic approach in the case of the concept of mass, Mach considered the following example. Imagine an isolated system consisting of two bodies "perfectly equal in all respects".(One ought to qualify that these traits are observational, for in order to substantiate the hypothesis of identity, one would have to measure, among other things, the masses of the bodies.) He expects, and finds by experiment, that "they will produce in each other in the direction of their line of conjunction equal and opposite accelerations." And this, according to Mach, is all we mean when we say that the bodies concerned have the same mass. In fact, he proposed "arbitrarily establishing the following definition: All those bodies are bodies of equal mass which, mutually acting on each other, produce in each other equal and opposite accelerations."[17]

---

16 H. Poincaré, *La Valeur de le science,* (Librairie Ernst Flammarion: Paris, 1902), p. 274. For the English version see: *The Value of Science,* Part III, Chapter 11, "Science and Reality", trans. G.B. Halsted, Science Press: New York, 1913, 1929; Dover ed., p. 354.

17 Mach (1942), op. cit., p. 266.

Thinking that he had *defined* the concept of mass equality and, indeed, without the help of Newton's *lex secunda,* he went on to metricize the mass concept by considering a slightly different example: a system composed of two arbitrary bodies which may be called 1 and 2. A comparison of their mutually induced accelerations $-a_1$ and $+a_2$ enables him to measure or else to compute their mass ratio. (Let us add that a suitable reference frame must be chosen.) And once he has confused "measuring" and "computing" with "defining", he can conclude that he has *defined* the mass concept in terms of observable (kinematic) properties. (A similar confusion lies at the root of Bridgman's operationalism.) In fact, pretending to ignore Newton's second and third axioms, Mach feigns that "experience alone can teach" that the ratio $-a_2/a_1$ is constant in time and positive, i.e., $-a_2/a_1 = m_{12}$, where m_{12} is just an abbreviation for that ratio. He finally writes the equation in the form

$$-a_2/a_1 = m_1/m_2 \qquad (5)$$

which is just a way of stating the "fact" that the ratio of accelerations is constant. (Mach did not inquire whether it remains constant under transformations of reference frames, in particular whether $-a_2/a_1$ is the same relative to an accelerated frame. In his piecemeal approach that jumps from one example to the next he seems to have forgotten the relativity of motion.)

Equation (5) was taken by Mach as the *definition* of relative mass, or mass ratio - namely as "the negative inverse ratio of the counteraccelerations".[17] Once the mass concept was so "defined" he then turned to the concept of force, which he "defined" as $m \bullet a$.[18] Consequently, Newton's *lex secunda,* far from expressing a law of nature, is degraded to a convention.

With this simple device, Mach believed to have attained his goal - "to remove all metaphysical obscurity"[19] - Since mass had now become an observable property of observable bodies and force a mere name for the product of two unobservable properties. (Atoms were considered by Mach as annoying fictions precisely because they were hypothesized to exist beyond sense experience. Hence there was no need to bother about atomic masses.) Moreover, the principle of action and reaction being absorbed in Eq. (5), did not have to be stated separately. Mach did not recall at this point that the principle is also needed in statics, where no accelerations occur. And he was not aware that his formula, far from having been established by ex-

---

18 Ibid., p. 304.

19 Ibid., p. 267. One wonders what the reaction of Mach to the mass operators in elementary particle theory would have been.

periment alone, was a consequence of the second and third laws of motion. In fact, he claimed that "in the concept of mass no theory of any kind whatever is contained, but simply a fact of experience."[20] Which, though untrue, was a definite advantage from an empiricist point of view: one more theory could be dispensed with.

Actually, Mach's "experimental proposition" (5) is not a direct expression of phenomena: from observation alone we could not infer that the components of a system move under the sole influence of their mutual actions: only an empirical checking of the consequences of such a *hypothesis,* when embedded in some theory, is possible. That proposition is in fact a deductive consequence of Newton's postulates for the very special case of an isolated system consisting of two interacting bodies. Being an application of a theory it does not allow us to dispense with the theory. It can moreover be anticipated that the extension or domain of applicability of Mach's "experimental proposition" (5) is smaller than Newton's theory - i.e., that it will fail in more complex cases. This is seen in a moment.

V. MASS STILL IN FORCE

Mach's formula (5) is much weaker than the Newton-Euler theory of mechanics, so much so, that it only holds for certain systems in equilibrium. The treatment of slightly more complicated problems requires the use of the full theory, which was to be expected, since no theory is equivalent to its lowest level consequences, let alone to one or two of them. In other words, a theory is much more (infinitely more) than a set of examples. But then Mach had no appreciation for theoretical physics: Indeed, for him the historical development of mechanics had established "only one great fact", namely the "experimental proposition" (5).[21]

But formula (5) breaks down when referred to an accelerated frame: in fact, relative to such a frame, a different acceleration ratio may obtain, although the mass values are, by hypothesis, invariant under arbitrary changes in reference frames. To realize this, consider again, Mach's example and compute the accelerations of its two constituents, first in a frame K linked to their center of mass - which is what Mach had unwittingly done - and then in another frame K' moving with accelerated motion relative to K and parallel to the line joining the two particles.[22] Calling a the acceleration of K' with respect to K, the new accelerations of the particles are

[20] Ibid., p. 271.

[21] Ibid., p. 307.

[22] See C.G. Pendse, *Philosophical Magazine,* **24** (1937), 1012.

$$a_1' = a_1 - a_1 \tag{6}$$

$$a_2' = a_2 + a_1 \tag{7}$$

whence

$$\frac{a_1'}{a_2'} = \frac{a_1 - a}{a_2 + a} = \frac{(1 - a/a_1)}{(1 + a/a_2)} \bullet \frac{a_1}{a_2} \tag{8}$$

If a_1/a_2 is, in fact, constant in time and positive, then Mach calls it $-m_{21}$ and interprets it as the mass of particle **2** relative to the mass of particle **1**. If, in addition, the coefficient of a_1/a_2 is constant in time and positive, he calls the left-hand side of (8) $-m_{21}'$, whence the previous equation becomes in Mach's interpretation,

$$m_{21}' = \frac{1 - a/a_1}{1 + a/a_2} \bullet m_{21}. \tag{9}$$

In particular, if K' moves alongside the first particle ($a = a_1$), then $m_{21}' = 0$ although m_{21} is not equal to 0. And by redoing the calculation for m_{12}' we can see that $m_{12}' = 0$ even though $m_{12} = 0$ if the acceleration of K' equals that of the second particle. Which shows that the primed quantities m_{12}' and m_{21}' cannot be interpreted as the relative masses since in classical mechanics, and also in Mach's thought, these are invariant with respect to the choice of frame.[23]

To that physical mistake stemming from elaborating a single elementary exercise rather than elucidating the whole theory, Mach added one logical and one methodological mistake. The logical mistake was to confuse equality with an identity and, in particular, with a definition. In fact, however Mach's pseudo-postulate (5), like most physical laws, established an equality between two expressions which differ in meaning and therefore cannot be regarded as the two sides of a definition. Indeed, while "m_1/m_2" means "the inertia of body **1** relative to the inertia of body **2**," the symbol "$-a_2/a_1$" stands for a purely kinematical quantity. The equality is numerical not logical: it does not authorize us to eliminate one of the sides in favor of the other.

Similarly, it is mistaken to regard "$f = ma$" as a definition of force in terms of m and a. It is not just a question of calling ma by the name f or conversely: the two concepts happen to be related in that way in classical

[23] C.G. Pendse, *Philosophical Magazine*, **27** (1939), 51.

mechanics but there is nothing either necessary or conventional in this. If the *lex secunda* were just a convention it would be impossible to subject it to empirical tests; in particular, it would be impossible to find fault with it. Yet, we know that it holds only relatively to inertial (Galilean) frames and, even so, only in the nonrelativistic limit and as a quantum-mechanical space average. Moreover, if it were a logical identity, masses and forces could not be assigned (hypothesized) independently. In particular, it would be impossible to assign masses to unaccelerated bodies.

In conclusion, Mach's proposed redefinitions of mass and force are a failure from every possible point of view: they are not independent of mechanical theory, they are not generally applicable, and they do not qualify as definitions (logical identities). Thus, their persistent lingering in textbooks may have to be explained as a result of philosophical loyalty and of neglect of foundations research among physicists.

VI. THE KINETIC APPROACH

The philosopher-scientist René Descartes (1596-1650) was the first among the moderns to attempt to reduce the whole of physics to figures and motions, these ideas being "clear and distinct" unlike those of potency, act, substance, and accident employed by the schoolmen. Of course, Descartes did not succeed in carrying out this reduction program: he introduced invisible atoms and kept the Aristotelian aether, which he endowed with complex vortex motions. Moreover, being a deductivist and not an inductivist, he had no intention of starting with observation. Like every other theoretical physicist he started from assumptions that would eventually explain observations. His program, interesting as it was, was unsuccessful because he chose the wrong axioms and because he was more anxious to justify them by metaphysical argument than to put them to the experimental test. The universal genius Leibniz (1646-1716), similarly inspired, failed for the same reasons, although he was able to utilize a more powerful mathematics. Newton criticized Cartesian physics - without mentioning it by name - for being aprioristic. He proposed instead to build a theory that would start with observation and stick to it: "In experimental philosophy we are to look upon propositions inferred by general induction from phenomena as accurately or very nearly true"; "the argument of induction may not be evaded by hypotheses."[24] Such the words, not the deeds: Newton never practiced the inductivist philosophy he preached. In fact, (a) he assumed objects or qualities not directly observable, such as mass, center of

[24] I. Newton in *Mathematical Principles of Natural Philosophy,* F. Cajori, ed. University of California Press: Berkeley, 1947, p. 400.

mass, and force; (b) he related such trans-empirical concepts to form hypotheses that he supposed stood for the invisible yet real patterns of motion, and (c) he assigned (hypothesized) forces and masses and worked out the consequences, which he finally checked against empirical data. In short, Newton was as much a practitioner of the hypothetico-deductive method as Descartes and Leibniz were - only he thought he was a faithful follower of Francis Bacon, the national philosophical hero.

Newton's anti-Newtonian philosophy was influential: people with no philosophical background who thought Newton practiced what he preached. But he did not fool Kirchhoff and Mach, two inductivists who felt Newton had overstepped the limits of experience. Without knowing it they tried to implement Descartes's kinetic program and the rather similar program of d'Alembert. Gustav Robert Kirchhoff (1824-1887), who seems to have exercized considerable influence on many of Mach's contemporaries, wrote in his treatise: "In my opinion mechanics can create out of motion alone the definitions of the concepts it is concerned with."[25] That is, observable events (phenomena) are to be the building blocks of mechanics and, in general, of physics. Every other idea is to be regarded as an auxillary construct, a useful fiction, as the philosopher Hans Vaihinger would put it later on. Thus, masses and forces are not to be found in reality, simply because they are not in human experience: they are just auxillary concepts (*Hilfsbegriffe*).[26] These ideas and the program of rescuing physics from the "metaphysical" ideas of Newton, had been pioneered by d'Alembert and have been in the air since the 1870's. Being the fashionable philosophy of physics, it has been adopted even by some of the creators of quantum mechanics - but only nominally, for when one speaks of a quantum mechanical "observable" one means a symbolic and high-brow representative of some physical property not directly accessible to the senses.

Physicists cannot work in the way recommended by inductivists: they cannot help introducing trans-observational concepts and they do not deduce theories from observations because this would be logically miraculous. In fact (a) experience does not supply those trans-observational concepts: (b) particular propositions can be deduced from general ones but the converse is logically impossible. In particular, there is no unique path from observed motions to laws of motion: the latter are hypothesized, their solutions are then found, and finally these solutions are contrasted with empirical data obtained with the help of a bunch of theories. One and the same class of motions can be derived from infinitely many equations of motion. By contrast, given an equation of motion, the class of motions remains

[25] G. Kirchhoff, *Vorlesungen über mathematische Physik, I, Mechanik,* B.G. Teubner: Leipzig, 1883, p. 11.

[26] Ibid., p. 1.

uniquely determined. For example, in particle mechanics a uniform rectilinear motion can be derived from any of the following equations of motion:

$$mx = f = \text{const (Aristotle's law for a constant force)}, \quad (10)$$

$$mx = 0 \text{ (Newton-Euler law for a null force)}, \quad (11)$$

$$mx = f+\lambda(\partial\phi/\partial x) = 0$$
$$\text{(Lagrange's law with constraint balancing applied force)}, \quad (12)$$

$$mx = f(t) + R(x,t), \quad (13)$$

R being a random force whose space average balances the instantaneous applied force.

Not only the laws of motion and the forces have to be hypothesized, but also the masses have to be prescribed, if only tentatively, if the trajectories of definite systems are to be derived from the equations of motion. In some cases - e.g., elastic collision - the mass ratios can be inferred from data concerning particle velocities in conjunction with the law of conservation of linear momentum. In a few other cases the mass values can be inferred from observed accelerations and Newton's second and third laws. The latter method, advocated by Mach as a universal procedure, has severe limitations: as shown by Pendse,[22 23] the method works for a system consisting of at most, 4 bodies when their accelerations are observed at one instant; and it works up to $n = 7$ if observed at different instants.

Consider an isolated system containing $n > 2$ particles. Call a_i the acceleration of the i th particle at a given instant (relative to a certain frame) and let a_{ik} be the contribution of the k th particle in the system to the acceleration of i. Calling *eik* the unit vector in the direction of i to k, we have

$$a_i = \sum_{k \neq i} e_{ik} \bullet a_{ik}. \quad (14)$$

Every such vector equation summarizes a system of three linear equations for the $(n - 1)$ unknown a_{ik} associated with the i th particle. For $n = 4$ there will be as many equations as unknowns - save if the particles are colinear or coplanar, in which case the number of equations will be less than that of

---

unknowns. If $n > 4$ then the equations are less than the unknowns which will therefore remain undetermined. In other words, the data a_i do not determine uniquely the relative accelerations a_{ik}, which are in turn proportional to the relative masses on condition that they are all referred to the same inertial frame. Similarly the method fails for $n > 7$ observations taken at different instants. The only way to proceed is to single out a subsystem with $n < 7$ particles or to assume that one of the components is dominant and apply a perturbation technique. But in such cases the computations are approximate, whence the mass ratios do not coincide with the acceleration ratios.

The unfeasibility of Mach's method becomes even more obvious in continuum mechanics, which, after all, is logically more basic and of course more realistic than particle mechanics. Indeed, consider, say, Cauchy's first law of motion,

$$\rho x = \mathrm{div} T + \rho f_b , \qquad (15)$$

where ρ, T, and f_b stand for the mass density, the stress tensor and the body force density, respectively, while x represents the position of an arbitrary particle in the body. Even assuming that the (infinitely many) accelerations were observable - which they obviously are not - nothing could be inferred from their knowledge alone. In practice, one hypothesizes ρ, f_b , and T are given (hypothesized), the stress state is not uniquely determined by the three equations of motion, as T has nine components. If we want to solve any real problems we must hypothesize the occult T - e.g., by adding some constitutive equation involving T. And T is not observable: it is a hidden property that manifests itself on the body boundary (as the stress vector $t = T \bullet n$). Not only in atomic physics but also in classical physics the overt behavior of matter is determined by hidden properties which are beyond observation, although they do show up in a roundabout fashion. In the best of cases, the observable behavior can be calculated on the basis of additional assumptions concerning inner structures and states. The inverse way, that is, from experiment to theory, simply does not exist. The way of science is not observation →theory, but rather observation →problem →theory →experimental checking →theory correction.

In brief, in a few simple cases it is possible to infer mass from observations fed to mechanical theory. But, in most cases, the mass values of the individual components of systems of bodies must be either hypothesized or just left as unknowns. Yet, although a mass value may not be computed from observed motions in most cases, it is assumed to be definite, i.e., to be an objective property of the body concerned. In other words, in mechanics we do not accept the positivist tenet that bodies acquire mass upon being subjected to mass measurement operations but assume instead that they have a definite mass value all along. Very often we assign mass values hypothetically, deduce consequences, confront the latter with data and, if necessary, correct the hypothesized mass values by successive approximations. This would be impossible without assuming that "mass" refers to (stands for) an objective property - which is all Mach wished to deny.

VII. HOW NOT TO BUILD MECHANICS

The preceeding considerations should suffice to show that the dynamics of a mechanical system can neither be ignored nor inferred from observation of the motions of every one of its parts - not even if these parts are observable. On the contrary, the motions are deducible from hypotheses concerning the laws of motion, the mass (or density) values, and the stresses and forces involved: the data only single out particular trajectories from bundles of possible trajectories. In short, as Newton had discovered, kinematics is deducible from dynamics but not, as d'Alembert, Kirchhoff, and Mach wanted conversely. In other words, the inverse problem of experimental mechanics - deriving masses and forces from a knowledge of motions alone - is in general as unsolvable as the problem of inferring postulates from theorems in a unique way.

Since this is true of the simplest possible physical systems - the mechanical ones - we may jump to the general conclusion that, while hypotheses concerning the nature of a system are, jointly with its laws of evolution, necessary and sufficient to determine its behavior in a broad way, given the behavior - in case it is observable - it is not possible to discover the nature and laws of the system. Consequently, every behavioristic (phenomenalist, empiricist) program, whether in mechanics or in elementary particles or in psychology, is bound to be superficial.[27] True, sometimes we are forced to remain on the surface: but let us not make a virtue of such poverty.

Despite the logical impossibility of building mechanics - or for that matter any other scientific theory - with observational concepts alone; despite the logical impossibility of inferring theories from data; despite the sterility of inductivism; despite the successes of the more general and abstract formulations of mechanics; despite all of this there have been further attempts to build mechanics on observational concepts. Those of Hermes[28]

---

[27] See M. Bunge, "Phenomenological theories", in M. Bunge ed., *The Critical Approach in Science and Philosophy,* The Free Press of Glencoe: New York, 1964.

[28] H. Hermes, "Zur Axiomatisierung der Mechanik" (reformulation of a system published in 1938), in L. Henkin, P. Suppes, and A. Tarski, eds. *The Axiomatic method,* North-Holland Publishing Company: Amsterdam, 1959.

Simon[29] and Eisenbud[30] are typical of that trend. Let us comment very briefly on the former, which is the earlier and the more elaborate.

The motivation of Hermes's work in the foundation of mechanics appears to be the same as Mach's, namely the elimination of unobservables: since what we measure are values of kinematic magnitudes - e.g., the position of a pointer in the case of weight measurements with a scale - we should be able to get along with purely kinematic concepts. The following objections can be raised. (a) Every measurement involves the use of a number of physical laws in which nonobservable concepts occur. (b) One of the characteristics of scientific theories is that they involve theoretical (nonobservational) concepts, this being why they can hope to explain experience. (c).The classical kinematics of the point particle is not a hypothetico-deductive system (theory) but a set of loosely related formulas: no wonder therfore that Hermes's system contains no law of motion and is therfore useless. (d) The concept of inertial or Galilean frame, which occurs as a primitive in this system, is not kinematical but dynamical, as is defined in terms of Newton's laws of motion or some similar set. (e) The conservation of linear momentum, which Hermes postulates and from which he derives the mass concept, is a dynamical, not a kinematical law - and moreover it cannot function as a definition if it is to be a law of nature. (f) Even if in simple cases we can determine mass ratios from velocity ratios with the help of that law, such a numerical determination does not amount to a definition.

In conclusion, the kinetic approach, which would reduce everything to observable figures and motions, was made obsolete by Newton three centuries ago. Moreover, it cannot be implemented, for if nothing is assumed in addition to what is observed, then nothing can be deduced. And if there is no deduction, there is no theory - which may be alright for empiricist philosophers but is lethal for science.

The morals for theory construction are clear: (a) Do not try to define the basic concepts, as they will be needed to define all others; (b) do not start with observables, otherwise you will be unable to explain them; (c) do not start with weak assumptions (theorems) unless you are willing to pile up infinitely many of them. In particular, in reconstructing classical mechanics one starts with a bunch of primitive (undefined) concepts such as those of body, configuration, mass, and force, and lays down basic requirements (axioms) to be satisfied by the primitives. Such a reconstruction is

---

29 H.A. Simon, "The Axioms of Newtonian Mechanics", *Philosophical Magazine,* **38** (1947), 888-905, and "Discussion: The Axioms of Classical Mechanics", *Philosophy of Science,* **21** (1954), 340-343; both reprinted in Herbert A. Simon, *Models of Discovery* (*Boston Studies in the Philosophy of Science,* vol. 54), Reidel: Dordrecht and Boston, 1977, pp. 349-375.

30 L. Eisenbud, *American Journal of Physics,* **26** (1958), 144.

under way, notably thanks to Suppes[31] and Noll.[32] Strange as it may seem, they are quite recent and are systematically ignored by those who think that Mach's foundational work was final or nearly so - although it was done at a time when mathematical logic was an esoteric science and scientific semantics had not been born.

The modern treatment of mechanics is not finished either. The available axiomatizations of mechanics are formally correct but they limp on the semantic side, in the sense that they correct the operationalist mania of attaching to every sign an operational meaning by failing to assign any objective meaning at all. The construction of formally correct and semantically adequate axiomatizations of the various branches of physics - not excluding mechanics - is an unfulfilled task of foundational research.[33] It does not need the assistance of automatic computers, and it will not lead to designing more effective lasers but it should help clarify ideas which, though capital and ever inspiring, are still half-baked or obscure.

VIII. CONCLUDING REMARKS

We have examined Mach's views on Newtonian mechanics and some of its descendents because, although they contain grave mistakes, they have contributed to freeing many minds from a blind faith in the Newton-Euler theory at a time when it seemed final. An additional reason for our examination is the amazing fortune Mach's mistakes have had with textbook writers (who on the other hand, usually ignore Mach's arguments against absolute time and absolute space). In fact, although Mach's ideas do not occur in the great treatises on mechanics - such as those of Levi-Civita, Whittaker, and Truesdell - hundreds of textbooks follow Mach in trying to reduce dynamics to kinematics by eliminating the concepts of mass and force via definitions in terms of observational concepts. of course, they just try. They do not realize that Mach's was not just a reinterpretation of the Newton-Euler theory but its utter destruction through its replacement by two formulas, the allegedly experimental proposition (5) - claimed to be at the same time a law and a definition - and the "definition" of force. And they do not know that the question whether a statement is a definition or a hypothesis is no longer a matter of opinion but can be settled in an exact way by

31 J.C.C. McKinsey, A.C. Sugar, and P. Suppes, *Journal of Rational Mechanical Analysis*, **2** (1953), 253.

32 W. Noll, "The Foundations of Classical Mechanics in the Light of Recent Advances in Continuum Mechanics" in L. Henkin, P. Suppes, and A. Tarski, eds., *The Axiomatic Method*, North-Holland Publishing Company: Amsterdam, 1959.

33 An attempt to fill formalisms with a physical content while respecting the axiomatic method is made in the author's *Foundations of Physics,* Springer Verlag: Berlin, 1967.

means of a logical technique (Padoa's method)[34] as long as it is applied in a closed context (an axiomatized theory).

Mach's mistakes were not accidental: they were due to his philosophy and to his inadequate knowledge of theoretical physics. (Nowhere in his work are four of the Big Five of theoretical physics in the 19th century mentioned: Cauchy's theory of elasticity, Maxwell's electromagnetism, Maxwell-Boltzmann's statistical mechanics, and Lorentz's electron theory. He only mentioned Fresnel's mechanical theory of light.) Mach was one of the very few scientists who evolved explicitly his philosophical ideas and tried to reform physics by adapting it to his philosophy, which continued the sensism of the 18th century British empiricists Berkeley and Hume.[35] He thought, in particular, that all physical concepts should be clear (Cartesianism), and that complete clarity is attained through definition (Aristotelianism) in terms of sense data (sensism). In addition, Mach was unaware of the logical impossibility of defining every concept; he seemed to regard every equality as qualified for functioning as a definition; he did not realize that the basic ideas taken in isolation cannot help being somewhat obscure if they are rich and deep, precisely because they are basic (explanatory and definitory) and pregnant with a number of relations, many of which are unexpected; and he could not accept that, in order to be rich, an idea must be far from concrete experience: indeed he was not after conceptual richness at all but, like Kirchhoff, a complete and simple description of experience (not of reality). In light of this philosophy Mach was able to criticize certain inscrutable ideas, but at the same time he fostered an antitheoretical attitude which continues to this day under the name of operationalism.

Mach's mistakes in his criticism of Newtonian mechanics - his most distinguished contribution to foundations research - can be corrected with the assistance of a bit of logic, a touch of semantics, and a dose of realism. A critical study of Mach's work in the foundations of mechanics should be helpful, if only to avoid repeating his mistakes, which were those of a philosophy distrustful of ideas. Ignore all philosophy and you will be the slave of one bad philosophy.[36]

---

34 See P. Suppes, *Introduction to Logic,* D. Van Nostrand: Princeton, 1957.

35 For Berkeley as a precursor of Mach, and for a criticism of both, see K.R. Popper, *British Journal for the Philosophy of Science,* **4** (1953), 21; and *Conjectures and Refutations: The Growth of Scientific Knowledge,* Basic Books Inc.: New York, 1963, chapter three.

36 For a comprehensive exposition of the philosophy of science, see the author's *Scientific Research,* two volumes, Springer-Verlag: Berlin, 1967.

CHAPTER 14

MORE ON MACH AND EINSTEIN

by Gerald Holton*
(Harvard University)

I

The appearance during the past several years of a number of books and articles devoted to Ernst Mach is perhaps an indication of the renewed interest in the topics and historical figures associated with various forms of empiricism. This is to be welcomed, for this area of studies in intellectual and cultural history had become relatively undervalued.

The two books with which this essay is concerned: Gereon Wolters, *Mach I, Mach II, Einstein und die Relativitätstheorie: Eine Fälschung und ihre Folgen,* (Walter de Gruyter: Berlin & New York, 1987) and John Blackmore & Klaus Hentschel (eds.), *Ernst Mach als Aussenseiter: Machs Briefwechsel über Philosophie und Relativitätstheorie mit Persönlichkeiten seiner Zeit,* (Wilhelm Braumüller: Wien, 1985) are very different, indeed diametrically opposite. One is in the new fashion of aggressive revisionism, where some of the most crucial documentation is absent; the other is chiefly in the category of the old-fashioned presentation of actual documents, with a minimum of editorializing. Wolters's 474 page volume, originally a *Habilitationsschrift* at the University of Konstanz, deals in turn with his estimate of the influence which Mach's approach to science appears to have had on Einstein's early work, Wolters's claim that Mach's forcefully stated rejection of relativity theory, published posthumously, was a 'forgery'; his lengthy and detailed attacks on much that has been previously written by almost all scholars dealing with the relation between Mach and Einstein, with a view to making more credible his claim of forgery; and some plausible and useful arguments for revising certain details, such as the likely date of Einstein's single visit to Mach, or that of one of the letters between them. In addition, there are also numerous '*Excursi* ', '*Beilagen* ', and other matters, mostly not bearing directly on the main topics. As we shall see, not until the final pages of the book does it become evident what may have chiefly motivated the author of this idiosyncratic and deliberately controversial work.

---

* This paper is a slightly revised version of one which appeared in *Methodology and Science,* **22** (1989), 67-81

J. Blackmore (ed.), Ernst Mach – A Deeper Look, 263–276.

Wolters's book starts on familiar ground. Building on his article "Topik der Forschung",[1] in which Wolters had adopted the concept of Themata for the analysis of scientific thought and its development. He now devotes the first chapter to the demonstration that "the influence of Mach on Einstein's development of the relativity theory may be described as a thematic influence of the first order." (page 14). The rest of the book, and its main novelty, is the argument that Mach, toward the end of his life, far from rejecting relativity theory, was "friendly and hopeful" toward it, and that Mach had "not the slightest reason" to act otherwise. (Ibid.)

II

As for Mach's influence on Einstein's scientific development, there is little question about it, as Wolters gladly agrees. The existing documentation is ample, not only through an analysis of Einstein's early work, but also, e.g., in Einstein's surviving correspondence to Mach, starting in 1909; remarks in Albert Einstein's letters to Michele Besso (6 January 1948), Carl Seelig (8 April 1952), Moritz Schlick (14 December 1915), and Mileva Maric (10 September 1899); as well as explicit passages in Einstein's Autobiographical Notes and other writings. There was indeed agreement between Mach and Einstein on certain fundamental thematic notions - including the importance given to the search for unity among scientific phenomena as well as among the branches of the sciences themselves and in their evolutionary rather than revolutionary model of scientific advance.[2]

Even while Einstein is the first to call seriously for deep-going modifications of the foundations of both mechanics and electronics, he says that "it will only be a matter of a modification of our present theories, and not a complete abandonment of them."[3] One must also remember that Mach's work, particularly the *Mechanik*, made a profound impression on many young scientists and intellectuals in addition to Einstein. Well-known examples include William James, Jacques Loeb, and Otto Neurath. In a confession typical of many others, Neurath wrote to Mach, "It was this conception [the Mach Principle] in your book *Mechanik*, which never left me ever since my first reading of it, and which, had on my own development of ideas . . . its influence in a curiously circuitous way. It is the tendency to derive the meaning, the sense of individualities, from the whole, not the

[1] Gereon Wolters, "Topik der Forschung" in C. Burrichter, R. Inhetveen, and R. Kotter (eds.), *Technische Rationalität und rationale Heuristic,* Paderborn, 1986, pp. 123-154.

[2] E.g., A. Einstein, "Zum gegenwärtigen Stand des Strahlungsproblems", *Physikalische Zeitschrift,* **10** (1909), 185-193.

[3] Loc. cit., p. 192. Emphasis in original.

whole as a summation of individualities . . . With respect to your work, I have always had a deep feeling of gratitude. Through your thought processes in physics, I learned not only the advancement in the subject of physics itself, but even more progress in other directions."[4]

Mach in turn was not only glad to keep in touch with his young admirers and correspondents, but encouraged and even pursued sympathetic readers, often enlisting them in his fight with more established opponents. In the European tradition, he was a builder both of a system of thought and also of a network of followers. Indeed, it is the combination of both of these that help make him such a significant figure in the history of modern thought. His influence on Einstein, and his expectations from Einstein, were part of that story.

III

Having reviewed these matters, and particularly Mach's few, brief, and mostly rather cryptic direct remarks which may be interpreted as showing a positive attitude toward Einstein's relativity theory, Wolters then reserves most of his space for the main claim to novelty, the supposed forgery of Mach's late disavowal of relativity. At first glance, this seems a curious preoccupation. In a sense, it really makes little difference to history or philosophy of science whether or not the ever-sceptical, but ever surprising giant did or did not turn his back on relativity in his last years, despite the few earlier bows toward it. Except for causing disappointment, the disavowal made little difference to Einstein himself. For it was published in 1921. Mach had been dead for five years, and Einstein had long passed beyond the stage in his evolving epistemology where Mach's interest or approval was of practical interest to him, or to anyone else.

Still, the inherent sensationalism of the claimed circumstances about the supposed forgery - Wolters himself has called it a *Schmutzwerk* and a 'soap opera', with designated culprit son Ludwig (dubbed Mach II) accused of having led a secret life of duplicity, diversion of research funds, false doctorate, psychical liability, drug addiction, etc.[5] - can have a redeeming aspect. It might draw attention to a set of larger problems, that of the historic circumstances favoring the reception or rejection of new theories. This is just where a source such as the Blackmore-Hentschel volume (which contains many of the letters quoted in this essay) becomes particularly valu-

---

[4] O. Neurath to E. Mach, no date [probably 1915], in Joachim Thiele, *Wissenschaftliche Kommunikation: Die Korrespondenz Ernst Machs,* A Henn Verlag: Kastellaun, 1978, p. 100.

[5] Elaborations were provided in Gereon Wolters, "Atome und Relativität - Was meinte Mach?" in R. Haller and F. Stadler (eds.), *Ernst Mach - Werk und Wirkung,* Verlag Hölder-Pichler-Tempsky: Vienna, 1988.

able. For the letters between Mach and his correspondents allow us to trace how he first became concerned about the apparent implications of Einstein's relativity theory.[6]

In this correspondence, one sees Mach's growing concern from 1909 on with trying to understand a number of rapidly evolving conceptions; Minkowski's widely discussed lecture, *Space and Time,* of 1908 had appeared in print shortly after Minkowski's death (January 1909). As noted, Mach had been attracted to some features of relativity earlier; but after Minkowski's paper he clearly felt the need to understand better the mathematically complex developments of the growing theory. He was then at age 71, for many years totally paralysed on his right side, and suffering from many other illnesses; but he tried to keep up with his large correspondence and other involvements - not least the fight with Max Planck that had been launched in a lecture by Planck in December 1908 and quickly grew into vigorous published attacks on Mach's conceptions, questioning even Mach's ability to serve the unity of the world picture. As Mach's correspondence shows, for the next few years Planck remained the darkest figure to be wrestled with.

On 28 March 1909, Mach wrote Friedrich Adler, a friend also of Einstein, to indicate that he had requested copies of Minkowski's talk. Others of the Mach circle were also mobilized with some urgency; Mach wrote Adler again on 16 November that his long-time disciple Joseph Petzoldt "is trying to find a man who can present the Einstein-Minkowski conceptions in a simple manner even for non-mathematicians. So far, he has not been able to find anyone. Perhaps you can direct him to somebody." Mach himself evidently could not.

Here we must stop to note a very important point which has been little discussed. It is what Mach knew and confessed to be his rather elementary knowlege of mathematics. Perhaps more than at any other time this was on Mach's mind in those years. He refers to it in letters to Hugo Dingler of 25 July 1911 and again on 20 November 1912 where he writes about the "weakness of my mathematical education as a youth, which unfortunately I have found no opportunity to repair."

Not until September 1910 did Mach receive - via Petzoldt - an article on relativity by Johannes Classen,[7] which was supposed to be of some use to Mach. Reading it, however, one sees that it did not even mention Minkowski's work. Worse, it dwelled on Planck's favorable reception of Ein-

---

[6] The 128 letters obtained by Blackmore and Hentschel since 1981, are thus complementary to Joachim Thiele's 1978 volume. An additional large set of documents has been recently published as Part 2 (pp. 167-305) of the book by Haller and Stadler, Ibid., (Ref. 5).

[7] J. Classen, "Über das Relativitätsprinzip in der modernen Physik", *Zeitschrift für Physikalischen und Chemischen Unterricht,* **23** (1910), 257-267. I am grateful to Dr. Dieter Hoffmann for his help in finding a copy of this publication.

stein's ideas and even merged the two in such phrases as "the revolutionary new material in the Einstein-Planck presentation."

Moreover, Petzoldt, the trusted disciple, had sent the article to Mach without his endorsement, and he added that while he thought Einstein's "fundamental idea is quite excellent", he questioned "whether he has freed himself altogether from the absolute." For example, Petzoldt said, he did not see why c and c' should be equal (Petzoldt and others, claiming to act in Mach's spirit, objected to the idea that through the postulate of a universally constant velocity of light an 'absolute' was being smuggled back into physics).

Petzoldt was not much wiser later on in a letter he wrote to Mach dated June 1 1911. But in it he made a very revealing remark: "You wrote me in your last letter [one that has not been found] that the relativity principle appears to you to lack much from the point of view of epistemological criticism. I believe this also." At the least, we must therefore suspect that Mach was having real reservations about relativity.

In the meantime, Adler had also been trying to be helpful. He informed Mach on 28 November 1909 that Einstein himself would soon publish[8] "an extensive explication in which the mathematics was insignificant [*unwesentlich*]". Mach thanked him at once cordially for this information (3 December 1909) and on 21 February 1910 asked again for the exact reference to it. In the meantime, on 11 January 1910, A. Föppl, also an acolyte, wrote to Mach, apparently also in reply to an inquiry about the Einstein-Minkowski thesis; Föppl said cautiously he had not yet formed a judgment about it, but regarded it as an unverified hypothesis, and even doubted it was in principle verifiable.

Mach's circle was evidently skeptical and worried about relativity, and his own search for authoritative illumination was not going well.

IV

It was in this setting that the first known personal meeting occured between Mach and a knowledgable informant - young Philipp Frank. In his biography of Mach, Blackmore noted[9] that Frank, a Privatdozent at the University of Vienna, had first been recommended to Mach in a response which was dated 5 June 1910 by the physicist Gustav Jaeger, to whom Mach had turned in puzzlement on another problematic theory in physics

[8] A. Einstein, "Le principe de relativité et ses conséquences dans la physique moderne", *Archives des Sciences Physiques et Naturelles,* **29** (1910), 5-28 & 125-134. The level is indeed very elementary; and Minkowski's work is barely touched on, in three paragraphs.

[9] John T. Blackmore, *Ernst Mach - His Life, Work, and Influence,* University of California Press: Berkeley, 1972, p. 263.

(one by Paul Gerber). Frank was praised as "a man who, I think, is the best qualified of the Vienna physicists to render a judgment on the matter." Eight days later, Philipp Frank gave his opinion to Mach in a letter (dismissing Gerber's work) and then visited Mach. As Frank later reported, Mach "especially wanted to obtain more specific information about the application of four-dimensional geometry . . . He requested that I supply him with a printed or handwritten statement [of my ideas]. I did that . . . "[10]

At last Mach had found his man. Another letter from Frank to Mach[11] has shown us details of the growing relationship. Frank wrote obligingly: "I would like to mention further that I am now working on a representation of the theory of relativity which is understandable to non-mathematicians, as you requested in your letter, Herr Hofrat, and as Herr Professor Lampa [another of Mach's associates, at Prague University] has also asked for. I will especially try to represent Minkowski's thoughts on space and time in an understandable way".

Whether Mach's scruples resolved in favor of Einstein-Minkowski at that time, and if so, how fully or for how long, we do not know directly from Mach. Frank himself informed Herneck[12] he "had the impression at the time" that Mach accepted Einstein's special theory of relativity and its philosophical basis, and even that it had been Frank's own interpretation "with which Mach agreed". Indeed, if we look at Frank's published papers on relativity of around that time,[13] we watch a masterful presenter of the elements of special relativity theory - but also a conciliant; for Frank stressed there the continuities with pre-Minkowskian sensibilities. For example, avoiding the use of the tell-tale square root of negative quantities involving t something which had startled so many who had encountered Minkowski's work directly.[14]

Moreover, Frank explicitly fashioned this article for readers "who do not master modern mathematical methods", to show that Minkowski's work brings out the "empiricial facts far more clearly by the use of four-dimensional world lines". Thus Frank managed to make it appear that Minkowski's treatment based itself not only on a functional and operational inter-

---

[10] Friedrich Herneck, "Die Beziehungen zwischen Einstein und Mach dokumentarisch dargestellt", *Wissenschaftliche Zeitschrift der Friedrich-Schiller-Universität Jena. Mathematisch-naturwissenschaftliche Reihe,* **15** (1966), 7. See Blackmore (1972), ibid., p. 183, for the comment that "Ernst Mach sent Frank two letters in 1910 at least partly in the hope that Frank could clarify the ideas of Einstein and Minkowski for him."

[11] Given by Blackmore (1972), ibid., p. 264 as undated "but clearly also in 1910".

[12] Herneck (1966), op. cit., p. 7.

[13] E.g., in *Zeitschrift für Physikalische Chemie,* **74** (1910), 466-495, which Herneck (1966), ibid., says is the most relevant, and of which Frank sent a reprint to Mach.

[14] Doing so was a point of pride for Frank, as he once told me. Moreover, in this respect he was following Einstein's lead in his 1910 survey article. See reference 8.

connection of space and time, but also - fully in accord with Mach's own views - on the primacy of ordinary, 'experienced' space and time in the relativistic description of phenomena. Thus was dissipated Minkowski's threat that, as Minkowski had put it at the head of his essay, "Space by itself and time by itself are doomed to fade away into mere shadows, and only a kind of union of the two will preserve an independent reality", Frank's version also kept quiet about the barely disguised idealistic boast at the end of Minkowski's article: that his point of view would conciliate even those to whom the abandonment of old-established views is unsympathetic or painful, by the idea of a preestablished harmony between pure mathematics and physics".[15]

Perhaps as a result of Philipp Frank's presentation, Mach mentioned the names of Lorentz, Einstein, and Minkowski in replying in 1910 to Planck's attack. But as is so typical in all Mach's published remarks on relativity and on those who fashioned it, the comment is cautiously brief, vague, and by no means an endorsement. He simply names these three in passing as "physicists who by-and-by are moving closer" to the problems of matter, space, & time.[16] Similarly, in his 1909 edition of *Conservation of Energy,* he had added a reference to Minkowski's lecture of 1908, but again in the context of a reservation: "Spaces of many dimensions seem to me not to be so essential for physics. I would only uphold them if things of thought such as atoms are maintained to be indispensable, and if, also the freedom of working hypotheses is upheld".

V

Wolters, however, never has a moment's hesitation in reading into every scrap an endorsement. For example, as part of his systematic attempt to discredit any thought that Mach may, after all, not have been enchanted by relativity in the last years of his life. Wolters argues at length that Mach could easily have accepted Einstein's theory in the form presented by Minkowski, because even in Minkowski's formulation the basic operational meaning of time and space intervals, taken separately, are those of conventional mechanics. However, this overlooks elementary facts of physics and of history.

First, even though Frank had taken care not to stress it, it had to dawn eventually on any reader of Minkowski's essay that the basic invariant interval, the "timelike vector element" *ds*, which Minkowski defined by 1/c

---

[15] For a discussion of the essentially Platonic undercurrent in Minkowski's article, see Peter L. Galison's essay: "Minkowski's Space-Time:From Visual Thinking to the Absolute World", *Historical Studies in Physical Science,* **10** (1979), 85-121.

[16] E. Mach, "Die Leitgedanken meiner naturwissenschaftlichen Erkenntnislehre und ihre Aufnahme durch die Zeitgenossen", *Physikalische Zeitschrift,* **11** (1910), 605.

$\sqrt{(c^2\, dt^2 - dx^2 - dy^2 - dz^2)}$, is not by any means on the same operational level as the space intervals and the time intervals defined in the ordinary world of meter sticks and clocks. The negative sign under the square root signals of course that *ds* contains imaginary components. It was a central insight of Minkowski to introduce in his 1908 paper a new quantity in place of ordinary time *t* namely the expression $\sqrt{(-1.t.)}$ The alarming implication this had at the time was made clear even in Minkowski's very next sentence: "Thus the essence of this postulate may be clothed mathematically in a very pregnant manner in the mystic formula 3 x 10^5 kilometers = $\sqrt{(-1)}$ seconds" (not cubic root as misprinted in the original journal article).

If the ever-sceptical Mach did not sooner or later perceive in Minkowski's article that Minkowski's spacetime conception was indeed an attack on the very root of a sensations-based physics, Einstein personally saw to it, by sending him on or shortly before 25 June 1913, what he called in his letter to Mach of that date "my new work concerning relativity and gravitation, which has now at last been finished after infinite labors and painful doubts". It was the first extensive version of general relativity which Einstein had begun in 1908, pursued during his years in Prague (1911-1912), and now advanced in collaboration with the mathematician Marcel Grossmann.[17]

It is a complex, lengthy, mathematically advanced publication, based on the tensor calculus. Reading the article will quickly force one to the conclusion that Mach, who had regarded himself as an innovator along 'relativistic' lines in mechanics, would have come away at best with a feeling of increasing helplessness concerning the fast-growing theory.[18] If he understood anything in it, it would have been that the contact with 'sense experience' was most tenuous.[19]

---

17 A. Einstein and M. Grossmann, "Entwurf einer verallgemeinerten Relativitätstheorie und einer Theorie der Gravitation", *Zeitschrift für Mathematik und Physik,* **62** (1914), 225-261. The title page of the journal says: "Issued on 30 January 1914"; but a separate, 38 page publication of the article had appeared earlier, B.G. Teubner: Lepzig, 1913. Wolters agrees that this article was the one Einstein mentioned in the letter of 25 June 1913 as having been sent to Mach.

18 Einstein had given a preview of that part of the article as a talk in Vienna during the 85th Naturforscherversammlung in Vienna. See *Physikalische Zeitschrift,* **14** (1913), 1249 ff.

19 Moreover, more evidence was accumulating that Einstein was being captured by - of all people! Max Planck, no matter how Einstein protested privately to Mach, in two of his letters, about Planck's attack on Mach. It could not have helped matters that at the outset of the Grossmann article it refers to a Planck publication as if it were the paradigmatic exposition of "the customary relativity theory", p. 226; that Einstein published a glowing article about Planck in 1913 in the first volume of the new *Naturwissenschaften*; and that he was negotiating with Planck to leave Zurich and join him in Berlin.

This is of course a crucial piece of the puzzle, if only because the vehement rejection of relativity,[20] appearing over Ernst Mach's name, was dated "July 1913", only a few weeks or days after the receipt of Einstein's mailing. In the Einstein-Grossmann article, the principle of relativity was extended to coordinate systems in nonuniform motion, and hence the inclusion of non-linear transformations became necessary. As Einstein himself confessed, this step "was inevitably fatal to the physical interpretation of the coordinates . . . It could no longer be required that coordinate differences should signify direct results of measurement with ideal scales or clocks. I was much bothered by this piece of knowledge."[21] To put it in modern terms, Einstein was forced to see that "a physical significance attaches not to the differentials of the co-ordinates, but only to the Riemannian metric corresponding to them."[22] It was indeed a wrenching experience during this part of his pilgrimage to what he called rational realism. As Einstein told his old friend Cornelius Lanczos (letter of January 1938): "Coming from skeptical empiricism of somewhat the kind of Mach's, I was made, by the problem of gravitation, into a believing rationalist, that is, one who seeks the only trustworthy source of truth in mathematical simplicity".

VI

Against this historical background, the content of the "July 1913" preface in Mach's *Principles of Physical Optics* becomes quite plausible. Because Wolters' book is at heart a gloss on this well-known passage, it is appropriate to quote here the most essential portions of the preface.

> I am compelled in what may be my last opportunity, to cancel my views [*Anschauungen*] of the relativity theory. I gather from the publications which have reached me, and especially from my correspondence, that I am gradually becoming regarded as the forerunner of relativity. I am able even now to picture approximately what new expressions and interpretations many of the ideas expressed in my book on Mechanics will receive in the future from this point of view. It was to be expected that philosophers and physicists should carry on a crusade against me, for as I have repeatedly observed, I was merely an unprejudiced rambler endowed with original ideas, in varied fields of knowledge. I must as as-

20 See the preface to E. Mach, *Die Prinzipien der physikalischen Optik,* Leipzig, 1921. Available in an English translation of 1926 as E. Mach, *The Principles of Physical Optics,* Dover Publications: New York, no date, pp. vii-viii.

21 A. Einstein, "Notes on the Origin of the General Theory of Relativity", lecture of 20 June 1933 at the University of Glasgow, reprinted frequently, e.g. in Albert Einstein, *Ideas and Opinions,* Dell Pub. Co., New York, 1954, p. 281; see also Einstein's *Autobiographical Notes,* p. 67.

22 Ibid (1954), p. 282.

> suredly disclaim to be a forerunner of the relativists as I personally reject the atomistic doctrine of the present-day school or church. The reason why, and the extent to which I reject [ablehne] the present-day relativity theory, which I find to be growing more and more dogmatical together with the particular reasons which have led me to such a view considerations based on the physiology of the senses, epistemological doubts, and above all the insight resulting from my experiments - must be treated in the sequel [a sequel which was never published].[23]

To Gereon Wolters, however, ascribing this rejection of relativity to Ernst Mach is a canard which, he charges toward the end of his book, was invented in an antipositivistic 'campaign', starting in the 1960's, to portray Mach unjustly as a 'philosophische *Dummkopf*'(page 403). Mach's preface would have had to be a "pathological break from Mach's personality" (page 405), or more probably a forgery. His finger points to son Ludwig even while he admits that there is no confession and no clear motive for this alleged crime though one might speculate about the influence of a friend, the philosopher Hugo Dingler, who Wolters says, had against relativity theory a "blind hatred" (page 402). Also, no documents seem to be available for independent scholarly study of Wolters's conjectures.

To be sure, technically, Ludwig had the opportunity after his father had died in 1916, to write or rewrite some or all of the preface that bears his father's name, to incorporate in it unauthorized inventions that were contrary to his father's views; to repeat unrepentantly later (e.g., in his letter to Petzoldt, 29 June 1920), that the abjuration is his father's; and to take the secret of his misdeeds to his grave. On the other hand, Wolters confesses at the end, after all "it could be that the *Optik* preface was written by Ernst Mach, or at least . . . in accord with his opinion [*in seinem Sinne*] . . . Truly, in this world much is possible" (page 405).[24] At any rate, convinced that Ernst Mach's reputation in philosophy somehow rests greatly on whether he did or did not write these passages, Wolters pleads that the 'antipositivism' which he says has been maligning Ernst Mach, should from now on target the son instead. Thus ends the book.

---

23 From the 1926 translation in the Dover edition. I have corrected some mistranslations.

24 He might also have added as others have shown, that the wording in the preface, and other internal evidence, shows it to be in accord with Ernst Mach's idiosyncratic expression style and ideas. See John Blackmore, "Mach Competes with Planck for Einstein's Favor", *Historia Scientiarum,* **35** (1988), 45-89 and John Blackmore, "Mach über Atome und Relativität - Neuste Forschungsergebnisse", in R. Haller & F. Stadler, op. cit., (Ref 5).

VII

Wolters would object at this point that two significant documents have not been mentioned: Einstein's undated letter to Mach (probably written at the end of 1913 or the beginning of 1914), and Mach's letter to Petzoldt that was dated April 1914. A few words in these are the centerpieces of Wolters 'evidences' that Mach could not, in July 1913, have been hostile to Einstein's work. (Passages in one letter, from Mach to Petzoldt, May 1, 1914, could be stretched to hint in that direction, but at least 3 other documents from the book by Blackmore and Hentschel say as much in opposition).* Let us therefore finally look at these two centerpieces in turn, for they illustrate the selective use of evidence in Wolters's book.

The first sentence in Einstein's undated reply (from late 1913 or early 1914) says merely, "I am very glad about your friendly interest which you bring toward the new theory". Mach's own letter, to which this sentence refers, has not been found. But let us assume that Mach wrote some explicitly pleasant phrases to this fast rising star, who had once called himself his "admiring student", and even assume further that such phrases referred to the Einstein-Grossmann paper. It would not have been the first or the last time that a scholar made polite remarks about a mailing that he really neither understood in detail nor approved of in general. In fact, one finds examples of this sort right in Mach's letters to others, and particularly to younger scientists of obvious quality who presented themselves as admirers. For example, Mach was generally supportive of William James in direct correspondence to him, but did not mind criticizing his work in letters with others.[25] Therefore, it is dangerous to let so much of a complex case rest on what can be read into Einstein's two words, "friendly interest".

This becomes even more compelling as soon as one continues the quotation, for in his letter Einstein added: "The mathematical difficulties which one comes up against in following these thoughts were unfortunately also for me very great". Mach appears to have made a comment on the grave mathematical complexities of the work, which in fact had necessitated Einstein's seeking the help of the mathematician Grossmann.

Wolters's other key piece of 'evidence' is in Mach's reply to Petzoldt of April 1914. At least there we are dealing with Mach's own words. But there, too, it is absurd to read into them an acceptance of Einstein's relati-

---

* Blackmore & Hentschel, op. cit., pp. 91, 114, and 156. - Ed.

25 E.g. Mach to Anton Thomsen, 4 September 1909. Mach writes that he likes James's *Principles of Psychology* well, but also, ever the skeptic, criticizes James's proclivity to "*Schwärmerei und Spiritismus* ". Similarly, in Mach's letter to Thomsen dated 21 January 1911, he refers to James's "dangerous argument" [about defending the truth of religion on pragmatic grounds, as if to the extent it seemed to 'work' it were true. - Ed.].

vity theory. Mach writes that he is pleased with Petzoldt's article,[26] sent to him by its author, first because "you have richly honored my modest merits with regard to this theme". (In fact, Petzoldt's article flatters Mach on page after page, in the sycophantic way Mach had come to expect from his acolytes.) Then Mach adds the vague phrase that Wolters takes to be a decisive indicator of acceptance: Mach says he likes the article "auch sonst" (also otherwise, or in general).[27] So the obvious thing to do, one would have thought, is to find out what lies behind the "*auch sonst*", i.e. to read Petzoldt's article in order to see what kind of relativity theory Petzoldt is writing about in 1914. Indeed, anyone beguiled with the notion of accepting Mach's opaque two words as an endorsement of relativity *Anno Domini* 1914 has to condemn himself actually to read that 56-page piece, in the official organ of the recently founded Society for Positivistic Philosophy, by its founding chief and member of its editorial board.

The article deserves a look to indicate the center of interest and level of understanding current in Mach's circle. In fact, Petzoldt's piece turns out to be not a report on relativity in the form it had then reached, but a quasi-philosophical attempt to show that an elementary version of relativity theory fits with the "relativistic positivism founded by E. Mach and R. Avenarius". The main problem discussed by Petzoldt is the relation of the physical and the psychological basis of experience, chiefly from Mach's point of view. The rejection of absolute space by both Mach and Einstein is highlighted, and the work of the latter is praised by him for showing that the Lorentz-Fitzgerald contraction is not a phenomenon requiring a physical 'explanation', but follows merely from a proper description.

The important point made here is "that physics in the end can deliver no more than the description of events; that complete description *is complete explanation*; and that at bottom physics has never delivered anything else." (page 16. Emphasis in original). Here Minkowski's essay is briefly mentioned as pointing to the same conclusions. Minkowski is occasionally referred to also later, e.g. by giving the first sentence of his paper, followed however by the observation that this 'marvelous' theory depends, as Einstein's does, entirely on the presupposition of the universal constancy of the velocity of light, and, like previous theories, may well fail with the eventual failure of that presupposition. What will then remain standing is Mach's anti-absolutist theory, "for it rests on sense-physiological foundations."

---

[26] Joseph Petzoldt, "Die Relativitätstheorie der Physik", *Zeitschrift fur positivistische Philosophie,* **2** (1914), 1-56.

[27] Petzoldt is another case where Mach rarely expressed an objection directly to his face, but nevertheless did so in letters to others, e.g., to Friedrich Adler, on 20 August 1909 and 23 January 1910.

Moreover, Petzoldt warns us that the relativity of Einstein-Minkowski has to be placed on a more '*anschauliche*' ground. And while Minkowski's four-dimensional world is acceptable as long as it is considered as "merely a conceptual system", Petzoldt cannot agree with Minkowski's remark that the four-dimensional world is given through the phenomena in it. Petzoldt points out other 'deficiencies' also. A few of them will suffice to indicate the general tone. Einstein's own formulation of the relativity principle is charged to be on the level of a "naive assumption, not of an assumption after critical examination". Einstein's conception of 'natural law', too, has elements of arbitrariness; the difference between spacetime determinations and natural laws cannot be tolerated. Therefore, "we see how important epistemological studies can be for a theoretical physicist and mathematician". Properly understood, relativity can teach us what is meant by a 'real' versus 'apparent' change, as in the length of a rapidly moving body, or even in observed shapes of objects as our vantage point is changed. Thus, observed shapes are not appearances based on hidden '*Dinge an sich*', but 'only psychological functions'. At its deepest, understanding itself consists "in the relativity of all actuality and its substancelessness" (page 46).

Out of the blue, there appears at this point - again only in the most elementary way - a discussion of the recently popularized clock paradox of relativity theory. Now Petzoldt reveals his ignorance in plainest terms. For he holds that while the traveler, on returning to his home-based fellow creatures, will seem to the latter to have aged less, the traveler in turn will have a *symmetrical* experience, and will perceive the home-based fellows also to have remained more youthful than he is himself. To think otherwise, says Petzoldt with solemn authority, is "an error and a regression back into absolutistic thinking" (page 50).

There is one additional noteworthy point in Petzoldt's paper, inserted as if without clear context, just before his doomed discussion of the clock paradox. It is a footnote reference to two of Einstein's papers extending relativity theory to include the "gravitation problem", one of which is the paper by Einstein and Grossmann. Petzoldt's text does not make clear why they are cited, for instead of attempting to discuss or describe the papers in any way, he says merely, in a way perhaps meant to express his disapproval: "Ordinarily the physicist in his *Anschauungen* does not hew closely to his formulas, but weaves around them a multiplicity of considerations which have no hold in the formulas themselves". And Petzoldt adds that "it is the function of epistemological critique, as carried out by Mach, to discover these addenda and their insupportability".

Conclusion

One can only conclude that if Mach - quite apart from being flattered - had indeed found pleasure and agreement with Petzoldt's 1914, as his letter indicates, it would necessarily have to be pleasure in and agreement with Petzoldt's *objections*. More importantly, Mach's agreement with Petzoldt's version of relativity would be a terrible indictment of Mach's state of ignorance about the current state of relativity theory. In sum, on this cumulative evidence - alas - it no longer matters *who* wrote Mach's disavowal dated July 1913. Whether he intended to accept or reject it, Ernst Mach would not have known at that point what relativity was about.

CHAPTER 15

THREE BATCHES OF REASONS FOR MACH'S REJECTION OF EINSTEIN'S THEORY OF RELATIVITY

by Ryoichi Itagaki*
(Tokai University)

1. ERNST MACH'S INTEREST IN EINSTEIN'S RELATIVITY THEORY:

Ernst Mach seems to have begun to take an interest in Albert Einstein's theory of relativity from about 1908 or so. Shortly after Herrmann Minkowski (1864-1909) had presented his epoch-making four-dimensional formulation of Einstein's theory in a colloquium in Göttingen and later in Cologne on September 21, 1908, Mach wanted to learn about it. In a letter dated March 28, 1909,[1] he mentioned to his friend and philosophical follower, Friedrich Adler (1881-1953), who was also a physicist and friend of Einstein, that he had ordered a copy of the *Rieckeschen Zeitschrift* which contained Minkowski's lecture. It was Mach during 1909 who began direct communication with Einstein. The seventy-one year-old Mach sent a copy of the new edition of *Erhaltung der Arbeit* (1872) to the thirty year-old genius. (Einstein's special theory of relativity seems to have been ordinarily called "the principle of relativity". Einstein himself apparently began to call it "the theory of relativity" from around 1909.) The book included a brief reference if not to Einstein at least to Minkowski in a supplementary note:

> Space and time are not comprehended as independent entities, but as forms of the dependence of the phenomena on one another. Therefore, I steer toward the *principle of relativity*, which is also firmly upheld in [my books] *Die Mechanik, Wärmelehre,* and [the chapter] "Space and Time - Physically Considered" in *Erkenntnis und Irrtum*, 1904, cf. H. Minkowski, *Raum und Zeit*, 1909.[2]

* This article has been written specifically for this anthology.

[1] J.T. Blackmore & K. Hentschel (eds.), *Ernst Mach als Aussenseiter: Mach's Briefwechsel über Philosophie und Relativitatstheorie mit Personlichkeiten seiner Zeit,* Wilhelm Braumüller: Vienna, 1985, pp. 55-56.

[2] E. Mach, *Die Geschichte und die Wurzel des Satzes von der Erhaltung der Arbeit,* zweite Auflage, Leipzig, 1909, p. 60.

J. Blackmore (ed.), Ernst Mach – A Deeper Look, 277–295.

Mach, especially in *Beiträge zur Analyse der Empfindungen* (1886), regarded both time and space as 'sensations', which were the most fundamental elements of the phenomenal world. Other elements, for example, included sound, smell, heat, and pressure. For him, material objects were just names designating a complex of these sensations. He considered learning as primarily an effort to arrange sensations. From his point of view, 'time' should not be based on Newton's 'absolute' time, but was only an abstraction drawn from the change of things. Time was usually determined by the positional interrelation of the Cosmos and the Earth in it, so that time could not be measured in itself at all. The preface of the Minkowski paper which Mach cited must have made a favorable impression on him, that is, when considered apart from the highly mathematical and abstract content of *Raum und Zeit*: "Henceforth space by itself, and time by itself, are doomed to fade away into mere shadows, and only a kind of union of the two will preserve an independent identity."[3]

Einstein must have been greatly pleased with the gift of a book from Mach, who was a well-known retired great scholar and an old man like his father. Einstein seems to have first studied Mach's works with his friend Michele Angelo Besso (1875-1955) while a second year student at the Federal Institute of Technology in Zürich together with other members of the "Olympia Academy". Einstein answered Mach in a postcard dated August 9, 1909 thanking him for the book, and sent another postcard on August 17. But perhaps without thinking, Einstein in the first card had naively asked Mach to consult his paper on Brownian motion even though he must have known how opposed Mach was to the reality of atoms.

This friendly exchange seems to have further aroused Mach's curiosity about Einstein's special theory of relativity such that he wrote letters asking some researchers about its character and possible soundness.[4] August Föppl (1854-1924) at the Institute of Technology in Munich answered on January 11, 1910: "I still can't seem to come to a definite conclusion on the work of Einstein and Minkowski. I can clearly see that provisionally the question is merely one of a hypothesis, whose confirmation by experience, that is, of

---

[3] H. Minkowski in "Raum und Zeit" in Albert Einstein [und andere Autoren] , *Das Relativitätsprinzip,* erste Auflage 1923, fünfte Auflage, Stuttgart, 1982, S. 54: "Von Stund an sollen Raum für sich und Zeit für sich völlig zu Schatten herabsinken und nur noch eine Art Union der beiden soll Selbstständigkeit bewahren." [For an English version see "Space and Time" in *The Principle of Relativity* by H.A. Lorentz, A. Einstein, H. Minkowski, and H. Weyl with notes by A. Sommerfeld, translated by W. Perrett and G.B. Jeffery (original English edition, London, 1923; Dover reprint N.Y., 1948 and later), pp. 73-91.

[4] Mach probably agreed with the definition of simultaneity by Einstein. Wataru Hiromatsu, a contemporary Japanese philosopher, has indicated that the operational definition was one of the few points which Mach and Einstein had in common. See Wataru Hiromatsu, "Mach's Philosophy and the Theory of Relativity" (in Japanese) in his *The Philosophy of the Theory of Relativity,* Tokyo, first edition 1981, second edition 1986, p. 169.

the movement of perceivable bodies, is completely lacking at least for the moment. The thought which I cannot suppress is directed more against their ability to carry the thing through at all..."[5] Philipp Frank (1884-1966), who was a young lecturer at the University of Vienna at the time, met Mach at his house and with his encouragement wrote a "readable" article concerning the four-dimensional representation of relativity theory.[6] As Gerald Holton has suggested, this representation was, far from being an exact expression of Minkowski's formulation, a revised edition from a rather Machian viewpoint. In particular, it excluded precisely those mathematical innovations by Minkowski most likely to disturb readers.[7] Frank's rendition should be kept in mind, since Mach's initial understanding or misunderstanding of the ideas of Einstein and Minkowski, like that of most people, seems to have been significantly filtered through secondary sources, and not all of them squarely faced the difficulties or novelties presented by the new theories.

Two articles written by Mach in 1910 as replies to his critics (Max Planck and Carl Stumpf) briefly mentioned the name of Einstein together with those of Minkowski and Lorentz:

> If the kinetic version of the physical world picture, which in any case I consider hypothetical without intending thereby to degrade it, were to 'explain' all physical appearances, I would still not consider the manifoldness of the world to be exhausted, since for me matter, time, and space would still be problems, a view which the physicists (Lorentz, Einstein, Minkowski) are gradually approaching.[8]

> Similarly, one will have to distinguish between metrical and physical (time containing) space, as has already been suggested in my *Conservation of*

[5] J.T. Blackmore, *Ernst Mach - His Work, Life, and Influence,* Berkeley, 1972, p. 263. For the original German see J.T. Blackmore & K. Hentschel (eds.), *Ernst Mach als Aussenseiter,* Wilhelm Braumüller: Wien, 1985, p. 66-67.

[6] Philipp Frank, "Das Relativitätsprinzip und die Darstellung der physikalischen Erscheinungen im vierdimensionalen Raum", *Annalen der Naturphilosophie,* **10** (1911), 161.

[7] Gerald Holton, "More on Mach and Einstein: Investigation of an Important Phase in the Rise of Logical Positivism/Empiricism", *Methodology and Science,* **22.2** (1989), 73. [See a slightly revised version of this article in this anthology (chapter 14). - Ed.]

[8] The quotation is from Blackmore (1972), op. cit., p. 253. For the original German see: E. Mach, "Leitgedanken meiner naturwissenschaftlichen Erkenntnislehre und ihre Aufnahme durch die Zeitgenossen", *Physikalische Zeitschrift,* **11** (1910), 605. Tetu Hirosige (1928-1975) insisted that Mach's refutation of the mechanistic world view was crucial for Einstein's new perspective and prepared the way for relativity theory, but his perspective must be reconsidered since Mach did not accept the "electromagnetic world view" which supposedly replaced the mechanistic approach, and instead, he laid stress on a biological world view which emphasized evolution. See Tetu Hirosige, "The Ether Problem, The Mechanical World View, and the Origins of the Theory of Relativity", *Historical Studies in the Physical Sciences,* **7** (1976), 81.

> *Energy* and in *Knowledge and Error*....Essential progress in this direction has been instituted and carried out by A. Einstein and H. Minkowski.[9]

Einstein's reaction to these articles is not known (nor is it even clear whether he read them). Interestingly, Einstein did pay a visit to Mach's Wohnung and met him, but the exact date of their only meeting is not yet certain.[10] Their discussion appears to have been short and probably concerned atomism, not the theory of relativity.[11] In spite of Einstein's attempts at persuasion, Mach appears not to have finally accepted atomism. Indeed, shortly before in April 1908, when a Japanese physicist Ayao Kuwaki (1878-1945) visited Mach, Mach said to him: "I do not pretend to be a philosopher, but if compelled to state my major field of study, I suppose I would say it is scientific epistemology. From an epistemological viewpoint I stand against atomism, but for energetics."[12]

Mach appears to have felt a measure of isolation from the academic world, judging from his following monologue in his notebook from about 1910: "Lost labor of love to the physicists. The old ones not yet, the young ones no more. Suddenly self-evident."[13] About at the same time Mach came to accept the viewpoint of the young mathematician and philosopher in Munich, Hugo Dingler (1881-1954), an anti-relativist, referring to Dingler's works in his *Mechanik* (7th German edition 1912). It was Dingler who would later (1924) insist that Mach never favored the theory of relativity and this rejection of it was a matter of course, because Mach saw from the outset that the theory was dogmatic. Dingler's position was as follows: "Mach's seemingly favorable attitude, shown on such actions as the presentation of his writings to Einstein and the maintenance of correspondence with Einstein, originated in his constant enthusiasm for new theories being developed by younger scientists [including Dingler himself]. He was, however, never actually in favor of the theory of relativity. One indication of

---

[9] Blackmore (1972), Ibid. For the original German see: E. Mach, "Sinnliche Elemente und naturwissenschaftliche Begriffe", *Pfluger's Archiv,* **136** (1910), 264.

[10] Gereon Wolters claims "shortly after September 25, 1910", but his calculation lacks strong evidence. See G. Wolters, *Mach I, Mach II, Einstein und die Relativitätstheorie - Eine Fälschung und ihre Folgen,* Walter de Gruyter: Berlin, 1987, p. 134.

[11] P. Frank, Einstein, *His Life and Times,* translated by G. Rosen (from German), edited and reviewed by S. Kusaka, New York, 1947, pp. 131-132.

[12] A. Kuwaki, *Einstein* (in Japanese), 1934, enlarged and supplemented by Tsutomu Kuwaki and Sigeko Nisio, Tokyo, 1979, p. 190. These passages were translated from the Japanese by the author.

[13] Manfred Sommer (ed.), "Auszüge aus den Notizbüchern 1871-1910" in *Ernst Mach - Werk und Wirkung*, edited by R. Haller and F. Stadler, Verlag Hölder-Pichler-Tempsky: Wien, 1988, p. 209. "Verlorene Liebesmühe den Physikern gegenüber. Alte noch nicht, Junge nicht mehr. Plötzlich selbstverständlich."

this is that Mach did not mention Einstein in his *Mechanik* (seventh edition)."[14] Dingler's standpoint is in accord with that of this study.

On June 25, 1913, Einstein wrote to Mach to tell him that a German observation party would be organized and sent to Russia to observe the bending of light rays of the stars behind the Sun, and that the success of this observation would confirm 'Mach's principle'.[15] Shortly after this letter reached Mach, Mach wrote the preface to *Die Prinzipien der physikalischen Optik* in which for the first time he openly* rejected the theory of relativity. Dingler's two visits to Mach in July 1913 are known from Dingler's diary to have included discussions about relativity theory and may well have had a great influence on Mach's rejection. According to Joachim Thiele, Mach probably consulted Dingler about this letter from Einstein.[16] In the preface to the *Optik* Mach not only discredited the theory of relativity, but also disclaimed the schema that made him a forerunner of this theory. The preface was dated July 1913, but its actual publication was delayed until 1921, five years after Mach's death (1916), and this delay has given rise to various speculations by several researchers.

In this study, Mach's actual rejection of Einstein's theory is presupposed. But one must acknowledge that at least one researcher, namely, Gereon Wolters at Konstanz University in Germany, has energetically insisted in several writings that Ernst Mach never rejected the theory of relativity.[17] His main allegation is that Ernst Mach's son Ludwig (1868-1951), who had helped his father for over thirty years, co-authored a number of articles, and edited the *Optik* for publication, 'forged' the anti-Einstein preface and in the process misrepresented his father's views. But while Wolters has done a lot of research, there is a clear gap between his documentation and both the 'forgery' thesis and the 'misrepresentation' argument. In short, both claims seem improbable. In point of fact, since no one had more direct acquaintance with his father's opinion on relativity than Ludwig and since he had nothing to gain by misrepresenting his father's posi-

---

[14] Hugo Dingler, *Die Grundgedanken der Machschen Philosophie*, Leipzig, 1924, pp. 76-79.

[15] Friedrich Herneck, *Einstein und sein Weltbild*, Berlin, 1976, pp. 143-144.

* Mach had written an autobiographical manuscript for Wilhelm Ostwald in June 1913 in which he referred to the first part of the *Optik* in the same manner as other published books, leading one to suppose that by the time the autobiographical manuscript would be published that the first part of the *Optik* would also be in print. See manuscript D in Chapter One of this book as well as the Mach-Ostwald correspondence in Chapter Four. - Ed.

[16] Joachim Thiele, *Wissenschaftliche Kommunikation - Die Korrespondenz Ernst Machs*, Kastellaun, 1978, p. 232.

[17] Gereon Wolters, *Mach I, Mach II, Einstein und die Relativitätstheorie - Eine Fälschung und ihre Folgen*, Walter de Gruyter: Berlin & New York, 1987 & "Atome und Relativität - Was meinte Mach?" in R. Haller and F. Stadler (eds.), *Ernst Mach - Werk und Wirkung*, Verlag Hölder-Pichler-Tempsky: Wien, 1988, p. 497.

tion on relativity, even if he had written part or all of the preface himself which so far does not seem likely, it almost certainly was an accurate representation of his father's beliefs. Let us now turn to the contents of the preface and attempt to give reasons why Ernst Mach rejected the theory of relativity.

According to a note added to the preface by Ludwig Mach the editor, although the printing of the *Optik* had begun in the summer of 1916 it was discontinued, ostensibly because of the need for further experiments, but other reasons may have been in connection with his being called into military service to work presumably in a medical capacity in an aircraft factory in or near Berlin, or perhaps simply due to unfortunate wartime conditions which made printing and publication rather difficult.[18] After having been proofread by Friedrich Adler in 1918, the manuscript and the existing printed pages were once again placed in the publisher's hands in 1921 when the book actually came out.

The preface is rather short and only contains nine paragraphs of which five concern relativity theory. The relevant parts of the preface to the *Optik* are as follows:

> ...the radiation, the process of the emission of light, Maxwell's theory, together with relativity, will be briefly dealt with in a subsequent part. The problem and doubts arising from the study of these chapters were the subject of lengthy researches undertaken conjointly with my son [Ludwig], my colleague for many years. And it would be desirable if the 2nd part which is connected with these researches were published soon, but I am compelled to touch on my viewpoint about the relativity theory on this probably last opportunity.[19]

Part 2 of the *Optik* [whose actual or future existence is indicated in Ernst Mach's autobiographical remarks written in June 1913] was never published. The paragraph quoted above mentions Mach's purpose to try to comment on the theory of relativity on the basis of optical experiments undertaken with Ludwig and his interferometer. Mach was probably interested in the problems of how light was emitted and transmitted and how Maxwell's equations could be reformulated from the viewpoint of relativity.

> I learn from the publications which have reached me, and above all from my correspondence that the role of forerunner of relativity theory is gradually assigned to me. Now I am able today to make an approximate picture of

[18] E. Mach, *Die Prinzipien der physikalischen Optik, historisch und erkenntnispsychologisch entwickelt*, Leipzig, 1921, p. vii.

[19] E. Mach, *The Principles of Physical Optics, a Historical and Philosophical Treatment*, translated by John S. Anderson and A.F.A. Young, Dover: 1926, p. vii. *Optik*, p. vii.

> what reinterpretations and explanations many of the ideas written down in my *Mechanics* will receive in the future from this point of view.[20]

The approach which regarded Mach as a forerunner of the theory of relativity can chiefly be found in letters by Albert Einstein and Joseph Petzoldt to Ernst Mach, dated respectively June 15, 1913 and June 25, 1913, that is, shortly before the preface was written down (July 1913).* Joseph Petzoldt (1862-1929) was a positivist philosopher on intimate terms with Mach, but it appears that without mentioning his name that Mach intended to include him as someone who had practiced reinterpretation or altered interpretation (*Umdeutungen*) of Mach's views on relativity. Mach had criticized Petzoldt's views before at least in correspondence with other people. Mach used the German word in the previous sentence to mean an arbitrary fiction. For example, Mach disapproved of atomism concerning the explanation of Newton's rings because of its various *Umdeutungen*, which means ad hoc remarks.[21]

> When philosophers and physicists preached a crusade against me, I found it natural and fully understandable. For, as I have repeatedly demonstrated, I was anyhow merely an unprejudiced rambler with his own ideas in various fields. I must, however, now as assuredly disclaim to be placed at the head of the relativists as I have disclaimed the atomistic doctrine of the present school or church for my part.[22]

Mach, who had experienced many controversies with other scholars during his lifetime, saw himself as an impartial rambler not belonging to any school of learning. He, therefore, felt no qualms in repeating his longheld opposition to atomism and in denying that he was a forerunner of the the-

---

20 Mach, *Optics*, pp. vii-viii & *Optik*, p. vii. In German: "Den mir zugangenen Publikationen und vor allem meiner Korrespondenz entnehme ich, das mir langsam die Rolle des Wegbereiters der Relativitätslehre zugedacht wird. Nun kann ich mir heute ein ungefahres Bild davon machen, welche Umdeutungen und Auslegungen manche der in meiner *Mechanik* niedergelegten Gedanken von dieser Seite in Zukunft erfahren werden."

* See Blackmore & Hentschel (1985), op. cit.., pp. 120-122.

21 Mach (1910), "Leitgedanken", p. 603.

22 Mach, *Optics*, p. viii; *Optik*, p. viii. In German: "Wenn Philosophen und Physiker den Kreuzzug gegen mich predigten, so musste ich dies natürlich finden, and war damit ganz einverstanden, denn ich war, wie ich dies wiederholt dargetan habe, auf den verschiedenen Gebieten doch nur ein unbefangener Spaziergänger mit eigenen Gedanken, muss es aber nun mit derselben Entschiedenheit ablehnen, den Relativisten verangestellt zu werden, mit welcher ich die atomistische Glaubenslehre der heutigen Schule oder Kirche für meine Person abgelehnt habe." [For similar remarks by Mach against "The Church of Physics" see his letters to The Austrian Academie of Sciences on April 28, 1913 and to Wilhelm Ostwald on July 23, 1913; See Chapter Four, section A, and Chapter Six, section D, in this book. - Ed.]

ory of relativity. This paragraph also indicates that the story that on looking through a spinthariscope that he had changed his mind and accepted the reality of atoms seems to have been false. He did not accept atomism.[23] In the preface above, Mach cited his 1910 article against Planck in which he maintained that there was no reason why it was necessary to accept the atom as a physical reality. Even though the results of experiments may appear to show the existence of the atom, the relationship between the atom and sensational reality was too vague and allowed for too many different interpretations.[24] Hence, we can easily imagine that Mach believed that any theory based on the dogmatic assertion that atoms existed had to be rejected. But he frankly confessed in his undated notebook from 1913 or early 1914 a measure of reluctant doubt: "Atoms not *occult?*"[25]

> But why and to what extent I reject for myself the present relativity theory which I feel is becoming more and more dogmatic, and which sensephysiological considerations, epistemological doubts, and above all experimentally-gained insights have led me to it [rejection of relativity theory], respectively, all that should be demonstrated in the next part of this work.[26]

Mach concluded his comments on relativity in the preface as follows:

> The ever-increasing amount of thought and work devoted to the study of relativity will indeed not be lost; it has already been fruitful and of permanent value for mathematics. Will it, however, be able to maintain its position in the physical conception of the world in the distant future, which will have to adjust itself to an enlarged manifold and furthermore to many new insights? Can it be maintained that it will become more than a witty remark in the history of this science? [27]

Mach points out three types of reasons for the rejection of the theory of relativity"*sense-physiological considerations, epistemological doubts, and experimentally-gained insights*. Since these types are not explained in detail

---

23 Blackmore (1972), op. cit., pp. 319-320.

24 Mach (1910), "Leitgedanken...", op. cit., p. 603.

25 Blackmore & Hentschel (1985), op. cit., p. 204.

26 Mach, *Optics,* p. viii and *Optik,* pp. viii-ix. In German: "Warum aber und inwiefern ich die heutige mich immer dogmatischer anmutende Relativitätslehre für mich ablehne, welche sinnesphysiologischen Erwägungen, erkenntnistheoretische Bedenken und vor allem experimentell gewonnene Einsichten mich hierzu im einzelnen veranlassten, das soll in der Fortsetzung dieses Werkes dargetan werden."

27 Mach, *Optics,* p. viii; *Optik,* p. ix. In German: "Gewiss wird die auf das Studium der Relativität verwendete immer mehr anschwellende Gedankenarbeit nicht verloren gehen, sie ist heute schon für die Mathematik fruchtbringend und von bleibendem Wert, wird sie sich aber in dem physikalischen Weltbild einer ferneren Zeit, das sich in eine durch mannigfache weitere neue Einsichten erweitereten Welt einzupassen hat, behaupten können, wird sie in der Geschichte dieser Wissenschaft mehr wie ein geistreiches Apercu bedeuten?"

or broken down into particular objections, in section 2 of this study we will examine them and clarify the true reasons for Mach's rejection.

2. THREE BATCHES OF REASONS WHY MACH REJECTED EINSTEIN'S THEORY OF RELATIVITY.

2.1 SENSE PHYSIOLOGICAL REASONS:

Physical Space Must Have Only Three Dimensions

Mach often mentioned that while one can think of a space having more than three dimensions which might indeed have value in mathematics and was possible in physics that it seemed improbable in physics, largely because it would misrepresent the empirical character of physical space and would be less economical than the use of three dimensions. While he favored the relativity of space and time and dependence of phenomena on each other from an epistemological perspective and would surely have had no objection about linking three spatial dimensions with one temporal dimension, provided that empirical and economic factors were fully preserved, to the extent that Minkowski's four-dimensional mathematical approach and Einstein's special and general theories interfered with these two criteria, he had no logical choice but to reject them.

According to Mach, the conception of four-dimensional space is only a product of thought, and has no basis in reality. Mach noted, ". . . [in physics] there is nothing in 1, 2, or 4 dimensions, but only in 3 dimensions."[28] And "The spaces of several dimensions don't seem to me to be so essential to physics. I would adhere to them, only if one takes things of thought such as atoms for indispensable."[29] Although Mach admitted the non-Euclidity of space from the physiological point of view, he insisted on its three-dimensionality. "The space of sight and touch is a *three-fold* manifold; it possesses three dimensions: and every point in it can be determined by three distinct and independent characteristics."[30] Dingler accepted only three-dimensional and Euclidean space on the ground that this concept

---

28 E. Mach, "Eine Betrachtung über Zeit und Raum" (1910), in his *Populär-wissenschaftliche Vorlesungen,* 5. Auflage, Leipzig, 1923, p. 505.

29 E. Mach, *Erhaltung der Arbeit,* zweite Auflage, 1909, p. 59.

30 E. Mach, *The Science of Mechanics,* translated by Thomas J. McCormack, sixth edition, La Salle, 1960, p. 590; *Die Mechanik in ihrer Entwicklung historisch-kritisch dargestellt,* achte Auflage, Leipzig, 1921, p. 487. In German: "Der Raum des Gesichts und Getastes ist eine *dreifache* Mannigfaltigkeit, er hat drei Dimensionen; jeder Ort in demselben kann durch drei voneinander unabhängige Merkmale bestimmt werden."

of space is supported by the movement of things. He pointed out that the definition of synchronism in the relativity theory leaves the problem of the velocity of light unsolved, destroys the mutual independence of time[31] and space, and distorts space in a non-Euclidean way.

The Distance Betwen Two Points in Four-Dimensional Space-Time Cannot Be Accepted as an Element of Sensation

From his scientific point of view, it was problematic for Mach to accept a four-dimensional space-time continuum (G. Holton stressed this reason).[32] Accepting a real four-dimensional continuum would have meant for Mach abolition of space and time as elements of sensation as well as their abolition as relations between other sensory or empirical phenomena. Mach's world was a continuum of interelated sensations, not a mathematical abstraction with select empirical support. Nature and abstract theory were not identical. Gerald Holton has stressed that if Ernst Mach had accepted a space-time continuum, his standpoint of sense-monism would have been undermined.[33] Holton's apparent insistence that acceptance of any space-time manifold on a physical level would have directly eroded Mach's philosophical basis seems to us overly harsh, since we can imagine Mach choosing to define the space-time continuum as a mere conceptual generalization which left the elements of space and time unchanged. But it has been Holton's very insight which has allowed us to see the need for this qualification.

Four-Dimensional Theory Encourages Psychic Speculation

Ernst Mach in his much revised seventh edition of *Die Mechanik* from 1912 did not mention Einstein or relativity theory, but he did praise Hugo Dingler strongly and did devote many lines to a psychic scandal which had broken out over thirty years before, one which had its origin in speculation by Hermann Helmholtz about a four-dimensional world (See J. Blackmore for details).[34] Mach was not a theist in religious matters and opposed all "superstition". He was also distressed by turn-of-the-century psychic societies in England and America which dabbled in parapsychology and included many prominent scientists and philosophers, including Sir Oliver

[31] H. Dingler, *Physik und Hypothese*, Berlin, 1921, pp. 164 ff & 200.

[32] G. Holton (1989), op. cit., p. 75-76f.

[33] Ibid.

[34] Blackmore (1972), op. cit., p. 261.

Lodge and William James. He thought such participation damaged science. Nor was he willing to forget how four-dimensional theorizing had seemingly encouraged the development of psychic fraud in Germany. Indeed, Mach criticized the psychic pretentions of several "scientists" in *Die Mechanik*.[35] In fact, Mach's fear of psychic mysticism may well have been a factor in doubting J. J. Thomson's notion of "electromagnetic mass" and with how it was supposed to increase with velocity, a controversy which surfaced around 1906-1907.[36] Although in the end, the idea of the electromagnetic mass was replaced by the $E = mc^2$ of relativity theory, it appears that Mach rejected both conceptions, which looked equally strange and mystical to him. Indeed, in his notebook from 1907 we find the following remark: "Mass seemingly increases with its velocity. . . . We need neither the mysticism of mass nor the mysticism of electricity."[37]

Einstein's Swiss biographer, Carl Seelig, mentioned in his fine book that around the time when Einstein had become famous as the creator of the theory of relativity, that many people still regarded four-dimensional space as mysterious and magical, and therefore tended to think of Einstein as a psychic himself.[38]

2. 2 *EPISTEMOLOGICAL REASONS:*

The Principle of Relativity Gives a Privileged Position to Inertial Systems

According to Mach, all coordinate system should be equivalent in physical theory, thus it is not allowed to treat only the inertial system as something special. Petzoldt wrote to Mach on June 1, 1911 when discussing the epistemological problems of relativity theory: "You wrote me last time that many things still seem to be lacking in the principle of relativity from the epistemologico-critical side."[39] Remarkably, Einstein himself also admitted this weak point in his relativity theory in his general survey up to the gen-

35 Mach, *Mechanics,* op. cit., pp. 590-591; *Mechanik,* pp. 467-469.

36 Abraham Pais, *Subtle is the Lord. . . The Science and the Life of Albert Einstein,* Oxford, 1982, pp. 155ff.

37 E. Mach as quoted by Manfred Sommer in Haller & Stadler, op. cit., p. 204. In German: "Die Masse wächst scheinbar mit Geschwindigkeit. . . .Man braucht weder Mystik der Masse, noch Mystik der Elektrizität."

38 Carl Seelig, *Albert Einstein, Leben und Werk - Eines Genies unserer Zeit,* Zürich, 1960, pp. 134-135.

39 Holton (1989), op. cit., p. 72; Blackmore (1985), p. 91. In German: "Sie schreiben mir zuletzt, das Ihnen am Relativitätsprinzip hinsichtlich der erkenntniskritischen Seite noch manches zu fehlen scheint."

eral theory of relativity (1916). With apparent reference to the privileged position of uniform velocity he wrote: "Classical mechanics and even the special theory of relativity included an epistemological defect which had been raised perhaps for the first time by Ernst Mach."[40] Petzoldt thought that this defect could be overcome by the basic principle behind the general theory of relativity, but as this theory adopted the special theory of relativity with respect to small regions - the so-called 'Lokal Lorentz System' - Mach probably objected to this development as well.

It Is Unacceptable that Observers and Light-Velocity Have No Connection with Each Other.

As is well-known, Mach refused to accept the notion of 'absolute zero' in thermodynamics. For Mach, absolute zero could never be reached by experimentation. In this sense, it was a product of metaphysics and had to be opposed.[41] Like the idea of absolute zero, Mach apparently reasoned that the constancy of light velocity might well be trans-empirical to the point of metaphysics, and perhaps even worse, strange to the point of mysticism. But in this case, Mach's son Ludwig as the co-inventor of the rather accurate Mach-Zehnder Interferometer [which is still widely used and much referred to in the literature, though commonly but falsely attributed to his father. - Ed.] might just be able to empirically test Einstein's allegation.

From the author's point of view, the conception of the light-principle is divided into two stages. The first stage (PCLI), which is the definition of it itself in Einstein's paper in 1905, states that the velocity of light is constant and unrelated to the velocity of the light source. The second stage (PCL2), which means the connection of the PCLI and the relativity principle, states that light velocity is also constant regardless of the velocity of the observer. PCL1 was easy to understand, for example by the analogy of wave-velocity in water, but it was difficult to grasp an everyday-life image resembling PCL2, because the observer plays no role in it. Thus, PCL2 seemed to be metaphysical and mystical to Mach.

The Postulation of Mach's Principle was Based on Einstein's Misunderstanding of Mach's Thought

40 A. Einstein, "Die Grundlagen der allgemeine Relativitätstheorie", *Annalen der Physik,* **49** (1916), 771. In German: "Der klassischen Mechanik und nicht minder der speziellen Relativitätstheorie haftet ein erkenntnistheoretischer Mangel an, der vielleicht zum ersten Male von E. Mach klar hervorgehoben wurde."

41 Mach, *Mechanics,* p. 345; *Mechanik,* p. 312. E. Mach, *Die Principien der Wärmelehre, historisch-kritisch entwickelt,* Leipzig, 1896, p. 342.

Mach's criticism about Newton's bucket experiment was based on the relativity of 'actual motion' (emphasis is the author's). Mach regarded all 'actual motion' as relative, but Einstein considered laws to be relative in the sense of being co-variant, that is, unconnected with ways of selecting coordinates. Mach reduced causes to mere mathematical functions, but Einstein apparently understood 'Mach's principle' as if the total distribution of mass in the universe had a literal effect on local inertia. Thus, two or more discrepancies are clear between the understandings of the two men, inspite of the fact that they both used the same word: 'relativity'. Einstein's theory basically stands on a four-dimensional time-space structure, which Mach, to the extent he was familiar with the theory, could only have viewed ironically at best. Indeed, when Einstein visited Japan in 1922, he even admitted in his Kyoto lecture that while Mach's influence on his thinking had been great he had not clearly understood Mach's point: "Although I did not exactly understand Mach's idea about inertia, his influence on my thought was enormous."*

There Is a Fatal Contradiction in the Conception of Time in Relativity Theory

Einstein's theory, which gave rise to what has been called 'the paradox of the clocks', apparently seemed absurd to Mach. Mach wrote a letter to Petzoldt on May 1, 1914, "A year ago the paradox of the clock hadn't even occured to Einstein."[42] Einstein stated in his paper "On the Electrodynamics of Moving Bodies" (1905) that one clock, which moves around along the closed line from one point, runs slower than another clock, which remains at that point. But this brings about the paradox. Why? Because the two clocks (coordinate) are symmetrically treated within the special theory. Einstein admitted this paradox and corrected it when he was considering the general theory of relativity.[43]

The following quotations below may be regarded as indirect evidence for Mach's rejection of relativity theory. This is because Mach apparently

* Y.A. Ono, "Einstein's Speech at Kyoto University, December 14, 1922" in *NTM. Schriftenreihe für Geschichte der Naturwissenschaften, Technik und Medizin,* **20** (1983), p. 27.

42 Blackmore & Hentschel (1985), op. cit., p. 135. In German: "Die Paradoxie mit der Uhr war vor einem Jahr Einstein gar nicht aufgefallen." See also Holton's discussion of Petzoldt's level of understanding of relativity theory which apparently provided the context for Mach's comment: Holton (1989), op. cit., p. 80-81.

43 E.g. A. Einstein, "Dialog über Einwände gegen die relativitätstheorie", *Die Naturwissenscaften,* **6** (1918), 697. Elie G. Zahar insists that the definition of time coordinate and the special theory of relativity are inconsistent with Mach's epistemology. E.G. Zahar, *Einstein's Revolution - A Study in Heuristic,* La Salle Ill., 1989, p. 127.

put this theory in the same category with electron theory which was essentially regarded as a product of atomism, which it is known he strongly opposed. Indeed, in 1908 Mach had told a Japanese physicist, Ayao Kuwaki, that he rejected atomism for epistemological reasons, which means that neither electrons nor atoms could be accepted as ultimate realities. This curious linkage between atomism and relativity and apparent rejection of both is present in the preface to the *Optik* as well as a pair of quotations from Ludwig Mach expressed shortly after his father's death but several years before publication of the *Optik*. Ludwig wrote to Hugo Dingler on May 6, 1917 ". . . he [Mach] sometimes talked a trifle ironically about electron theory and relativity theory."[44] Ludwig made almost exactly the same point a bit later in a letter to Friedrich Adler on March 3, 1918: "He was until his death somewhat ironic in a gentle way about ions and the new views of the relativists."[45] In short, one may well have to agree with an observation which John Blackmore stressed that "No one who believes in the reality of physical atoms can have the same epistemological position as a phenomenalist".[46] Mach's philosophical position simply did not allow him logically to accept either the reality of atoms or the soundness of Einstein's theory of relativity.

2.3 EXPERIMENT-BASED REASONS:

Michelson-Morley Null Results Do Not Give Enough Experimental Support to Justify Accepting Einstein's Theory of Relativity

Mach the experimental physicist was well-aware that experimental failure to detect non-obvious phenomena need never be conclusive as evidence for persuading those people who believed in the reality of the non-obvious, as most supporters of Einstein's theory of relativity apparently did. This fact apparently tempted him to think that his son Ludwig with his interferometer for objectively measuring light could perhaps find incontrovertible 'enlightening' results which were either more positive about the ether than Michelson-Morley results or negative about the alleged curvature of space, thus compromising either the special or general theory of relativity or both.

44 Wolters (1987), op. cit., pp. 336-337. In German: ". . . über die Elektronentheorie, die Relativität sprach er manchmal leise ironisirend, . . ."

45 Ibid., pp. 338-340. In German: "Er hatte bis zu seinem Tode etwas leises Ironisirendes für die Ionen und die neuen Anschauungen der Relativisten. . ."

46 Blackmore (1972), p. 83.

Mach wrote down his primary reason for moving from Vienna to Bavaria in a letter to Anton Thomsen dated June 1 1913, that is, a month before writing the anti-Einstein preface, that he needed to carry out further experiments at less cost and with the help of his son who had prepared a house and laboratory for him in Vaterstetten, Bavaria[47] According to Anna Karma Mach, Ludwig Mach's second wife, they began to carry out optical experiments, indeed, sometimes even locking themselves in their laboratory all night to help solve a difficult problem with only chocolate to sustain them.[48] Mach's reference to experiments in the preface to the *Optik* almost certainly referred to initial work with Ludwig and his interferometer and to more extensive work anticipated in the future, which presumably was or would be the heart of Volume II of the *Optik*.

Ludwig mentioned in a letter to Friedrich Adler (March 3, 1918): "In his [Mach's] time, I brought his attention to certain insufficiencies in the Michelson experiment, and he acknowledged that he was uncertain about it - and I promised him that I would carry out a counterexperiment - The resources are there; only the time and rest are lacking."[49] We do not know the results of these experiments, but that Ernst Mach, like many reputable scientists of the time, had serious reservations about the sufficiency of the Michelson-Morley experiments is certain.

Logically, but not historically, the zero-effect of the Michelson-Morley experiment (hereafter cited as the M-experiment) guided the conception of the non-existence of the light ether, and this idea led to the assertion that

47 Blackmore & Hentschel (1985), pp. 117-118. "Why have I resettled at the age of 75 shortly before the end of my life?. . . . Because of the paralysis on my right side, I could not carry out experiments, but I still had the need to do so. So I often had to wait a long time for my son, who was occupied by technical work in Jena, Berlin, or elsewhere until he could get away in order to help me. On top of that, I could no longer manage to pay the rent for sufficiently well-lighted rooms which were needed for my books and laboratory equipment. I therefore let my son draw me to his place in the neighborhood of Munich." In German: "Warum ich noch mit 75 Jahren, also kurz vor Lebenschluss unziehe?. . . . Experimentiren konnte ich wegen meiner rechtseitigen Lähmung nicht mehr, hatte aber doch das Bedürfnis darnach. So musste ich oft lange auf meinen Sohn warten, bis dieser, in Jena, Berlin und anderwärts technisch beschäftigt, endlich abkommen konnte, um mir zu helfen. Es kam nun hinzu, dass ich den Zins für ausgiebige lichte Räume, welche die Aufstellung für Bücher und Apparate gestatteten, nicht mehr erschwingen konnte.Ich stimmte also meinem Sohn zu, zu ihm in die Umgebung von München zu ziehen..."

48 Anna Karma Mach, "Ernst Mach", *Bibliothek Ernst Mach*, Theodor Ackermann Antequariat, Katalog #634, München, 1959, p. 1. [Also private communication from A.K. Mach to John Blackmore (1967). - Ed.]

49 The translation is taken from Chapter Five of this anthology. A copy of the original letter can be found in J. Blackmore, "Mach über Atome und Relativität - Neuste Forschungsergebnisse" in Haller & Stadler, op. cit., p. 479. In German: "Ich habe ihn seinerzeit auf gewisse Unzulänglichkeiten des Michelson Versuches aufmerksam gemacht, und er gab die Unsicherheit zu; ja - und ich versprach ihm, den Gegenversuch aufzuführen - die Behelfe sind da, es fehlt noch die Zeit und die Ruhe."

there was no special coordinate system, which suggested the equality of the various inertial systems, namely, the principle of relativity. This consideration has been briefly mentioned by Friedrich Herneck and Gerald Holton. Mach clearly announced in the preface to the *Optik* that he hoped to disprove the theory of relativity through experimentation, and that this disproof was important and conclusive "above all" as a factor in his rejection of Einstein's theory. He did not accept Einstein's theory of relativity.

Michelson-Morley Null Results Are Not Even Sufficient to Conclusively Refute the Ether hypothesis

Even though Mach made few explicit references to the words "ether" or "light-ether" and on a conscious level seems to have been rather sceptical of the ether theory as a mere hypothesis, it appears that unconsciously like many physicists of the period he often presupposed it. He not only thought of physical change as studied by physicists as always continuous as he mentioned in his book on heat theory in 1896,[50] but a number of his contemporaries, especially Emil Wiechert [whose article about Mach and Einstein is partly included in this anthology], believed that not only did Mach presuppose an ether theory but that it was at the root of the difference between Machian relativity and Einsteinian relativity and the clear incompatibility between the two.[51]

Concerning the propagation of light, Mach said: "in physically homogeneous-filled space, light proceeds from luminous object in a straight line with a definite (bound by the medium) velocity."[52] Mach's apparent belief in the existence of such a medium reminds us that the advocates of 'light-ether' could not establish a definite interpretation of the M-experiment even as late as 1925. Although Mach thought of the essence of light in wave terms, following the works of C. Huygens, one must repeat that he was not

---

50 E. Mach, "Das Verhältniss physikalischer und chemischer Vorgänge", *Die Principien der Wärmelehre,* vierte Auflage,Leipzig, 1923, pp. 354-361. See also Heinrich Löwy, "Historisches zur Quantentheorie", *Naturwissenschaften,* **21** (1933), 302-303.

51 E. Wiechert, "Der Äther im Weltbild der Physik", *Königliche Gesellschaft d.Wissenschaftlichen Nachrichten (Göttingen),* Math-phys. Klasse, Heft 1, 1921, pp. 57-62.

52 Mach, *Optics,* op. cit., p. 5; *Optik,* op. cit., p. 7. In German: "Im physikalisch homogen erfüllten Raum schreitet das Licht vom leuchtenden Körper aus mit einer bestim-mten (an das Medium gebundenen) Geschwindigkeit geradlinig fort." [On the other hand, one must keep in mind that numerous parts of the *Optik* had apparently been written decades before publication, and it is not certain how much of the contents he still accepted during his last years. - Ed.]

an advocate of light-ether;[53] nor does he seem to have worried about whether vacuums with electromagnetic properties ought to be considered as a kind or variety of matter-ether.

On the other hand, the null effect of experiments which Michelson himself called a failure[54] did trouble many wave theorists,* and seemingly even Mach, which is indirect evidence that accepting wave mechanics may have involved assuming a position about ether, whether one was conscious of that assumption or not. One may recall that Mach's most famous physical invention was a wave machine, which was used in laboratories around the world for many years. Also, his son's biggest invention, his interferometer, was designed to help measure light waves.

From 1925 to 1930, Dayton C. Miller (1888-1941), President of the American Physical Society at the time, continued to reexamine the work of Michelson, looking for flaws and ways of improving upon his approach.[55] In other words, it was not 'beyond the pale' for Ernst and Ludwig Mach to pursue their tests further. Hindsight is not legitimate in this case. According to Ludwig and his wife,[56] he burned the manuscript for Part II of the *Optik* during the Second World War in frustration at the German military setting up high tension wires so near his experimental apparatus as to make it impossible for him to obtain reliable results (presumably about the alleged curvature of space in the general theory and how it affected light). Nevertheless, some data on how he planned to carry out his experiments has survived.

In one of Ludwig's extant notebooks, there still exists a reference to Ernst Mach's opinion about Michelson's experiment: "In E. Mach's opinion Michelson's experiment is in principle right, and it should not be repeated at all in the old form, but there exist doubts about the subtlety of how well he set up [and used his apparatus], whether the experiment is not wrong quantitatively. Also, the experiment should be repeated by all means with

---

53 Arthur I. Miller pointed out that Mach took a cautious attitude about the reality of the ether. A.I. Miller, *Albert Einstein's Special Theory of Relativity,* Reading, Mass., 1981, p. 166.

54 A. Michelson, *Studies in Optics,* Chicago, 1927, p. 111.

* In his later years, Mach corresponded many times with Otto Wiener in Leipzig, perhaps one of the more extreme ether theorists, who like Mach also opposed Einstein's theory of relativity. After Mach's death, Ludwig continued the correspondence. See *Ernst Mach - Studien und Dokumente zu Leben und Werk,* edited by D. Hoffmann & H. Laitko, Deutscher Verlag der Wissenschaften: Berlin, 1991, pp. 404-427. - Ed.

55 Lloyd S. Swenson, *The Ethereal Ether,* Austin, 1972, p. 228.

56 See Blackmore (1972), op. cit., pp. 284-285.

fixed sources of light, sunlight, or light from a fixed star."[57] When Einstein spoke at Kyoto University (1922) Jun Ishiwara recorded: "...when I had these thoughts in my mind, still as a student, I got acquainted with the unaccountable result of the Michelson experiment..." Though Einstein would state to R.S. Shankland (October 24, 1952) that "I guess I just took it for granted that it [the Michelson experiment] was true",[58] it is clear that Dayton Miller, Hugo Dingler, and the Machs were no longer young students and took a more critical attitude.

There is a rather interesting interpretation of the M-experiment in G. Gleich's book *Einstein's Relativitätstheorien und Physikalische Wirklichkeit,* which Ludwig mentioned in the preface to the *Mechanik* (9th edition, 1933), commenting that Mach himself would have shared Gleich's point of view, if he had lived to read the book. Gleich insisted that although the M-experiment was merely an experiment of interference and was not related directly to the theory of relativity, if the result of this experiment had been positive, the theory of relativity would have been refuted, and there could have been the possibility that light rays changed their frequency or direction when reflected in a mirror which moved around with the Earth.[58]

Indeed, Einstein himself admitted that the M-experiment was essential to understanding the theory of relativity. He stated that ". . .the physicists would have abandoned the principle of relativity without hesitation, if an important experimental result (such as the experiment of Michelson) had not been put forward [to support it]" (1915).[59] Given the importance of Michelson's result and Mach's knowledge as an experimentalist of how easy

---

57 Ludwig Mach, "Experimentalreihe als Einleitung für den II. Teil der Optik gedacht", (Aus einem Notizbuch Ludwig Machs), Wolters (1987), p. 416. In German: "Nach E. Mach's Ansicht ist Michelson's Versuch principiell richtig und soll in der alten Form durchaus nicht wiederholt werden, aber es bestehen bei der Subtilität der Aufstellung bedenken, ob er quantitativ nicht verfehlt ist. Auch soll der Versuch unbedingt mit stehender Lichtquelle, Sonnen- oder Fixsternlicht, wiederholt werden."

58 G. Holton, "Einstein, Michelson, and the 'Crucial Experiment'" in *Thematic Origins of Scientific Thought - Kepler to Einstein,* second printing, Cambridge, Mass., 1974, p. 283. Concerning Einstein's early acquaintance with the Michelson experiment, see Tsuyoshi Ogawa, "Japanese Evidence for Einstein's Knowledge of the Michelson-Morley Experiment", *Japanese Studies in the History of Science* (from September 1980 this journal changed its name to *Historia Scientiarum*), **18** (1979), 79. Holton's research shows that although Einstein probably knew about the Michelson experiment and the result of it before 1905, he did not directly rely on this experiment in generating relativity theory. See Holton above, "Crucial Experiment", p. 327.

58 Gerold von Gleich, *Einsteins Relativitätstheorien und Physikalische Wirklichkeit,* Leipzig, 1930, pp. 5-6.

59 Holton (1974), op. cit., 274; see also A. Einstein, "Die relativitätstheorie" in E. Warburg (ed.), *Physik* (volume 1 of section 3 of part 3 of *Die Kultur der gegenwart,* Leipzig, 1915, p. 706.) In German: "...daß die Physiker unbedenklich das Relativitätsprinzip fallen gelassen hätten, wenn nicht ein wichtiges experimentelles Resultat vergelegen wäre...."

it is to make mistakes, is it possible to blame Mach for being 'overcautious' and wanting to see other experimental results? Indeed, until he studied other results, can he be blamed for withholding assent from Einstein's theory of relativity?

On the other hand, while his experimental reservations might have been good science, can the same be said about his sense-physiological and epistemological objections? Indeed, could they have allowed him to accept solid experimental backing for Einstein's theory could it have been forthcoming? That is perhaps the big question. Many theoretical physicists jumped on Einstein's bandwagon early, starting with Max Planck, but is it fair or just to expect experimentalists, regardless of their philosophical orientation, to do or to have done the same? And if the experimentalist is also a phenomenalist in epistemology, ontology, or both, as Ernst Mach surely was, *is it even possible?*

CHAPTER 16

MACH, EINSTEIN, AND KUWAKI

by Setsuko Tanaka*
(Aoyama Gakuin University)

EARLY CONTRIBUTIONS

During the rush toward modernization which took place in Japan during the Meiji period, that is, the last decades of the 19th century and the first twelve years of the new century, it was perhaps natural that the emphasis in the new universities which were being established should be on applied disciplines such as engineering and medicine.[1] Nevertheless, "pure" fields such as physics were not entirely neglected. Aikitsu Tanakadate and Hantaro Nagaoka led the way during the early 1890's by studying Newtonian physics and 19th century additions such as thermodynamics and what would later be called statistical mechanics. The Japanese government in its efforts to catch up to European countries financed the overseas study of numerous students, scholars, and professors. For example, Hantaro Nagaoka studied under Helmholtz in Berlin and later under Boltzmann at Munich.[2]

But around the turn of the century, just when Japanese physicists began to feel - thin as their numbers were - that at least Newtonian physics was in their grasp, a virtual or actual revolution broke out which clearly transcended what they had learned. The altered situation was an immense challenge. To meet it, Japanese physicists would have to master English, French, German, dificult mathematics, complex experimentation, and the full range of conflicting ideological and philosophical intrusions into the interpretation of "modern physics".

Ayao Kuwaki (1878-1945) appears to have been the only Japanese physicist of his day with the training and education to understand the most recent developments in physics well-enough to accurately describe and explain them both to other Japanese physicists and to the interested general public in his country. European and American scientists often struggled to contribute something new to their discipline as did an increasing number of

---

* This paper was especially prepared for this volume.

1 See James R. Bartholomew, *The Formation of Science in Japan,* Yale University Press: New Haven and London, 1989 for a study of the general development of Japanese science during this period as well as for useful information about Japanese physics.

2 Ibid., pp. 68-82.

J. Blackmore (ed.), Ernst Mach – A Deeper Look, 297–332.

Japanese, but Kuwaki saw that what was most needed at least in Japan was greater reliable understanding of what was going on. His main task was to learn well so that he could become a better educator. Knowledge had to be circulated before largescale efforts could be made to significantly improve upon it. Kuwaki the educational pioneer had to take precedence.

Ayao Kuwaki was born in Tokyo in 1878 into a book-focused, intellectual family. His father, like so many of the samurai class, was interested in scholarship, and Ayao's elder brother, Genyoku Kuwaki, who was four years senior, would become a recognized philosopher and Kant-scholar. Also, Ayao's son Tsutomu Kuwaki would become a philosopher, one noted for his translation of Martin Heidegger into Japanese.

Ayao Kuwaki entered the Physics Department of the University of Tokyo in 1896 and graduated in 1899. His strongest subject apart from languages appeared to be mathematics, an ability which gave him open sesame to the new developments in theoretical physics. He published his first article very shortly after graduating.[3] It involved a discussion of the notion of force with another scholar, and the discussion continued back and forth between them.[4] He also introduced Mach's definition about mass. The exchange demonstrated how wide his range of reading was, including books by Tait, Thomson, Helmholtz, Hertz, and Boltzmann, that is, men whose ideas his teachers Nagaoka and Tanakadate had largely studied first hand during the early 1890's. Especially interesting for our purposes, however, was the fact that that he showed considerable familiarity with Mach's *Mechanik.*

Without presuming to compare the abilities of the two men, Kuwaki who was a half-year older than Einstein seems to have read basically the same books on physics which Einstein had read, and a major result was Kuwaki's relatively quick sympathy for Einstein's ideas on relativity. It is possible that Mach's *Mechanik* with its criticisms of Newton's ideas on absolute space, time, and motion helped smooth the way for both men. But if Kuwaki lacked Einstein's originality, he partly compensated with rapid apprehension and acceptance of his colleague's work on relativity.

In 1901, Kuwaki published a translation of Henri Poincaré's article "Sur les rapports de la physique experimentale et de la physique mathematique"[5] which was included in Poincaré's book *La Science et l'hypothese.* Already in this article Poincaré had raised the question whether the ether was real, a point not lost on Kuwaki. As is well-known Poincaré would la-

---

[3] Ayao Kuwaki, "Goto-shi no 'chikara ni tsuite' o yomite" (On Reading Mr. Goto's Article 'On Force'), *Toyo Gakugei Zasshi*, No. 223, 1900, p. 173 & No. 224, 1900, p. 209.

[4] Ibid., "Goto-shi ni kotau" (To Answer Mr. Goto), No. 226, 1900, p. 311.

[5] Ayao Kuwaki, "Jikken butsurigaku to suri butsurigaku to no kankei", *Toyo Gakugei Zasshi,* **18** (1901), No. 234 pp. 99-103, No. 235 pp. 151-157, No. 242 pp. 455-464, & No. 243 pp. 501-512.

ter develop his own theory of relativity. One may add, that Kuwaki's translation seems much better than the much later one of the whole book by a different scholar in 1938. Kuwaki's facility at languages so different from Japanese can only provoke envy and admiration among those of us from a later period.

Kuwaki became a lecturer on graduating from the University of Tokyo and an assistant professor shortly afterwards. His numerous papers reflected his wide interests. In one article he distinguished between Ernst Mach's use of the word "description" and what Kirchhoff seems to have originally meant by it[6] and in another, also published in 1906, he addressed the issue of absolute versus relative motion head on.[7] The latter question had become a very active issue even before the appearance of Einstein's revolutionary article in the year 1905. Mach, Neumann, Poincaré, Boltzmann, Abraham, Russell, Föppl, and Schuster had all addressed it. Kuwaki began by posing the question whether "the idea of absolute motion could be meaningful in science or not". Newton had argued for the affirmative and Mach in his *Mechanik* for the negative side. "Ernst Mach in 1883 rejected absolute motion which was not relative with respect to another body as metaphysical and useless in science."[8]

Kuwaki showed his extensive background in the history and philosophy of science by carrying the discussion back several centuries and by transforming the question into an epistemological one. He addressed in turn: the ideas of Newton, those of his followers, Leibniz's approach, standard textbook arguments, Mach's opinion, ideas on space, the notion of ether, and absolute motion in electron theory. Above all, he focused on Mach: "Mach said that all empirical knowledge was relative: that there were only relative space and relative motion as basic facts, that it was a kind of illusion that perception of absolute motion could be proved mechanically since human perception of things was itself relative."[9] Kuwaki concluded his paper in terms of the then fashionable *Bildtheorie* of Hertz and Boltzmann, as if physics was primarily concened about "pictures" rather than what the pictures were supposed to represent: "To analyse absolute motion and absolute rest, it might be well to step into the picture of electron theory. But the picture of electron theory can be replaced by other pictures based on relative motion".[10]

---

6 Ayao Kuwaki, "Setsumei to kisai" (Explanation and Description), *Rigakukai,* **4** (1906), 7-10.

7 Ayao Kuwaki, "Zettai undo ron" (On Absolute Motion), *Tokyo Butsuri Gakko Zasshi,* No. 180, 1906, pp. 494-508. The article was included in his book *Butsurigaku to Ninshiki* (Physics and Recognition) published in 1920.

8 Ibid., p. 494.

9 Ibid., pp. 500-501.

10 Ibid., p. 507.

If the last line is read carefully, one may notice that it suggests Einstein's theory as presented in his epochal work "On the Electrodynamics of a Moving Body" of the previous year, that is, 1905. Kuwaki later wrote as follows (1921): "When I published this lecture more than ten years ago, it was just after Einstein's paper on special relativity, and at that time the evaluation of his paper was not yet settled. I had been conscious of the importance of the problem and had written a paper 'On Absolute Motion' and discussed the meaning of the problem from an epistemological point of view."[11]

Though it seems probable that Kuwaki had read Einstein's original paper in the *Annalen der Physik* and been impressed by it, one of the few scientists apart from Planck to take it very seriously that early, he did not openly refer to it in the 1906 paper. But he did point out that the necessary solution of the problem about absolute motion in ether could be found by eliminating the distinction between static charge and electric current in electromagnetic theory, which had apparently been a basic motivation behind Einstein's paper. Since Kuwaki had already studied Mach's *Mechanik,* Poincaré's paper, and Lorentz's electron theory, he was prepared for a natural development of the theory such that he apparently saw that the difficulties found in electron theory could be resolved by following the direction of Einstein's argument.

Kuwaki's next paper[12] which appeared in 1907 was more explicit about Einstein. Kuwaki began his article about electrons with the question of their spatial distribution within atoms, and understandably, he mentioned his teacher Hantaro Nagaoka's "Saturn model". One will recall that there was competition between "the raison-cake model", "planetary model", and other fashionable "pictures" of the time. Then he mentioned the dependence of electron mass on the velocity of electrons and the difference between longitudinal and transverse masses when they traveled with velocities of at least a tenth of the velocity of light, something known from experiment on beta rays or cathode rays. On the other hand, the famous Michelson-Morley experiment showed a negative result about motion with respect to the ether (1887). In order to interpret the result, Lorentz proposed a theory of contraction of bodies in the direction of motion. In the theory of Lorentz, he preserved absolute motion with respect to the ether. Then Kuwaki introduced Einstein's theory:

> Mr. Einstein has treated the problem in a different way than Lorentz by eliminating the idea of absolute motion from the foundation of the theory. To define simultaneity at two different points, Einstein used the time of the pro-

---

[11] Ayao Kuwaki, *Butsurigaku Joron* (Introductory Course in Physics), Tokyo 1921, p. 1.

[12] Ayao Kuwaki, "Denshi no keijo ni tsuite" (On the Shape of Electrons), *Tokyo Butsuri Gakko Zasshi,* No. 183, 1907, p. 73-80.

> pagation of light to exchange signals between two points. Einstein accepted the local time introduced by Lorentz and in order to measure the length of a body from different systems moving relative to the body, one should record both ends of the body at the same time of the system as defined above. Then the measurement of the length of the body should depend on the velocity of light and the velocity of the body with respect to the system. Finally, the results of Einstein coincide with that of Lorentz.[13]

Next, Kuwaki discussed the theories proposed by Abraham and Bucherer about the shape of electrons. He also described Kaufmann's experiment of 1906 on beta rays from radium. The Einstein-Lorentz approach appeared to deviate about 10% from what one ought to expect while the Abraham-Bucherer theory only deviated by 3 or 4 percent.

But at the 78th Congress of German Scientists and Physicians in 1906, Max Planck had commented that the experiment of Kaufmann was not a final one and had proposed a theory for new experiments to decide the validity of the two theories. Also, Kuwaki mentioned the famous controversy between Planck, Abraham, Bucherer, and Sommerfeld at that Congress. The main issue of the debate was not only the shape of electrons but also the validity of the idea of absolute motion in the ether. Kuwaki quoted Planck's words from the conference as follows:

> I can admit a non-uniform absolute motion such as rotation, but uniform absolute motion can not be admitted in generalized mechanics. Therefore, the latter should not be allowed in electrodynamics either.[14]

Kuwaki commented on Planck's declaration:

> Planck mentioned relativity of motion but he also defended absolute motion. So he seems to be an absolutist on mechanics. The controversy [at the Congress of 1906] did not lead to any conclusion about the correctness of the criticism about the fundamental idea.[15]

We see from Kuwaki's 1907 paper his ability to focus on fundamental issues almost at the very moment they appeared in public or in print and his speed in transmitting this knowledge to the educated Japanese public by means of graphic, non-mathematical descriptions. But if Kuwaki the educator is prominent in his papers, and especially his remarkable ability to take an objective or detached position, another deeper tendency also makes its appearance, namely, Kuwaki the Mach-influenced philosopher and reflective thinker:

---

[13] Ibid., p. 76.

[14] Ibid., p. 78.

[15] Ibid.

> The notion of atoms, molecules, and electrons derived from Democritus and Lucretius has become more and more uncertain, and still it is not easy to master the notion of them even in the recent progress of physics. These objects escape from our perception, so they cannot ever be understood completely. Sometimes, phenomenology or energeticism, which have rejected the atomic theory, have been thought to be better opinions, but these are also arbitrary concepts. Should the purpose of a hypothesis be to help discover some unknown results and to help unify the known results, it does appear that the theory of electrons has clarified many facts and stimulated the study of basic concepts of space and time. Thus we have been able to understand the great influence of the theories of electrons in physics. In this paper, I have only explained that part of the problem related to the controversy about the shape of electrons.[16]

During 1906 and 1907 Kuwaki wrote a series of twelve articles on recent and contemporary physics which were apparently to be read as if they were a book.[17] His intent was clearly to help interested but non-professional Japanese readers penetrate the non-mathematical essentials of recent developments during one of the most turbulent periods of the discipline. Perhaps without realizing it, Kuwaki may also have been well-ahead of any comparable attempt in Europe or America. The published articles finally appeared in book form years later in 1921.[18] But the broad scope of the work also allowed Kuwaki's natural tendencies toward the history and philosophy of science to become prominent. More and more he preferred to *understand* the basic assumptions of physics rather than *do* physics.

> My program is planned in a similar way to that of Helmholtz's *Vorträge und Reden,* and the first chapter 'Experience' is based on a chapter from Thomson and Tait's *Treatise on Natural Philosophy.* But this book is neither a translation of those books nor abstracts from them. My concluding idea is rather different from those books . . . . I have tried to describe some tendencies of recent theoretical physicists.[19]

The reflective nature of Kuwaki's understanding seems clear from the following passage:

> Newton thought that mathematical study could lead to results which could master the reality of nature, that is, absolute truth. This kind of thought which allows the existence of the reality of nature and the possibility of reaching reality is thought to be a view of transcendent realism. On the other hand, Kant said that axioms of mathematics were given a priori, but our experience about nature only consisted of pictures which appeared in our

---

16 Ibid., p. 80.

17 See *Butsurigaku Kogi* (Lectures in Physics) in each issue from June 1906 to May 1907.

18 Ayao Kuwaki, *Butsurigaku Joron* (Introductory Course on Physics), Tokyo, 1921.

19 Ibid., p. 2.

> mind. Thus, we could not reach nature itself. The idea was opposite to that of Newton and said to be transcendental idealism. Both kinds of scholars considered that mathematical proof was unique. But after non-Euclidean geometry was established, it is thought that mathematical axioms are not determined uniquely but can be chosen from many possibilities arbitrarily.
>
> The next question is what kind of choice is valid in physics. Physical laws can be derived from the results of experiments and observation. The laws of physics are expressed quantitatively by algebraic symbols which represent physical quantities. When functional relations change according to differential equations or integral equations, mathematical study of physics is valid.[20]

Nor did he always agree with Mach:

> One of the roles of hypothesis is to unify the explanation of physical laws. Here the word explanation means to fix the relation between cause and results. From ancient times, causality has been the great issue among philosophers, Plato, Aristotle, occult theory in the Middle Ages, Hume, and Mill.
>
> Among them, Mach followed Hume's opinion. Mach said that knowledge was obtained from sensation, and he thought that experience was absolutely important and that Newton's idea which seemed to transcend experience was not scientific . . . . In short, the propositions of natural laws already contains the causal relation, that is, in order to propose natural laws an expectation of unification of nature is necessary and at the same time the proposition means a prediction by causality . . . . Science could not be constituted without so-called hypotheses.[21]

In the last part of the 1907 book, Kuwaki included an annotated bibliography with rather detailed descriptions and interpretations of selected publications by Helmholtz, Hartmann, Mach, Maxwell, Poincaré, Stallo, and Ward. The longest interpretation focused on Poincaré's *La Science et l'hypothèse* (1903). But while Kuwaki surely sympathized with his ideas, especially on hypotheses and conventions, he appears to have been most stimulated by Mach. Nor is it inappropriate to recall the extent to which Poincaré himself appears to have been indebted to the Austrian philosopher.

In the last passage of his work, Kuwaki not only seems to share Mach's position on simplicity and the nature of theories but appears to take a relativistic position on truth and theoretical physics as well, one which passes beyond the physical relativism of Einstein and the epistemological relativism of Mach. Objective and detached as Kuwaki almost always was in describing the ideas of *other* people, there is a sense in which his *own* extreme relativism, at least in terms of his apparent philosophical position, could be seen as perhaps unwittingly approaching subjectivism:

---

20 Ibid., pp. 5-6.

21 Ibid., pp. 10-12.

> The choice of atomic theory, energeticism, or other hypothesis is arbitrary and depends on the way each scholar considers it to be simple.[22]

2. KUWAKI'S EARLY TRIP TO EUROPE

Ayao Kuwaki went to Europe to study under Max Planck at the University of Berlin from October 1907 until July 1909. His curiosity about Europe in general and recent developments in physics in particular seem to have motivated him as well as the presence of his elder brother, Gen-Yoku, who was studying philosophy in Berlin and later Paris. The fact that his visit began shortly before the bitter controversy between Planck and Mach but lasted well into the dispute and while physicists were only just beginning to make up their minds about Einstein's special theory of relativity give it a special interest. Kuwaki's vivid impressions based on direct contact with Max Planck in Berlin, Mach in Vienna, Einstein in Bern, Poincaré in Paris, and Lorentz in Leyden plus quotations from them were apparently written down in notes at the time which he published a few months after his return to Japan.[23] The fact that these impressions do not seem to be known in the West plus the possible importance they may have for biographers and historians has encouraged me to quote from them directly and extensively. I have also added a few additional remarks he made about his trip from later writings. With allowances from the reader for the difficulty in determining the exact spelling of little-known European and American names, and hence identification of at least some of the people from the Japanese symbols and calligraphy which Kuwaki used, here is a selection of his comments:

A. About Max Planck:

> "The main figure of the physics department of the University of Berlin is still Planck. When the University of Vienna lost Ludwig Boltzmann in 1906, they courteously invited him to teach there, but Planck did not accept the offer. Finally, I heard that at that time a union of physics and mathematics students at Berlin University expressed their gratitude with a candlelight parade which gathered in front of Planck's house. Also, the Physical Society of Berlin held a banquet to celebrate that they could keep Planck in Berlin. A

---

22 Ibid., p. 40.

23 Ayao Kuwaki, "Ryugaku Zakki" (Notes on My Study Abroad), *Toyo Gakugei Zasshi,* 1910, No. 345 p. 283, No. 347 p. 369, No. 348 p. 427, and No. 349 p. 481.

person made a poem at the banquet: 'Sie kennen unsern Planck. Noch ist er Unserer, Gott sei Dank!'

During the first semester in which I heard Planck's lectures, there were two American students. One of them was an assistant professor of Columbia University in New York called Pegram. Another was called Harrington from Clark University. Dr. Lise Meitner who has received a doctor's degree at the University of Vienna has studied radiation with Otto Hahn. I have often seen their papers in *Naturwissenschaftliche Rundschau.* Another Japanese student was a military officer called Fujino."[24]

"Altogether, there were over a hundred and twenty students in the lecturehall. Among them were more than ten female students. I heard that ten years ago Planck said that Amazons were against nature, even in intellectual matters. In Berlin, women were treated as special students. But since last year they have become regular students."[25]

"Planck's lectures on theoretical physics cover the whole field in seven semesters. The lectures are given four hours a week with an hour *Prakticum.* I heard that he once gave a general lecture on music and that he himself was a good piano player. But now he only gives his one lecture series. The level of his lecturing was appropriate for students who have graduated from Gymnasium. Planck's lectures were well-prepared and he gave calculation of formulas on the blackboard. He wanted to give comprehensive lectures and his expression was lucid. His lectures were systematic and suggested deep reflection. American students were surprised that Planck's lectures were very philosophical."[26]

"Usually, Planck saved the sixth semester for lectures on something general or on a new topic. This took place during the summer term 1909. He had been invited to Columbia University in New York where he had given a lecture on the current situation in theoretical physics. He gave us a similar lecture. He explained about the general theory of physics, the principle of least action, and the principle of relativity. Planck ended his lecture by saying that especially the latter principle still played only a *bescheidene Rolle,* yet it [Einstein's theory] would be finally recognized as the main principle of nature concerning the epistemological meaning of physics and astronomy. Concerning astronomy, if the principle were applied to

---

24 Ibid., No. 345, pp. 286-287.

25 Ibid., p. 287.

26 Ibid.

planetary motion, the orbits of the planets would be calculated as being slightly different from elliptical orbits.

For the one-hour-a-week *Prakticum,* one or two questions normally would be given as homework for the students and they would answer them and give them back to Planck. Then the assistants would check them and give them to Planck. At the next *Prakticum* Planck returned them to each student with comments about any mistakes. One of the students would then correct a mistake at the blackboard.

Planck's house was in Grünewald outside of Berlin and he used a subway with third-class monthly tickets. He was thin and taller than ordinary German people. His head was almost bald with a few hairs. Once his portrait which was painted by a painter in Berlin two-hours-a-day was sold and in it he looked like a peasant scholar. Ordinarily, he wore a hat and a shirt with a stand collar and a black bow and never changed them when he came to the university."[27]

"When I was studying under Planck, there was a spell when dinner parties were held once a week for four consecutive weeks. At the first dinner party, I went with Abraham. More than ten people joined the party every time, and the participants were almost always the same. They were young and old, ladies and gentlemen, and many young doctors. An American and I were the foreigners [Professor Pegram of Columbia University].

At one of the parties we had been told that some kind of show would be held. Everyone was waiting for it. Then the main doors were opened on both sides, and we could see two figures with devil's masks dressed like ladies with long skirts standing in the neighboring room. The professor played piano in the corner, two people walked elegantly with the music backward and forward. After several times of such performance, suddenly they turned to us and we found two of Planck's daughters there. The two daughters had worn masks of devils on the back of their heads and astonished us. The daughters were twins and completely similar to each other. When they wore the same dresses at the party, we could not distinguish them. Sometimes it happened that someone greeted the same lady twice."[28]

B. About Ernst Mach:

27 Ibid., pp. 287-288.

28 Ayao Kuwaki, *Einstein-Den,* Tokyo, 1934, pp. 79-80.

"I visited Ernst Mach in his residence in a Vienna suburb in April 1908 while on my way through Austria and Italy [in order to attend the Fourth International Conference in Mathematics in Rome]. Mach's house was in the so-called *Villenviertel.* There were houses along the street and the neighborhood was rather noisy with the traffic of streetcars. Mach's residence was on the upper floor of the two-floored building in which the groundfloor was a shop which sold miscellaneous items. He met me in his study room. He was over seventy years old and his hair was all white. I heard that he had had disease in his legs for a long time and even in the room he used a cane. He said that he lived on a pension. [I remember that once] he took out a volume by Jaumann on mechanics and he said to me that vector analysis was used widely in that book [by his former student and assistant]."[29]

"There are many kinds of criticism of Mach's point of view. He often explained that he did not like to be called a philosopher. His opinion was also criticized because it did not consist of a systematic philosophy. He claimed that all metaphysical hypotheses, that is, hypotheses which could not be controlled by experience, should be avoided, and that sensation alone existed. He was also of the opinion that the *Ding an sich* and absolute space and time were all useless and obstacles. His historical and critical study of mechanics is famous; he said to me that mechanics was his favorite study. But from his own perspective he opposed the atomic theory which was thought to be an example of mechanical explanation of nature by so-called orthodox physicists. Boltzmann in his inauguration address [for his series of lectures in philosophy] in the University of Vienna [1903] had criticized Mach's claim 'I do not believe in the existence of atoms'. He thought that something which could have an exact *Vorstellung* could be said to exist beside sensation."[30]

"Recently, Planck opposed himself against Mach and argued that it was contrary to the progress of science to treat natural science in an anthropomorphic way as if it rested on the unique foundation of sensations. Rather, natural science should emancipate itself from the anthropomorphic factor while on the way of progress. For example, earlier physics was classified into mechanics, heat theory, acoustics, optics, and electromagnetism by the kind of sensations studied, but thermal radiation was sometimes treated in the same chapter with light waves, and finally, physics has been simpli-

29 op. cit., "Ryugaku zakki", 1910, No. 347, p. 369. See also *Zettai to Sotai*, Tokyo, 1921, p. 127.

30 Ibid., p. 370. See also *Zettai to Sotei,* pp. 127-128.

fied into mechanics, thermodynamics, and electrodynamics. And now, physical phenomena in motion have been reduced to just two categories, reversible and irreversible, a distinction far indeed from the original ones between sensations. As mentioned above, Planck opposed Mach more from the metaphysical perspective of separating physical phenomena from subjective factors than in a way related to phenomenology.

But it is clear that Ostwald, who is said to have been influenced by Mach, in proposing energeticism was also doing metaphysics. Even Mach's opinion is thought to be a kind of metaphysics based on human intent, because Mach claimed that knowledge was [mentally] selected from sense data.

Wilhelm Wundt thought Mach's opinion was a critical form of metaphysics and discussed the way Mach had tried to base his theory of economy of thought on sense data, which was an inverse approach from that of Kant. Mach's idea, however, was simpler than the complicated categories of Kant, who had wanted to derive all he could from the general action of thinking. But it must be admitted that it is not clear where Mach's principle of economy of thinking actually comes from. It looks like something suddenly shot from a pistol, indefinite, arbitrary, and similar to teleology.

There are some polemical disagreements about the soundness of Mach's idea that it is possible to have natural science apart from metaphysics, but I think that Mach has had great influence on scientists because he tried to introduce and carry through an epistemological *Wendung*. When I mentioned to Mach that I had been attending Planck's lectures, he said to me 'German scholars are speculative; you may like English scholars better'."[31]

"Mach died in Munich eight years after I visited him. He was generally thought of as an empiricist, sensationist, or positivist, but in some circles as a materialist and in a few even as a spiritualist or idealist. He did not think of himself as a philosopher but merely as a scientific epistemologist. He opposed the atomic theory from the epistemological point of view and he agreed with energeticism. His criticism of Newton's mechanics was empirical, relativistic, and sceptical, and he was thought to be heretical by the more orthodox physicists of that time."[32]

"When I met Mach ten years ago, it was after Einstein's theory of special relativity had been published. He merely said to me that Einstein's principle was just a theory. At that time I saw that Mach

31 Ibid.

32 Ayao Kuwaki, *Einstein-Den*, Tokyo, 1934, pp. 282-284.

was already old, so I did not want to ask him in further detail. I thought to myself that I could imagine Mach's thought."[33]

C. About Albert Einstein:

"In March of 1909, I went to the *Eidgenossiges Amt für geistigen Eigentum,* that is, the patent office in Bern and visited *Herr Offizier* Albert Einstein. He had proposed the theory of relativity by himself. When I visited him, he came out from his office and said to me that he could have free time for three hours. Then he took me downtown. The city of Bern had sidewalks with roofs on both sides of the street. Those sidewalks were called *Lauben* and made the special character of the town. It was almost lunchtime. Einstein said to me: 'I cannot take lunch with you, because I have a wife. If I take you to my house, my wife would be surprised with a sudden guest. How much would you like to pay for lunch? Such and such a restaurant can serve good food with reasonable price.' Then he guided me to the neighborhood.

In the afternoon, I went to his house. His wife and daughter who was almost ten years old came to the porch and they welcomed me.[34] His wife showed me lacquerware and asked me if it might not be genuine. Mr. Einstein spoke to me as follows:

> It is the first time that I have talked with a Japanese person. It is interesting that there are various kinds of civilization on the earth. It might have been happier for our life if the steam engine would not have been invented. I am working in the patent office eight hours every day. Applicants for patents propose possible or impossible plans. I have graduated from the Polytechnicum in Zürich. I have not met German scholars yet. Recently, I have been corresponding with Planck. What kind of person is he? I am sorry to oppose such a great scholar, but I could not agree with his theory of radiation.[35]
>
> I cannot agree with Planck on his attempt to withdraw his quantum theory of energy with respect to absorption.[36]

---

[33] Ayau Kuwaki, "Mach to Sotaisei Genri" (Mach and the Theory of relativity) (1922) in *Kagakushi-Ko* (Essays on History of Science), Tokyo, 1944, p. 455.

[34] This passage appeared in Kuwaki's "Ryugaku Zakki" in his *Toyo Gakugei Zasshi* of 1910 (volume 27, No. 349 p. 481) and in *Zettai to Sotai* in 1921 but was changed in *Einstein-Den* in 1934. The matter will be discussed at greater length in the next section.

[35] Ibid.

[36] This sentence is from a later publication by Kuwaki, and is a somewhat more concrete version of the preceeding sentence. See *Einstein-Den,* Tokyo, 1934, p. 59.

"When I came back to Berlin and told Professor Planck that I had met Einstein while on my trip, Professor Planck just asked me what was the personality of Einstein and whether he was a Jew."[37] [How strange that they had not met and that both men asked the same question about each other. - Ed.] Einstein continued by saying the following:

> Abraham is a good physicist, but it is not correct for him to insist that the motion of electrons is 'absolute'. Hertz might have heard some very incomplete lectures on mechanics [a thing which might have stirred him to write his famous book]. Planck's lecture in Leyden had a nice ending. Mach's argument is logical but physicists cannot be satisfied with it. I think the same about Kant. If someone could develop a sound atomic theory of light, his fortune would be made [*Denkmal gebaut*].[38]

Some years later, Kuwaki wrote a letter to Einstein hoping that he still remembered his visit of 1909. Einstein answered in a very flattering way:

> I remember your visit to Bern very well. At that time you were the first Japanese person, indeed, the first East-Asian of any kind that I had gotten to know. You amazed me at that time with your vast theoretical knowledge.[39]

Kuwaki added further "I had asked Einstein then if the relativity of uniformly accelerated motion could be extended to a negation of absolute rotation." Returning to Kuwaki's visit:

"When I told him that I was going to Paris, he thought that I might have trouble in an unknown country and he wrote a letter to introduce me to his friend. After I arrived at the accomodation of my elder brother [who was then staying in France], I visited M. Solovine who was a friend of Einstein in his accomodation in the Quartier Latin in Paris. His residence reminded me of the word 'Attic philosopher'. He was a specialist on the history of science and recommended that I hear the lectures of 'Wyroboff' in the College de France. He also mentioned that it was very convenient to study the history of science in France because from well in the past there had been many scholars who had made a rather systematic study of science or philosophy such as Descartes, Comte, d'Alembert, Ampere, etc.. At the Sorbonne, I heard the lectures of Mme Curie and Picard. When the professor came into the classroom or went out of

---

37 op. cit. "Ryugaku Zakki", p. 481.

38 Ibid.

39 Ayao Kuwaki, *Butsurigaku to Ninshiki* (Physics and Cognition), Tokyo, 1922, pp. 122-123.

the room, the students clapped hands in Paris, but in Germany they made noise with their feet."[40]

D. About Henri Poincaré:

"During my stay in Paris, one day I went to hear the regular meeting of the Academy of Science. It was the day when 'Kolbe' of Göttingen gave a lecture. I later asked him who the scholar was down front. At first, because I had seen Picard and Darboux who had a dignified appearance, I felt surprised on hearing that the short man with a bent back who had been walking around was Poincaré. Gradually, however, a feeling developed which I could not hold back to visit the really great person. I wanted to meet and talk with him. I visited his house at 63 Rue Claude-Bernard in Quartier Latin. But he was absent and I heard that he was staying in his villa in Lozère, so I followed him there. It was a southern suburb of Paris, and I thought that it would take about twenty or thirty minutes from the Sorbonne by train. His villa was located about two hundred meters from the closest station. It was on a slope with a forest, and I could see a hill in the distance across a field. After the visit when I came back from his house, Poincaré came with me to the station, saying that he would like to take a little walk. Afterwards, I sent him a picture of him which I had bought and asked him for his signature on it. He accepted my request and sent it back to me."[41]

"[During the visit to] his house which was on the middle of a steep slope with the second floor on the same level as the backyard, Poincaré and I went out into the garden and talked sitting under the trees. While Poincaré was talking, he changed his language from French to German. He said to me 'I am going to give a lecture in Göttingen in the near future, so I am practicing on you.' Also he said to me that among recent works he had found that Ritz's paper was interesting. (Ritz treated only retarded potential and did not accept the theory of ether as connected with electromagnetism). But he said that ether was '*noch etwas*' (still something)."[42]

E. About Hendrik Antoon Lorentz:

40 op. cit., "Ryugaku Zakki", p. 482.

41 Ayao Kuwaki, "Henri Poincaré", *Toyo Gakugei Zasshi,* **29** (1912), No. 373 pp. 459-460. See also *Zettai to Sotei,* op. cit., pp. 96-97.

42 Ayao Kuwaki, *Toyo Gakugei Zasshi,* **27** (1910), p. 483.

"Leyden was an old city. There were no electric streetcars as yet but only horsecars. The experimental apparatus of high pressure and low temperature [M.H. Kamerlingh Onnes] in the physics department was very famous. I saw some technician blowing glass. I even heard that there was a school within the physics department for training technicians. I met Professor Lorentz. What he said was as follows: 'The experiment of Michelson and Morley has an accuracy of the second order. It is too bold to think that it offers final results. About philosophical arguments we should be very careful, but [the idea of] absolute time could still be thought of in a variety of ways [at least] for me. I would like to think that ether is still 'substantielle'."[43]

III. EINSTEIN'S LOVE CHILD

Since the publication of Volume I of *The Collected Papers of Albert Einstein* in 1987,[44] it has been known that Albert Einstein and Mileva Maric had been the parents of a baby girl before they were married, what *Webster's Unabridged Third International Dictionary* (1976), p. 1340 calls a "love child". But what happened to the girl still seems to be a mystery. Perhaps it will be solved in a few years such that the comments by Kuwaki will prove to be superfluous, but they do seem to be relevant now. But even after the fate of the girl is better known, a second mystery while easier to plausibly explain may be even more difficult to definitively resolve, namely, why much later editions of Kuwaki's work *appear* to change the sex of the child he saw in Bern.

As I have quoted above from the 1910 publication,[45] Kuwaki was greeted at the door by "Einstein's wife and his daughter who was almost ten year's old". In this the earliest version, Kuwaki used the Japanese expression for daughter "*reijo*", a term for a daughter of respectable people and which is a very polite expression. The phrase "almost ten year's old" has to be understood in the Japanese context of the period in which a child was thought to be one year old at the time of its birth. For example, a child born in December of one year would have been considered two years old a month after birth in January of the next year. In other words, the Japanese of that time added at least one year and in some cases almost two years to

---

43 Ibid.

44 John Stachel (ed.), *Albert Einstein - The Early Years 1879-1902,* with David C. Cassidy and Robert Schulmann associate editors, Jürgen Renn assistant editor, and Robert Summerfield editorial assistant, Princeton University Press: Princeton, 1987, p. 332.

45 Ayao Kuwaki, "Ryugaku Zakki",*Toyo Gakugei Zasshi,* **27** (1910), No. 349 p. 481; also included in *Zettai to Sotai*, Tokyo, 1921.

the "real" age of the child. In short, in saying that the Einstein girl was "almost ten", Kuwaki actually intended in terms of Japanese usage of today that the child seemed eight or almost nine, which would fit almost exactly the age of the "love child". If the mentioned age still appears a little high, one might do well to keep in mind height differences between Japanese and European children of the same age, such that it would be natural for a Japanese person to overestimate the age of a European child.

But even if there is an apparent possibility that Kuwaki saw the girl, and as we shall see there is some evidence to the contrary, we are still faced with the apparent backtracking in Kuwaki's later reference. When he published his biography of Einstein in 1934, *Einstein-Den,* the passage was rewritten to the effect that he was met by the wife and then incidentally saw a child inside: "In the afternoon, I went to his house because he invited me. Then I met his wife and also saw '*sho A shi*'."[46] The word "*sho*" means little in this context, the symbol A is an abbreviation of Einstein, and "*shi*" is put after the name of a person just like Mr. is put before the name in English. It is used almost always for gentlemen but sometimes for ladies. So the expression "*sho A shi*" is rather ambiguous and can be translated as "junior Einstein". Kuwaki's book *Einstein-Den* was partly re-written into modern Japanese by Ayao Kuwaki's son Tsutomu Kuwaki the philosopher and Sigeko Nisio the historian of science in 1979. In the partly rewritten text, the same passage refers to a meeting in the house and is expressed: "In the afternoon, I went to his house because he invited me. I met his wife and also saw '*shisoku*',"[47] a word which signifies boy. From the original publication of 1910 until the revised edition of his Einstein biography in the year 1979, a girl had *seemingly* become a boy. What possible explanations come to mind?

First, the original reference is to having been greeted by the mother *and* daughter. The second and third references are about Kuwaki meeting the mother but also incidentally seeing a child, very possibly a boy in the 1934 version and explicitly a boy in the 1979 version. The important point is that the 1910 verb *mukaeta* means greeted by or welcomed by (the wife and daughter) whereas the 1934 and 1979 verbs are *ai* (Kuwaki *met* the wife) and *mita* (Kuwaki *inadvertently saw* the child or boy). In short, perhaps there is no real contradiction. He may have been greeted by the daughter at the door, as mentioned in the 1910 version, and inadvertently saw the boy inside, as mentioned in the 1934 and 1979 versions. *He may*

---

[46] Ayao Kuwaki, *Einstein-Den,* Tokyo, 1934, p. 59. The 1910 Japanese translated into romaji was: "Fujin to 10 bakari no reijo ga mukaeta." (I was greeted by the wife and the almost-ten-year-old daughter). The 1934 version was "Fujin ni mo ai sho A shi o mo mita" (I met the wife and also saw Einstein Jr.) The 1979 version is the same as the 1934 one except that instead of incidentally seeing an Einstein Jr. of undetermined sex, he saw a boy.

[47] Ibid., 1979 edition, p. 41.

well have been describing two successive but different experiences. But be this explanation ever so plausible, it admittedly still falls far short of proof.

Second, concerning the identity of the girl, could Kuwaki have confused a girl for a boy in general on his original visit? Under normal circumstances, this seems unlikely. Third, could he have confused the girl with one of the much younger Einstein boys in particular? At least initially, this seems even harder to believe. The 'love child' called "Liserl" or "Lieserl" was apparently born in January 1902,[48] which would have made her seven when Kuwaki visited the household. The oldest boy was only five. How could Kuwaki, a man gifted at clear observation of the people and objects around him, have confused him with "Liserl"?

One may add that in March 1909 Einstein was still working at the Patent Office in Bern and was far from being famous. He seemingly had nothing to hide which could injure his still meager "public" reputation. He was quite straightforward with Kuwaki and was presumably pleased to show him his family, including "Liserl". Perhaps it may have seemed prudent to keep the girl with Mileva's relatives in Hungary. But this is speculation. No one seems to have published the final answer yet, if there is one.

But there is another side to the story. It has come to our ear from a knowledgeable source, namely, Robert S. Cohen, managing editor of this series of publications, that a number of leading Einstein scholars situated down the hall from his office at Boston University currently appear to believe that Einstein's son Hans Albert, in spite of a well-known photograph of him in boy's clothes, was often dressed as a girl until he was 5, 6, or 7, that he was big for his age, and that Kuwaki, contrary to apparent probability, almost certainly did confuse being greeted by a seven year old girl, who he thought of as almost ten years old, with also inadvertently seeing a five year old boy. Furthermore, the experts have surmised that the existence of Liserl was probably concealed as early as the time when Einstein first applied for a position at the Patent Office in Bern. In the absence of additional supporting evidence, one can only repeat the current interpretation and withhold judgment. But one may expect at least an initial published conclusion soon from one or more Einstein scholars which can be supported by significantly more evidence than is available to this author now.*

Concerning Kuwaki's 1910 claim of having been greeted by the daughter and its replacement by the 1934 phrase that he had incidentally seen Einstein Jr., by then, Einstein had become the world's most famous scien-

---

[48] Stachel, op. cit., p. 332.

* The following book which is scheduled to come out in 1992 may help shed light on the situation: *Albert Einstein and Mileva Maric: The Love Letters,* ed. with an introduction by Robert Schulman and Jürgen Renn, Princeton University Press: Princeton. - Ed.

tist and it was presumably known by all educated people that while Einstein had two sons he had no daughter. The mention of a daughter who was "almost ten year's old" in early 1909 could well have raised questions which could have seriously embarrassed Einstein, Kuwaki, or both men *if* republished during the 1930's. On the other hand, *had* Kuwaki repeated his earlier reference to a girl, enough publicity might have been raised in the newspapers and elsewhere which eventually could have hounded Einstein or Mileva into explaining just what had happened to the "love child".

Another possibility, of course, is that Kuwaki now believed that he had been mistaken, that he had not been greeted by a girl, let alone Einstein's daughter, but there is no evidence for this opinion, and his careful phrasing in 1934 argues against it. Also, it would have been imprudent for him to have publically mentioned the possibility of a mistake, since it would only have brought attention to the issue, which by itself could have embarrassed Einstein. Nevertheless, Kuwaki's subtle alterations and prudent way of expressing himself in 1934, regardless of the reason, may well have contributed to our relative ignorance about the daughter now, i.e. by not socially urging Einstein to reveal basic details. Yet, on reflection, it seems very likely that Kuwaki did the kind-hearted and respectful thing precisely by not causing personal embarrassment for the Einstein he so much admired. It would be ungenerous to blame him for that.

Let me close this section of my paper by thanking Professor Cohen for his informative comments.

IV. KUWAKI AND MACH'S PHILOSOPHY

The period between Kuwaki's return to Japan in 1909 until Einstein's visit to the island nation in 1922 fell largely into what is called the Taisho Era in Japan (1912-1926). There was much industrial development and democratic government appeared to be promising. Even the Tokyo earthquake could not blunt the optimism. Though Kuwaki was soon academically buttressed by his appointment as professor at Kyushu Imperial University in 1911, the period actually saw his increasing switch from physics to philosophy.

Kyushu Imperial University was a largely one-faculty university which had developed out of an engineering college such that Kuwaki's primary contribution was apparently to teach mathematics and mechanics, though one may assume that he also sometimes taught more recent theoretical physics. One should add that it was far from Kyoto and Tokyo and the centers of active research in physics. Kuwaki's first article after coming back from Europe was in a sense his last as a physicist. It was published in three successive issues of a journal and demonstrated his mastery of a difficult subject. He explained the theory of relativity in mathematical terms and introduced the four-dimensional theory of Minkowski, but even here he could not resist a philosophical interpretation of what it meant:

> The expression 'three dimensional geometry is one chapter of four dimensional physics' by Minkowski can be thought of as a confusion of formalism and empiricism, and the term 'four-dimensional world' is already mathematical abstraction apart from experience discussed formally. Another expression by Minkowski that only unification of space and time is self-consistent and that sensation cannot always distinguish space from time brings out the thought that sensation and ideas are different. He also said that the two notions, space and time, can be obtained by separating motion in intensive and extensive ways, but are the expressions the same as so-called separation? The only thing which distinguishes Minkowski's 'world' from an ordinary theory of transcendent reality about space and time is that what he considers to be separation between them is not unique.[49]

Kuwaki paid his debt to the Japanese need for translations of important European scientific works by co-translating with Nagaoka the most recent German edition of Lorentz's *Physics* in 1913 and by himself Lagrange's classic work *Méchanique analytique* in 1916. Hantaro Nagaoka wrote in 1912 about his 1910 trip to Europe where he discussed possible translation with Lorentz:

> [Lorentz said to me] 'When I went to Göttingen and met Professor [Eduard] Riecke, I saw that his book had been translated into Japanese. I have written a physics textbook, but mine has a different character than most. Recently, it was translated into other languages. Is it possible to find someone who can do it in Japanese?' I answered that if I could have enough time, I would like to do it myself. Lorentz was glad to hear it. I also said to him that I might ask someone to help by translating parts of it, should that prove necessary. Lorentz told me that Mr. Kuwaki who had visited him some years previously had a good comprehensive ability. If only he could help. I answered that Kuwaki might be happy to do it. Lorentz then added: 'For the latest German version, I ought to add something. If you are going to translate it soon, I would like to tack on 50 or 60 more pages. . . ' The translation is being printed now.[50]

But Kuwaki's main interest had clearly shifted from physics to philosophy of science. In his article "Epistemological Issues in Physics" (1912), he compared the philosophical approaches of Mach and Planck and once again

---

49 Ayao Kuwaki, "Sotaisei gensoku ni okeru jikan oyobi kukan no kannen" (The Concept of Time and Space in the Theory of Relativity), *Tokyo Butsuri Zasshi,* 1911, No. 234, p. 203. The full article appeared in No. 232, pp. 117-131, No. 233 pp. 157-165, and No. 234, pp. 197-206.

50 Hantaro Nagaoka, *Tokyo Butsuri Gakko Zasshi,* 1912, No. 250 P. 364.

demonstrated his ability to be relatively neutral and fair to different points of view:

> According to so-called *Naturwissenschaftlichen Erkenntnislehre*, there are no facts and no reality besides sensation. It is the purpose of scientific recognition to describe (*beschreiben, nachbilden*) those sense data, that is, phenomena, in a manner which economizes thinking best. The *Erkenntnis* obtained like this can not be separated from sensation . . . On the other hand, wanting to know what is independent of our sensation is metaphysical inquiry, wanting to know reality and not just phenomena. Professor Planck and others admit this inquiry and they think that it is the purpose of scientists to obtain constant *Erkenntnis* invariant with respect to time and race, independent of our personal character and which is *emancipiert* from sensation. According to this idea scientific knowledge does not only describe phenomena, but can also know the correlation of facts which transcend [sensory] phenomena, that is, the relation between cause and effect, or in other words, to explain (*erklären*) the phenomena is the purpose of science.[51]

In his 1915 article "Physics and Recognition", which would be included in and provide the title for his 1922 book, like many positivists he interpreted early versions of Einstein's general theory as a repudiation of the absolute character of the light constancy axiom and as introducing the relativization of inertia. he also stressed the compatibility of Mach's philosophy with his own understanding of Einstein's contributions:

> The epistemological side of Einstein's theory of relativity proposed in 1905 negated recognition of the absolute motion with respect to the ether, which was thought to be physical space. Then the physical space and ether which could not be perceived were rejected and space was thought of again as a real vacuum. Also, direction in space and time was to be relative and this negated the idea of absolute space and time. After 1911, the assumption of constant light velocity was removed from the new theory of gravitation, and also the relativity of inertial mass was found.... Hertz defined mass as number of mass points but Mach defined it by a ratio between acceleration of two masses which interacted with each other such that mass could be changed according to the accumulation of mass around the mass, that is, in addition to the change of mass according to the state of motion which had been obtained by the old electron theory. The idea of relativity was inherent to what was obtained empirically. But the hypotheses which are included in the idea of relativity vary.[52]

Kuwaki concluded his paper on the following note:

---

51 Ayao Kuwaki, *Butsurigaku to nishiki* (Physics and Recognition), Tokyo, 1922, p. 33. For the original article see: "Butsurigakuju no ninshiki no mondai" (Epistemological Problems in Physics), *Rigakukai*, **9**, No. 9 (1912), 1-8.

52 Ibid., pp. 11-12.

> As an empirical science, physics cannot escape from the range of perception, still, it gives a system of deduction for the accumulated knowledge of perception. And for the purpose of making it easy to make new discoveries, physics produces world views according to the conditions required. However, world views which fulfill the given conditions can be thought of as being infinitely many, just as indefinitely many curves can be drawn which pass through given points . . . . The physical world view should be a world view under physical observation. The problem of reality is not a physical problem.[53]

Kuwaki published numerous philosophical articles together with a few historical ones shortly before, during, and after World War I. They were all connected with science and several showed his mastery of parts of it, but the focus of concern was increasingly with what he regarded as the philosophical foundations of science. Nor was he alone in this investigation. He was able to attract his friend Torahiko Terada as well. Terada had graduated from Tokyo University a little later than Kuwaki and had become an experimental physicist, mostly in geophysics, but his interests were also broad. Terada's talent in writing essays attracted many young students to study science. Let me give exerpts from two of Terada's letters to Kuwaki during this period. The first is dated January 12, 1916:

> After I read the books of Poincaré and Mach last year, I have become interested in the basic problems in physics. When I have free time, I read books in that area. I am also reading books written by your brother. Please guide me also along this path in the future.[54]

And Terada to Kuwaki, January 19, 1917:

> Thank you very much for mentioning the book by Pierre Duhem. I would like to buy it soon and read it. I have read your lecture 'Descriptivism and Science', and am interested in it very much. I have been stimulated by it in various ways. If you are going to give lectures on those thema, I would like to know more about your approach. Recently, I have been reading an American translation of Mach's *Analysis of Sensations*. I think that the book is very interesting. I still have some points where I disagree with him, but even then his thoughts are suggestive and I admire the book. In my childhood I felt a kind of premonition before my arrow would hit the target. Mach's book suggests that it might have been a kind of illusion about the sensation of time.
>
> When I was a student in the fifth year of middle school, a teacher of history recommended to me that I should study psychology in the faculty of humanities. But I entered science and never did learn what psychology was. But after reading Mach's book, I feel I would like to study psychology. But

---

53 Ibid., pp. 13-14.

54 *Terada Torahiko Zenshu* (The Collected Works of Torahiko Terada), Volume 15, Tokyo, p. 62.

> I have to be patient so as not to be distracted from my work too much. Though apologizing for my enjoyment, I bought a book by William James at Maruzen bookstore. Would you like to tell me about some interesting books in psychology?[55]

One might add that Terada once wanted to write a book (*Introduction to Physics*), but he could not finish it. However, the written part was published. One can see Mach's influence on Terada in it. Another philosophically-oriented friend of Kuwaki was Jun Ishiwara, who later translated Einstein's lectures when he visited Japan. Inspite of his philosophical interests, Ishiwara was also known as a physicist, having studied with Einstein about the theory of relativity. It is common knowledge that Kuwaki and Ishiwara often discussed Einstein's theory together. It is possible that Kuwaki's deep understanding of physics helped Ishiwara a lot.

I have mentioned before that Hantaro Nagaoka, Kuwaki's revered teacher at Tokyo University, thought highly of his former student. In general, he did not like his pupils to study history and philosophy of science. Kuwaki seems to have been the only exception.

Most of Kuwaki's philosophical articles were gathered together in 1921 and 1922 into two books *Absoluteness and relativity* (1921) and *Physics and recognition* (1922). The preface to the latter spelled out his hopes for the renewed union of physics and philosophy:

> The close relation between physics and philosophy has not been universally acknowledged until the recent period. And of course it is derived from Einstein's theory of relativity. The change of the idea of space and time which has been caused by the new theory is thought to have stimulated alterations in philosophical ideas. But scientists did not anticipate that this change in philosophical outlook could also influence scientific problems. Originally, science and philosophy shared the same search for universal truth, but their methods were completely different. Therefore, it is only lately that correlations [even in this area] have begun to be noticed. . . . Recently, concerning the problem of whether the Lorentz-Fitzgerald contraction is real or not, scientists have not been able to remain indifferent to the question about what it means to be real.[56]

Kuwaki also began to become a little more explicit about his own philosophical position:

> There may be several philosophical standpoints from which to treat scientific problems. If I may not be misunderstood, my position is thought to be that of criticism. But in detailed arguments, there may be several distinctions one can make between different kinds of criticism, such as between idealism and empiricism. In the articles collected in this book the reader may recog-

---

55 Ibid., pp. 72-73.

56 op. cit., *Butsurigaku to Ninshiki,* pp. 1-3.

> nize that my philosophical standpoint tends to be that of empiricism. Also, the author's voluntaristic tendency has come from his relativistic, genetic standpoint. . . . The epistemological study of physics began around 1870 and was caused by the controversy over whether physics was more concerned with explanation or with description as Kirchhoff and Mach had maintained. I would like to pursue this development somewhat further in this book.[57]

Kuwaki ended the preface with acknowledgements to Hantaro Nagaoka, Kitaro Nishida who was the most famous philosopher in Japan at that time, Hajime Tanabe a fellow philosopher of science, and his brother Gen-yoku Kuwaki. They indicate how far his interests had strayed from doing professional physics.

V. KUWAKI AS A HISTORIAN OF SCIENCE

Kuwaki often wrote about Einstein's physics and especially what he regarded as its philosophical implications. He took it for granted that Mach had strongly influenced Einstein, but as the following 1921 paper "The Philosophy of Einstein" indicates, he was not really sure what Einstein's philosophy was:

> Since Einstein has not yet written about his own philosophy or which philosophical assumptions made his theory possible, we cannot know the answer. We can see that in several papers he appreciated Ernst Mach, who favored positivism and relativism. As I am going to mention later, no one can deny that Mach's *Erkenntnislehre* had seriously influenced him. But it is not clear that Einstein's philosophical opinion, if he has one, is completely the same as Mach's.[58]

Nor was he willing to fall victim to confusing physical with philosophical relativity:

> According to Auguste Comte, the only absolute truth is that all truth is relative. But Comte was talking about relativity in a philosophical sense. What Einstein argued for was relativity of motion.[59]

Learning about the philosophical assumptions of other people is not always easy, but Kuwaki's curiosity had been aroused, and during the last decades of his life his interest in philosophy would be substantially augmented by historical research, not least about Albert Einstein himself. One of the earliest signs of Kuwaki the historian, apart from his early reading

57 Ibid. pp. 4-5 & 7.

58 Ibid., p. 112.

59 Ibid., p. 116.

and impressionistic reports on his first trip to Europe, was his 1916 paper titled "D'Alembert's Dynamics" in which he mainly discussed d'Alembert's so-called principle.[60] This was followed in 1917 by a paper "History and Development of the Notion of Force"[61] in which he discussed the ideas of Mach, Kirchhoff, and Hertz, but his attitude was even more neutral than in his philosophical articles.

From 1911 until 1938 Kuwaki was a professor at Kyushu Imperial University in Fukuoka. The city is not far from Nagasaki where even in the period of Japanese national isolation the Japanese Shogunate government had diplomatic relations with Holland. So Nagasaki was the only window to see the Western World for Japanese people during the Edo period. The translators of the Dutch language in Nagasaki could be the pioneers of Western science. They introduced and translated some books on astronomy, physics, chemistry, and medicine. Some of the translators had a reasonably good ability to understand Western science and introduced and explained what they could about Western science. These old books were largely left in Nagasaki.

Isolated as Kyushu Imperial University might have seemed, its presence in the city of Fukuoka had this one very big advantage for an historian. It was close to Nagasaki. Kuwaki began hunting old documents down, and not just Western ones or even Japanese attempts to learn from them, but manuscripts going back centuries, and many on Japanese efforts to develop science and especially mathematics by themselves.

In 1920, Ayao Kuwaki wrote an article "Pioneers of Science in Kyushu"[62] in which he outlined the efforts of some 18th century Japanese pioneers in Western science. For example, Jindayu Motoki, translated a book on astronomy in 1774 which was the earliest translation in Japan and which introduced the Copernican theory to the country for the first time. Kuwaki had found the original translation and studied it. He continued to collect documents on the Edo period in Nagasaki until he was retired from Kyushu Imperial University. His collection "The Kuwaki Archive" is very important for researchers on that era of Japanese science.

During the Showa period, that is from 1926 on, Kuwaki's interest in history of science continued to grow. He may well have been the first person in Japan to study history of science based on original documents. Nevertheless, I would like to focus on what is perhaps his best known book in the history of science: *Einstein-Den* his 1934 biography.

---

60 Ayao Kuwaki, "D'Alembert no Rikigaku", *Butsurigaku to Nishiki,* Tokyo, 1922.

61 "Chikara no Kannen no Hattatsu no Rekishi", ibid.

62 "Kyushu ni okeru no senku", Appendix II, ibid.

Concerning Einstein, Kuwaki met him again on Einstein's visit to Japan in 1922 and still again in 1927 when Kuwaki visited Europe for the second time. Both events are briefly mentioned in his 1934 biography:

"Einstein and his wife who had taken the *Kitano Maru* landed at Kobe on November 17, 1922 . . . . Dr. Hantaro Nagaoka from the Japanese Imperial Academy, Mr. Yamamoto from Kaizosha [publishing house which had invited Einstein to Japan], Dr. [Jun] Ishiwara, Dr. Aichi from Tohoku Imperial University, and myself went to the boat to welcome him. That night he stayed in Kyoto and the following evening arrived at Tokyo station.[63] Mr. Inagaki who had accompanied Einstein and his wife from Shanghai described their arrival at Tokyo station:

> We felt almost crushed by the people who came to welcome him. His wife was shortsighted and was afraid of crowds. After being shoved into an elevator and reaching a [reception] room in the station hotel, we noticed that Einstein's necktie and his wife's dress had been mangled. After receiving many flower bouquets, flower rings, cheers of 'Banzai!', and shaking hands - real turmoil - they could go to the Imperial Hotel.[64]

The very first day he arrived in Tokyo, Einstein talked with us about recent developments in physics. Einstein spoke about his theory of gravitation, but electromagnetic force was not included in his theory. He then discussed a number of attempts to combine the two:

> Weyl's theory is skillful, but it is merely mathematical and not physical. Eddington's theory is mathematically too complicated. I agree with Kaluza's five-dimensional theory, but I am not satisfied with it. The theory of the universe is now being developed. Quantum theory is at a turning point.[65]

Einstein's lectures in Tokyo and in other big cities were mostly translated by Dr. Ishiwara. Concerning auditoriums, only the largest halls in those cities were used, but they were still full. I do not think that many people could understand his lectures, but his naive and scholarly attitude strongly impressed his listeners, and the lucid translation by Dr. Ishiwara was much admired by the audience.

One day during his two week stay in Tokyo he was invited to see a Noh play. . . . When he arrived at Hosho Noh theater in Kanda, he was introduced into a dressing room and I explained about

---

63 Ayao Kuwaki, *Einstein-Den,* Tokyo, 1934, p. 106, and 1979 edition, p. 72.

64 Ibid., (1934), p. 107, and (1979), p. 73.

65 Ibid., (1934), p. 112, and (1979), p. 75.

Noh plays in general and the one being presented that evening using the book by [Ernest] Fenalosa. . . . Dr. Torahiko Terada and my brother supplemented my explanation. At that time it was quite rare for foreign people to see Noh plays. The theater prepared a short version of *Hagoromo* and played it for him. But he wanted to see other plays too. I thought that he might have become tired, because he could not understand the verse of the play. Still he watched it very earnestly. He especially enjoyed watching the fast dance of the dragon god in the second half of the play.[66]

Einstein came to Fukuoka in Kyushu [where Professor Kuwaki lived] after he visited Tokyo, Sendai, Nagoya, Kyoto, Osaka, and had stayed a night at Miyajima. He gave a lecture in a theater in Fukuoka, but the theater did not have a blackboard and we borrowed one from a middle school. Einstein used it and the middle school has kept the blackboard with his writings on it ever since.

In the city of Fukuoka, he stayed in a traditional Japanese inn called '*Sakaeya*'. It was the first experience for him to stay in a non-Western hostel. He observed the *gaku* [Japanese calligraphy formed into a frame], the *kakemono* [scrolls of Japanese paintings hung as ornaments], and the *fusuma* [sliding doors made of wood and paper]. He said that the room conveyed him to a land of poems and dreams. He was surprised by absence of keys in the room and by the beautiful quilts . . . . During his stay in Sakaeya Inn, he was asked to draw something on a silk cloth with a brush, and he drew the letters S A K A E Y A."[67]

Kuwaki also wrote down a few remarks about his visit to Einstein in Berlin in 1927, after he had attended the international conference in Como. Kuwaki talked with Einstein about physics:

"His hair had become whiter than the time I saw him in Japan. When I said to him that I had expected to meet him at Como he answered that he didn't like to go to places where so many people had gathered. He asked me: 'What are you studying now?' I answered 'I am only trying to follow [the development] of other theories'. He murmured: 'Recently, it has become quite a task to do that.' And he said the following about a number of topics:

There is reconciliation between wave theory and discrete theory in a new experiment in the United States; Schrödinger only represented a part of the idea of Louis de Broglie. Uncertainty by Heisenberg is uncertainty of obser-

---

66 Ibid., (1934), pp. 111-113, (1979), pp. 76-77.

67 Ibid., (1934), pp. 114-115 and (1979), pp. 77-79.

vation, and the idea is interesting but I don't agree with it.There is difficulty in the development of field theory. There is extension of parallelism. The relation between energy and frequency is basic, or is it rather wave length? In this theory duality between space and time is broken. I feel the future of physics is *dunkel*. I have studied physics for twenty three years. I feel I'll go mad even thinking about it.

While we were talking, his wife came into the room. She said to me: 'Do you remember me still?' We talked about their trip to Japan. He asked me how Japanese literature was doing outside the traditional approach. I said to him that Russian literature had been introduced somewhat early. We talked about Dostoevski's Raskolnikov. Einstein said to me that the present situation in Russia was a big experiment, as was the new movement in psychology. He also mentioned that the recent arguments of physicists about causality were a kind of fashion.

Einstein talked in such a quiet and relaxed way that I could not leave his house for a long time. Afterwards, I would sometimes run into him in a subway, in a bus, or on returning from a colloquium. On such an occasion, he told me that 'the colloquium was *hübsch*, the theory of relativity was not welcome in Berlin, the talk of many young physicists who do not consider position and velocity may change soon, *Realität* was extracted from causality before, but now causality is doubted and still *Realität* is not doubted'. Afterward, I went to Paris and London and looked for old books. Then I returned to Berlin and left the city for Japan at the end of February 1928."[68]

Kuwaki's interest in collecting information about Einstein also focused on the occasional biographical attempts which had been made by friends and scholars of the still relatively-young genius. In order to explain the character of his 1934 biography of Einstein I would like to quote from his preface:

Two biographies of Einstein have already been published. One was written by Alexander Moszkowski in Berlin in 1920, and another one has been written by Anton Reiser [Rudolf Kayser] which was published in London in 1931. According to *Wer ist,* Mr. Moszkowski is a journalist born in Poland and a chief editor of the magazine *Lustige Blätter* in Berlin. About Reiser I do not know anything. Both books were written on the basis of stories told by Einstein and his wife. In the former book, the author wrote that Einstein had not checked it before publication, but in the latter Einstein wrote a preface and approved the book as almost correct. What I could learn about his boyhood, student years, and private life comes from these two works. I have also examined Ostwald's collected papers, *Wer ist,* several kinds of

68 Ibid., (1934), pp. 220-223, and (1979), pp. 148-151.

> encyclopedias, and Einstein's short biography added in French or English translations of his articles. In those references, I could find when he moved to Munich etc., and added something. New and old copies of *Minerva,* collections of catalogues of lectures in universities, *Dictionary of Biographies of Scientists* by Poggendorf was used to examine his student years, his years in Switzerland, etc..
>
> Recently, unexpected difficulties have occurred in his life, which I described using his *Mein Weltbild* published in Amsterdam in 1934 plus magazines and newspapers in England, France, and the United States.
>
> In order to know Einstein as a scientist, it is important to analyse his scientific works and quite a few books interpreting his theories have been published, but a book where other physicists have reviewed the progress of his scientific study does not seem to have been published yet. Even though my book is short and fragmentary, I have tried to analyse his scientific works and describe the circumstances of scientific study relating Einstein's works. I have also added geographical and historical explanation, biographical data on scholars, and have presented an internal and external account of the circumstances under which his work was done. Nevertheless, I am afraid that I have allotted too much space to side stories.[69]

As a rule, it is not wise to clutter articles with either quotations or lists, but given the inaccessibility of Kuwaki's major writings for so many European and American scholars, and especially his biography of Einstein, I have thought that it is the only practical option left. At this point, I would like to present a list of the contents of Kuwaki's *Einstein-Den* in the hope that some aspects may move Einstein scholars sufficiently to request that some of their Japanese colleagues provide more detailed information on the parts of the book which they are most curious about. The book has twelve chapters and their contents are as follows:

1. From Boyhood to Graduation from ETH:

   His birth - Who was his father - Prediction by his mother - Primary school period - Middle school period - Newton - The Pythagorean theorem - Young - Hamilton - His taste for natural science - A half year in Milano - Choice of profession - Entrance examination - ETH - His friends - Student life.

2. From Graduation to the Period in Bern:

   Assistant - Technical School - The Patent Office in Bern - Director Haller - What he did in the Patent Office - His family - His friends.

3. Einstein's Work on Physics in Bern:

   His papers after 1900 - Four papers in 1905 - His friend

[69] Ibid., (1934), pp. 2-4, and (1979), pp. ii-iv.

Besso - Privatdozent at the University of Bern - Extraordentlich Professor in the University of Zürich.

4. Responses to his Earlier Work:

Einstein's complete discontinuous theory - Improvement of Lorentz's electron theory - The Lorentz contraction - Kaufmann's experiment - The Congress of Physicians and Scientists in Stuttgart - The Congress in Cologne - Einstein in Berlin.

5. From 1909 to 1912:

The University of Zürich - The meeting in Salzburg - The University of Prague - Poincaré's Recommendation.

6. The Progress of Relativity Theory after 1911:

Propagation of light - The observation of solar eclipse - Abraham - The dinner party at Professor Planck's house - Minkowski - Laue - The debate with Abraham - Webster - The Congress of Physicians and Scientists in Vienna - The reports of Abraham - The reports by Laue,

7. Einstein Moves to Berlin, Demonstration of the General Theory Theory of Relativity, and trips Abroad:

The invitation from Berlin - The quantum theory of radiation - Observation of the solar eclipse in 1919 - The response - His lectures in Paris and London - Sir Haldane - Einstein comes to Japan.

8. Einstein's Subsequent Career I:

Opposition to the theory of relativity - Society of Jewish people - The concert - Einstein's illness - The field theory - Berlin and humor - The fiftieth anniversary of his birth - Gift from the city of Berlin - His study room in Berlin -

9. Einstein's Subsequent Career II:

Einstein and Bernard Shaw - Einstein's wife - His views on the United States of America - The Institute of Advanced Study in Princeton - Flexner's opinion about the university - Johns Hopkins University - The Flexner brothers - Compton's address - The Compton brothers - Lord Rayleigh - Shapeley - Einstein's lecture - Einstein's confession - His withdrawal from the Prussian Academy - The British Association for helping German refugees - The Congress in Paris - Methods of theoretical physics - Einstein's escape to London - His London lecture - Laue's lecture.

10. Einstein and Lorentz:

Acquaintance with Einstein - Death of Lorentz - About Lorentz by Larmor - Planck's address on Lorentz - Solvey Conference - Volta - International Conference on Physics - Como - Pavia - Rome International Conference - Einstein in Berlin - Einstein's obituary on Lorentz's death.

11. Einstein and Planck (1):

Planck's first letter to Einstein - Biography of Planck - Hertz's resignation and his last notes - Planck's rejection of the invitation to Vienna - Planck's successor - Planck's doctoral thesis - Boltzmann's travel notes - The University of Berlin - Helmholtz - Reichsanstalt - The radiation theory - The birth of quantum theory - Klassiker und Romantiker - Planck's dinner party.

12. Einstein and Planck (2):

Bohr - Heisenberg - Axioms of atomic theory (basic principles of the atomic world) - The uncertainty principle - Abolition of causality - de Broglie waves - Mechanical causality - Mach - Meyerson - The realism of Planck and Einstein.

Appendix: List of Einstein's papers.[70]

VI. MORE ON KUWAKI'S INTEREST IN MACH

From 1920 until the publication of his *Essays in the History of Science* in 1944, a year before his death, Ayao Kuwaki continued to write and publish a vast number of articles of which forty-five were included in that large book of almost 600 pages.[71] The first 195 pages were on Meiji and especially pre-Meiji Japanese science, but the rest was generally on aspects of Western science, its history, and philosophy. He included articles on the science of Aristotle, Archimedes, Leonardo da Vinci, Francis Bacon, Newton, Immanuel Kant, Faraday, Maxwell, Gibbs, Poincaré, and Rutherford. In philosophy, he wrote on Meyerson, Moritz Schlick, and especially Ernst Mach. As might have been expected, he kept abreast of the development of the Vienna Circle during the 1920's and of Logical Positivism during the 1930's. The list of books in his library not only reveals numerous publications from Bertrand Russell and A. N. Whitehead which might have been expected but also a large number by Hugo Dingler and his pupil Rudolf Carnap during a period when neither were household names anywhere.

---

70 Ibid., (1934), pp. 1-4. (Only the earlier edition lists the detailed contents of each section.)

71 *Kagakushi-Ko*, Tokyo, 1944.

Kuwaki wrote two articles on Ernst Mach's philosophy and view of science which were included in the book, and he or his ideas are mentioned in thirteen other papers. The two specific articles were titled: "Mach and the Theory of Relativity" (1922) and "Ernst Mach's World View" (1938).[72] In the earlier paper, one sees Kuwaki's lingering identification of Einstein's epistemological views with Mach's approach, but it is possible he may have been increasingly disturbed by the feeling that the situation was actually more complicated:

> No one doubts nowadays that Einstein's theory of relativity coincides with Mach's Erkenntnislehre. Einstein himself often said so. But it can be imagined that concerning basic philosophical [i.e.metaphysical?] opinions that Einstein is apart from Mach, that is, in so far as Einstein agrees with Max Planck. Still, it is quite a sensation to hear that Mach rejected the theory of relativity strongly before his death in 1916. Last year the book *Die Prinzipien der physikalischen Optik - historisch und erkenntnispsychologisch entwickelt* was published. I have not seen the book yet, but I saw a review of it by Max von Laue in the recent *Die Naturwissenschaften.*[73]

Kuwaki then introduced an outline of von Laue's review and mentioned once again his own experience when he had met Mach in Vienna. He followed that reference with conjecture about how he thought Mach might have changed his opinion had he been more familiar with general relativity and Einstein's obituary of him:

> When I met Mach ten years ago, it was after Einstein's special theory of relativity had been published. Mach said to me that Einstein's principle was merely a theory. At that time Mach already was so old that I did not think to ask in further detail, and I thought then that I could almost understand what Mach's idea was. Einstein himself thought that his theory was heuristic and a conventional hypothesis. At that time, it was thought that there was an essential difference between Einstein's theory of relativity and Mach's epistemological relativism.
>
> But when Mach wrote the preface [to the *Optik*] in 1913, Einstein had already begun moving from the special theory of relativity to the general theory where he had derived the relativity of mass from a suggestion by Mach with the help of the mathematician Grossmann. I think it would be natural for Mach to have kept the same attitude toward the [Max Planck dominated?] society of physics as when I had met him. But if Mach could have known Einstein's paper of 1916 and Einstein's obituary on Mach's death, I think that the preface would have been considerably rewritten.
>
> Mach's opposition to atomism had brought him into conflict with Boltzmann and opened the way for Planck's challenge, but the opinion of Mach was thought to be too divergent from the physics of our day. Historically, the situation looks similar to when Auguste Comte had opposed the

---

72 "Mach to sotaisei genri" and "Mach no sekaikan".

73 *Kagakushi-Ko,* op. cit., pp. 455-456. [See chapter 3, section C of this anthology for a translation of von Laue's review. -Ed.]

> ether hypothesis at the beginning of the 19th century. Nevertheless, concerning theory of physical space, I think Mach was closer to Riemann than to Comte. Indeed, Einstein said in his Leyden lecture that he would be willing to keep the notion of ether if it was understood in Mach's sense.[74]

Kuwaki ended the article with an expression of hope:

> Anyone who knows Mach's extraordinary character must wait earnestly for the publication of the continuation of the book, even in an unfinished version.[75]

In the 1938 article, which appears to have been connected with celebration of the 100th anniversary of Mach's birth, Kuwaki sketched Mach's biography but also referred to a subject dear to Japanese at the time by mentioning that Mach's father had been able to successfully raise oak-leaf-eating silk worms on his farm in Slovenia, but had not been able to find commercial buyers.[76] Kuwaki mentioned that he thought that Mach's early book *Compendium der Phyik für Mediziner* (1863) had been a turning point in Mach's move from the atomic theory to sense monism. In this article, Kuwaki quoted Einstein's famous obituary on Mach's death. Kuwaki wrote at the end of the article:

> Mach's rejection of 'absoluteness' and his criticism of Newtonian mechanics only threw doubt on one type of physics and did not construct anything positive. Nevertheless, it is rather natural that orthodox physicists did not welcome it. Mach's idea plus the idea that the velocity of light is constant could be thought of as a main theme of physics.
>
> It should be acknowledged that Mach's philosophical ideas reflected his personal character and originality, but this has not stopped people from thinking that his ideas could be covered under a general label such as 'positivism' as the opinions of Hume and Comte have been. In the history of philosophy, the opposite of positivism has been realism. It may be difficult or even useless to argue between realism and positivism over which is true.[77]

Kuwaki then mentioned how both theories had changed through time. He concluded his article as follows:

> The postulate of the unity of nature is the ultimate one on which we base knowledge and experience. Mach's positivism looked as if it could not be

---

[74] Ibid., pp. 457-458.

[75] Ibid., p. 458.

[76] "Mach no sekaikan", ibid., pp. 385-391. [Marie Mach described her father's attempt to raise silk worms in some detail in her book *Errinerungen einer Erzieherin*, Vienna, 1913, pp. 33-50. - Ed.]

[77] Ibid., pp. 389-390.

> compatible with the atomic theory, and it has been widely thought that such a philosophical opinion which was so immersed in sense experience was useless and bad. [On the other hand], the negation of causality in the new quantum physics while derived from atomic theory has been held to have had positivism as its leading principle. However, to the extent that physics looks for models and needs some hypothesis to unify it, then one might say that it has a tendency toward realism.[78]

Even later, in an article first published in 1941, Kuwaki gave what might be regarded as his final position on Mach:

> There are of course other points of view besides that of Mach about the history and philosophy of science, such as those of Planck and Meyerson, but one cannot overlook that Mach had a particular character as a pioneer and original thinker. Were someone to ask me what is my favorite book, I would not hesitate to answer that since my student days the best one has been Mach's *Mechanik.*[79]

VII. CONCLUSION:

It would perhaps be appropriate to finish this paper with remarks about Kuwaki's last years, two of his students, and a pair of general observations. After he retired from the University of Kyushu in 1938 he became the principal of a high school called "*Matsumoto Koko*". Concerning the old system of *Koto Gakko* such as *Matsumoto* which were then in effect, these institutions had been established to prepare students to enter universities. Most of them had been established by the government, and they were relatively few in number, though later, some municipal and private schools were established which also called themselves *koto gakko*. The original or "real" *koto gakko* had a curriculum focused on humanities, sciences, and languages. Individual classes were organized in the European fashion, that is, where the same group of students basically stayed together until they graduated some years later. But in Japan, the different classes or groups of students were also classified, first, by science or humanities, and second, by which language was given the most emphasis, English, French, or German, though students also had to learn other languages. Nevertheless, the schools closely resembled German Gymnasiums in numerous ways. Students were taught to read foreign books in the original languages but they were also given enough free time to think about philosophical issues. Most graduates remembered their young days in the school with their respected teachers and good friends. They believed that school days in *koto gakko* created their cultural background. The schools were much more than mere prep

---

78 Ibid., p. 391.

79 Ayao Kuwaki, "Shi, tomo, shoseki", *Kagakushi-Ko,* Tokyo, 1944, p. 531.

schools for universities or even a method for developing an intellectual elite. Unfortunately, however, women were not allowed to enter such institutions, and the whole system was altered after the war, outwardly to fit an American approach, and inwardly to put an extraordinary emphasis on doing well in middle school and high school examinations. It was a considerable honor at least before the war to be appointed principal of a real *koto gakko,* though in Kuwaki's case his reign was short, since he died in 1945.

In 1941, Kuwaki became the first president of the Japanese Society for the History of Science. He wrote an article for the first edition of the society's journal *Kagakushi Kenkyu* titled simply: "On the Study of History of Science". Kuwaki picked his own brain for fine models from European and American science of the past emphasizing the importance of the discipline, even quoting the words of George Sarton, the longtime editor of *Isis*. He ended the article with optimism:

> I hope that an ideal institute might be established in Japan which would have the size suggested by Sarton to be able to study the cultures of both East and West with the magnitude which their importance deserves.[80]

The most outstanding scholar among Kuwaki's students seems to have been Kiyoshi Amano (1907-1945) who accomplished a great deal in a short life. He worked as assistant under Kuwaki for two years from 1932 to 1934 where he helped his mentor collect material for *Einstein-Den*. Amano then moved to the National Research Laboratory of Meteorology where he worked on the standard temperature scale of thermometers in high temperature. His main books were *Origins of the Theory of Thermal Radiation and Quantum Theory* (1943) and a posthumous work published in 1973 by Seiji Takata *History of Quantum Theory*. The 1943 volume was primarily on the work of Max Planck, but he also described much technical and experimental background.

Another disciple of Kuwaki, who, however, had not been one of his students, was Suketoshi Yajima. He was interested in Japanese history of science and not only carried on Edo period research in general which Kuwaki had opened up, but continued on with the very investigation Kuwaki was engaged in when he died. Yajima wrote an obituary on Kuwaki in *Kagakushi Kenkyu* in 1949 when the journal in the history of science was revived after the war years. Yajima also studied the history of Islamic science and produced several biographies of Western scientists who had previously been little studied or known in Japan.

Mach's influence in Japanese physics has been strengthened by publication of virtually all of his major works, especially his *Mechanik* in 1931,

---

80 "Kagakushi no kenkyu", ibid., p. 11.

and by the tendency of many textbook writers to employ his definition of mass. On the other hand, Kuwaki's interest in Mach never seems to have interacted either with Left Wing concern over "Machism" or with the profound but esoteric Nishida school of philosophers in Kyoto. Kuwaki was not just independent of these streams, always preferring to go his own way, but he seemed loyal to what he wanted to do, which was to be a generalist in physics, philosophy, and history of science while mastering the basic assumptions of the first two while opening up serious historical research into Japanese science. Above all, I think he enjoyed his life. He truly loved learning.

CHAPTER 17

ERNST MACH'S INFLUENCE ON FOUR JAPANESE THINKERS

By Michael A. Santone Jr.*
(Temple University, Japan)

PART ONE / A.

Starting with a brief account of early Japanese interest in Western philosophy, part one of this paper will focus on the influence of Ernst Mach's ideas on a small group of philosophers of science. We will begin with *Ishiwara Jun,* follow with *Tanabe Hajime* from the influential Kyoto School of Nishida Kitaro, and conclude this section with Mach's philosophical influence on *Taketani Mitsuo,* a thinker of some importance about methodological questions in science and a colleague of the Nobel prize winner, Yukawa Hideki.

Part Two will be devoted entirely to the influence of Ernst Mach on the carefully thought-out, relational philosophy of *Hiromatsu Wataru,* who is Professor of the History and Philosophy of Science at Tokyo University. (Following Japanese custom, family names are given first and then personal names second.)

The introduction of Western thought into Japan, particularly scientific and philosophical thought, was conditioned by its role in the modernization of the state.[1] Though British and American instructors at the University of Tokyo, the first and generally most important National University, introduced their students to Comte, Mill, Spencer, and Kant, the emphasis would change by the 1890's. Prussia, which had recently made spectacular progress with its own industrialization, became the preferred model and German philosophy soon took the forefront, with students sent to the Second Reich, first, largely to study Hegel, and second, given the strength of Neo-Kantianism in German universities, to study the major works of the sage of Königsberg himself, who soon dwarfed his rivals in terms of major influence on Japanese professional philosophers, and to a surprising extent even

---

* This article was especially prepared for this anthology. The author studied at Tokyo University and is a philosophy instructor at Temple University Japan.

1 All translations except for Taketani (1951) in Nakamura (footnote 15) are those of the author. The book by Piovesana mentioned in footnote 8 is to my knowledge the only history of Western philosophy in Japan in English, and it only reaches the early 1960's.

J. Blackmore (ed.), Ernst Mach – A Deeper Look, 333–361.

on physical scientists with an interest in philosophy. One should add, however, that even today, it is normal to compartmentalize philosophy as a discipline, first, away from science in general,[2] and second, as divided between Japanese, Chinese, and Western thought frequently taught in different buildings and institutes, and often in their original languages or older versions of current ones. This practice has resulted in a noticible tendency to think of all school philosophy as somewhat remote, including Western points of view.

The reception of European science was taken much more seriously, especially applied science.[3] Physics tended to be a relative latecomer, but was almost immediately confronted with a significant transition from Newtonian mechanics to the modern physics of Planck and Einstein. It seemed that just as some Japanese physicists were on the verge of mastering the earlier approach that the advent of quantum mechanics and relativity theory produced considerable mental turmoil and not a little conflict within Western philosophy of nature, conflict which spilled over into Japan. On the other hand, the introduction of science has probably been somewhat facilitated by the sophisticated intellectual culture which had developed over many centuries, based especially in Kyoto, the old Imperial capital, culture understandably derived largely from Shinto, Buddhism, and Confucianism. The vocabularies of these older traditions could be exploited in order to create the language of the new science. In addition, there was a well-developed native mathematical culture (wasan) originally stimulated by ideas from China and the need to fix the yearly calender more exactly but which had gradually advanced to the level of pure mathematics pursued for its own sake. Western mathematics, apparently unlike physics, had been introduced from the end of the seventeenth century, though it is not clear how widely such ideas were circulated.

While some of Ernst Mach's contributions to science and the publication of his influential book *The Science of Mechanics* had aroused some interest in Japan even during the 1880's and 1890's, his ideas on the philosophy of science appear to have first made an impact during the first decade of the 20th century, precisely when the transition from Newtonian to modern physics was becoming serious and major scientific and philosophical conflicts were developing into the order of the day.

Mach's philosophical influence in Japan seems to have followed at least two different paths, the first was led before 1910 by the physicist Kuwaki Ayao who also seems to have introduced Einstein's ideas into Japan. Kuwa-

---

2 Samuel Coleman, "*Riken* from 1945 to 1948", *Technology and Culture*, **31** (1990), 228-250.

3 See James Bartholemew,*The Formation of Science in Japan,* Yale University Press: New Haven, 1989.

ki also made a tour of Europe in 1908 and 1909 where he met Planck, Einstein, Poincaré, and Mach. The second group, based in Kyoto would start with professional philosophers largely focused on the ideas of Kant, Hermann Cohen, and Rickert. But some of their students would be physicists who would study aspects of Mach's contributions to science often together with Vladimir Lenin's 1908 book directed against the "Empirio-Criticism" of Avenarius and Mach. The result would be greater interest in the controversial Austrian combined with often heated discussions about the soundness and applicability of his philosophy of science.

B

Our account begins with the physicist Ishiwara Jun (1881-1947) who in a general sense may be considered to have worked in the half-scientific, half-philosophical path of Kuwaki and who in his Neo-Kantian way would do his very best to reconcile Mach's ideas with those of Planck. Ishiwara graduated from the physics department of the University of Tokyo (which was then called the Imperial University of Tokyo) and continued on to graduate school. Unlike almost all of his colleagues, he was Protestant and the son of a Protestant minister.[4] From 1912 to 1914, he studied relativity theory and quantum mechanics under Sommerfeld. He also visited Einstein in Zürich. In fact, Ishiwara would serve as Einstein's translator during the latter's trip to Japan in 1922.

Ishiwara obtained a position as a professor at the National University of Tohoku in northern Japan. He also published many papers for the Tokyo Society for Mathematics and Physics which began the task of attracting interest in the new theoretical physics. But his position was cut short in 1921, that is, immediately before Einstein's visit to Japan. Apparently, a love scandal broke out at his university which forced Ishiwara to resign (perhaps helping to set back the introduction of important aspects of theoretical physics in Japan by several years in spite of the newspaper attention given to Einstein during his much publicized lectures up and down the country). Ishiwara then worked for private companies, but starting in 1931 he became editor of the important journal *Kagaku* (Science) published by Iwanami Shoten.

His philosophical development which as already suggested leaned toward Neo-Kantian epistemology found expression in 1929 in his book *Shizen kagaku gairon* (Outline of Natural Science). He later published a collection of papers under the title *Shizen kagakuteki sekaizo* (A Natural Scientific Worldview) which appeared posthumously in 1948. In his 1929 book,

4 Hirokawa Shunkiti, "Ishiwara Jun", *Kagaku* (Science), **50** (1980), 768.

Ishiwara recognized a fundamental mystery of nature, but unlike the Kyoto scholars - who we will soon discuss - attempted to fit science into a larger metaphysical or religious scheme, he clearly separated their domains, science to the natural and religion to the supernatural. All knowledge of natural fact was limited to science, and Ishiwara wrote he only felt the mystery of nature when events were explained in terms of laws expressed in natural science: There should be no conflict between religion and science.

In what may seem an odd juxtaposition, he appealed to both Planck and Mach for support on the same page of *Outline*.[5] He drew on Planck to support his view that we preserve a unitary scientific view of the world by excluding sense elements by means of quantifying them and describing their relations with universal mathematical functions. But at the same time, he quoted from Mach to support his assertion that in a mathematical equation, variables can be freely moved across the equal sign, making assignments of cause and effect in terms of laws meaningless, and that such a concept of cause has made no substantial contribution to the development of physics. He ended by asserting that cause and effect are projections of the human will into natural processes.

We can perhaps better understand his attitude toward Mach and Planck by looking at an article written in 1933 called "Butsurigakujo no gainen ni tai suru yuibutsusei no imi ni tsuite" (The Meaning of Materiality with respect to Physical Concepts).[6] He attempted to understand the conflict between the two physicists in what might seem a dialectical way. He agreed with Planck's belief that there was ultimately one physical world view which was unique and absolute, and in terms of which all phenomena could be explained. However, he could not accept what he regarded as the basis of Planck's belief, his focus on an ideal world view of the future toward which recent physics converged. Since the period Planck criticized Mach (1908-1910), quantum mechanics had gone through a revolution which made such prediction unrealistic.

In order to know anything about accessible facts, which are the basis of our knowledge, he agreed with Mach that we need to emphasize sensation, without which we cannot even begin the process of acquiring knowledge. It was reasonable to reduce materiality to what seemed most manifest, which, again following Mach, consisted of sensations of resistance to touch, heat in the sense of temperature, light as vision, and sound as hearing. Physics surely ought to be the description of what one can directly observe, whether it was consistent with or converged toward an ideal worldview of the

[5] Ishiwara Jun, *Shizen Kagaku gairon* (Outline of Natural Science), Hyoronsha: Tokyo, 1929, p. 245.

[6] See Ishiwara Jun, *Shizen kagakuteki sekaizo* (A Natural Scientific World View), Hyoronsha: Tokyo, 1948, pp. 259-265.

future or not. We could not directly observe, for example, the orbit of an electron, only changes in the quantity of its energy or motion. For Ishiwara, the basic quantities were the quantum of action, motion, and energy, and these produced the whole of natural phenomena, including the stimulations of sensation by which these basic quantities were known. Materiality existed only in the activity received in observation which could be measured (the basis of the theory of measurement).

Finally, he concluded that while we must aim for a unique physical world view as Planck did, it must be based on sensation (and thereby observation) as Mach asserted. Application of physical observation must be the basis of materialistic physics, and to the extent that a unitary world view was not attained, we should assume that the theory still possessed pure or "metaphysical" ideas not connected with observation, and such pure ideas must eventually, as Mach suggested, be purged from the system to make it unitary.

It is doubtful whether contemporary philosophers (in particular of the Anglo-American persuasion) would find these arguments interesting in any more than a historical sense. Our brief picture of an attempted synthesis of ideas from Kant, Mach, Planck, and the Copenhagen interpretation of quantum mechanics lacks sufficient clarity and detail to allow proper judgment, but Ishiwara's approach does deserve respect as a bold, thought-provoking attempt to reconcile what some observers might consider irreconcilable. He did his level best to come to terms with obtuse or even contradictory notions as sensibly as possible. But his solution does not seem unreasonable given the background conditions and the state of science and philosophy in the Japan of his day.

C

Beginning in the 1910's, a group of philosophers (meaning those familiar with the Western tradition starting with the Greeks) gathered at the University of Kyoto, a school whose prestige at least in physics and philosophy would soon eclipse that of the University of Tokyo. Most of the members had been educated at the University of Tokyo, and were inspired to continue its research tradition aimed at achieving a historical understanding of Western philosophy. Nishida Kitaro (1870-1945), the central figure in this school who originally taught religion and moved to "pure" philosophy, marked the beginning of the attempt to initiate an interpretation of Western philosophy which would make possible its assimilation into Japanese intellectual culture. Perhaps because religion, not science, was common to East and West, this assimilation focused on trying to create a logical system to express the Eastern religious experience but patterned largely after German Idealists like Hegel, who had attempted to find a language and logical system in which to organize the data of revelation in Christianity.

One of the first references to Ernst Mach in a purely philosophical context was made in the preface to Nishida's maiden work, *Zen no Kenkyu* (A Study of Good) in 1911. Nishida wrote that:

> For a long time I had entertained the thought of trying to explain everything using pure experience as the only existent (*jitsuzai*). At first, I tried reading Mach, but was not at all satisfied. Subsequently....I came to believe that one could escape solipsism thinking that it is not the case that the individual comes before experience, but that experience is prior to the individual and that experience is more fundamental than the distinction of the individual. Additionally, by thinking that experience is active, I was able to agree with post-Fichtean transcendental philosophy.[7]

But while it is not yet clear which books of Mach that Nishida may have read, the influence of Fichte and Hegel on the Kyoto philosopher can be seen everywhere in a work which as a whole attempted to establish a complete system of ultimate metaphysical categories founded within religion, which, as Nishida wrote, would "explain everything". Nishida's initial "axiom" was pure experience, which by means of self-differentiation produced epistemology, ethics, and a philosophy of religion in order to incorporate the experience of both Christianity and Buddhism. A thinker under Fichtean or Hegelian influence, with his concomitant metaphysical and religious commitments (even though agreeing about abolition of the thing-in-itself, though for different reasons) would find Mach's phenomenalism rather unsatisfying. In subsequent works influenced mainly by Neo-Kantians, Nishida elaborated his approach into a system whose main axiom would be "the place of absolute nothingness", and whose logic would be that of the "self-unity of absolute contradictories".[8] Although perhaps declining in influence in some parts of Europe during the 1920's, German Neo-Kantian philosophy, in particular the perspectives of the university professors Hermann Cohen and Heinrich Rickert, became influential among other Kyoto School members as well. This was very possibly because of its emphasis on ethical reason, a major concern of the time.

While deeply influenced by religious concerns, which provided their basic framework, members of the Kyoto School also introduced Western philosophy of science. Second in stature to Nishida among the Kyoto Scho-

---

[7] Nishida Kitaro, *Zen no kenyu* (A Study of Good), Iwanami Bunko: Tokyo, 1979 (original 1911), p. 4.

[8] Gino Piovesana, *Recent Japanese Philosophical Thought: 1862-1962,* Enderie Bookstore: Tokyo, 1968, pp. 145-158 & James Heisig, "The Religious Philosophy of the Kyoto School: an Overview", *Japanese Journal of Religious Studies,* **17** (1990), 65-71.

lars was Tanabe Hajime (1885-1962),[9] who began his studies in mathematics but changed to philosophy and graduated from the University of Tokyo in 1908. He started his career in the sciences (writing his dissertation on the philosophy of mathematics), but became a lecturer in philosophy at Tohoku University in northern Japan in 1913. He studied in Germany from 1922 to 1924 and began teaching at the University of Kyoto in 1927, taking over Nishida's position upon his retirement. He himself retired in 1945.

Early in his career, while still well under the influence of Nishida and the Neo-Kantians, Tanabe wrote several books introducing Western philosophy of science to Japanese readers. As a student, Nobel physicist Yukawa Hideki, for example, was influenced by Tanabe's works as well as having attended Nishida's philosophy seminar during his senior year.[10]

As early as 1915, Tanabe had published *Saikin no shizen tetsugaku* (Recent Philosophy of Nature), which deals with the development of Western thought from the Greeks to Kant and the Neo-Kantians while overlooking Descartes. He emphasized what he believed were the main characteristics, goals, and methods of European natural science with physics as a representative example.[11] He was particularly sympathetic to Cohen, perhaps because they shared a background in mathematics. In the 1915 book, Tanabe wrote that the final goal of human civilization which included science, art, and religion, was the incarnation of ideal values which were present in existence: actual life was a means to this end. Contrary to pragmatism, it was not the case that life was the goal and truth the means, but that truth was the goal and life was the means. Tanabe followed the principle of Neo-Kantian value theory giving preference to ethical over theoretical values. Science in particular purified disordered *a priori* principles which were present in existence and known in experience through inutition. This experience was trans-individual (thus allowing science to claim universal validity), and after differentiating experience into subject and object, scientific laws expressed the relations between objects. Natural science was an aspect of the self-appearance of existence, and hypotheses and theories naturally changed in the process of its development.

Tanabe objected to Mach's idea that natural law was merely a product of economy of thought and was complex only to the degree required by the needs of guiding life's activities. Nor did he accept that hypotheses are merely guides to future experimentation.

According to Tanabe, his scheme excluded all epistemologies based on copy theories, whether the object "copied" was within or beyond exper-

---

9 Piovesana, ibid., pp. 145-158 & *Encyclopedia of Philosophy,* vol. 8, p. 77.

10 Yukawa Hideki, *Tabibito* (Traveler), Kadokawa Bunko: Tokyo, 1960, pp. 153 & 191.

11 Tanabe Hajime, *Saikin no shizen kagaku* (Recent Natural Science), Iwanami Shoten: Tokyo, 1915, pp. 199-320.

ience. And in his opinion, this excluded Mach's element-based monism as well:

> The theories of science are neither related to the cognitive ability of the humans who know them nor connected to whether or not humans know them. They express the unchanging principles which have an inner relation to the reality which is the basis of experience, or to speak in the language of modern mathematics, express an invariant.[12]

Tanabe was following the preoccupation of German philosophy with awareness as "knownness" (*Bewusstheit*), and not so much as the individual knowing consciousness (*Bewusstsein*), a trend, for example, conspicuously followed by Hermann Cohen. The word "invariant" in the quote was given in German and was taken from Planck, a Neo-Kantian who gradually became a realist of the indirect or representationalist type, and who like Tanabe was sympathetic with religion. Tanabe used Planck for a measure of support in his arguments against Mach.

In his discussion of Mach, Tanabe did not criticize his assertion that sensations were not individual (a point common to Tanabe's notion of experience), but argued that his sensations were simply abstract elements analysed out of direct experience. True science must be built on concrete living experience. Direct experience was not a collection of elements, but was given form, and was a system unified in its inner relations. Space and time, which Mach tried to reduce to kinds of sensation, were for Tanabe (and the Neo-Kantians) not sensations but constructive forms based on inner relations of the original experience which developed internally. "Facts" were a product of cognitive construction, not a collection of sensations.

Again, following Planck, Tanabe asserted that while Ernst Mach's view contributed much to understanding the origin of physics, his epistemology was contrary to the spirit of physics. Physical theories grew by a process of excluding sense elements, and reducing all phenomena to quantitative or mathematical relations or non-sensory concepts. In short, Sensations were not quantitative; laws were not copies of sensations, but cognitive constructions which act according to their own norms; "Truth" as demanded by science existed in the sense of a process in which existence developed internally and manifested itself; and the truth realized by science, Tanabe asserted, was an absolute truth.

Tanabe's characterization of Mach's theory will strike many readers as oversimplified, and it is difficult to ascertain whether the views were based on a reading of primary or secondary sources. His view of Mach was determined in general by his Neo-Kantianism, in particular by Cohen, while particular criticisms were derived from Planck. But one feels that the

---

12 Ibid., pp. 306-307.

critique went astray in that it was addressed to technical aspects of Mach's epistemology, and failed to come to terms with more fundamental ontological and epistemological differences. For example, while Tanabe criticized Mach as mentioned above he did not refer to Descartes in a significant way, which suggests that Mach unlike Descartes was at least still within the pale of what he considered to be important and reputable thought. Many philosophers tend to ignore extreme opponents, saving most criticism for "friendly" half-allies. Both Mach and Tanabe, for example, favored science, favored reason, and favored the use of both observation and mathematics. Most of their differences were well within the general tradition of Berkeley, Hume, and Kant, but even within that tradition Tanabe neglected to make what are normally considered to be rather fundamental distinctions between Neo-Kantianism and Mach's basic Neo-Humean position.

D

The spread of Marxist ideas in Japan during the 1920's and 1930's while increasingly opposed by the Japanese government, especially during the latter decade, put its mark on the thinking of numerous intellectuals, including many scientists and philosophers. Like Neo-Kantianism, it was very critical of Ernst Mach's ideas, but not because it was so distant from Marxist advocacy of science and opposition to Christian theology, which it was not, but because "Machism" had an appeal largely to the same intellectual audience which the Marxists wanted to influence. "Machism" and "Marxism" were competitors for intellectual allegiance, but even more distressing from a Marxist perspective, many Machists attempted to argue that phenomenalism and economy of thought were compatible with Dialectical Materialism, as if one could and should combine both philosophies. This heretical tendency helped drive Vladimir Lenin into writing his 1908 book *Materialism and Empirio-Criticism* which was largely a criticism of Ernst Mach's world view.

On the academic scene, many of the ideas of Karl Marx gained a real platform with the establishment of economics departments in the 1920's at Japanese universities. The interests were chiefly economic (the interpretation of surplus value theory), but philosophical concerns were eventually integrated into economic studies. Marxist-influenced philosophers such as Tosaka Jun (1900-1945) and Saigusa Hiroto (1892-1963) tended to follow Lenin's critique of Mach. Tosaka wrote in his book *Kagakuron* (Theory of Science):

> This (Mach's) approach, in fact, tears positivist natural science away from so-called critical philosophy (Kantianism and Neo-Kantianism) and causes a self-inflicted wound, as Lenin has correctly analysed and criticized.[13]

Saigusa in his book *Tetsugaku Nyumon* (Introduction to Philosophy)[14] also cites Lenin, and in a typical move assimilated Mach to Berkeley as an explanation of subjective "taking in" of experience, and insisted that those who rejected the scientific nature of materialism leaned toward Mach and Avenarius in accusing it of "mysticism" for accepting things which lay outside of experience, such as matter.

Turning to a figure of greater importance for philosophy of science, Taketani Mitsuo (b. 1911) was a Marxist-influenced physicist who worked extensively with Yukawa Hideki as an advisor on methodology, that is, with the man who became Japan's best-known scientist. In addition to numerous contributions of his own, Taketani co-authored in 1938 the last two of the series of four papers in which meson theory was developed with Yukawa and Sakata Shoichi. He had graduated from the physics department at the University of Kyoto (also Yukawa's alma mater) in 1934 and taught at Rikkyo University, later becoming a freelance researcher. His range of interests has been very wide, extending from theoretical physics through technology and history of science[15] to and including philosophy.

After the war in 1946, he published a collection of papers called *Benshoho no shomondai* (Problems of Dialectic) which had originally been written during the 30's and 40's. The following presentation of his point of view is based on the papers collected in this publication plus a 1951 paper which appeared in *Shinri no ba ni tachite* (In the Course of our Study).[16] Takatani wrote that his own methodology in physics which he labeled "dialectical analysis" had been inspired by Marx's multidimensional logical analysis of what he called "commodity", a substance whose phenomenal form consisted simultaneously of use and exchange values, while at the essential level, value and labor power explained the necessity of the phenomenal form.[17] Expanding on the conventional two-fold view of science as consisting of phenomenal and essential levels, Taketani developed this into a me-

---

13 Tosaka Jun, *Tosaka Jun zenshu I* (Tosaka Jun: Collected Works, vol. I), Keiso Shobo: Tokyo, 1966, p. 125 & for Marxism in Japan see: Piovesana, op. cit., pp.170-190.

14 Saigusa Hiroto, *Saigusa Hiroto Chosakushu,* VIII (Saigusa Hiroto: Selected Works, vol. 8) Chuo Koronsha: Tokyo: 1972, pp. 243-253.

15 Nakamura, S. et al., *Science and Society in Modern Japan,* University of Tokyo Press: Tokyo, 1974, p. xx.

16 Taketani Mitsuo, *Benshoho no Shomondai* (The Problems of Dialectic), Keiso Shobo: Tokyo, 1968 & Yukawa, H., Sakata, S., and Taketani, M., *Shinri no ba ni Tachite* (In the Course of Our Study), Mainichi Shimbun: Tokyo, 1951.

17 Taketani (1968), ibid., p. 29.

thod that pealed away layers first from sensory phenomena revealing a substantial layer and then pealing away from that until one finally reached an essential layer, the process being repeatable as the essential became the main focus of phenomenal analysis. He applied this three-stage, semi-dialectical theory to the development of physics, emphasizing that it did not represent necessary historical progress, but rather the progressive stripping away of simultaneously existing layers.

Assuming that our knowledge of nature consisted of a collection of systematized judgments, Taketani grounded his three-stage theory on a theory of judgment. The type of judgment must reflect the character of each stage, and Taketani followed Hegel's dialectical scheme of the *Science of Logic*, applying "in-itself" singular judgments at the phenomenal level, particular "for-itself" judgments at the substantial level, and universal "in-for-itself" judgments at the essential level.[18] The first stage, *the phenomenal*, consisted of the description of singular observations made without the intention of finding a law. Methodologically, this level was associated with Machism, and historically, for example, with Tycho Brahe's astronomical observations, his emphasis on empirical description. The second stage, *the substantial*, ascertained the structures which were manifest at the phenomenal level, making particular judgments showing law-like correlations between some observations. This was characteristic of Kepler's three laws of planetary motion, introducing a substantial model (heliocentrism) which had given order to Brahe's observations. These laws were phenomenal, not necessary. The final *essential stage* used universal judgments to help comprehend why the structures of the substantial were necessary, and was represented by Newton's demonstration of the necessity of motion by introducing the essential notion of force or gravity as cause.[19]

According to Taketani, the proof of necessity was found in the activities, experiments, and labor of human beings, a facet introduced into physics by Galileo, a physics was born during the Renaissance-development of production technology. This concept was related closely to his Marxist-based theory of technology as praxis. This methodology also served to orient the direction of research in physics, as will be explained below.

Since Taketani considered Galileo to be a transitional figure in the movement of physics from the phenomenal to the substantial stage, it is not surprising that he criticized Mach's representation of Galileo as a phenomenalist in *The Science of Mechanics*. His criticism of Mach's analysis of the relation between Aristotelian-Scholastic physics and Galileo according to his schema began with the relation of force and acceleration. Mach had

---

[18] G.F.W. Hegel, *Science of Logic* (tr. A.V. Miller), George Allen and Unwin: London, 1969, pp. 643-650.

[19] Taketani (1968), op. cit., p. 65.

wrtten that when Aristotle examined the motion of a falling stone in its alleged return to its "natural place", he was using explanatory methods, whereas Galileo used simple description.[20] Mach considered Galileo's mechanics to be complete in so far as it was merely a description of motion, but Taketani, who unlike Mach refused to reduce dynamics to kinematics, insisted that for Galileo mechanics was much more than description of motion, even if Galileo himself could not provide the dynamic explanation which Newton was able to put forward. In his dialogues on mechanics, Galileo, regardless of how unconsciously, was moving toward the substantial level in mechanics when he considered acceleration to mediate between force and motion. He distinguished between the case when force was present (acceleration) and when it was absent (uniform velocity), thus linking force to acceleration, whereas Aristotle simply linked force with velocity as one commonly does in everyday life where one tends to judge force by its potential impact against an object or counterforce.

Taketani's criticism of Mach's analysis turned to his exclusion of the notion of force as a cause which Ernst Mach considered "metaphysical". Taketani agreed with Mach that the relation between force and acceleration was empirical, but Mach and Machists had tried to exclude the concept of force as a cause, dismissing force linked with cause as a metaphysical abuse. This was because according to Taketani, Mach lacked a multidimensional physics for properly analysing the mediation of force and motion by acceleration, reducing science to a mere description of phenomena. Mach avoided use of the term "force", in favor of the expression "circumstances determinative of motion" (*Bewegung bestimmende Umstände*). Taketani reminded us that force did not determine motion (an Aristotelian idea), but rather with Galileo, force determines acceleration, a point which must be understood to understand Newton's multi-dimensional theory which covers both "substance and phenomenon".

Aristotle's theory was one-dimensionally substantial, but Dr. Mach's was one-dimensionally phenomenal. According to Taketani, motion was not directly determined by force, but by initial conditions, which were contingent conditions at the phenomenal level. Force as cause was not something hidden behind phenomena, but worked at the essential level. It could introduce a necessity to the substantial level telling us why, given force and contingent initial and boundary conditions, the necessary interactions at the substantial level produced the phenomenal motion which appeared as it did. Newton's method could relate all levels (phenomenal, substantial, and essential), and force could not simply be understood as a "circumstance determinative of motion".

---

[20] Ernst Mach, *The Science of Mechanics* (tr. T. McCormack), Open Court Publishing Company: Chicago, 1902, p. 131.

Finally, the reason Takatani understod force as a cause is found in his general theory of technology as praxis. In praxis, to put an object into a state of motion, we must apply appropriate initial and boundary conditions, and force, which produces a concept of cause, is what actually makes praxis possible, and is not simply a subjective criterion. In attempting to avoid metaphysics, Mach simply adopted its antipode, offering something on the same level, but lacking a notion of praxis.[21]

If this were all there were to the matter, then Mach could simply be dismissed as having put forward a mistaken understanding of the relation between Aristotle, Galileo, and Newton, or a misguided analysis of the history of mechanics generally. However, a proper worldview (which the approach of Mach was not even in the opinion of its author) was necessary according to Taketani because it helped orient the direction of future research, and in this respect, Machism was still regrettably producing a pernicious influence on the development of physics.[22]

In his 1951 essay, he wrote that around the time of the development of meson theory in 1936, physics was still dominated at least methodologically by Machism and Neo-Kantianism. Concerning Machism he wrote that:

> It was thought that physics discovers laws by manipulating empirical facts, and only sense impressions are real; all the rest is constructed by the subject who arranges experienced facts. Moreover, it was maintained that the essence of science lies in function, and therefore, in the course of scientific development, substance will be absorbed in function... If applied to physics, such thinking made it nonsense to consider the internal structure of the nucleus - with a diameter of only about 10^{-12} cm -- or its structural elements. It is not the role of science to arrange mathematically the various phenomenal aspects that concern the nucleus.[23]

"At the time", he continued, "the problem of nuclear force and beta decay were dealt with mainly from the phenomenological side."[24] The methodological implication was that concepts were to organize empirical data, and the introduction of a notion not supported by data was contrary to valid scientific procedure.[25] He then cited Niels Bohr as having been unhappy

---

[21] Taketani (1968), op. cit., pp. 110-116.

[22] Ibid., p. 34. [What Taketani apparently meant was that Mach's philosophy when not understood as a first stage to a higher philosophy could be pernicious. It was Taketani who tried to provide those higher stages and by doing so helped in a certain sense to legitimate Machism. - Ed.]

[23] Taketani Mitsuo, "Methodological Approaches in the Development of the Meson Theory of Yukawa in Japan", in Nakamura et al, op. cit., pp. 24-28 (a translation of an article originally appearing in Yukawa et al., op. cit., 1951, p. 27).

[24] Ibid., p. 29.

[25] Taketani (1968), op. cit., p. 6.

with Fermi's proposal of the neutrino to preserve the law of conservation of energy in beta decay because such a particle had never been observed, preferring a statistical interpretation of the law or jettisoning it for new laws.[26]

Yet, in introducing his notion of the meson (the U-quantum at the time) to explain nuclear forces, a theory based on inference and calculation, not observation, Yukawa would be following exactly the method of Fermi (and before him, Heisenberg and Pauli): molecular analogy.

When Niels Bohr came to Japan in 1937 and discussed Yukawa's proposal, he rejected it: "Why do you want to create such a new particle?" he wondered.[27] Yukawa and his associates, including Taketani, continued to develop the theory despite Bohr's opposition, and the fact that Yukawa's earlier publication of 1936, introducing the meson, as Segré wrote, "...did not cause a great stir. At the time nobody had yet seen particles similar to those proposed by Yukawa, and thus they appeared mainly as an interesting speculation."[28] Yukawa was aware that his arguments had "merely speculative character",[29] and he even wrote in his 1935 paper introducing what would become known as the meson, "As such a quantum with large mass and positive or negative charge has never been found by experiment, the above theory appears to be on a wrong line."[30] He then showed that a meson would not be observed outside of the nucleus unless the absolute value of the difference in transition energy between the neutron and proton state was much greater than the mass of the meson times the speed of light if squared, a condition which never obtains.[31] In his paper of 1937, he stated that with the observation of a particle in cosmic radiation which appeared to match the meson, then "...the only serious drawback to our theory disappears."[32]

---

26 Ibid., p. 28.

27 Taketani (1951), op. cit., p. 33.

28 Emilio Segré, *From X-rays to Quarks,* W.H. Freeman and Company: New York, 1980, p. 247 & Abraham Pais, *Inward Bound,* Clarendon Press: Oxford, 1986, pp. 429-436.

29 Yukawa Hideki, "On the Interaction of Elementary Particles, I", in Tanikawa Yasutaka (ed.), *Hideki Yukawa: Scientific Works,* Iwanami Shoten: Tokyo, 1979, p. 12. (original article by Yukawa Hideki appeared in *Proceedings of the Physical and Mathematical Society of Japan,* **17** (1935), p. 57.)

30 Ibid., p. 8 (original, p. 53).

31 Ibid., p. 9 (original, p. 54).

32 Yukawa Hideki and Sakata S., "On the Interaction of Elementary Particles, II" in Tanikawa Yasutaka (ed.), Hideki Yukawa: Scientific Works, Iwanami Shoten: Tokyo, 1979, p. 193 (original: Yukawa Hideki, *Proceedings of the Physical and mathematical Society of Japan,* **19** (1937), 1085.)

Taketani attributed Bohr's rejection of Fermi's neutrino and Yukawa's meson to his observation-obsessed Machism.[33] In 1940, during the later development of meson theory (the symmetrical meson theory), he and Sakata found themselves on the defensive, co-authoring a paper in which they suggested that neutral mesons had escaped observation because of their extremely short lifetime.[34]

Yukawa, Sakata, and Taketani persisted on in the development of meson theory in spite of such adverse circumstances because Taketani had concluded that at the time quantum mechanics had progressed past the phenomenal stage (at which Bohr and the Machists had stalled) and was at the substantial stage. At this stage, it was still concerned with organization at the level of substance, that is, to show that substances (or particles) were producing the observations. This was not the time for a radical change in the theory as it stood (as had been asserted, for example, by Bohr and Schrödinger), and physicists therefore should continue using substantialist methods. Taketani wrote in 1951:

> Throughout these thrusts into the dark, it was the three-stage theory of materialistic dialectics that guided our research and fortified our resolution to overcome difficulties. We maintained the following viewpoint of our methodology: In contrast to current opinion that difficulties of nuclear theory lie in some deficiency in quantum mechanics itself, the problems of the meson theory should be attacked within the framework of quantum mechanics by using the correspondence principle. Our method focussed on substance. That is, we thought that nuclear physics was at the stage where its main problems were the investigation of what the constituent substances might be, and what might be the properties of these substances. We felt that we were still far from the stage at which quantum mechanics itself should be fundamentally modified.[35]

That they depended on the correspondence principle may show that they were not dogmatically anti-Bohr, but this related to a calculation technique rather than an epistemological problem per se. Nevertheless, Yukawa's method (following Pauli, Heisenberg, and Fermi) of introducing an unobserved particle to resolve the problem of nuclear force was justified in spite of the rebuff from such a notable physicist as Bohr. Taketani concluded that:

> This was the merit of our methodological attack based on the three-stage theory of material dialectics. With idealistic Machism we might have become lost or even seduced into an overall denial of the theory.[36]

---

33 Tanikawa (1968), ibid., p. 33

34 Pais, op. cit., p. 435.

35 Yukawa, Sakata, & Taketani (1951), op. cit., p. 34.

36 Ibid., p. 35.

This appears to show the real importance of having a proper methodology. When the scientist is presented with a contradiction or dilemma, the correct methodology should make it much easier to determine the correct path to take: During the 1930's this may have meant either change the theory and laws or introduce new substances (particles). Machism in all cases, that is, when used as an exclusive methodology, seemed to lead to the wrong choice and to have produced wasted effort.

The problems of the 1930's were largely resolved within the context of the time by the successive introduction of new particles: the neutron, neutrino, meson, and positron. This strongly suggests that Taketani's methodology - largely developed during the period and before its results were able to be fairly judged - was basically sound and had explanatory value, perhaps considerably more sound and more enlightening than some recent methodological approaches, such as that of Imre Lakatos and his theory of progressive and degenerating problem shifts.

Writing in 1946, Taketani asserted that quantum mechanics was beginning to enter the essential stage, which would demonstrate the unity of particles and their function and the reason why this substantial level structure was necessary.[37] We note that Yukawa's description of the situation of meson theory in 1949 was very much in the style of Taketani, discussing phenomenological observations, then moving to his substantial stage, "...we are still at the stage of accepting different kinds of mesons...", to an essential stage, giving "...fundamental laws determining the masses of various kinds of particles which appear to us to be so irregularly distributed", and "...the law which guarantees the stability of the whole system of elementary particles interacting with each other and changing from one to another."[38]

A comprehensive understanding of Taketani's three stage theory is dependent on his theory of technology, but this would take us too far away from Mach. Nevertheless, in spite of the impressive character and clear utility of Taketani's methodology of science a few comments should be made about possible limitations to his approach. For example, there is an eclectic aspect to his work which may not be able to withstand close analysis, and one may also legitimately wonder whether his methodology applies to other fields as well as it apparently does to physics. Nor is it clear why he so strongly emphasizes ideas drawn from Hegel's philosophy, especially from the latter's theory of judgment, when, first, Taketani's own epistemology seems incompatible with that of Hegel, and second, when there seem to be other more recent sources closer to his own epistemology available. Concerning the history of science, there is grave doubt whether Bohr's views

---

37 Taketani (1968), op. cit., p. 17.

38 Tanikawa, op. cit., 193 (original: Yukawa [1949], op. cit., 479.)

can be so readily assimilated to Machism, since that implies that Bohr was more consistent on the one side and narrower in outlook on the other than appears likely.[39] Finally, Taketani's own curious admission that, "Once the results prove correct and natural, then the methodology used has less importance",[40] while perhaps emphasizing the need for an open mind and possible influence from Heraclitus or Hegel seems to undermine his own claim expressed elsewhere on the importance of an external method to regulate one's thinking while doing science, a claim which his three-stage theory appears to solidly substantiate. But on the other side, Hirosige Tetsu has argued that Taketani's three-stage theory bore no relation to the development of meson theory, and that Taketani missed predicting a number of important events in physics that he should have been able to predict.[41] *We cannot settle the dispute here, but what does seem evident is that Taketani incorporated part of Mach's philosophy into a three-stage methodology of science which is so strikingly appropriate to understanding recent developments in physics when seen within a larger context as to deserve much wider attention from scientists, methodologists, and philosophers.*

The rest of our story from the middle 1930's until the end of World War II is not a happy one. Most work in physics and philosophy apparently came to a halt. The *Tokko,* or "thought police", suppressed philosophy not in accord with military ideology. Leftist philosophers like Tosaka and Miki, who worked as popular writers at that time because they had lost their university positions, died in prison. The repression was directly or indirectly supported by several members of the Kyoto school. Nishida's "unity of absolute contradictories" was employed perhaps not always with his full approval in order to give a metaphysical justification for the necessity of conflict between the ostensible ideals of the Greater East Asian Co-Prosperity Sphere and less savory realities such as the Rape of Nanking. Many contradictions were presumably overcome (to speak in dialectical or Neo-Hegelian terms) in the historical avatar of the "place of absolute nothingness", which some people identified with the house of the Japanese emperor himself, keeping in view that in much Oriental thought "nothingness" has a positive connotation and often means peace of mind or freedom from bad things such as desire.

Even Tanabe, who we have already discussed, was influenced by the spirit of the times. He wrote in the preface to his book of 1936 *Tetsugaku to Kagaku to no aida* (Between Philosophy and Science):

---

39 For an argument against "mistaken association" between Mach and Bohr, see Henry Folse, *The Philosophy of Niels Bohr,* North Holland: Amsterdam, 1985, pp. 224-227.

40 Taketani (1951), op. cit., p. 32.

41 Hiromatsu Wataru et al. "'Kindai no chokoku' to Nishida tetsugaku" ("'Overcoming Modernity' and Nishida's Philosophy") in *Shicho*, Shichosha: Tokyo, no. 4, (1989), pp. 16-17.

> ...overcoming international and domestic crises and achieving our country's historical mission cannot be done in a brief fit of excitemen. Disinterested and meticulous plans guided by the consciousness of this mission must be given a correct basis in knowledge and suffused with a strong will. The passion of the love of the country must be accompanied by the scientific spirit...Extolling love of country, today there is a great pressing need for the evoking of the scientific spirit.[42]

The book was dedicated to the end of bringing the two together. Nishida, the longtime mentor of the Kyoto School, would die a few months before the war came to a close.

Work on meson theory also stopped as physics became directed towards improving military technology for the war effort. Taketani was imprisoned for "...my analyses of quantum mechanics, my analyses of the development of nuclear physics, and my methodological approach to meson theory - in short, my research activities on natural dialectics."[43] Taketani's work diminished at the time because his health had declined due to imprisonment.

PART TWO

After the war, the Kyoto School's philosophy declined because of its perceived complicity in the war effort.[44] Tanabe Hajime went into a life of seclusion to develop his philosophy of repentance. Marxism became influential as the philosophy of powerful labor groups involved in postwar reconstruction of the peacetime economy,[45] as well as because the Marxists were one of the only groups to consistently resist fascism during the war. (Reflecting a perceived failure of many prewar values, the other influential movement was existentialism, in particular Heidegger and Sartre.)

The generally Marxist orientation in philosophy of science was represented by the Democratic Science Society (DSS). This group was concerned with methodological praxis in theory of science, and was widely supported by scientists. On the other hand, there were the "academic" philosophers of science, represented by groups such as the Association for the Foundations of Science and the Association for the Philosophy of Science. They were

---

42 Tanabe, Hajime, *Tetsugaku to kagaku to no aida,* Iwanami Shoten: Tokyo, 1937, p. 1.

43 Taketani, (1951), op. cit., p. 51.

44 Ibid.

45 See Noé Kei'ichi, "Kindai Gorishugi no chokoku: Hiromatsu tetsugaku to gendai butsurigaku" (Overcoming Modern Rationalism: Hiromatsu's Philosophy and Modern Physics.) in Ohba Ken et al., *Hiromatsu-ron* (Reviews of Hiromatsu), Unite: Nagoya, 1982, pp. 197-246.

concerned with the technical theory of science, and largely followed logical positivism.

Neither of the groups sharply questioned the meaning of "science" or "rationality", both maintaining a scientistic attitude assimilating knowledge to scientific knowledge. During the 1960's, Hiromatsu Wataru, along with Hirosige Tetu, were the first in Japan to begin the questioning of modern rationalism as represented in Western philosophy. Hiromatsu was born in 1933, and after graduating from the University of Tokyo philosophy department would teach at the University of Nagoya before returning to the University of Tokyo as professor in the Department of the History and Philosophy of Science. He has been most strongly influenced by Marx and worked to develop his own epistemology and ontology, but also elements of Neo-Kantianism, Husserl, and several Continental philosophers can be seen in his work. He also analyses parallels between Western and Buddhist metaphysics. He has done extensive analyses of reification and intersubjectivity grounded in role theory. He was the co-translator of the Japanese translations of Mach's *Analysis of Sensations* and *Knowledge and Error*, and contributed introductory essays to the translations. To understand Hiromatsu's view of Mach's significance, I will briefly characterize Hiromatsu's analysis of his position in the history of Western philosophy as it was in Mach's time.

Very roughly, Hiromatsu has developed the characterization of the formation of modern Western philosophy as being bound by the subject/object schema which has its origin in dualist Cartesian substantialism. Within epistemology, this dualism is worked out based upon the following three propositions, stated along with their criticism and means of resolution.[46]

The first proposition is the "individuality" of the subject (*Satz der Persönlichkeit*). Epistemologically, the subject is understood as the personal consciousness of each individual, and all subjects are isomorphic. This proposition has been rendered untenable by modern research in the anthropology of nonliterate societies and the psychology of psychiatric patients. Its resolution requires recognition that consciousness is socially intersubjectivized and that there is a role of the other in knowledge (the categories of culture learned from the other mediate my relation with nature and society: I never encounter an object directly).

The second proposition is the "trialism" of knowledge (*Schema der Triarität*). The content of consciousness that is directly given to the knowing subject is separated from the object itself such that knowing consists of three parts: conscious activity, the content of consciousness, and the object itself. In addition, at least part of the content of consciousness is actively

[46] Hiromatsu, Wataru, *Sekai no kyodoshukanteki sonsaikozo* (The Intersubjective Ontic Structure of the World), Keiso Shobo: Tokyo, 1972, pp. 3-20.

contributed by the subject. In overturning the constancy hypothesis, Gestalt psychology renders this proposition suspect, and resolution requires recognition of the Gestalt articulation of sensation as intersubjectively determined, not a form assembled from the elements of the contents of consciousness by the activity of consciousness.

Finally, the third proposition is the interiority of the given (*Satz der Immanenz oder Satz des Bewusstseins*). What is known is only the content of consciousness: perceptual images, ideas, representations, etc. The existence of collective representations cannot be explained under this assumption, and its resolution requires an analysis of reification.

More particular to the natural sciences, the dualistic framework and its three propositions could accomodate Newtonian mechanism and atomism and its epistemology, but has failed to assimilate the two pillars of modern physics, the theory of relativity and quantum mechanics. Hiromatsu argues that the problems of epistemology and ontology that first appeared in the context of this new physics are not actually unique to it. They are merely the most dramatic and acute symptoms of a systematically impaired ontology and epistemology that no amount of therapy will cure. Overcoming these problems requires nothing less than a jettisoning of substantialist Cartesian dualism and substance-attribute view of matter, and as indicated by the example of physics: the development of a relationalist world view in which the relation is given ontological priority.

Although Mach was a physicist and Hiromatsu regards his philosophy as insufficiently mature, he argues that while his philosophy did not itself escape Western philosophy's dualistic framework, in replacing Cartesian substantialist dualism with a neutral monism, his approach provided a decisive inner critique of dualism and the subject-object schema, and was an advance towards a relationist epistemology and ontology. The hint from Mach is best clarified by an analysis of how the relation between his philosophy (epistemology and ontology) and its influence on Einstein, whose theory of relativity and quantum mechanics broke out of the old framework and has provided the new one for modern physics. Finally, from the point of view of his relationist ontology, Hiromatsu offers a diagnosis and cure for the problems which originate in realist understanding of relativity theory.

Hiromatsu argues that by using functional relations, Mach avoided both substantialism and the concepts of cause and effect, which contain a hidden concept of personification.[47] Materialism and idealism simply suppress one side of the subject/object dualism while accepting the framework. Mach avoided this framework because, instead of seeing the subject/object struc-

---

[47] Hiromatsu Wataru, *Kototeki sekaiken e no zensho* (Outposts of the Relational World-view), Keiso Shobo: Tokyo, 1975, pp. 51-75.

ture as primary, elements are "prepersonal", and thus the first proposition can be mitigated. He posited a monism of neutral sensations as primary and the "mental" subject and "physical" object as secondary complexes of functionally related sensations: color, heat, pressure, space, time. As subject and object are secondary constructions or complexes of sensations, they cannot be the "cause" of sensations, and therefore an "object itself" and "subject itself" are excluded from epistemological consideration, mitigating problems entailed by propositions two and three.

Aiming to clarify Mach's role in the development of modern science, Hiromatsu begins his analysis of the relation between Mach and Einstein with the logical structure of their concepts as they relate both the special and general theories of relativity. Analysis of the similarity in logical structure shows that Einstein's concepts can be derived from Mach's with the addition of a few elements, indicating the closeness of their thought. Mach's operationalist criticism of Newton's physical concepts provided Einstein with both a means for making explicit Newton's conceptual system, and a technique for concept formation that allowed him to reform fundamental physical concepts that became the basis of the theories of relativity. Hiromatsu does not pretend to know what transpired in Einstein's mind while he was writing the paper of 1905, but attempts to show as an outside observer a filiation in the logical structure between Mach's and Einstein's concepts.

In his book *Sotaiseiriron no tetsugaku* (The Philosophy of Relativity),[48] Hiromatsu's analysis of this conceptual filiation between Mach and Einstein centers on the concepts of mass, space, time, and dynamics. Mach pointed out that Newton's definition of mass only works with respect to homogeneous substances. (Mach does not mention that Newton entertained a homogeneity produced by "inner substance".) He redefined mass in terms of an observable or measurable property which determines a relationship between two bodies in terms of inertial mass. Hiromatsu believes that this concept was a direct precursor to Einstein's concept of mass, but understanding the link that he proposes between the two concepts depends on his analysis of Mach's concept of space as an acceleration-determining medium and its inconsistency within his definition of mass.

Hiromatsu argues that even more than Mach's critical analysis of Newton's concept of space, it was his reconstruction of it that influenced Einstein. Newton's space had no physical effect relationship on matter, but only a relation of geometrical position. Mach suggested the possibility of a space being filled with a medium that might determine an object's position. Mach concluded by saying that this medium would be more valuable to

[48] Hiromatsu Wataru, *Sotaisei riron no tetsugaku* (The Philosophy of the Theory of Relativity), Keiso Shobo: Tokyo, 1986, 90-178.

science than Newton's "forlorn idea of absolute space",[49] but that at present, nothing was known about such a hypothetical medium and the only way known to measure position was with respect to other fixed bodies. Hiromatsu then asserts the similarity of logical structure of their two conceptions is demonstrated in Einstein's quote about Mach's medium in *Ether and Relativity*: ". . . this conception of the ether to which we are led by Mach's way of thinking differs essentially from the ether as conceived by Newton, by Fresnel, and by Lorentz. Mach's ether not only *conditions* the behavior of inert masses, but *is also determined* by them . . . . Mach's idea finds its full development in the ether of the general relativity. . . . Empty space in its physical relation is neither homogeneous nor isotropic..."[50] In general relativity, space determines the path of matter and matter determines the shape of space.

Then, turning to this geometry, Hiromatsu points out that Mach, contrary to those who considered it to be an arbitrary fiction of the mathematical imagination, believed physical space was probably non-Euclidean.[51] It exists as a positive object (with its property as an acceleration determining medium in the background). Like any theory, a geometrical theory is an economical description of reality, and in Mach's opinion, theories are all equally valid until it is determined which one most economically describes experience.

Hiromatsu indicates that a further reason for Mach to accept non-Euclidean geometry was his distinction between "metrical space" and "physiological space", the first being the derived Euclidean space of classical physics and the second the primitive space of the empirically directly given.[52] "Physiological space" is neither homogenous nor symmetrical, and is most appropriately described by non-Euclidean geometry. If we introduce the concept of metrical as an inseparable part of space, the constancy and rigidity of rulers must be reconsidered. While not writing in terms of coordinates, Mach recognized that if a ruler and object undergo expansion or contraction while moving together through space, they may not appear to change in size if they are in the same frame of reference. Mach moved from the relativity of spacial metrical to what we could call the relativity of space itself. If "space were a movement-determining medium" and this concept of metrical were under-stood together, asserts Hiromatsu, we could understand them to be the precursor of Einstein's concept of space.

---

49 Ernst Mach, *The Science of Mechanics,* (trans. T. McCormack), Open Court: Chicago, 1902, p. 231.

50 Albert Einstein, *Sidelights on Relativity,* Dover: New York, 1983, pp. 17-19.

51 Ernst Mach, *Knowledge and Error,* Reidel: Boston, 1976, chapter 20.

52 Ernst Mach, *The Analysis of Sensations and the Relation of the Physical to the Psychical,* Open Court: Chicago, 1914, chapters 7 and 9.

Mach's rejection of Newton's notion of absolute time as a metaphysical superfluity is well-known, but with respect to Einstein, what draws Hiromatsu's attention is Mach's operationalist definition of simultaneity, which he argues prefigures Einstein's, and his attempt to eliminate time, reducing it to a spacial determination. Even if we observe an eclipse of two stars and an event on earth occurs simultaneously, operationally we cannot speak of the eclipse as having "actually" occured millions of years ago.[53] In an operationalist definition, "simultaneous" must be oriented to observed simultaneity -- simultaneity within the observer's systems of time sensations -- and thus the observer cannot be excluded from the concept of time any more than from that of space. However, Hiromatsu argues, because the constancy of the speed of light had yet to be discovered, this had no practical significance. But if we include this constancy of the speed of light, then we arrive at special relativity with logical necessity.

Hiromatsu concludes that all three of Einstein's claims about time (that two events will be a priori simultaneous in two different systems: concepts and concept systems are only valid in their utility for *Orientierung*: absolute time cannot be measured, and time must be defined so as to be measurable) have all been previously asserted by Mach (Einstein follows Mach's usage of the word "Orientierung" from *Die Analyse der Empfindungen*), and indeed if time is defined as suggested by Mach's procedure, Einstein's time dilation follows with logical necessity.

Hiromatsu suggests that Mach's reduction of time to spacial determination of a sense index and his use of functional description instead of cause-effect explanations allow elimination of time as an independent variable in the description of a phenomenon as a functionally dependent relation between sense indices. While Mach recognized a specific sensation of time, it cannot be incorporated into an intersubjective system of order. Reducing time to a spacial determination allows movement previously described by a three-dimensional space-coordinate system and a time-coordinate system to be described by one four-dimensional spacial-coordinate system, with the fourth-coordinate according with $s = vt$ oriented to the constant movement of a watch or celestial body. Thus Hiromatsu concludes, "...(this) four-dimensional description has essentially the same structure as the four-dimensional description in which the fourth-coordinate of light-time $1 = ct$ is introduced".[54] Of course, "Mach himself was not oriented toward the idea or thought of a four-dimensional continuous body, nor did he link the non-Euclidean character of physical space with a four-dimensional coor-

53 Hiromatsu (1986), op. cit., pp. 112-113.

54 Ibid., p. 115.

dinate-system, but the path to four-dimensional space-time was opened up for Einstein by Mach's critique of Newton."[55]

Finally, Hiromatsu analyses the relationship between Mach and Einstein in terms of dynamics, in particular, Mach's critique of Newton's laws of motion and the alternatives to them that he proposed. As is well known, Mach not only criticized Newton's definitions and laws, but also offered alternatives. One of Mach's experimental propositions read:

> The mass-ratios of bodies are independent of the character of the physical states (of the bodies) that condition the mutual accelerations, be those states electrical, magnetic, or what not.[56]

Returning to the concept first mentioned, Mach viewed it as an experimental fact that qualities such as color, smell, the electro-chemical, etc., do not affect mass, and if so, transitivity is preserved. However, mass is dependent upon the condition of motion of the body, a thing which implies for Hiromatsu the collapse of Mach's definition of mass. Motion has a dynamic relation with respect to space as medium, and thus the accleration-determining relations are also related to space, which again is inconsistent with the above proposition, where the mass-ratio of two objects is being determined without considering space.[57] Such inconsistencies, Hiromatsu maintains, were resolved by Einstein.

Hiromatsu quotes Einstein to the effect that he agreed with the points predicted by Mach (that the inertia of an object would increase if the amount of mass of objects in the vicinity increased, and acceleration would increase if neighboring bodies accelerated)[58] were incorporated in general relativity as the relativity of inertial actions. Einstein's general relativity also supported Mach's assertion that this interaction would hold only in a finite universe. The one point which separated Mach and Einstein was that Mach assumed that, using a *quasi* -Euclidean space, inertia was a type of interaction between bodies like universal gravitation, which, Einstein maintained, was mistaken: Mach's conception required *non*-Euclidean space. Hiromatsu writes that Einstein was able to remove this inconsistency between Mach's principles by orienting his physics around the constancy of light coupled with his theory of space and time.

Moving to the next level of analysis, Hiromatsu considers the epistemological and ontological influence of Mach on Einstein.[59] He maintains that

---

[55] Ibid., pp. 117-118.

[56] Mach (1902), op. cit., p. 241.

[57] Ibid., (in experimental proposition *a*).

[58] A. Einstein, *Vier Vorlesungen über Relativitätstheorie,* Vieweg: Braunschweig, 1922, p. 64.

[59] Hiromatsu (1986), op. cit., pp. 129-178.

Einstein himself acknowledged the former (though later rejecting it) while according to Hiromatsu there was no ontological influence. Hiromatsu makes two important points about this difference. First, as evidence for it, he notes that in all of Einstein's writing concerning Machist ideas which influenced or were rejected by him (which one may add only appeared in popular publications or correspondence), only epistemological ideas are mentioned, not ontological ones. Second, Hiromatsu points out that the epistemological ideas accepted by young Einstein, for example, the operationalist definition of concepts, were all compatible with both realist and phenomenalist ontologies. Early Einstein was not a realist, if by this label one would indicate a Newtonian sense of realism, which suggests that he could have accepted most of Mach's critique of Isaac Newton's definitions of space, time, and mass, including their operationalist definitions without ever having been a phenomenalist in Mach's sense either (provided that the early Einstein completely comprehended the meaning and implications of Mach's phenomenalism). Ontologically central to Einstein's physics was the concept of field, while force and matter appeared to be central for Newton.

One example of a popular text in which Einstein dealt with Mach was in his obituary of him, written in early 1916, a few weeks after Einstein's major work on the general theory of relativity had come out.[60] In this text, Einstein remarked on the usefulness of the critical analysis of established concepts in physics whose origins may have been forgotten and may have come to seem natural. In particular, he mentioned Mach's critique of Newton's concepts of space, time, and mass (treated above) and bucket experiment, and quoted extensively from Mach's *Science of Mechanics*. Einstein concluded with the now well-known remark that if Mach had been a little younger and his mind more flexible, and had the constancy of the speed of light been recognized as a problem among physicists during his youth, it would not have been unlikely that he could have discovered relativity himself.

Hiromatsu uses this suggestion that Mach *could* have discovered relativity as the starting point for his own analysis of its implications as to why this did not happen and what it means concerning the similarities and differences between Mach's and Einstein's thinking. The statement itself suggests to Hiromatsu that Einstein considered his thinking to be close to that of Mach, at least at that time, and that the obituary was not merely a formal exercize in praising the dead.

Einstein pointed out that Mach had analysed the weak points in classical mechanics, including the concepts of space and time, indicating that strictly speaking time was abstract and could not be measured. Einstein did not state it explicitly, but Hiromatsu conjectures that he was referring to doubts

---

60 Albert Einstein, "Ernst Mach", *Physikalische Zeitschrift,* **17** (1916), 101-104.

held at the time concerning the possibility of successfully employing an operationist definition of time. In terms of Mach's method, concepts were abstracted and formed into laws in such a way that concepts and laws were formed together, and it is this which constitutes the basic point in Mach's theory of concept formation (and which entailed his rejection of absolute space and time). Mach came near to relativity theory when he pointed out in his critique of absolute motion that movement was determined with respect to the universe as a whole and that force was determined with respect to the entire mass of the universe. But along with an operationalist definition of time, Mach needed the concept of ether and the physical fact of constancy of the speed of light. In addition, because Mach was not manifestly aware that the equivalence of inertial and gravitational mass implied a principle of relativity, thus it was unclear whether what appeared like falling through space was caused by a gravitational field or by an accelerating frame of reference.

In later writings, Einstein began to delineate his developing differences with Mach, and once again, these differences related to epistemology. Hiromatsu cites, for example, a 1917 letter to Besso (first mentioned by Gerald Holton), saying that Mach could destroy but not build; he could only produce a catalogue, not a system. In short, Einstein rejected Mach's phenomenalistic physics and indeed any descriptivism based on monism, in favor of realism.[61] It was with Einstein's assertion of his realism that the distinction which Hiromatsu draws out between Mach's epistemological influence and ontological lack of influence becomes important in helping to clarify Einstein's intellectual debt to Mach.

As suggested earlier, what Einstein gradually came to conspicuously reject was the epistemology of Mach. Hiromatsu observes that being influenced by Hume or Mach entails an epistemological investigation into the origin of concepts, the mechanism of their formation, or their justification, and none of this contradicts any aspect of a "realistic position". Epistemology as a type of historical investigation is compatible with either a phenomenalistic or realist *ontology*. In particular, there is no inconsistency between realism and, for example, operationalist definitions. It would have been it seems quite natural for Einstein to have sympathized with Mach's critique of Newton's concepts of space, time, and mass if we do not assimilate realism to a Newtonian realism. It is possible that Einstein's view was always at least implicitly realist from the beginning.

The first epistemological difference between Mach and Einstein comes with regard to system construction.[62] Although initially Mach and Einstein

[61] Gerald Holton, *Thematic Origins of Scientific Thought,* Harvard University Press: Cambridge, 1973, citation 239-240.

[62] Hiromatsu (1986), op. cit., pp. 169-170.

may have used the same method of system construction, it seems that with the development of the mature Einstein (post 1916) their methods became different. Hiromatsu argues that Einstein's statement that Mach was a mere catalogue-maker is an unfair evaluation, and simply shows that Einstein's understanding of Mach's methods was not deep enough. Mach, Hiromatsu observes, departs from concrete facts and ends up with abstract formalizations. Einstein, on the other hand, works from abstract principles to concrete facts. Einstein considers this approach to be "deductivist". This does not mean a formal logical deductivist system, but rather making an implicit principle of a generalized form of an empirical fact, establishing it as an explicit principle by determining its conditions of possibility, and then developing from the ground up a system using this principle as a starting or initial point.

To some extent this may seem to resemble, for example, Mach's making the impossibility of eternal motion a principle, but Mach never consciously traced this principle back to determine the conditions of its possibility. This difference in system construction was insufficiently realized by Einstein. But, Hiromatsu points out, even these methodological differences in system construction would not have prevented Mach from accepting the special theory of relativity if the constancy of the speed of light had been recognized as an experimental fact. That Mach recognized this to be a testable matter, Hiromatsu notes, is demonstrated by the fact that Mach and his son conducted several experiments to try to ascertain this. Had they proven to their satisfaction that the results of the experiments showed the constancy of the velocity of light, Hiromatsu believes that Mach would have had no qualms about accepting special relativity *as a physical theory*. At any rate, whether one believes that if equipment failure or misuse could be avoided such an experiment would prove the constancy of the speed of light is a pre-theoretical affair, and cannot be used to argue for the incompatibility of Mach's philosophy with either relativity theory or its philosophical foundations.

Hiromatsu believes that Einstein oversimplified matters when he alleged that Mach in terms of his preferred methodology of science was simply a cataloguemaker. Contrary to Einstein, Mach was not just noticing and recording what was directly given in the appearances, but doing "something more", "something other". Hiromatsu in the context is referring to the "as" structure of perception, that is, emphasizing that it is meaning-laden. Even the idea of "pure perception" is an intellectualization. In Hiromatsu's epistemology, this "something more" is intersubjectively determined and is conditioned by socio-historical factors in its formation.[63]

Mach realized that sensory appearance contained "something more", but tried to exclude it as metaphysical. Mach was in fact half-consciously building the "something more" in its intersubjetively valid aspect as a *Gestalt, hinzudenken, Gedankending,* or *idealizieren.* So were the positivists, who included *Gestalten* among phenomena. What Mach and the positivists wanted, according to Hiromatsu, was not only to avoid any wrong or mistaken *Gedankending* which in the received view could be passed off as the "real thing" (like Einstein's enduring real existent), but also to replace it with the new "evidential" *Gedankending,* that is, in order to provide the new "something more" strictly connected to evidence. Thus Mach's phenomenalist descriptivism was more complicated than Einstein considered it to be. For Einstein, the "something more, something other" is the traditionally existent independent of the content of consciousness, known intellectually. Further, Einstein asserted that science was the study of such "objective elements". For these ontological and disciplinary reasons, Einstein could never accept Mach's basic views.

In any case, this "something more" is a necessary structural moment for knowledge of description. Mach was not engaged in a mere descriptivism, as Einstein thought, and to the extent that Mach and the positivists believed that their thinking excluded a higher level "something more", they fell into self-deception. But Einstein and Planck on the other hand, fell into the short-sighted error of reducing the "something more" to an "existent" in the traditional sense. Hiromatsu writes that in the end both came to grief:

> The origin of this tragi-comedy is that they did not correctly grasp the dual structure of empirically given knowledge consisting of the "directly given" and the "something more", and the existence characteristics (what is termed *Idealität*) and the intersubjective validity of the latter moment. As a result, they gradually fell into a situation wherein one became an advocate of empirical, phenomenal descriptivism and the other became an advocate of a worn-out realistic explanationism.[64]

Hiromatsu feels that Einstein's realism brought about a deformation of relativity theory itself. While one cannot go so far as to say that this led to the failure of the unified field theory of later years, the assumption of a pre-ordained harmony between mathematical structure and the empirical world seems to Hiromatsu to be a poor philosophy. For Hiromatsu, Einstein's realism led him to grant special privilege to the measurement of an

---

[63] Hiromatsu has developed his epistemology with which he evaluates Mach and Einstein in a series of volumes (including Hiromatsu 1975), none of which are presently available in English.

[64] Hiromatsu (1986), op. cit., p. 177.

object in its inertial frame of reference. Thus, contraction phenomena became "appearances" and space and time became absolute existents. This is a distortion of relativity theory. While Mach directed decisive criticism at subject/object framework epistemology and ontology, Einstein had one foot in and one foot out of it. A complete interpretation of relativity theory requires escaping both Machist phenomenalism and traditional realism by introducing a relationist ontology:

> ...the complete relativity and equivalence of observed phenomena cannot be developed by a dualistic separation of "observed object" and "observing subject" but by understanding the dialectical "interpenetration" of knower and known (more appropriately, in the aspect of a "functional relation" on the basis of an ontological view <derived from> the primacy of the relation).[65]

65 Ibid., p. 81.

CHAPTER 18

THE "VEREIN ERNST MACH" - WHAT WAS IT REALLY?

by Friedrich Stadler*
(Ministry of Education, Vienna)

It seemed plausible to most members of the Vienna Circle, the philosophical group most responsible for the rise of Logical Positivism, who also saw themselves as representatives of a scientific movement with tendencies based on the Enlightenment, to do their best to help spread and popularize the heart of their own teachings.[1] They also stood in the adult-education tradition of empirically-minded philosophers and scientists of the recently dismembered Austro-Hungarian Monarchy such as Ernst Mach, Ludwig Boltzmann, Friedrich Jodl, Wilhelm Jerusalem, and Adolf Stöhr. The leading members of the Vienna Circle decided during the1920's in a manner consistent with this earlier tradition and their own clear intentions to begin institutionalizing the adult-education half of their movement, bring more popular attention to their version of philosophical positivism, especially to its recent enrichment by Mach, Wittgenstein, and Russell.

What is so remarkable is that while this step seemed obvious and natural to the Circle, that is, broadening the mode of communicating philosophy of science, it has represented something of a rarity. Most philosophers in dominant cliques and schools seem quite happy to expound to each other within academic ghettos, utterly without any kind of outside social anchor, political stance, or feedback mechanism from the general population.

This rarity plus ignorance of the new institution's activities by scholars in other countries may help explain why most histories of the Vienna Circle simply pass over the Verein Ernst Mach as if it were just another name

* This paper has been shortened, translated, and partly re-written by the editor in order to be more easily understandable to an English-speaking audience. It originally appeared as "Popularisierungsbestrebungen im Wiener Kreis und 'Verein Ernst Mach'" which appeared in the book *Philosophie, Wissenschaft, Aufklärung - Beiträge zur Geschichte und Wirkung des Wiener Kreises,* Ed. Hans-Joachim Dahms, Walter de Gruyter: Berlin & New York. For a much more extensive and detailed version see Friedrich Stadler: *Vom Positivismus zur 'Wissenschaftlichen Weltauffassung',* Löcker Verlag: Wien, 1982.

[1] *Wissenschaftliche Weltauffassung. Der Wiener Kreis.* Hrsg. vom Verein Ernst Mach, [edited and written by Hans Hahn, Otto Neurath, Rudolf Carnap], Vienna, 1929. (For the English version see: "The Scientific Conception of the World: The Vienna Circle" in Otto Neurath, *Empiricism and Sociology,* ed. Marie Neurath and R.S. Cohen, (*Vienna Circle Collection,* vol. one, chapter nine); Reidel: Dordrecht and Boston, 1973, pp. 299-318.

J. Blackmore (ed.), Ernst Mach – A Deeper Look, 363–377.

for the Vienna Circle and had no significant existence of its own at all. *But it was a legally separate organization and had its own unique character and history, which this paper will describe.*

As is reasonably familiar to numerous readers, the second[2] and more familar Vienna Circle was founded not long after Moritz Schlick's appointment in 1922[3] to Ernst Mach's old chair at the University of Vienna, but which, however, in Mach's day (1895-1901) had been titled: "The History and Philosophy of the Inductive Sciences". Schlick's once-a-week private philosophical sessions became popular during the middle 1920's with a number of Viennese scientist-philosophers. The addition of Rudolf Carnap and his friend Otto Neurath gave further dynamism to the group.

The pre-history of the Verein Ernst Mach from 1926 to 1928 falls together with the earlier or non-public phase of the 2nd Vienna Circle, while the establishment of the Verein at the turn of the year 1928-1929 was buttressed during the same period by the publication of the programmatic essay by the Verein titled *Wissenschaftliche Weltauffassung. Der Wiener Kreis* (1929), which began the public phase of the Vienna Circle and soon led to its internationalization from 1929 to 1936[4] under the names "Logical Positivism" and "Logical Empiricism", a period which also saw the Circle come under considerable political pressure from rightwing groups within Austria and together with internal differences helped lead to the emigration of many members from Vienna starting about 1930.

At the end of the non-public period, the Vienna Circle, which now considered itself to be an independent movement of some importance, took over a journal, *Annalen der Philosophie,* which they renamed *Erkenntnis,* though during much of that turbulent decade it would actually be compiled in Turkey. Another related organization was the "International Congress for the Unity of Science" which Otto Neurath a few years later in Paris helped put together.

As early as April 11, 1927, a sizable and interesting local group of the period called the "Union of Free Thinkers" [*Freidenkerbund*] had *proposed* to the Vienna magistrates the establishment of the *Allgemeinen Naturwissenschaftlichen Bildungsverein Ernst Mach* which would help "spread scientific knowledge and technical literature by setting up courses, giving lectures, and arranging excursions and tours."

---

2 The first Vienna Circle may be identified with a discussion group which lasted from 1907 until 1914. Its members included Philipp Frank, Hans Hahn, Otto Neurath, and Richard von Mises, who all became associated with the second circle as well.

3 The thursday-evening meetings owed much to Hans Hahn as initiator and seem to have literally begun in 1924 under Moritz Schlick's direction.

4 Moritz Schlick was shot and murdered in 1936 and the discussion group in Vienna apparently ceased in 1938, the year Hitler marched into Austria.

At that time, no member of the Vienna Circle was also a member of the *Bildungsverein Ernst Mach*, but later with the *factual* founding of the group *Verein Ernst Mach - Verein zur Verbreitung von Erkenntnissen der exakten Wissenschaften* on November 23, 1928 when a large meeting took place, Otto Neurath already appeared as a member of the *Proponentenkomitee* and gave a lecture "Ernst Mach und die exakte Weltauffassung".[5]

Moritz Schlick took over the leadership, which led to greater participation by members of the Vienna Circle, but he also expressed reservations on its [leftwing] partyline and propagandistic activity. Nevertheless, he remained in charge of the organization until it was dissolved in 1934. Soon after the actual or effective founding, we find among the functionaries Hans Hahn as deputy chief, Otto Neurath as secretary, and Rudolf Carnap as his deputy, and among the committee members Edgar Zilsel and Heinrich Löwy. Although the Union of Free Thinkers had been responsible for framing the original statutes as well as organizing the founding assembly of participants under the Baconian slogan "Knowledge is Power" with its emphasis on the control of nature through science, progressive ideas, and "*Technophilie*", a perceived decline in dominance by the *Freidenkerbund* took place to the benefit of a (neo-positivist) profile whereby the philosophical discussions about world views gradually began to shift to concern with interdisciplines and the unity of science.

The history of the Verein Ernst Mach with respect to character and function, apart simply from inviting speakers and providing a forum for lectures, can be said to have started like the perceived transition from the Vienna Circle to Logical Positivism itself with the publication of the programmatic essay *Wissenschaftliche Weltauffassung. Der Wiener Kreis*, Hrsg. Verein Ernst Mach, Artur Wolf Verlag: Vienna, 1929, by Carnap, Hahn, and Neurath, a philosophical call to arms which seems to have been the product of two different working sessions.[6] But the document clearly reveals how these neopositivists saw their own movement. In the appendix of this rather thin "book", the purpose and future program of the Verein Ernst Mach is spelled out; it also inadvertently mirrors the pluralistic and antagonistic internal character of the Vienna Circle itself by means of the basis on which a minimal consensus was reached. That consensus sharply distinguished a scientific conception of the world from "*geisteswissenschaftlicher*" world views [i.e. The Vienna Circle rejected all mental powers as unscientific except what could be reduced to empirical or formal factors.]. On the positive side, the Circle supported the following high

[5] *Verein Ernst Mach. Vereinsbüro. Bundespolizeidirektion Wien.*

[6] See H.L. Mulder, "Wissenschaftliche Weltauffassung. Der Wiener Kreis" in *Journal of the History of Philosophy*, **6** (1968), 368-390 and also correspondence between Schlick and Neurath in the Wiener Kreis-Archiv in Amsterdam.

principles: humanism [*Diesseitigkeit*], human cooperation, interdisciplinarianism, anti-metaphysical study of facts, the tradition of liberal and social adult-education in Austria, and "efforts to place economic and social relations on a new basis [leading] to a renewal of schools, education, and the unity of mankind".[7] In short, their scientific conception of the world was to be a means leading to the emancipation of intellectuals and the masses or as the proposed program of work for the Verein Ernst Mach in the1929 essay puts it:

> ... to provide lectures and publications on the current stage of the scientific conception of the world, and thereby to show the significance of exact investigation for the social and natural sciences. Thus, conceptual tools of modern empiricism will be formed which are needed for the public and private conduct of life.[8]

The fact that this phraseology recalling the 18th century Enlightenment went unappreciated by that wing of the movement around Schlick, Waismann, and Wittgenstein already showed cracks in the profile, which would come to extreme expression in the philosophical and political antagonism between Schlick and Neurath.[9]

The *Verein Ernst Mach* first appeared to come under significant criticism from academic philosophers when it held a joint meeting in Prague in September 1929 with the Berlin-based *Gesellschaft für Empirisiche Philosophie* of Petzoldt and Reichenbach. The meeting was titled: *1.Tagung für Erkenntnislehre der exakten Wissenschaften.*[10] Lectures were on subjects such as: logic, foundations of mathematics, causality, and probability. Some of the gradually developing differences between members of the Vienna Circle also began to become conspicuous about this time with a proliferation of mostly private offshoots such as the so-called "Schlick-Wittgenstein Kreis"[11], Karl Menger's "Mathematische Kolloquium"[12], and on the peripheries what might be designated the "Heinrich Gomperz-Zirkel". Among

[7] *Wissenschaftliche Weltauffassung. Der Wiener Kreis,* Hrsg. Verein Ernst Mach, Vienna, 1929, p. 14.

[8] Ibid.

[9] See F. Stadler, "Otto Neurath - Moritz Schlick: Zum philosophischen und weltanschaulich-politischen Antagonismus im Wiener Kreis" in *Schlick und Neurath - Ein Symposion,* Hrsg. von R. Haller, Amsterdam, 1982, pp. 451-463.

[10] *Erkenntnis,* Vol. I, 1930/31, pp. 93-340.

[11] *Witgenstein und der Wiener Kreis - Gespräche,* Aufgezeichnet von F. Waismann, (*L. Wittgenstein Schriften,* 3), Hrsg. von B.F. McGuinness, Frankfurt/M, 1967 .

[12] K. Menger, *The New Logic - Selected Papers in Logic and Foundations, Didactics, Economics,* (*Vienna Circle Collection,* vol. 10), Dordrecht/Boston/London, 1979.

Neo-Kantians, a circle which gathered around Robert Reininger should also be mentioned, especially given its rather long life.

Viennese intellectual life was truly active in those days, and the Verein Ernst Mach was in the forefront. Here is a list of their lectures from 1929 until 1934. Understandably, most were given by members of the Vienna Circle and their sympathizers, but unassociated scientists and empirically-oriented philosophers, and a wide variety of scholars and theorists from in and out of the country also contributed. An unusual feature was the large number of lectures on interdisciplinary topics.

LECTURES GIVEN UNDER THE AUSPICES OF THE VEREIN ERNST MACH:

1929: J. Frank: *Moderne Weltauffassung und moderne Architektur.*
[Modern Conception of the World and Modern Architecture]
H. Hahn: *Überflüssige Wesenheiten* (*Occams Rasiermesser.*)
[Superfluous Substances (Occam's Razor)]
H. Vokolek: *Begabungsproblem und Vererbungslehre.*
[Problems of Aptitude and Theory of Inheritance]
R. Carnap: *Von Gott und Seele. Scheinfragen der Metaphysik und Theologie.*
[Of God and Soul. Pseudo-Questions of Metaphysics and Theology]

1930: O. Neurath: *Einheitswissenschaft und Marxismus.*
[Unity of Science and Marxism]
M. Schlick: *Über wissenschaftliche Weltauffassung in den USA.*
[On the Scientific Conception of the World in the United States]
Wl. Misar: *Probleme der Astronomie.*
[Problems of Astronomy]
H. Feigl: *Naturgesetz und Willensfreiheit.*
[Natural Law and Freedom of the Will]
R. Carnap: *Einheitswissenschaft auf physischer Basis.*
[Putting the Unity of Science on a Physical Basis]
E. Zilsel: *Der Geniekult, ein soziologisches Problem.*
[On the Cult of Genius, a Sociological Problem]
J.K. Friedjung: *Unwissenschaftliches in der Erziehung.*
[On the Unscientific in Education]
O. Bauer: *Industrielle Rationalisierung und Wissenschaft.*
[Industrial Efficiency and Science]

(The following lectures held during 1930 were sponsored jointly by the Verein Ernst Mach and an affiliated association: "Studiengruppe für wissenschaftliche Zusammenarbeit". The talks were organized under the leadership of R. Carnap.)

H. Feigl: *Statistische Gesetzlichkeit.*
[Statistical Lawfulness]
L. Bertalanffy: *Entropieproblem und Normbegriff in der Biologie.*
[The Problem of Entropy and the Concept of Norms in Biology]
W. Marinelli: *Statistische Methoden in der Biologie.*
[Statistical Methods in Biology]

E. Zilsel: *Über Induktion.*
[On Induction]
H. Zeisel: *Konjunkturstatistik.*
[Statistics on Business Trends]
K. Polanyi: *Wirtschaftsstatistik.*
[Economic Statistics]
H. Hartmann: *Die Psychoanalyse und das Problem der Illusionen.*
[Psychoanalysis and the Problem of Illusions]
R. Strigl: *Die ökonomischen Kategorien.*
[Economic Categories]
E. Brunswik: *Gestaltpsychologie mit Demonstrationen.*
[Gestalt Psychology with Photographs]
W. Reich: *Trieblehre der Psychoanalyse.*
[Theory of Human Drives in Psychoanalysis]
G. Halpern: *Zum Kausalbegriff der Quantentheorie.*
[On the Causal Concept in Quantum Theory]

1931: O. Neurath: *Magie und Technik*
[Magic and Technology].
R. Gicklhorn: *Zellphysik.*
[The Physics of Cells]
Ph. Frank: *Physikalische und biologische Gesetzmässigkeit.*
[Physical and Biological Conformity with Laws]
O. Neurath: *Empirismus in der Pedagogik.*
[Empiricism in Teaching]
M. Schlick: *Probleme der Kausalität.*
[Problems of Relation-Based Causality]*

Lectures on *Probleme der Einheitswissenschaft* [Problems in the Unity of Science] were held under the auspices of the *Wiener Volksbildungsverein* [Vienna Adult Education Union] in connection with the *Verein Ernst Mach* as follows:

O. Neurath: *Einheitswissenschaft und Empirismus der Gegenwart. Soziologie in physikalischer Sprache.*
[Unity of Science and Empiricism of the Present. Sociology in Physical Language]
R. Carnap: *Die Sprache der Physik. Psychologie in physikalischer Sprache.*
[The Language of Physics. Psychology in Physical Language]
F. Waismann: *Logik, Sprache, Philosophie.*
[Logic, Language, and Philosophy]
H. Hahn: *Mathematik und Wissenschaft.*
[Mathematics and Science]

* The word "cause" in English is normally translated as "Ursache" in German. But the German term "Kausalität", especially when Kantians and positivists use it, does not mean causality in English but that which is relation or law-determined. Except for some scientists and philosophers, the English word "causality" refers to causes as powers or forces. To cause something to happen is to help force or make it happen. - Ed.

Under the auspices of the Verein Ernst Mach, lectures were also given for a weekly seminar on "Moderne Wissenschaft":

H. Hahn: *Einleitungsvortrag.*
[Introductory Lecture]
F. Waismann: (Leiter der Abteilung "Physik")
[Director of the Division on Physics]

Additional lectures were held in the series *Physikalismus* [Physicalism] under the auspices of the Verein Ernst Mach:

O. Neurath: *Empirismus in Vergangenheit und Gegenwart. Psychologie und Soziologie in physikalischer Sprache.*
[Empiricism in the Past and Present. Psychology and Sociology in Physical Language]
H. Hahn: *Sprache der Physik.*
[Language of Physics]
F. Waismann: *Logik und Sprache.*
[Logic and Language]
H. Hahn: *Mathematik und Wissenschaft.*
[Mathematics and Science]

1932: Ph. Frank: *Philosophische Strömingen in der Sovjetunion.*
[Philosophical Currents in the Soviet Union]
O. Neurath: *Das Fremdpsychische in der Psychologie.*
[The Psychically Alien in Psychology]
H. Hahn: *Sein und Schein.*
[Being and Appearance]
M. Schlick: *Philosophische Strömungen in den USA.*
[Philosophical Currents in the USA]
H. Hahn: *Logik und Wirklichkeit.*
[Logic and Reality]

There were also lectures which were planned to be offered in common with the *Verein für angewandte Psychopathologie und Psychologie* [Union for Applied Psychopathologie and Psychology], but which it is not certain were actually given:

H. Hartmann: *Empirismus in der Psychologie.*
[Empiricism in Psychology]
M. Pappenheim: *Reflexologie.*
[Reflexology]
O. Neurath: *Einheitswissenschaft und Psychologie.*
[Unity of Science and Psychology]
H. Gomperz: *Entstehung der Atomistik.*
[On the Origins and Development of Atomism]

1933/34: The journal *Erkenntnis* gives no report on lectures held during these two years, although according to documents in the Archive of the Verein Ernst Mach, lectures are referred to as if having been given up until the forced dissolution of the society in 1934.

The list of themes and lectures clearly show the interdisciplinary focus and collective working style. For example, their scientific conception of the world is viewed as even being linked with modern architecture, something which appeared especially conspicuous with the new functionalism of the Bauhaus where Otto Neurath, Rudolf Carnap, and Herbert Feigl gave lectures.[13]

According to information from contemporaries, the number of people who listened to lectures in the Verein Ernst Mach did not exceed those who normally attended lectures given in Viennese adult schools.[14] In general, the "masses" scarcely beat down the doors to get in, nevertheless, Neurath and Carnap were often able to fill their lecturehall. During the year 1930, the governing board of the Verein had twenty-two people, including several "free-thinkers" along with Neurath, Carnap, Zilsel, Julius Tandler, and Hans Thirring. Schlick remained head man, although, perhaps anticipating the future, he increasingly tried to distance himself from political and propaganda aspects.*

Starting in 1930, in addition to attempting to popularize the unity of science on the basis of what the movement called "physicalism", an effort was begun to integrate psychoanalysis into the total concept of science as if it were compatible with behaviorism, but, understandably, the forced marriage had very little success. Nevertheless, Heinz Hartmann and Josef Karl Friedjung, for example, put forth their point of view within a context as if they were a cooperative working part of the scientific community.[15] Otto Bauer, the Austrian socialist theoretician and leader, in a like interdisciplinary spirit, lectured on efficiency and modernization [*Rationalisier-*

---

13 See H.M. Wingler, *Das Bauhaus: Weimar - Dessau - Berlin 1919-1933,* Wiesbaden, 1975, p. 170.

14 The information is from Professor Paul Neurath, Vienna/New York. There are also indications of this from the Carnap diaries in the Carnap Archive in Pittsburgh.

* Schlick correspondence. Wiener Kreis-Archiv, Amsterdam.

15 Concerning attempts to integrate psychoanalysis with other kinds of psychology, see: M. Jahoda, "Aus den Anfängen der sozialwissenschaftlichen Forschung in Österreich", *Zeitgeschichte,* 8th year, January 1981, Heft 4, pp. 133-141. The same article can also be found in F. Stadler (Hrsg.), *Arbeiterbildung in der Zwischenkriefzeit,* Löcker Verlag: Vienna, 1982, p. 43ff. For the interpretation of psychoanalysis in terms of natural science see: H. Hartmann, *Die Grundlagen der Psychoanalyse,* Vienna, 1927 and O. Neurath, *Einheitswissenschaft und Psychologie,* Vienna, 1933 (*Einheitswissenschaft,* 1). For the English version see: "Unified Science and Psychology" in *Unified Science,* editor Brian McGuinness, trans. Hans Kaal with an introduction by Rainer Hegselmann (*Vienna Circle Collection,* vol. 19, Kluwer: Dordrecht and Boston, 1987, ch. 1, pp. 1-23.

ung][16], which manifestly had been stimulated by discussion among Bauer, Neurath, and Zilsel.[17]

Not only undervalued but scarcely noticed at all by subsequent scholars has been the existence of another organization established by the Verein Ernst Mach, namely, the *Studiengruppe für wissenschaftliche Zusammenarbeit*, which starting in 1930 contained a regular working circle of significant scientists under the leadership of Rudolf Carnap, which met in the Vienna *Arbeiterkammer* and had as its ostensible goal: "To bring about a closer relation between individual specialized sciences . . . concerning new methods, problems, and conceptual notions by means of reports, lectures, and connected discussions as well as to help clarify their position within the context of science as a whole."[18] This group manifestly included together with the positivists (Neurath, Frank, Carnap, Zilsel, Feigl, and Brunswik) the biologists Ludwig Bertalanffy and Marinelli, the sociologist Hans Zeisel, the economists Karl Polanyi and Richard Strigl, as well as the psychologists i.e. psychoanalysts Egon Brunswik, Else Frenkel-Brunswik and Wilhelm Reich. Some of the latter also appeared as authors in the "Foundations of the Unity of Science" series. Nor should one overlook the Verein's connection with several of the founders of empirical social research in Austria: Paul Lazarsfeld, Maria Jahoda, and Hans Zeisel. While only Zeisel seems to have given lectures with the Verein, all of them had personal connections with it or the Circle as well as epistemological and methodological affinities.[19]

A remarkable activity was the series of lectures presented by the Verein Ernst Mach in conjunction with the *Wiener Volksbildungsverein* in February/March 1931 on *Probleme der Einheitswissenschaft*. In his own lecture, Otto Neurath added in his fiery way the following popular slogans: "The last consequence of empiricism is science without philosophy!" "Freedom from metaphysics and theology demand freedom from bad social conditions too!" "Empiricism and Unity of Science together with social behaviorism and social epicureanism symbolize the demands of the present!" "Let

---

16 See O. Bauer, *Kapitalismus und Sozialismus nach dem Weltkrieg*, 1 Band, *Rationalisierung - Fehlrationalisierung*, Wien, 1931, in Otto Bauer, *Werkausgabe*, Bd. 3, Wien, 1976.

17 Conversations with Prof. Eduard März and Margarete Schütte-Lihotsky, both from the city of Vienna.

18 *Erkenntnis*, Bd. I, 1930/31, p. 79.

19 There is a problem in delimiting movements, since social adhesion does not always mean ideological cooperation; nor is it always clear to what extent even a vigorous lecture series is part and parcel of a movement. Nevertheless, the leaders of the Verein Ernst Mach were undeniably active.

us have clarity, strictness, and practicality!" "All scientific work is to be understood as a function of the social situation . . . !"[20]

With the advantage of hindsight, it seems to us very possible that perhaps the most lasting influence of the Verein Ernst Mach has been the encouragement which Neurath's rhetoric and other lectures on unity of science gave to the publication of a long series of books, which were followed by more than one subsequent series, and which continued directly well into the 1960's and when coupled with the Vienna Circle Collection put out by Reidel/Kluwer has continued up to the present.

The Verein Ernst Mach began pumping out short works in 1933 under the title of *Einheitswissenschaft.* They were edited and largely written by O. Neurath, R. Carnap, P. Frank, H. Hahn, and later J. Joergensen and Ch. Morris, that is, basically the left-oriented non-Schlick, non-Waismann, and largely non-Wittgenstein wing of the Vienna Circle plus affiliates overseas.

This series of Verein Ernst Mach booklets was apparently originally thought of as a popularized version of Moritz Schlick's somewhat conservative 1929 publishing venture, but which became more and more independent and radical. (All seven of the Verein Ernst Mach *Hefte* listed below plus two more, i,e, articles published as separate pamphlets or short books, have recently been published together in a single book in English,** hence we will give German and English titles plus the English page numbers.)

H. 1. O. Neurath, *Einheitswissenschaft und Psychologie,* 1933.
[Unified Science and Psychology", pp. 1-23.]
H. 2. H. Hahn, *Logik, Mathematik, und Naturkennen,* 1933.
["Logic, Mathematics and Knowledge of Nature", pp. 24-45.]
H. 3. R. Carnap, *Die Aufgabe derWissenschaftslogik,* 1934.
["The Task of the Logic of Science", pp. 46-66]
H. 4. O. Neurath, *Was Bedeutet rationale Wirtschaftsbetrachtung?,* 1935.
["What is Meant by a Rational Economic Theory? pp. 67-109]
H. 5. Ph. Frank, *Das Ende der mechanistischen Physik,* 1935.
["The Fall of Mechanistic Physics", pp. 110-119]
H. 6. Neurath/Brunswik/Hull/Mannoury/ Woodger, *Zur Enzyklopädie der Einheitswissenschaft,* 1938.
["Towards an Encyclopedia of Unified Science", pp. 130-165]
H. 7. R. v. Mises, *Ernst Mach und die empiristische Wissenschaftsauffass*sung, 1938.
["Ernst Mach & the Scientific Conception of the World" pp. 166-190]
[H. 8-9. H. Gomperz, "Interpretation: Logic Analysis of a Method of Historical Research", pp. 191-272.]

---

[20] O. Neurath, in *Erkenntnis,* Bd. 2, P. 311.

** *Unified Science,* ed. Brian McGuinness, trans. Hans Kaal with an introduction by Rainer Hegselmann, (*Vienna Circle Collection,* vol. 19), Kluwer: Dordrecht and Boston, 1989.

As mentioned above, Moritz Schlick and Philipp Frank had already begun publishing their own series of books under the title: *Schriften zur Wissenschaftlichen Weltauffassung*. These books while generally representing the point of view of the Vienna Circle were, apart from the works by Carnap, Neurath, and Frank themselves, not as radical, Machist, or "unity of science" focused as those put out in the Verein Ernst Mach series by Carnap and Neurath. Schlick's collection would contain two volumes on ethics, one on value theory, and even one from a severe critic of the Circle, who would later allege that it was he who had killed logical positivism.

Bd. 2. R. Carnap, *Abriss der logistik,* 1929.
[*Introduction to Symbolic Logic and Its Applications,* trans. W.H. Meyer and John Wilkinson from the enlarged and revised 2nd German edition of 1954, Dover: N,Y, 1958.]
Bd. 3. R. v. Mises, *Wahrscheinlichkeit, Statistik, und Wahrheit,* 1928.
[*Probability, Statistics and Truth,* trans. J. Neyman, D. Scholl and E. Rabinowitsch, William Hodge: London, 1939; and the same title, ed. Hilda Geiringer from revised 3rd German edition of 1951, Allen & Unwin: London, 1957.]
Bd. 4. M. Schlick, *Fragen der Ethik,* 1930.
[*Problems of Ethics,* trans. D. Rynin, Prentice-Hall, N.Y., 1939.]
Bd. 5. O. Neurath, *Empirische Soziologie,* 1931.
["Empirical Sociology. The scientific Content of History and Political Economy", in *Empiricism and Sociology,* ed. Marie Neurath & R. Cohen, (*Vienna Circle Collection,* vol. 1), Reidel: Dordrecht and Boston, 1973.]
Bd. 6. Ph. Frank, *Das Kausalgesetz und seine Grenzen,* 1932.
[*The Law of Causality and Its limits,* ed. R.S. Cohen from translation by Marie Neurath, (*Vienna Circle Collection*) , Kluwer: Dordrecht and Boston, 1992.]
Bd. 7. O. Kant, *Zur Biologie der Ethik,* 1932.
Bd. 8. R. Carnap, *Logische Syntax der Sprache,* 1934.
[*The Logical Syntax of Language,* trans. by Amethe Smeaton from a text revised and enlarged by Carnap, Kegan Paul: London, 1937.]
Bd. 9. K. Popper, *Logik der Forschung,* 1935.
[*The Logic of Scientific Discovery,* trans. by author with J. & L. Freed of 1935 text plus 12 new appendices, Hutchinson: London, 1959.]
Bd. 10. J. Schächter, *Prolegomena zu einer kritischen Grammatik,* 1935.
[*Prolegomena to a Critical Grammar,* trans. Paul Foulkes (*Vienna Circle Collection,* vol. 2), Reidel: Dordrecht and Boston, 1973.]
Bd. 11. V. Kraft, *Die Grundlagen einer wissenschaftlichen Wertlehre,* 1937.
[*Foundations for a Scientific Analysis of Value,* ed. Henk L. Mulder, trans. E.H. Schneewind, based on the revised 2nd German edition of 1951, (*Vienna Circle Collection,* vol. 15), Reidel: Dordrecht and Boston, 1981.]

But it was the Verein Ernst Mach's title of another series of books openly called *Einheitswissenschaft*, or Unity of Science, which would set the tone for the immediate future, though in terms of publication date they would be partly anticipated by an American series, which we will discuss presently. The first two books of this German language series which had become expatriate in that it could not be continued in Austria or Germany were titled:

R. v. Mises, *Kleines Lehrbuch des Positivismus - Einführung in die empiristische Wissenschaftsauffassung,* 1939.
[*Positivism: A Study in Human Understanding,* trans. Jeremy Bernstein and Roger G. Newton, Harvard University press: Cambridge Mass., 1951, and with corrections, Dover: N.Y., 1968.]
H. Kelsen, *Vergeltung und Kausalitat,* 1941.
[*Society and Nature*, Univ. Chicago Press: Chicago, 1943, and Kegan Paul: London, 1946.]

But there was a lot more far from Central Europe. A much larger series of books came out under the imprint: "International Encyclopedia of Unified Science". The editers were Neurath and Carnap again, plus Charles Morris, but the New World had become El Dorado and all of the books would appear in English, which reflected the emigration of most prominent members of the Verein and the growing popularity of logical positivism in the United States, but in fairness, one would have to allow that some of the books published at least partly escaped the positivist line, especially those by John Dewey and in the subsequent section Thomas Kuhn:

V.F. Lenzen, *Procedures of Empirical Sciences,* 1938.
Ch. Morris, *Foundations of the Theory of Signs,* 1938.
L. Bloomfield, *Linguistic Aspects of Science,* 1939.
R. Carnap, *Foundations of Logic and Mathematics,* 1939.
J. Dewey, *Theory of Valuation,* 1939.
E. Nagel, *Principles of the Theory of Probability,* 1939.
J.H. Woodger, *The Technique of Theory Construction,* 1939.
G. de Santillana/E. Zilsel, *The Development of Rationalism and Empiricism,* 1941.
O. Neurath, *Foundations of the Social Sciences,* 1944.
P. Frank, *Foundations of Physics,* 1946.

After World War II, even more books appeared, but apart from Hempel and Joergensen, there seemed to be less overt positivist ideology in the series, though perhaps much was simply taken for granted and did not have to be repeated ad nauseum. The following is a brief and rather summary list:

E. Finley-Freundlich, *Cosmology*.
F. Mainx, *Foundations of Biology*.
E. Brunswik, *The Conceptual Framework of Psychology.*
T. S. Kuhn, *The Structure of Scientific Revolutions* .
A. Edel, *Science and the Structure of Ethics* .
G. Tintner, *Methodology of Mathematical Economics and Econometrics* .

C.G. Hempel, *Fundamentals of Concept Formation in Empirical Science*.
J. Joergensen, *The Development of Logical Positivism*.

Concerning the lectures given by the Verein Ernst Mach during 1933 and 1934 back in Vienna, we simply lack information. The house-journal of the time, *Erkenntnis,* doesn't list anything, though as we have mentioned before, there is indeed evidence from the Verein's archive that activity of some sort continued up until the group was outlawed.

The basic problem was political. Adolf Hitler had risen to power in Germany in early 1933 and his influence cast a long shadow over Austria, even in "Red Vienna". There were three major political groupings in Austria: The Social Democrats on the left led by Otto Bauer, which through Neurath and some others had close connections with the Verein, the right-wing Christian Socialists which consisted of Austrian nationalists, Roman Catholics, and rural interests, and even further to the right German nationalists and Nazis who wanted *Anschluss* with the Third Reich. All three factions had their own quasi-military "support groups", and rioting rapidly became the order of the day. Moritz Schlick, to the extent he may have had a political opinion, like many middle-class Austrians, apparently leaned toward the hope that under the circumstances that the Christian Socialists in spite of their increasingly authoritarian approach were the least of three evils. But the leader of the Christian Socialists and the government of the time, Engelbert Dollfuss, in order to win the support of Mussolini's Italy to counterbalance Hitler had to agree to the abolition of all political parties and the introduction of at least the trappings of a fascist state. In June 1933, the Nazis were briefly suppressed only to murder Dollfuss a year later, and after Mussolini abandoned his successor Schuschnigg in 1937, the Germans grabbed power the next year.

In February 1934, still attempting to eliminate all political opposition by force in order to keep Mussolini's personal backing, as if preserving Austrian independence were more important than protecting human rights or democratic government, Dollfuss's troops using artillery removed the Social Democrats from all remnants of power in Vienna by blasting a huge public housing project called the "Karl Marx Hof" where a large number of the socialist leaders were. The closing of numerous groups and societies suspected of favoring the socialists [SDAP] soon followed, including the Verein Ernst Mach, which took place on February 23, 1934. The official reason was not very specific: "Verein Ernst Mach in Wien, von dem Amtsbekannt ist, dass es fur diese Partei [SDAP] tätig war."[21] (Or put in

21 Vereinsbüro BPDion 1152/34.

English: "It is known by this office that the Ernst Mach Union has been an active supporter of the Social Democratic Workers' Party.")

Moritz Schlick as head of the dissolved Verein sent two letters of protest which corresponded to his conservative-liberalism and his generally apolitical attitude. He objected to the closing on the ground of the Verein's supposedly pure scientific activity, the absolutely unpolitical lectures. He denied every connection with the Social Democratic Party and even emphasized his own personal reserve with respect to it.[22] This position converged with Moritz Schlick's support of Dollfuss's *Vaterländischen Front* - which in his opinion - served as an effective barrier against Hitler's National Socialism.[23] Schlick was no doubt sincere in denying that either he or the Verein "actively supported" the Social Democrats, but the officials, like the government itself, could hardly accept that the Verein Ernst Mach had played an objective role given Otto Bauer's presence as a lecturer and the known left-wing orientation of Carnap, Neurath, and many if not most of the other speakers. Thus, they condemned the Verein just as they had previously done with the Vienna Circle in the spirit of their unliberal and undemocratic ideology about a "New Austria". They categorically rejected Schlick's protest.

Schlick's second letter[24] spelled out even more firmly than before his insistence that regardless of who the Verein invited as speakers it was politically and religiously neutral; he also tendered his sympathy for the difficult task facing the Dollfuss government, but he was rebuffed perhaps even more strongly, but this time he was also criticized by Neurath and Carnap who did not feel comfortable having either themselves or the Verein labled "politically neutral". As activists they had already felt compelled to leave Austria and live abroad.[25] As a victim of Dollfuss's anti-left policy, Neurath expressed his annoyance at Schlick's behavior using the expression: "With Dollfuss and against the Unity of Science."[26]

Schlick's dilemma was grounded in his Socratic ethos of pure science and philosophy on the one hand and his wish to preserve this ideal under impossible conditions on the other, where the very political ideology he felt forced to support was trying to stamp out the "valuefree" empiricism and scientific conception of the world which he normally favored. In the larger context, even apart from left-right political considerations, Neurath and Carnap were saying that the need ιo establish and popularize their philoso-

---

22 Schlick to HR Ganz, March 2, 1934, Wiener Stadtarchiv und Vereinsbüro.

23 Schlick-Korrespondenz, Wiener-Kreis-Archiv Amsterdam. See also Schlick's posthumously published small book *Natur und Kultur,* Vienna/Stuttgart, 1952.

24 Schlick to the Sicherheitskommissar Vienna, March 23, 1934.

25 Carnap-diaries, University of Pittsburgh.

26 Neurath to Carnap, August 18, 1934, *Wiener Kreis-Archiv,* Amsterdam.

phical movement required that freedom of speech, press, and assembly be put above nationalistic considerations. Dollfuss and Schlick thought that the nation and blocking Hitler came first. In the event, because of Mussolini's flip-flop, Austria had little or no chance to protect its independence against Hitler in any case, but before Dollfuss's assassination and before German military power became too ominous for Mussolini, this was not obvious, hence, legitimate differences of opinion were still possible. On the other hand, for people on the left such as Neurath and Carnap, the differences between Dollfuss and Hitler did not always seem incommensurable.

Put in other terms, this meant that even if Moritz Schlick had not been assassinated (June 22, 1936), that because of the success of Neurath and Carnap in finding teaching and especially publishing opportunities abroad, particularly in the United States, that their version of logical positivism as based on the notions of "physicalism" and "unified science" as originally popularized in the Verein Ernst Mach had a better chance to be influential in the world philosophical community. When Schlick chose Austria over America or nationalism over freedom, though he also kept fighting for the latter, he may have helped eclipse his own philosophical wings.

Concerning the assassination of Schlick,[27] which happened right in the university itself in 1936 and which was of course most unfortunate and tragic in its own right, what made it philosophically worse from the perspective of logical positivism in Vienna was the phasing out of the Mach-Schlick chair in the philosophy of science and the triumph of anti-positivist tendencies in the university itself.[28] The Circle, Verein, and professorship had disappeared. On the larger scene, logical positivism seemed to be expanding its influence abroad, thanks to vigorous publishing by Neurath and Carnap. But in fascist and communist Europe, and especially in Vienna, the situation appeared bleak. Only well after World War II would things look up again, and even then more in Salzburg and Graz than in the capital.

But hope springs eternal, and even the University of Vienna has very recently started thinking about establishing an Institute for the History and Philosophy of Science [Mach had held the original chair but without an institute as such as early as 1895], an Institute which may well be active by the time this paper is published. *But will it become as bold as either Mach or the Verein Ernst Mach?*

---

27 M. Siegert, "Mit der Browning philosophiert. Der Mord an Moritz Schlick am 22. Juni 1936", *Forum,* Juli/August 1981, Heft 331/332. A shortened version is in L. Spira (Hrsg.), *Attentat, die Osterreich erschutterten,* Vienna, 1981.

28 F. Stadler, "Aspekte des gesellschaftlichen Hintergrunde und Standorts der Wiener Kreises am Beispiel der Universitat Wien", in *Wittgenstein, der Wiener Kreis und der kritische Rationalismus,* (Akten des 3. Internationalen Wittgenstein-Symposiums Kirchberg, Hrsg. von H. Berghel/A. Hübner/E. Köhler, Wien, 1979, pp. 41-59.)

CHAPTER 19

THE INFLUENCE OF MACH'S PHILOSOPHY IN YUGOSLAVIA 1900-1941

by Miodrag Cekic*
(University of Nis, Belgrade)

I

Let us start this paper with an account** by Branislav Petronijevic, who is widely considered to have been Yugoslavia's most original philosopher, of several visits he made to Ernst Mach between 1907 and 1912:

> The third significant German philosopher with whom I came into close contact was Ernst Mach, who had been professor at the University of Vienna....While active, he taught courses in the methodology of the inductive sciences and was known for siding with students during oral examinations. Because of a serious stroke, which in no way reduced his intellectual capacity, he had been forced to go into retirement in 1901. Since that time he lived in Vienna as a private scholar, but during his final years he settled in a village by Munich. He died in 1916.
>
> I visited Mach for the first time in 1907 in Vienna during a long stay of several months. Earlier, I had already sent him the first section of my *Prinzipien der Metaphysik*+ and a treatise on discrete

* This article was originally taken from the chapter of a book written in Serbo-Croatian. The editor has added biographical and background information plus the article by Petronijevic (which was translated into German by Professor Cekic) to this article in order to make it easier to understand for those English-speaking readers not familiar with Yugoslav history or philosophy. An attempt has been made to translate some of the book and article titles from Serbo-Croatian into English in the footnotes. - Ed.

** This article by Petronijevic originally appeared in the Serbian newspaper *Srpski narod*, Beograd, 2/1943, br. 30.7, Avgust, str. 9.

+ Ernst Mach's colleague at the University of Vienna, the physicist-philosopher Ludwig Boltzmann, also seems to have read Petronijevic's book on metaphysics, whose contents he partly described in his lecture notes of 1904. He mentioned him again in early 1906. Boltzmann is perhaps most famous for his work on the statistical interpretation of the second law of thermodynamics, but, at least among Boltzmann scholars, he is also well-known for his remark that metaphysics gave him a "migraine headache". Hopefully, he did not have Petronijevic's rather original book in mind. See Ilse M. Fasol-Boltzmann (ed.), *Ludwig Boltzmann 'Principien der Naturfilosofi' - Lectures on Natural Philosophy 1903-1905*, Springer Verlag: Berlin, 1990, pp. 110 & 145. - Ed.

J. Blackmore (ed.), Ernst Mach – A Deeper Look, 379–402.

geometry, which had been published in Ostwald's *Annalen der Naturphilosophie* in 1905. I reminded him of these and recalled his interest in the construction of discrete space which had been clearly expressed in his *Erkenntnis und Irrtum* (second edition, page 447f). He said he had read both of my writings and requested that I visit him again. Although he was antimetaphysically-minded, Mach read my work through carefully, concentrating on the mathematical sections, and was even ready to accept my new geometrical doctrine, saying to me one day: 'If I could only overcome a few difficulties, I would completely agree with you.'

The greatest difficulty which Mach found in the new doctrine was the dependence of the Pythagorean theorem [*Satz*] on superficial propositions [*Lehrsätzen*]. In discrete geometry, the Pythagorean theorem also exists as a pure linear proposition, which presupposes grounding the proposition of proportion quite independently from superficial considerations. I remarked that the grounding was quite possible and had already been carried out in my work, but I could not persuade him.

Although this difficulty hindered him from publically calling himself a follower of the new doctrine, he was still curious enough to ask what kind of reception it had received in mathematical and philosophical circles. I replied to him: 'mainly negative'. I cited two French philosopher-mathematicians, the first alleged that the new doctrine had only indirect-negative significance, as a *'réduction `a l'absurde du finitisme'*, and the second said it was only a *'singuliere curiosité '*. Mach was indignant with the second opinion and called it frivolous. But when I visited him in the following years and told him that the American journal *Monist* had called the new doctrine 'acute', 'original', and 'bold', he remarked that I ought to be satisfied with that. Generally speaking, he would always ask me during each succeeding visit: 'How is the new geometry coming?' 'Are people writing about it?' But all the while he was becoming more and more reserved, and on my last visit in the summer of 1912 he explained his position as follows: 'Don't put your hopes in us old people; you must turn to a younger generation.'

While talking about geometry, let me mention another interesting observation from Mach. Since discrete geometry is a much more radical departure from classical Euclidean geometry than Riemannian and Lobachevskian geometries are, Mach was so impressed by my intellectual boldness that he once literally said the following: 'Differences between individuals of one and the same race could be greater than those between races.'

Concerning his book on mechanics and those of other authors I had studied, especially basic concepts such as force, mass, action, reaction, and so forth, I once asked him whether there was another book which described them according to his historical and critical conception. He pointed out an English book on mechanics which I later studied and which really did help me understand basic concepts and principles better. I couldn't help but noticing, however, the the English book had been dedicated to Mach himself 'whose genius has enlightened the historical and philosophical development of mechanics together with many other branches of physical science.'

In the narrower sense, we rarely engaged in philosophical discussion. Nevertheless, one day I brought forward difficulties which I believed had made his standpoint of naive realism** impossible. The difficulties were later precisely formulated, enlarged, and systematically represented in the second part of my *Prinzipien der Metaphysik* which appeared in 1912 and which I sent to him at once after its appearance. When I visited him for the last time in summer of that year, I directed his attention to that chapter.

More often than general philosophical discussions were conversations about philosophy of nature. During one of them, Mach praised the book of a professor in the University of Vienna, which had just appeared and which was concerned with the problem of matter. I became quite involved with the book and studied it thoroughly. Nevertheless, I finally came to the conclusion that the author's theory of matter suffered from irremediable difficulties. I went to Mach and laid out the difficulties before him. He completely agreed with me then. But I could not refrain from reminding him of his earlier praise of the book. Whereupon, Mach literally said the following: 'My dear colleague, You are still a young man and too strict. There are so many bad books written, that at my age I am happy if out of five hundred pages I can find a single correct idea.'

Mach gladly discussed many things. He was particularly happy to relate to me different anecdotes about the great physicists. He was an uncommonly freethinking and independent spirit, a persuaded atheist and anti-clerical. He didn't like noble titles. When the conversation turned to the English physicist, Lord Kelvin, he said he thought 'William Thomson' was good enough.

In conclusion, it might be of interest to certain readers to learn about the following detail as well. While discussing his profound

** Petronijevic is classifying Mach's world view in terms of his own highly original and interesting definition of a label rarely applied to Mach, who called his own philosophical position "physical phenomenology" which in most dictionaries and encyclopedias of philosophy it is listed as a type of "phenomenalism" or "sensationism". - Ed.

books on mechanics and heat theory, I asked Mach why he had not written a volume of the same kind on optics. He replied: 'Even before I wrote *Die Mechanik,* I did work out the principles of optics in a special treatise, but I was not satisfied with the results, and the work has remained a manuscript.' This (at least the first part) was actually published after Mach's death in 1921 under the title of *Die Prinzipien der physikalischen Optik.*

II

SOME YUGOSLAV PHILOSOPHERS

Partly Anticipated Mach's Ideas:	Largely Friendly Toward Mach's Ideas:	Largely Critical of Mach's Ideas:
Uros Milankovic	Dragisa Djuric	Branislav Petronijevic
	Radmilo Vucic	Dusan Nedeljkovic

Before presenting the views of those Yugoslav philosophers who either embraced or rejected Mach's position, I would like to point out that in his work *World Organism and the System of Entire Life* (Vienna, 1845) Uros Milankovic (1800-1849) rejected Isaac Newton's doctrine of absolute space and absolute world-coordinates and thus anticipated Mach's (and restated Berkeley's) criticism of Newtonian classical mechanics. It seems that Milankovic was under a considerable debt to Schelling's natural philosophy, but I have been unable to obtain a copy of the Milankovic book, hence I would need to examine the original work and rely less on secondary sources in order to clarify in full detail the exact extent to which the Yugoslav philosopher's conceptions anticipated those of Mach.[1]

The philosophy of Dragisa Djuric (1871-1941), reflects some of Mach's basic ideas as well as other fundamental positivist notions at the turn of the century. He had originally earned his Ph. D. in philosophy in Germany and later taught at the University of Belgrade. Politically, he was a social democrat. Some of his books have included *The Problem of the Subject and Object* (1912), *The Philosophy of the Phenomenon 'Now'* (1925), *The Philosophy of the History of Philosophy* (1925), and *The Origins of Philosophy* (1928). Djuric repudiated metaphysics as impossible, adopted gnoseological phenomenalism (sensation or "psychognoseological element" is the

[1] Andrija Stojkovic, *Razvitsk filozofije u Srba 1804-1944* (The Development of Philosophy in Serbia 1804-1944), Beograd, 1972, str. 91.

primary element of knowledge) and he believed that knowledge has a biological basis and explanation. Djuric thought that the struggle for life underlies all knowledge, and regarded truth as a biological and gnoseological tool. He came closest to Mach's views in his acceptance of an economy principle. The law of least action, i.e. the law of the motion of bodies in the direction of least resistance is manifest in nature, both animate and inanimate, as well as in society and thinking. To economize is the principle of organic matter: the psychognoseo-logical tendency is exhibited in every living being in that it seeks to obtain maximum psychognoseological profit by minimum expenditure of psycho-gnoseological energy or labor. Society is governed by the principle of economy of labor while thinking is governed by the principle of economy of gnoseological labor. Abstraction, judgment, reaching conclusions, etc., owe their existence to the principle of the economy of thinking. The same principle applies in both material and spiritual aspects of cultural life; it is valid both in science and philosophy; and it is employed or should be employed by every thinker.

Although Djuric often refers to the principle of economy of thought, nowhere in his work is it spelled out in detail, explained, or applied. Moreover, there is a vagueness, obscurity, and diffuseness which suggests rather decisively that this philosopher rarely practiced what he preached, as if he failed to understand that practical theories without practice can be empty. In my opinion, there is great confusion in many of his books, and some of his basic definitions of terms such as "philosophy', "metaphysics", and "science" are so difficult to make out that it is hard to imagine that any of his writings will have lasting value.*

Radmilo Vucic, an educator and psychologist born in 1898, properly understood Mach's principle of the economy of thought and applied it successfully to an interpretation of the pedagogical problem of cultivating the will. In his view, the principle of economy applied to nature as well as to man, history, and society. Vucic drew on Serbian philosophers such as Bozidar Knezevic and Dragisa Djuric, especially the latter, for an interpretation and acceptance of this principle. He also appealed to many other philosophers and scholars from antiquity to this century. He believed that all changes in the inanimate world rely on the principle of the least consumption of energy; this holds true for changes in the animate world as well. Concerning psychic life, economy was its fundamental principle. Intellect, being the highest form of psychic life is also capable of becoming the most economical tool for man in the struggle for life. In the course of history, man has increasingly become aware that his ability to accomplish his nu-

* Part of the problem may have been his attempt to combine the direct world view of the positivist Ernst Laas with the indirect world view of traditional materialist philosophy. - Ed.

merous cultural and social tasks depends on his ability to economize the forces at his disposal. The principle of economy becomes all the more important the more cultural life becomes complex in its development.

Science and philosophy are the highest realizations of psychic economy; They are man's most successful attempt to give the simplest explanation of reality possible. Unguided by the principle of economy in human action, the human mind would be faced by infinite possibilities and become hopelessly bogged down. The same applies to many operations in nature as a whole as well as history. Our mind is so constituted that it seeks to express whole realms of life and universe by a single symbol, notion, or law. Vucic understood that the principle of economizing when applied to a man's action and thinking is not the same as the principle of the conservation of energy, but is, rather, its proper utilization.

The fundamental problem of how best to cultivate our use of the human will is also one of economy, how to be effective and sparing. In developing the will an increasingly better utilization of the principle of economy can be observed. Instinct, training, and intellect, the three forms or developmental stages of the will, all show its increasing importance. In short, for the wisest cultivation of the will people need to master instinctive functioning of the will, the development of the will through exercizing particular skills and abilities, and the acquaintance with and control over natural factors basic to cultivation of the will. All this is part of the economy of the will, which in a larger context can help establish in people a close cooperation between mind, will, and emotions.[2]

Turning to opponents of Mach's philosophy, perhaps the most important non-Marxist one was ironically his good friend Branislav Petronijevic (1875-1954). Though 37 years his senior, Mach was impressed by his intellectual audacity. If reluctant to dispute with Petronijevic over the criticism which the Yugoslav philosopher raised against Mach's "naive realism", yet he gladly discussed problems of natural philosophy with him. Petronijevic not only gave the first volume of his book *The Principles of Metaphysics* to the older man but also his treatise on his interpretation of discrete geometry "Über die Grösse der unmittelbaren Berührung zweier Punkte" (On the Size of Two Points which are in Immediate Contact), an offprint from Ostwald's *Annalen der Naturphilosophie* (1905). Though opposed to metaphysics, Mach patiently read these works carefully, especially the sections dealing with mathematical problems. He also wrote a personal letter of in-

---

2 Radmilo Vucic, *Obrazovanje volje* (Educational Will) Beograd, 1939, str. 5,9, 11, 13-14, 15, 20-25, 42, 115-117.

troduction to an English philosopher for Petronijevic.[3] Concerning correspondence, the Ernst-Mach-Institute in Freiburg, West Germany contains several letters from Petronijevic to Mach. Unfortunately, letters by Mach to him are not likely to be found since Petronijevic´s house in Belgrade was burned in 1941 and a second house in 1944. One should add that this bold thinker had extensive contacts with other German philosophers as well and appeared to be highly respected by them. Let us turn to his philosophy at this time, which is often thought to have been influenced by the ideas of Gottfried Wilhelm Leibniz, the famous German polymath.

In Petronijevic's view, Newton's ideas of empty space and time do not exist in direct experience. However, he accepted the existence of absolute space and absolute time as well as the existence of absolute rest and absolute motion. That is why he found Mach's ideas of relative space and relative motion absurd. Granted, Petronijevic accepted Mach's and Einstein's thesis about the relativity of space and motion for the physical sphere of nature; in the metaphysical sphere, however, there exist absolute rest and absolute motion, as can be decisively concluded from the facts in direct experience, as direct experience is absolutely real. According to Petronijevic, rest and motion are the original states of bodies; motion entails objective features - direction and velocity, which thus keeps the moving and resting points absolutely apart.

Petronijevic's rejection of Mach's thesis of the relativity of time, space, and motion was clearly the result of his theoretical-cognitive and metaphysical conceptions.[4] The same conceptions led him to reject Mach's views of causal relations, the principle of inertia and the principle of the conservation of energy. Causality according to Mach, is induced; therefore, argues Petronijevic, it is contingent truth. Consequently, such a conception of causality must be rejected, since causal relation is necessary truth. Though admitting that Mach had demonstrated the unrationality of the fundamental propositions of mechanics, Petronijevic nevertheless rejected Mach's theses of the empirical origin of the principle of inertia and the principle of the conservation of energy. Mach's theses of an empirical origin of these two principles were derived from his suppositions that all motion is relative and the the universe is infinite, while Petronijevic's theses that the principles of inertia and conservation of energy are rational were grounded on his sup-

[3] Branislav Petronijevic, "Iz mojih uspomena na filozofe" (From My Recollections of Philosophers), *Vreme,*, Beograd, 27-30, IV. 1940, str. 18. Branislav Petronijevic´, "Iz mojih uspomena na nemacke filosofe" {From My Recollections of German Philosophers), *Srpski narod* (Serbian Volk), Beograd, 7. VIII, 1943, str. 9.

[4] Branislav Petronievics, *Principien der Metaphysik,* I, Heidelberg, 1904, S. 313-320, also *Osnovi teorije saznanja* (Foundations of the Theory of Knowledge), Beograd, 1923, str. 131, 164, and *Matematicki rad K Rudjera Boskovic'a, (The Mathematical Work of Rudjera Boskovic*), u knjizi: *Olanci i studije,* III, Beograd, 1922, str. 106.

positions that motion is absolute and the universe finite. Petronijevic admitted that his theses were not verifiable within the field of natural science, but could be justified metaphysically, which Mach was reluctant to do.[5] Adducing the second principal law of thermodynamics (the law of entropy), Mach claimed that energy is not constant in the world: Part of energy lost in the form of thermal energy (part of which is not convertible into other forms of energy) has in fact vanished. In Petronijevic´s opinion, this brings Mach's empiricist theory into conflict with itself, as in this case he disregards his basic tenet that sensory evidence is the ultimate criterion of truthfulness. According to this Yugoslav philosopher, energy which has "vanished" into thermal energy is available to our senses; therefore, it still exists. The principle of the conservation of energy is grounded deep in the constitution of matter and motion, and it is not accessible through sensory experience but only through metaphysical contemplation.[6] Petronijevic gladly pointed out that Mach felt the need to construct discrete space,[7] as this was a significant innovation contributed by the metaphysical system of this Yugoslav philosopher. Talking to Petronijevic on one occasion, Mach expressed willingness to accept the discrete geometry proposed by his Yugoslav colleague provided he could succeed in eliminating some troublesome points. Later, however, he had serious reservations about that conception.

In his principal work, *The Principles of Metaphysics,* in which are fully expounded his fundamental gnoseological and ontological doctrines, Petronijevic subjected to critical evaluation other doctrines different and opposed to his own and rejected them. However, instead of considering them in the form in which they were presented by their own authors, he argued against their ideal form, i.e. the form in which, according to Petronjevic's scheme, they ought to appear. In the notes he even cites the names of authors whose doctrines are closest to these idealizations. In this way the criticism raised by this philosopher against various teachings is aimed primarily at those idealized constructs and only secondarily at the teachings as factually formulated in the history of science and philosophy; the latter was the case only when a given doctrine (or parts thereof) happened to coincide

---

[5] Branislav Petronijevic, *Osnovi teorije saznanja* (Foundations of the Theory of Knowledge), Beograd, 1923, str. 118-119.

[6] Ibid., str. 112-113.

[7] Branislav Petronijevic, "Über Herbart's Lehre von Intelligiblem Raume", Sonderdruck aus dem *Archiv für Geschichte der Philosophie,* 1914, Band 27, Heft 2, S. 170. Ernst Mach, *Erkenntnis und Irrtum,* Leipzig, 1920, S. 447-448.

with the idealized constructs subjected to a critical evaluation by this Yugoslav philosopher.[8]

According to Petronijevic, four metaphysical paths to reality are possible: *extremely naive realism, moderately naive realism, absolute consciousnessism,* and *relative consciousnessism.* The first approach was mainly advocated by Mach, Avenarius, Schuppe, and Rehmke, with Mach closest to that view; Aristotle was closest to the second approach, though some of Mach's propositions fit the theory of moderately naive realism; Berkeley and Wundt were close to the third approach, while Petronijevic adhered to the fourth one, in its pure form, of course. Accordingly, an analysis of Petronijevic's attitude towards Mach's conceptions should take into account the conceptions of extremely naive realism and moderately naive realism.

Extremely naive realism claims that the immediate datum is not to be regarded as the internal world different in principle from the external world, but rather as the internal world partly grasping in a direct way the external world as such. It claims that the world consists of elements qualitatively and numerically identical. Its contents can be summed up in the following four propositions: (1) sensations also exist *outside* the perceiving subject and they make an integral part of the objective outside world, (2) sensations can exist simultaneously as the *common* contents of consciousness of several subjects, (3) emotions and memorized images make up a *specific* reality content of each individual subject, (4) the world of sensations is constant and its existence is absolutely real, while the world of emotions and memorized images is changeable and unreal; in other words, the objective set of sensations of the world is *eternal,* while the subjective sets of subjects are transient.[7]

Citations under (1) and (4) are false, if the reference is to Mach; in his view, sensations are not an integral part of the objective outside world, and anybody attributing such a conception to Mach is raising an old metaphysical dilemma "whether in the last resort the world is made up of matter or

[8] It is known from the history of philosophy that Aristotle first provided a critique of Plato's teaching, i.e. the teaching advocated by a particular philosopher, and then expounded his own. Spinoza first produced a critical evaluation of Descartes' dualism, and after that founded his monism. Kant criticized Wolff's metaphysics in order to point to the difficulties with which it is confronted. Petronijevic´ did not follow their example. He constructed not only his own teachings about the world and knowledge but also the teachings different from and opposed to his own, in order that he may submit them to critical analysis. Thus he did not only define the absolute reality of the world, like other philosophers before him, but also suggested how the metaphysical and gnoseological misconceptions must be formulated. Finally, by subjecting these misconceptions to a critical evaluation he arrived at his own, metaphysically and gnoseologically correect truth about the world and knowledge. It would be of interest to find our whether a similar procedure has been followed by some other philosopher.

[7] Branislav Petronievics, *Principien der Metaphysik,* II, Heidelberg, 1912, S. 1-2, 4.

spirit"; according to Mach, any set of elements, whether consisting of physical or psychic elements, or both, is changeable; consequently, there are no eternal sets of elements. My criticism, however, is not aimed at Petronijevic's criticism of Mach, because he did not level his criticism against Mach's conception directly, but rather against the conception of an idealized construct of extremely naive realism. Petronijevic gave a detailed analysis of all four basic propositions of extremely naive realism and then rejected them as untenable along with its solutions of the problem of the reality of the world, problem of relations within being, problem of change and problem of infinity; finally he rejected the whole approach.[8]

In analysing moderately naive realism, Petronijevic proceeded along the same lines, reducing it to two propositions; he rejected it following a critical evaluation; the moderately naive realism, too, is not able to solve metaphysical problems; it only succeeds in solving the problem of the reality of the world. This, however, does not suffice to make it acceptable as a definite metaphysical solution to the riddle of the world.[9]

Contemplating the necessity of metaphysics, Petronijevic was justified in rejecting Mach's proposition that this branch of philosophy was superfluous. In Petronijevic's view, metaphysics is a science which deals with reality as a whole; therefore, it comprises both immanent and transcendent elements. Anyone rejecting the possibilities of and need for metaphysics, would at the same time be rejecting both its immanent and transcendent elements. Yet, no one does it, since this cannot be done. All opponents of metaphysics reject its transcendent elements without at the same time rejecting metaphysics as a whole, since they accept its immanent elements. It follows that adversaries of metaphysics are metaphysicists too, but of a poor kind: their metaphysics consists of immanent elements only. In this way, they renounce too soon what may prove at a later time to be indispensable in constructing the picture of the would as a whole, i.e. they renounce transcendent elements.[10] This argumentation is directly aimed at Mach's conviction that metaphysics is superfluous.

In his theory of knowledge, Petronijevic subjected to criticism Mach's phenomenalism, which he called naive realism, using more or less the same argumentation as in his metaphysics. He claimed that naive realism in its pure form must split the immediate datum into two parts: the subjective one (emotions and memorized images) and the objective one (sensations). This split of the immediate datum has a corroding effect which ultimately destroys naive realism. In contrast to naive realism, Petronijevic argues for

---

[8] Ibid., S. 5-46.

[9] Ibid., S. 46-52.

[10] Branislav Petronijevic, *Osnovi teorije saznanja,* (Foundations of the Theory of Knowledge), Beograd, 1923, str. 23.

the individual nature of the immediate datum as a whole. He infers this conclusion from all of sensory experience i.e. from the following sense data: resistance, effort, temperature, taste, smelling, hearing, and seeing; he concludes that the immediate datum can be given to one subject only, not to several subjects. He goes on to argue in favour of the rejection of the thesis maintained by naive realism about the external world being an integral part of the immediate datum. According to naive realism, a subject's contents of consciousness can become the content of the external world even when there is not any subject whose content it would be. But this claim, in Petronijevic's opinion, can only be realized by means of reflection, which was known by Hume rather long ago. It follows from this that the immediate datum (namely, the external world) is the product of reflection; after all, this cannot be denied even by naive realism. Petronijevic therefore concludes that Mach's naive realism and his own objective idealism do not significantly differ on this point, the difference being only in degree. Namely, both theories use reflection in order to prove the existence of the external world, but in the objective idealism it is a conscious act, whereas in the naive realism it is not. Finally, an adherent of naive realism, according to Petronijevic, has to presuppose absolute reality of the objective part of the immeduate datum, while the subjective part is unreal; that is the thesis on which naive realism depends for survival, yet in Petronijevic´s view it is untenable. In this way, the three fundamental theses characterizing naive realism were rejected by this philosopher as untenable - from the point of view of his objective idealism or relative conscioussism - to be sure. This rejection, however, does not hit Mach's gnosiological teaching as it was, since Mach did not divide the immediate data into a subjective and an objective part, neither did he deny the importance of reflection in the processing of sensory experience or invoke the concept of illusion in support of his basic thesis. All these elements of naive realism were constructed by Petronijevic himself. True, the argumentation he employed in their rejection was coherent enough, yet it could not refute Mach's gnosiological views as presented in his own works. Nevertheless, the criticism expressed by Petronijevic was not mere fiddling around. By refuting naive realism (and other gnoseological theories, differing from his own) Petronijevic attempted to help evolve and find support for his own gnosiological views. Thus, for instance, he explicitly claimed that his objective idealism derives its force from this naive realism, which as a movement (Mach, Avenarius, Schuppe, Rehmke) constitutes an acceptable reaction against both spiritualism and materialism, i.e. against the conviction that the world, as perceived by our senses, is unreal. Petronijevic accepts that viable core pertaining to naive realism: the perceived things have an absolute reality. He goes beyond that: all contents of consciousness are absolutely real. This is how Petronijevic arrived at the basic proposition about the absolute reality of consciousness, namely at the conviction that all data in consciousness are absolutely real. To this he added his second basic proposition (rejected by naive realism) that that reality is dependent on the subject himself.[11]

Another critic, this time from a Marxist point of view, was Dusan Nedeljkovic (1899-1984) who was professor of philosophy at Skoplje before World War II. After the conflict, he taught at Belgrade until 1954. He is known for works on Pascal, Boscovic, and Hegel and for a book titled *Anti-Bergson* (1939). He was very interested in the history of philosophy. Nedeljkovic criticized Mach's conceptions during the 1920's as not being grounded on facts but rather on uncritical constructivism. Against Mach's thesis that thinking is one of the tools which man as a living being uses to adjust himself to his environment in order to survive, Nedeljkovic constructed, for the purpose of dispute, the thesis that thought has created life for itself in order to expand and develop its experience of reality and that, therefore, life is just the outward manifestation of man's spiritual existence. Mach's as well as Nedeljkovic's construction of the relationship between biological and psychic facts are allegedly probable or improbable, showing that constructions as such cannot lay claim to truthfulness. Nedeljkovic rejects as uncertain Mach's assertion that thought gets rid of its internal contradictions by adjusting to facts, because reality may be in contradiction with itself. Similarly, he rejects Mach's notion of the economy of thought thereby committing the same error as so many other critics of that principle by attributing to Mach the idea that the economy of knowledge is the ultimate criterion of truth, something which Mach did not claim.[12] Nedeljkovic accused Mach of constructivism, which is a thing this Austrian physicist and philosopher endeavored to avoid - largely successfully - in all his scientific and philosophical activity.

---

[11] Ibid., str. 7, 8-14, 19-21, 25-27, 31-32.

[12] Dusan Nedeljkovic, "Nemacki filozofski konstruktivizam i Ernst Mach, (German Philosophical Constructivism and Ernst Mach)" *Letopis Matice srpske,* 1925, br. 4-5, str. 82-87.

III

CONTROVERSIES BETWEEN MARXISTS OVER "MACHISM" DURING THE 1920'S AND 1930'S

"Revisionists" (pro-Mach):	"Moderates" (a little anti-Mach):	"Dogmatists" (very anti-Mach):
Mirko Kus-Nikolajev	Richard Podhorsky	Ognjen Prica
Tomislav Pinter	Zvonimir Richtmann	Milos Popovic
	Sima Markovic	

It is of special interest to give here an account of how Mach's epistemology was received among Yugoslav scientists and philosophers of a socialist orientation before the Second World War and the triumph of the Partisan movement. There was much interest in and controversy over Mach's philosophical ideas, even though few of the particpants are currently mentioned in dictionaries and encyclopedias which include articles on Yugoslav philosophy.

The ethnologist Mirko Kus-Nikolajev (1896-1961) sought to fill out cognitive gaps in Marx by supplementing them with the philosophical principles of Mach and Avenarius. Therefore, he can be regarded as a typical "revisionist". In his opinion, Marxian historical materialism must in the course of time be modified so as to accomodate new scientific results, yet this must be carried out without upsetting the essential propositions of historical materialism. Like the Austrian Max Adler, from whom he drew ideas, Kus-Nikolajev held that Marx's historical materialism marked the beginning of an exact approach to the interpretation of society and history, and yet, that it was not a theory of the world as an integrated whole; Mach and Avenarius perceived nature in the same way as Marx perceived society. This explains why Mach's conception of nature and Marx's conception of society complement each other in an integrated (Marxist) conception of the world. Physical, biological, and psychological laws are not applicable to history in any simple way; history and nature must instead be subsumed under one law, integrated at a higher level; the path of discovering that law was covered by Mach and Avenarius. Besides, according to Adler and Kus-Nikolajev, Marxian theory is not materialistic. It would be more appropriate to call it "socio-economic determinism" now that it has been expanded and deepened. Feuerbach, too, was no materialist; he just returned to the entire human experience introducing the line of thought terminated by

Mach and Avenarius. Besides, the terms "materialism" and "dialectics" as used by Marx and Engels correspond to the modern positivistic (as well as Mach's) terms "experience" and "evolution". In this way, Mach's conceptions become the theoretical-cognitive foundation of Marxism. Marx's requirement that social processes be judged by a determinist model of natural processes is reflected in Mach's positivism, i.e. in his antimetaphysical conception of pure experience. Adler and Kus-Nikolajev drew an analogy between the Marxian requirement that in determining social phenomena distinction must be made between real processes and their ideological forms in which they are experienced by man - on the one hand, and Mach's requirement that thoughts must adjust to facts, leading to an exclusion of all subjective additions to social laws - on the other. Adler and Kus-Nikolajev drew an analogy between the Marxian thesis that ideological superstructure must be derived from strictly defined material relations within the social being and Mach's thesis that ordinary reasoning differs from complex speculation only in degree, not in essence. In Kus-Nikolajev's opinion Marx's thesis that social consciousness is determined by social being resembles Mach's thesis that the essence of the world is neither physical nor psychical but rather made up of sets of elements. It must be admitted that these analogies between Marx and Mach are rather far-fetched, yet they were constructed by our author to enable him to conclude that Marxian doctrine must join up with modern positivism. According to Kus-Nikolajev, Marxism must not remain a social theory alone; it must build up an integrated theory of the world and must therefore give up its one-sided economism and embrace the historical-evolutionary conceptions propounded by Mach and Avenarius.[13] Although these "revisionist" ideas of Mirko Kus-Nikolajev are untenable, the fact remains that this Yugoslav author (like Max Adler) at least voiced an independent opinion about Mach's epistemology. This in itself is not of particular importance, yet deserves to be noted because in this country some other interpreters and critics of Mach's conceptions, who had a socialist orientation, failed to do so.

Tomislav Pinter (born in 1899) gave in one of his studies[14] an accurate account of some of Mach's theoretical-cognitive conceptions, expressing a favorable judgment about them. In Pinter's view, Mach must be given credit for the proper understanding of modern natural science, though he was wrong about restricting science to sensations, considering that Planck's quantum theory had proved the existence of things unavailable to them.

---

[13] Mirko Kus-Nikolajev, "Marx-Mach-Avenarius", *Bankarstvo,* V, 1928, br. 2, str. 71-72; br. 3, str. 132-133; br. 6, str. 268-271.

[14] Tomislav Pinter, "Machovi spoznajno-teoretski nazori i razvoj nove fizike" (Mach's Theoretical Views on the Development of Modern Physics), *Almanah savreminih problema,* Zagreb, 1933, str. 5-16.

Pinter, whose approach to Mach's physics and epistemology was independent and rather scientific, was attacked by Richard Podhorsky on *ideological* grounds. Podhorsky criticized Pinter's study for taking Mach's belief for granted, for failing to accuse Mach of idealism, and for allegedly depriving natural science of the necessary assumption about the existence of an objective world beyond our consciousness. Following Lenin's *Materialism and Empirio-criticism* of 1909, which had been largely directed against the philosophies of Berkeley, Avenarius, and Mach, Podhorsky criticized Pinter for being unable to resolve the (big!) problem whether or not Mach was an idealist and for not even attempting to answer the (very clever!) question - How come Mach, in spite of his mistaken philosophical views, succeeded in contributing with his experimental and theoretical works to the development of physics? Pinter is reprimanded for allegedly bypassing Planck, the materialist (or for doing it inconsistently) in order to justify the rejection of Mach, the idealist. In his "scientific" activity, Podhorsky undertook the *political* and *ideological* task of refuting at all costs Mach's hostile epistemology. This led him to reject in his article[15] Pinter's unbiased *scientific* approach to Mach as unacceptable.

Sima Marcovic (1888-1937), a mathematician, knew how to bring his socialist beliefs into harmony with his scientific criteria. He know how to maintain a scientific approach to both Lenin and Mach (as well as to all other scientists and philosophers) and so was able to accept (or reject) their views on the basis of scientific argumentation. He quoted from Mach's and Schlick's opinions freely and with approval in seeking to support his dialectical-materialist convictions.[16] He did not adopt Mach's conception of elements nor did he approve of his rejection of atomism; he derived these views from his own reasoning and arguments,[17] without looking to Lenin for proof. Markovic disagreed with Mach's conception of lawfulness and causal relations, yet he admitted that Mach was one of the forerunners of Einstein's theory of relativity.[18] This Yugoslav Marxist also rejected the suggestion about an organic link between Mach's philosophy and epistemology on the one hand and the theory of relativity on the other, arguing that physical theories, including the theory of relativity, were not derived from some philosophical theory but from experience and experimental facts. Markovic gave credit to Mach for rejecting Newtonian concepts of absolute

15 Richard Podhorsky, "Razmatranje o problemima prirodne filozofije povodom 'Almanah savremenih problema' 1933 (Observations on Problems of Natural Philosophy...)", *Kultura*, 1933, br. 5, str. 344-362.

16 Sima Markovic, *Iz nauke i filozofije,* (On Science and Philosophy), Beograd, 1924, str. 43-46.

17 Ibid.

18 Ibid., str. 115-116. See also: Sima Markovic, *Teorija relativiteta,* Beograd 1924, str. 12, 47-48.

space, absolute time, and absolute motion and for concluding that all motion is relative, but he criticized his inability to relativize distance in time and space. This was done by Einstein who carried Mach's audacious ideas all the way to the end, and dispensing with the Newtonian concept of absolute acceleration and Newton's gravitational theory brought about a great scientific revolution by his theory of relativity. Markovic gave Mach credit for establishing formal equality between geocentric and heliocentric systems (the latter system being accepted nowadays - at least in the opinion of many - merely because it is simpler and more practical), but it was Einstein who has appeared to give it a dynamic explanation.[19]

Sima Markovic rejected the principle of economy of thought in science (without relating it to Mach). He thought that insistence on scientific theories belong simple, clear, and obvious was appropriate to the former mechanistic view of nature. We know now that nature is anything but simple; it would be more appropriate to regard it as extravagant. True, insistence on simplicity, clarity, and obviousness is a subjective principle which is of some importance in the first phase of acquirung scientific knowledge, but which loses significance with the advancement of Knowledge about the understanding of the essence of science, whose structure is infinitely complex. Besides, if scientific theories and laws are to provide adequate reflection of complex objective reality it would be absurd to insist on their simplicity, clarity, and obviousness. A case in point is the increasing complexity of mathematical symbolic language which is designed to account for an increasing number and variety of natural phenomena whose essence is gradually being comprehended.[20] Since differential equations of classical mechanics had long been regarded as the only means by which all natural phenomena can be mathematically formulated, mathematical formulations by means of these classical schemata has come to be identified with a causal explanation of natural phenomena. However, since the phenomena in the field of quantum physics were not accountable for by means of that mathematical apparatus, some physicists - according to Markovic - jumped rashly to the conclusion that the principle of causality does not hold for the phenomena in quantum and atomic physics. Bearing in mind that the principle of causality is surely the foundation on which the whole edifice of science depends, the whole edifice began to sway: the crisis of causality in modern physics was the cause of the crisis of modern science, its essence and totality. Markovic believed that the principle of causality must be rejected in its metaphysical and mechanistic form. This is not a rejection of the principle

---

[19] Ibid., str. 10-11.

[20] Sima Markovic, *Princip kauzaliteta i moderna fizika,* Beograd, 1935, str, 179, 180, 183.

itself; it remains valid, but receives a new, dialectical materialistic interpretation.[21]

Let us now turn to Yugoslav socialists, who while indeed critical of "revisionism", nevertheless, at least theoretically and in principle have possessed enough knowledge, talent, and courage to pursue both scientific and philosophical studies successfully. Among this group, I would like to mention: Zvonimir Richtmann (1901-1941) and Richard Podhorsky (born 1902).[22] Richtmann, thoroughly acquainted with the problems of modern natural science, especially classical and recent physics, firmly believed that the problems of traditional philosophy would either be rejected as pseudo-problems, or else would be tackled by strictly scientific methodology. Unless reduced to logic and dialectics, philosophy will merely have an ideological function: to conceal social opposites. Like logic and dialectics, philosophy too according to Engels and Richtmann, will be reduced to investigating scientific concepts and their consequences, and the historical processes of thinking and knowledge. Philosophy will be preserved by realizing itself in the field of the very science. Though the origins of these ideas can be traced back to Marx, Engels, and the logical empiricism of the Vienna philosophical circle, they may also be regarded as Richtmann's own; he reached them by his own intellectual efforts, i.e. he did not act like a religious zealot who blindly accepts the dogmas of his faith.

Richtmann gave a thorough, detailed account of both Mach's and Max Planck's achievements in physics, of Einstein's theory of relativity, of Heisenberg's uncertainty principle, etc., and especially of the philosophical problem of determinism as shaped under the influence of the latest results in natural science. By introducing Heisenberg's uncertainty principle and Schrödinger's wave mechanics to support his claims, Richtmann asserted that physics had relinquished insistence on obvious evidence and causal relations in describing natural phenomena. This has brought about great successes in the investigation of microprocesses. Simple mechanical laws have

---

21 Ibid., str. 186-187.

22 The following works by Richtmann have been considered: "Kriza determinizma u danasnjoj fizici" (The Deterministic Crisis in Contemporary Physics), *Almanah savremenih problema*, Zagreb 1932, str. 66-77. "Borba oko Einsteinove teorije relativnosti" (The Fight around Einstein's Theory of Relativity), *Kultura*, 1933, br. 2, str. 115-133. "Savremena fizika i filozofski materijalizam", *Almanah savremenih problema*, Zagreb 1933, str. 154-160. *Znacenje teorije relativnosti za razvoj spoznaje*. Pogovor Einsteinovom delu: *O specijalnoj i dpstoj teoriji relativiteta*. Izdanje Biblioteke "Socijalni i naucni problemi" (Social and Scientific Problems), Beograd, 1935, str. 113-155. "Prevrat u naucnoj slici o evijetu", *Pecat*, 1939, br. 1-2, str. 90-102 i 1939, br. 4, str. 235-245. With Podhorsky as co-author, Richtmann published: "Utjecaj prirodnih nauka na modernu filozofiju" (The Influence of Natural Science on Modern Philosophy), *Znanost i zivot*, 1937, br. 2-3, str. 49-73. "Nauka ili dogmatika?" (Science or Dogmatism?), *Znanost i zivot*, 1938, br. 1, str. 51-76. In my account when speaking of Richtmann I am referring to Podhorsky at the same time.

been replaced by a new kind of laws - the statistical laws. Statistical laws do not exclude determinism altogether. Modern physics does not reject all causal relations and lawfulness (since that would be the end of science); but it admits the existence of new, previously unknown forms of causal relations and lawfulness in nature. Richtman felt that the discoveries in the new physics were not in favour of the idealistic view of the free will of the electrons or the theological belief in the existence of the divine creator governing the world of occurences; on the contrary, in his view they are a confirmation of the laws of dialectical materialism, i.e. the laws of the unity of opposites, of the reduction of qualities into quantities, of universal interconnection, etc. Among other things, the discoveries made in modern natural sciences should be accepted by dialectical materialists because in them are revealed the same general laws that scientific sociology (or historical materialism) had already discovered in social life: namely, just as the social science can comprehend social phenomena only if it views itself as an integral part of those phenomena, so modern physics proves that man cannot comprehend natural phenom,ena unless he views himself as their integral part.

Richtmann based his interpretations of the results of the new physics on Lenin and the fundamental propositions of dialectical materialism. He was, however, prepared to take over from Mach and the Viennese logical positivists any ideas which, in his judgment, were applicable to a scientific interpretation of nature. Richtmann criticicized Mach for his rejection of atomism and the neglect of hypothesizing in scientific investigations, yet his attitude towards Mach's views (as well as towards the views of the Vienna philosophical circle) was in principle the same as towards any other scientific and philosophical theories: namely, he believed that whatever is of value to science should be accepted, while all things hindering the the scientific interpretation of the world should be rejected. Richtmann was resolutely and soberly on the side of science, willing to become thoroughly acquainted with any bourgeois theories which might stimulate scientific research.

Richtmann believed that Marxism must be applied to new findings in natural sciences. This is practised by bourgeois scientific philosophy too; therefore its results are not to be rejected together with its metaphysicial interpretations, though they are not to be accepted uncritically. The adherence to dialectical materialism, a standpoint which reveals clearly class limitations of the bourgeois scientific philosophy, enables one to critically evaluate all its results and accept those which are compatible with dialectical materialism rejecting at the same time the false ones. By incorporating the latest results of the natural sciences into its body of teaching, the form of dialectical materialism will become changed, but its principles will remain the same. Thus Richtmann was convinced that logistics (or Carnap's logical syntax of scientific language) could effectively eliminate from scientific language pseudo-notions and pseudo-problems of traditional metaphysical philosophy. He also believed that the true unity of science could be achieved through a unified scientific language, as had been suggested by the Vienna philosophical circle.

Such conceptions were condemned by Ognjen Prica who in the name of orthodox Marxism rejected all achievements of bourgeois science and philosophy as reactionary and false. In his response to him, Richtmann showed that Prica was a dogmatic Marxist who hindered science and critical thinking from developing. In the name of science which he pursued, or wished to pursue, Richtmann rejected Prica's "clichéd pseudo-radicalism", emphasizing that Marxists, too, must selectively employ conclusions pertaining to science in its totality, including bourgeois science. Richtmann could not accept Prica's conceptions as they embody negation of both science and Marxism. He refused to condemn without genuine knowledge, to automatically condemn as reactionary any bourgeois science and philosophy, realizing that some of their achievements could be incorporated into dialectical materialism. Prica did not distinguish between Richtmann and bourgeois scientific philosophers, so he reproached the Zagreb "revisionist" for identifying dialectical materialism with neopositivist idealism, - a criticism rightly rebutted by Richtmann. This gifted "revisionist" insisted on accepting the achievements of bourgeois science and philosophy on a selective basis, the criterion being a true contribution to the solution of particular scientific and philosophical problems. The open-minded Richtmann characterized Prica's attitude as a pogrom, and expressed a firm belief that among Marxists argued criticism was possible. Bowing to authorities was unacceptable to him. In a naive way, he believed that the "progressive movement" would not turn into a religious sect whose adherents, including intellectuals, would have to accept all proclaimed truths in good faith, without exception and without questioning.

Richtmann knew all too well that his polemic with Prica involved him in fighting on two fronts: first, against the real opponents of dialectical materialism, and second, against other party dogmatists, who in turn identified Richtmann with his true opponents. Nevertheless, he was forced to admit that the majority of the educated public did side with Ognjen Prica. That is why it is indeed strange that Richtmann, who died young, i.e. in the same year Yugoslavia was invaded by the Germans and the Partisan movement began, and who was also a talented scientist, failed to see that ideological arguments are not reconcilable with scientific ones, that one cannot profess adherence to dialectical materialism and Lenin and at the same time be prepared to take over from the physics and epistemology of Mach and Vienna logical positivists all things that would be acceptable by scientific criteria. Richtmann was wrong in that he wished to belong to a movement preparing to carry out radical social change and at the same time to keep an open mind for the scientific achievements of bourgeois science. By combining these two irreconcilable things, he made it impossible for himself to be a successful scientist, while incurring the risk of being regarded as a socialist renegade. Richtmann failed to see what Prica understood well, that those in power and those who are getting ready to assume power are anxious to extract from scientists the confession that science is on their side (or, in other words, that they are anxious to ideologize science) whereas scientists do not care whether or not the rulers are on their side or against them. Had he been more sober, Richtmann could have dedicated himself wholly to science and gone far. As it was, given the increasingly socialist orientation in some Yugoslav universities even before World War II, he was compelled to struggle for the barest essentials without ever securing the proper conditions for his scientific work. Nevertheless, considering all that has been said, we admit that in a sense there was little difference between Richtmann and Prica in either science or philosophy. The only significant difference, even in their polemic, was only of an ideological and political nature, with which I am not concerned here. Richtmann's case should serve as a warning to all intellectuals who engage in the socialist movement.[23]

And finally, let us turn to a pair of Yugoslav socialists who were unalterably opposed to Mach's philosophy and who criticized it extensively. Ongnjen Prica, who we have already begun to discuss, pleaded in his writings[24] for drawing a clear dividing line between Machism and dialectical materialism in order to prevent Machism from being smuggled into dialectical materialism, the only scientifically-based philosophy. He regarded Machism as a reactionary, idealistic, and metaphysical philosophy leading to fideism and theology (and the same was true of the logical empiricism of the Vienna philosophical circle) which should be unmasked as such. This would prevent the spoiling of pure Marxism by interpolating those false philosophical views into dialectical materialism. In Prica's opinion that task had already been completed by Vladimir Lenin in his work *Materialism and Empirio-Criticism* (1908). So it remains for the Marxists after Lenin to apply his conclusions to new situations. It never occured to this Yugoslav Marxist to evaluate the scientific and philosophical aspects of Mach's physics and epistemology and their results which might prove relevant for fur-

---

23 Stanko Lasic viewed the conflict between Prica and Richtmann differently. He criticized Richtmann only for failing to distinguish between analytical and dialectical reason, which was later done by Sartres (*Sukob na knjizevnoj ljevici 1928-1952,* (Literary Conflicts Among Leftists 1928-1952), Liber, Zagreb, 1970, str. 164-178). I find this explanation of Richtmann's failure both strange and untenable, and feel compelled to reject it.

24 Ognjem Prica, "Savremena filozofija i prirodne znanosti", *Znanost i zivot,* 1937, br. 1, str. 4-14. "Takozvana naucna filozofija" (So-Called Natural Philosophy), *Znanost i zivot,* 1937, br. 5-6, str. 207-227. "Jos o neopozitivizmu kod nas" (More on Neopositivism and How it Concerns Us), *Znanost i zivot,* 1938, br. 1, str. 37-51.

ther scientific investigations. Instead, he uncritically adopted the judgments of other authors as his own and allowed himself to be guided by them. In my view, it shows that Prica lacked the essential intellectual qualities that would enable him to accept Mach's teaching or provide a true criticism of it, so his contact with Mach was bound to be totally barren. This Marxist rejected the very thought of re-examining the opinion that Machism was an idealistic and metaphysical brand of philosophy incapable of being incorporated into Marxism; Lenin had definitely excluded Mach's idealism from dialectical materialism; therefore, no one can claim that idealism is "the actual result of science or scientific philosophizing" (for example, as Zvonimir Richtmann had thought). Prica had no doubt that dialectical materialism is the only scientifically-based philosophy, which cannot be undermined, let alone overturned by the further advance of science. To uphold the truly scientific philosophy means to believe that further development of science will bring about further strengthening and perfecting of materialism and dialectics, dispensing at the same time with idealism and metaphysics. Prica himself gave an account of the most recent results in natural science from the perspective of dialectical materialism (in the article "The So-Called Scientific Philosophy") as they were explicable only in terms of this philosophy and not in terms of Machism or neopositivism. Accused by Richtmann and Podhorsky of providing "a silly caricature of modern physics seen through the eyes of a dogmatic dilletante", Prica defended himself by revealing the fact (which for him was conclusive proof) that the account was quoted almost verbatim from the article written by Rudas, a physicist and theoretician of dialectical materialism, which was published in Moscow in the Soviet periodical *Unter dem Banner des Marxismus* [as well as in the German edition published in Berlin - Ed.]. This "revealing" remark is in fact doubly detrimental for Prica: 1) He quoted from Rudas but failed to acknowledge it (this is plagiarism that would have gone unnoticed but for the polemic between him and Richtmann and Podhorsky), and (2) What if Rudas was wrong? - It had never even occurred to him to question the validity of Rudas's views (He went by the old slogan: *De Roma locuta causa finita*). To finish this part, it would not be precisely correct to call Prica a dogmatist in philosophy: he simply tried to use philosophy for political purposes.

As a violent political opponent of Machism, neopositivism, and all other wrong trends in bourgeois philosophy, Ognjen Prica entered the polemic against Richtmann and Podhorsky, accusing these two "revisionists" of embracing the logical empiricism of the Vienna philosophical circle (essentially a mere continuation of Machism) as a truly scientific philosophy for the purpose of incorporating it with the aid of quotations from Marx and Engels, into dialectical materialism. In Prica's view, this very purpose was tantamount to sin, so any further polemic was superfluous. Richtmann and

Podhorsky (further in the text: Richtmann for the sake of simplicity) allegedly believed that logical empiricism, aided by the latest findings in natural science, would prove capable of resolving important philosophical problems about space, time, matter, and causal relations. Logical empiricism would accomplish this by replacing the notion "objective" by the notion "intersubjective", by creating logistics (symbolic logic) as a new scientific methodology, and by viewing the unity of science as a unity of scientific language. According to Prica, these neopositivist theses show unmistakeably that that philosophy is a variant of Machism, i.e. no more than an idealistic distortion of the results of modern natural science; to accept it would mean to supplement dialectical materialism with neo-Machism - which would suffice to disqualify anyone attempting to do it. Richtmann was accused by Prica of planning the expulsion of dialectics from natural science and its replacement by logistics; he reduced Marxism to historical materialism, which is wrong because without the theory of reflection neither Marxist materialism nor dialectics could survive. Richtmann was accused of having caused great confusion by identifying (1) two diametrically opposed theories - dialectical materialism with neopositivism, and (2) neopositivist idealism with the results of modern natural science. His aim was to improve or supplement dialectical materialism with neopositivism or "scientific philosophy" - which is to be strongly condemned. None of his polemic against Richtmann was based on scientific or philosophical argumentation. In other words, Prica repudiated Richtmann's conceptions not because they were incompatible with the state of affairs, i.e. because they failed partially or wholly to explain old and new facts pertaining to natural science or because they offered an inadequate interpretation of the methodology and results of natural science, but because they did not fit in well with his own preconceived and borrowed theses. By judging Prica so harshly I am not trying to defend Richtmann and his views: I am not even sure that Richtmann had the intentions ascribed to him by Prica (although Prica was probably justified in doing it because he was very well acquainted with Richtmann's conceptions from the spoken polemics they had had). All I am trying to do is point out that Prica did not much care about either science or philosophy. He was interested in rejecting and thwarting - by ideological and political argumentation - the plan that Richtmann may have had - to accept Marxism as the scientific philosophy and socialism as the true future of mankind while at the same time reserving the right to act as a scientist, i.e. to make evaluative judgments about Mach's ideas, about Viennese neopositivism and above all the other movements in science and philosophy on the basis of their actual contribution to an interpretation of nature, society, and the world in general, and not on the basis of the interpretations imposed by authoritative socialist prophets.

In this account at this point, I would also like to mention Milos Popovic [Djurin] (1911-1943), an orthodox Marxist who assailed his opponents with arguments similar to those of Prica and was even more violent, more fanatical and unscrupulous than Prica. In his three articles,[25] Popovic attacked Kajica Milanov from Belgrade and Richtmann and Podhorsky from Zagreb for adhering to the neo-positivism of the Vienna philosophical circle that had continued to develop Mach's idealistic philosophy. The followers in this country of that pseudo-scientific philosophy from Vienna together with the neopositivistic philosophy itself were regarded by him as a philosophical assault on modern natural science aimed at aggravating the crisis in modern science and preventing its resolution by turning logical thinking by means of logistics into meaningless juggling with empty notions. To attain this goal the logical positivists used tricks and maneuvers so skillfully that they succeeded in distracting the attention of scientists away from the natural and social processes they were investigating, pushing them into pure sophistry. That Popovic was poorly versed in science and philosophy can be seen from the fact that he criticized Richtmann for attempting to solve some scientific problems which had long ago become obsolete. However, the drastically false conceptions that he held about science and philosophy did not prevent Popovic from advocating the views of the only true science and its name to claim that Richtmann's solutions were based neither on science nor on scientifically-based philosophy, but that they were an expression of purely subjective idealism. In Popovic opinion, a clue to explaining the "fact" that the results of natural science are in contradiction with the philosophical interpretation of these results in bourgeois science and philosophy simply lies in the restraints imposed by the bourgeois social system on bourgeois scientists and philosophers, which prevents them from creating a truthful picture of the would as a whole. Besides, the results obtained by Planck, Einstein, and Heisenberg have no connection whatsoever with the scientific and philosophical generalizations made by those scientists. To support these assertions, Popovic only quotes Lenin's name, without giving his arguments. If to this we add that for Popovic formal logic is not preserved in dialectical logic as a special case, but that it is completely discarded as a type of false reasoning and that Mach's philosophy of nature is simply reduced to Kantianism - then we can safely conclude that here we have an ideological fanatic who in the name of science and philosophy, which he did not even comprehend, condemned and hindered any real efforts made in those spiritual activities. Popovic was entrusted with a politi-

25 Milos Popovic Djurin, "Neopozitivisticka teorija saznanja" (Neo-Positivist Theory of Knowledge), *Zivot i rad,* 1938, nov. -dec., dtr. 212-225. "Neopozitivizam u nasoj filosofija", *Pregled,* mart 1940, str. 180-184. "Naucna slika svijeta prema zagrebbackim neopozitivisma", *Pregled,* Juni-Juli 1940, str. 318-324.

cal task (Why did he accept it?) of repudiating Machism and neopositivism at all costs, of proving that they were idealistic and reactionary, of condemning them and thus putting them under the taboo for all impartial researchers, - all of which has little to do with science and philosophy. To carry out that task, Popovic looked upon science and philosophy as a battlefield on which apart from true scientists and philosophers, i.e. dialectical materialists, there appear also a number of suspicious persons, enemies, saboteurs, and spies (some of them pretending to be dialectical materialists hoping that it might be profitable), whose aim is to hinder the advancement of knowledge, of science and philosophy, and whose dishonest dealings should therefore be exposed and prevented. In my opinion, these kinds of accusations have nothing to do with science and philosophy or even legitimate politics. The fanatic who succumbs to them should be sent to a psychiatrist who will decide whether he can be saved. Some might quote the words: "Let bygones be bygones." But I do not go along with it. We must prepare ourselves for a confrontation with similar attacks on science and philosophy in the future, in order to avert them more effectively.

In conclusion, as touched on earlier in connection with Zvonimir Richtmann, I said that it is not possible to reconcile ideology with science. He is an example of what can happen to well-meaning people when they try to do science while advocating philosophical views which they believe follow from their scientific work. On the other hand, Sima Markovic showed that it is possible to integrate socialist conviction with the postulates of scientific research. That rare man, who was capable of successfully combining two different vocations - of a revolutionary and a scientist - was a thorn in the flesh of many people in the socialist movement. He was finally punished by execution by those whose beliefs he shared. Like Zvonimir Richtmann, Sima Markovic is a momento to all those intellectuals who may join the socialist movement in the future.

CHAPTER 20

MACH, UTRECHT, AND DUTCH PHILOSOPHY

by Henk Visser*
(Tilburg University)

I

A study of Ernst Mach's philosophical influence in the Netherlands and of ideas similar to his can scarcely avoid an initial conclusion that with comparatively few exceptions significant impact has been centered at a single university, though it has seemed to survive well there. It appears that virtually all the philosophy chairs at other Dutch universities from the late 19th century well into the 20th century were occupied by Neo-Kantians,[1] who considered themselves to be supporters of what they called "critical philosophy", or by psychologists and Hegelians.[2]

In physics and chemistry there were numerous very outstanding scientists of whom several won Nobel prizes, but unlike Mach they either accepted the reality of atoms and molecules or at least considered the utility of the atomic theory to be indispensable in their discipline. Neither phenomenalism nor epistemological positivism appealed to them. Nevertheless, many basic books by Mach were widely read by physicists, philosophers, and psychologists and gradually his ideas did exercize influence. This paper is the story of that development.

Part of Mach's experimental work on "Mach Bands" which had been carried out during the 1860's would be rediscovered over thirty years later by Hermanus Haga at Groningen and a colleague C.H. Wind (who would be the first to realize that Mach had indeed anticipated them).[3] Gerardus Heymans (1857-1930) at the same university would share many psychological

---

* This paper has been especially prepared for this anthology. Henk Visser teaches philosophy at Tilburg University in the Netherlands.

1 Jan Pieter Nicolaas Land (1834-1897) was Professor of Philosophy and Oriental Studies at Amsterdam from 1864 to 1872 and of Philosophy at Leyden from 1872 to 1895. Cornelis Bellaar Spruyt (1842-1901) taught at Amsterdam from 1877 until 1901. Allard Pierson (1831-1896) held a position in philosophy at Amsterdam from 1877 to 1895. Gerardus Bolland was Professor in Philosophy at Leyden from 1896 until 1922. And Bernard van der Wyck (1836-1925) taught philosophy at Groningen from 1863 to 1890 and at Utrecht from 1890 to 1906.

2 Gerardus Heymans (1857-1930) was successor of Van der Wyck at Groningen and taught philosophy and psychology there from 1890 to 1927. Most of his works, however, were written in German.

3 See Floyd Ratliff, *Mach Bands*, Holden-Day: San Francisco, 1965, pp. 211-214.

J. Blackmore (ed.), Ernst Mach – A Deeper Look, 403–430.

views with Mach which both men had derived from Gustav Fechner; both would be criticized by opponents as "psychomonists" and would even correspond with and cite each other's works in friendly fashion,[4] but since Heymans verbally favored "inductive metaphysics" and Mach claimed to reject all metaphysics, they apparently did not see each other as philosophical allies. But our concern is with a number of positivists and philosophers of science at a different university, which was long a haven for the kind of philosophy which Mach preferred, and where several Dutchmen gradually became interested in Mach's philosophy of science. I am talking about Utrecht.

Max Planck presented his famous anti-Mach lecture at Leyden in December 1908, the university of the Nobel prize winners Hendrik Antoon Lorentz, Pieter Zeeman, and somewhat later Heike Kamerlingh-Onnes. Other outstanding physical scientists such as Johannes Diderik Van Der Waals and Jacobus Henricus Van't Hoff taught and did research at the University of Amsterdam, but to find Dutch philosophers and scientists attracted to positivism and Mach's philosophy, Utrecht was the place.

II

But it was not always so. Auguste Comte the French founder of Positivism who attempted to replace the Catholic Church with his own religion of humanity found a relatively early epistemological disciple in the Protestant theologian Josué l'Ange Huet (1840-1924) whose ideas have many resemblances to those of Mach. He was only active as a philosopher until 1876, but to understand Dutch academic life one has to recognize that in practice one could also write a doctoral dissertation on philosophy in the theology faculty which is what Huet did. His work was titled *The Method of Positivistic Philosophy According to Auguste Comte* (in Dutch). It was defended in 1866 before theology professors at the University of Leyden and presented a plea for a "pure" Comtean Positivism. Mill, Spencer, and the Utrecht philosopher Opzoomer were accused of having moved away from it: Mill for his views on cause and effect, Spencer for his principle of the equivalence of forces, and Opzoomer for his idea that philosophy should put "the cooperation of forces in the Universe" in the proper light.[5] All three authors according to J. Huet abandoned "the relativity of human knowledge". His conclusion was that Positivism was definitely not a phi-

[4] See a letter from Heymans to Mach, Groningen, dated October 15, 1891 in Joachim Thiele's book *Wissenschaftliche Kommunikation,* Aloys Henn Verlag, Kastellaun, 1978, pp. 210-211 and index citations in Mach's *Analyse der Empfindungen.*

[5] J. l'Ange Huet, *De methode der positieve filosofie volgens Auguste Comte,* Leiden, 1866, p. 312.

losophy of causal connections, but the philosophy of "the development of humanity". It was by showing the way to the reality of human development in each field that Positivism had the right to call itself a truly scientific philosophy and to turn its back on the unreal speculations of metaphysics.[6]

In his *New Solution of an Old Problem. The Method of Positivism Applied to the Concept of Cause and Effect* (in Dutch) of 1872,[7] Huet developed an idea of causality akin to Mach's, and indeed in the same year in which Mach first published a book on philosophy.[8] According to Huet even philosophers who regarded themselves as empiricists were in error when they believed (1) awareness of causation could arise from repeated observation, and (2) the nature of explanation was to be found through an indication of causes.[9] Scientists would have solved the problems of physics and chemistry even without a notion of causality, that is, solely by paying proper attention to "what belonged to each other".[10]

Huet's publications were reviewed by the Neo-Kantian philosopher, C.B. Spruyt, who became a professor at the University of Amsterdam in 1877 and who was also interested in science. Spruyt admitted that they "without doubt belong to the best that has been written in the field of philosophy in the last years",[11] but he also saw clearly that J. Huet believed that he could "explain the origin of all human science without the help of aprioristic forms". He disagreed with his theory about the history of the development of the notion of cause - as an anthropological concept of force - since it failed to appreciate the Kantian "conviction that one appearance follows the other according to a rule".[12] Moreover, Spruyt was disturbed by Huet's lack of clarity about the demarcation between sensations and observations.[13] He also found Huet's interpretation of an "I" as "that which the changing sensations leave behind" insuficient for establishing a *self-identical self*". Huet should have acknowledged an *a priori* faculty or, still better, accepted Kant's doctrine of the synthetic unity of consciousness, in order to assign another kind of existence to that unity than to the sensations.[14] Finally, Spruyt contended tht Positivism served no useful purpose for a philosophy of natural science: "The theory of ether and the theory of

[6] Ibid., p. 316.

[7] J. l'Ange Huet, *Nieuwe oplossing een oud vragstuk. De methode van het positivisme toegepast op het begrip oorzaak en gevolg*, Leiden, 1872.

[8] Ernst Mach, *Erhaltung der Arbeit*, Prag, 1872.

[9] Huet (1872), op. cit., p. 132.

[10] Ibid., p. 137.

[11] C.B. Spruyt, "Oude dwalingen in een nieuw kleed", *De Gide*, **37.1** (1873), 226.

[12] Ibid., p. 239.

[13] Ibid., p. 248.

[14] Ibid., pp. 231 ff.

atoms may not be positivistic, but they seem to be indispensable for natural science."[15]

After Huet stopped writing philosophy in 1876, what remained of positivistic tendencies in the Netherlands appeared to retreat to the University of Utrecht, which we will turn to now.

III

The city of Utrecht is very old and what remains from earlier centuries is quite picturesque. The town is situated in the center of the Netherlands and the university founded in 1636 has at times, including recently, been the largest of the state-run educational institutions. The city became largely Calvinist during the Reformation but is also the center of a Catholic archbishopric and was long famous for its Old Catholic and Jansenist communities. It is not clear what singularities helped set the university at least somewhat philosophically apart from the rest of the Netherlands, but one is reminded of an early precedent, namely, when Gisbert Voet, president of the newly founded university bitterly attacked René Descartes and his world view, perhaps the sharpest criticism he would receive during his long stay in the Lowlands. In any case, a Positivist but anti-dualist and anti-Cartesian tradition could be found two centuries later during C. W. Opzoomer's long reign as professor of philosophy.

Cornelius Willem Opzoomer (1821-1892) held the chair of philosophy at Utrecht from 1846 until 1890. He was called the "Dutch John Stuart Mill" and had developed his "philosophy of experience" under the influence of Gerhardus Johannes Mulder (1802-1880) who had been a professor of chemistry in the same university and was known as the discoverer of protein. Opzoomer was highly respected by scientific colleagues such as Donders and Buys Ballot who were pleased that the chair of philosophy was not occupied by a proponent of "speculative philosophy", but by an advocate of scientific method, at least as many Positivists understood it. Understandably, however, philosophers at other Dutch universities, especially Neo-Kantians wondered how they could recapture "their" chair. They had to wait a long time, but in 1890 they finally had their way. A student of Opzoomer's, Bernard van der Wyck (1836-1925) was appointed to the position but soon defected toward Neo-Kantianism. But while this meant that all the major professorships in philosophy seemed under "safe" control at Dutch universities, Opzoomer's influence survived precisely where one would have expected it to linger longest, namely, among several scientific colleagues at Utrecht and a number of science students.

---

15 Ibid., p. 262.

The first apparent Dutch reference to Mach's philosophy came quite early but was not particularly friendly. One of Opzoomer's renegade disciples, Allard Pierson (1831-1896) mentioned Mach's *Erhaltung der Arbeit* (1872) in 1875, but as a theologian he praised it rather obliquely by quoting from Mach to the effect that electrical phenomena were still mysterious and hard to reduce to material terms.[16] In 1896, the Hegelian philosopher, Gerardus Bolland (1854-1922) included a review of Mach's *Beiträge zur Analyse der Empfindungen* (1886) in the back pages of his own book of collected papers under the title "Ein Unbewusster Metaphysiker", where he maintained that Mach's "lack of speculative education" had kept him from seeing the contradictions in his own empirical approach.[16a] The book review had originally been written at Batavia in the Dutch East Indies and was signed "May 1890".

The "official" introduction of Mach to the Dutch public was due to Willem Koster (1834-1907), who had been professor of anatomy, general pathology, pathological anatomy, and forensic medicine from 1866 until 1888 in Utrecht, but who had had to retire because of an infectious disease. But in 1892 he became editor-in-chief of the influential *Wetenschappeltjke Bladen* (Scientific Papers), a Dutch scientific journal which contained translations of outstanding foreign articles, book reviews, and sometimes original publications. Near the end of his life he also wrote three philosophy books.[17]

Koster in his role as editor gave a favorable review of Mach's *Populärwissenschaftliche Vorlesungen* (the third edition of 1903), in which he also drew attention to *Die Mechanik, Die Principien der Wärmelehre,* and *Analyse der Empfindungen* (the fourth edition).[18] Koster pointed out that Mach was even more radical than Helmholtz, who "kept recognizing an aprioristic activity of the mind" in the urge for causal explanation and the explanation of the "localization" of the visual sensations:

> For Mach, on the contrary, there is no question of an *a priori* or synthetic apperception. The endless stream of sensations and representations, which under certain circumstances produce temporarily little changeable combina-

[16] A. Pierson, "Eene Geschiedenis van het Materialisme", *De Gide,* **39.2** (1877), 433.

[16a] See G.J.P.J. Bolland, *Het wereldraadsel. Wijsgeeräge verhandelingen,* Leiden, 1896, pp. 502-510.

[17] W. Koster, *De Ontkenning van het bestaan der materie en de moderne physiologiche psychologie,* Haarlem, 1904; *Subjekt en objekt bij zintuigelijke waarneming (Naar aanleiding der Metaphysica van Prof. Heymans),* Haarlem, 1906; and *Kant's noumenale wereld en de zinnelijke waarnemingen*, Haarlem, 1907.

[18] W. Koster, "Mach als populair-wetenschappelijk schrijver", *Wetenschappelijke Bladen,* **48.2** (1903), 149-153.

> tions (...), is exclusively the basis and the source of our conscious mental life.[19]

But, instead of elaborating the differences between the "idealistic" and the "positivist" philosophy of nature, Koster referred the interested reader to *Die Krisis in der Psychologie* by Rudolf Willy.[20] Willy had argued that Mach took things too easily, when he (1) considered time and space to be sensations comparable with color and sound, and (2) solved the problem of other people's minds by the assertion that "in thinking we add consciousness to certain phenomena" (*dass wir das Bewusstsein hinzudenken*). Koster concluded that every physicist who goes further than a limited professional study, becomes willy-nilly a metaphysician. Nevertheless, he declared himself "a great admirer of Mach's physical and physiological investigations" and he praised his book of popular scientific lectures for their conciseness and clarity. For the rest, Koster called Mach a precursor and co-founder of Ostwald's *Energetics* and the progress of physiological psychology. He observed that Ostwald's well-known *Vorlesungen über Naturphilosophie*[21] had been dedicated to Ernst Mach and remarked that both energetics and physiological psychology had a *monistic* view of nature. For Mach, the concept of "matter" was only an abstraction, and the word nothing but an expedient of language. His review ended with a quotation from "Die Ökonomische Natur der physikalischen Forschung"[22] about the uncertain status of "matter" in the future.

This extensive review of *Populär-wissenschaftliche Vorlesungen* was, however, by no means the last word by Koster on Mach. But familiarity would breed increased doubts. In the very next year, 1904, he published a book, *The Negation of the Existence of Matter and Modern Physiological Psychology* (in Dutch), which dealt extensively with Mach's philosophical investigations.[23] Though a part of the book was devoted to a discussion of Mach's younger contemporaries and correspondents, Wilhelm Ostwald and Theodor Ziehen (the latter being a German positivist philosopher who was

---

19 Ibid., pp. 150 ff.

20 R. Willy, *Die Krisis in der Psychologie,* Leipzig, 1899. See also E. Mach, *Die Analyse der Empfindungen,* zweite Auflage, Jena, 1900, (preface).

21 W. Ostwald, *Vorlesungen über Naturphilosophie*, Leipzig, 1901.

22 E. Mach, *Populär-wissenschaftliche Vorlesungen,* dritte Auflage, Leipzig, 1903, p. 242.

23 Koster (1904), op. cit.

known largely as a psychologist while at Utrecht from 1892 to 1903),[24] the main focus was on Mach as a central figure in physiological psychology because he had indicated the "general philosophical consequences" of this nascent branch of science more than the other thinkers.[25]

Koster's main source on Mach this time was the third edition of his *Analyse der Empfindungen* (the fourth edition with its curious preface appeared during the preparation of Koster's own book). He did not assume acquaintance with Mach's writings on the part of readers, so he gave several quotations such as "Matter is for us not something given. The elements (sensations)[26] come first" and the more radical claim: "Everything that we can know about the world is necessarily expressed in sense impressions" etc..[27] Special attention was paid to Mach's amazement how anyone could think it was possible to derive or explain sensations from the movement of atoms.[28] This allowed Koster to refer to the inaugural speech in 1890 of Van der Wyck, his longtime colleague at Utrecht, who as we have already mentioned replaced Opzoomer while deserting positivism for a Neo-Kantianism, and who even added some traces of realist epistemology. Van der Wyck wrote:

> We must not ask how can we go from moving atoms to mere sensations? If it is clear that there is no way, then no way exists. Rather, we should ask the reverse question: How can we get from sensations to atoms?[29]

But this did not mean literal agreement with Mach, but was more a case of *les extremes se touchent,* since Van der Wyck had smuggled a form of ontological dualism into his Neo-Kantianism, acknowledging the existence of matter as well as mental phenomena.[30]

Though Koster's sympathy is clearly on the side of Van der Wyck, he does not attack the monistic positivist view directly, but instead analyses its consequences. For whereas Mach's approach which reduces apparent matter and most of the mental faculties to sensations appears to simplify onto-

---

[24] Both Ziehen's well-known *Leitfaden der physiologischen Psychologie* and his popular booklet *Ueber die allgemeinen Beziehungen zwischen Gehirn- und Seelenleben* were translated into Dutch. Though not a Dutch philosopher, his presence at Utrecht, the closeness of his philosophy to Mach with whom he later corresponded, and his possible influence on Dutch students deserve attention, which unfortunately cannot be given here in this already rather lengthy paper. Though his philosophy books were published after he left Utrecht, nevertheless, he may have been a contributing factor in Utrecht sympathy for Machism.

[25] Koster (1904), op. cit., p. 20.

[26] Ibid., pp. 21f.

[27] Ibid., p. 22.

[28] Ibid., pp. 23f.

[29] B.H.C.K. Van der Wyck, *Nog eens: oorsprong en grenzen der kennis,* Utrecht, 1890.

[30] Koster (1904), op. cit., pp. 31f.

logy as if there were only one kind of basic reality instead of two or several, it runs into both epistemological and ontological trouble when it encounters the relations between subject and object, consciousness and unconsciousness, and particularly the minds of different people. If only my mind exists, I fall into solipsism, but if other minds indeed exist, then where are they and how can they exist unless connected with reasonably well-fed bodies separated from each other by space? If there is no ego, self, or mind but only sensations, then why should there be consciousness of one group of sensations but not groups normally identified with or attributed to what other people are presumably conscious of? In fact what is consciousness? Indeed, what are relations? Does it make any sense to consider consciousness and relations to be sensations? If they are not real why do they seem indispensable? And if they are real but not sensations, then how can one retain ontological monism or reductionism? These are some of the types of questions plus Mach's reluctance to answer any of them which apparently made it hard for Koster to accept the Austrian thinker's treatment of philosophical and psychological issues.

Other difficulties which Koster found in Mach's philosophy were connected with the notion of "sensation" itself and concern at least three issues: (1) the breadth of the notion, (2) the "localization" of sensations, and (3) the building of complexes of sensations. Koster remarked that Mach already dangerously simplified things by treating pains and feelings of pleasure as sensations, but his inclusion of "space" and "time" under sensations was a surprisingly new point of view which seemed to neglect "our intuition in the form of time", and which identified the notion of time with the determination of it as *duration*.[31] In this connection, Mach's treatment of the irreversibility of the course of time seemed characteristic of the light-hearted way he dealt with important philosophical problems.

Finally, Koster formulated his "standing main objection" to Mach's monistic positivism: *the lack of any indication of how different sensations if thought of as atoms, elements, or isolated things could unite or be united into a genuine psychical or psychological complex.*[32] In physics and physical chemistry there are binding laws in the form of mathematically expressable relationships as "the important work of Van't Hoff, Bakhuis Roozeboom, and Cohen shows", but in the absence of such laws about sensations how can one even discover and confirm the very functional relationships between elements considered as sensations which Mach so strongly advocated? Was it possible to establish either mathematical laws or functions for components of the Proteus-like life of consciousness?[33]

---

31 Ibid., p. 112.

32 Ibid., p. 124.

33 Ibid., pp. 125f.

When we look at Mach's arguments in support of his ontological monism as an alternative to "confusing dualism"[34] such as could be found at the end of chapter five in his *Analyse der Empfindungen*, how meagre they appeared. Indeed, Koster concluded that identification of the physical with the psychical was hazardous and speculative, such that the accusation of unneeded metaphysics was inescapable, rather ironic testimony on a philosophy which aimed at replacing metaphysics. But such observations did not prevent Koster from expressing his admiration for "Mach's genial mind, broad knowledge, noble personality, and masterly contributions to the field of science".

Such kind words, however, were coupled with still one more defection from Positivism to what Kantians and neo-Kantians called "critical philosophy". According to the Dutch historian of philosophy F. Sassen writing as late as 1941 (i.e. during the German occupation): Wynaendts Francken, a former disciple of Opzoomer, had been "the last representative of Positivism in The Netherlands"[35] and in 1905 even Francken in his book *Introduction to Philosophy* (in Dutch) had deserted the cause, though retaining Mach's insight to the effect that "the highest philosophy consisted in tolerating an incomplete world view and preferring it to a seemingly closed but inadequate one".[36]

This development may help raise two questions in the reader's mind, first, how could positivism be dead in The Netherlands less than two years after Mach's philosophy had been so conspicuously introduced by Koster? And second, why had Mach's philosophy penetrated the country so slowly up to then, Koster's comments coming as late as twenty years after Mach's two most famous books had been published?

In reply to the first question, three points, first, Sassen was apparently only thinking of professional philosophers, second, one has to remember that what Auguste Comte had meant by "Positivism" was not fully the same as Mach's philosophy, which one might best designate as "positivism" with a small "p" or perhaps even better, "neo-positivism". Sassen was referring very largely to the demise of Auguste Comte's Positivism, But the newer version while largely ignored by professional philosophers or strongly criticized as Bolland and Van der Wyck had done would find initial advocates among a number of philosophically inclined scientists who had studied under some of the earlier Positivists. And third, Sassen was writing during the war in an overrun nation, where optimistic, scientific philosophy was not exactly in vogue. Concerning the second question about the slowness of Dutch response to Mach and his philosophy and a possible reason for it, it

34 E. Mach, *Die Analyse der Empfindungen*, dritte Auflage, Jena, 1902, p. 237.

35 F. Sassen, *Wijsgerig leven in Nederland in de twintigste eeuw*, Amsterdam, 1941.

36 E. Mach, *Die Mechanik*, Leipzig, 1883, p. 437.

was slow even in Austria. His major philosophical influence only began with the turn of the century everywhere, but please read the next section for more information.

IV

Perhaps the Utrecht *scientist* whose general world view had come closest to that of Mach had been the mathematical and theoretical physicist Victor August Julius (1851-1902),[37] who will be called with no disrespect intended "Uncle Julius" in this paper to distinguish him from his nephew Willem Henri Julius (1860-1925) who taught experimental physics at the same university and who we will discuss later. Uncle Julius's early work *Remarks on the Foundations of Physics* (in Dutch, 1880) had made him the outstanding Dutch philosopher of science of his era. This book demands discussion, since it seems to present an explanation why Dutch *physicists* with philosophical interests *also* may have appeared to drag their feet in paying attention to Mach: *They had their own Ernst Mach!*

Uncle Julius's book showed many Machian traits. With respect to Newtonian mechanics it had reached the same conclusion as Mach: Newton's first two laws had been only definitions in disguise, but the theorem of the parallelogram of forces had been a result of experience, when one stuck to the hypothesis that two arbitrary material points exerted an influence on each other, and as a result of this each of these points had a movement with respect to fixed points, etc. But when one abandoned Newton's hypothesis, the parallelogram rule fell off, as well as d'Alembert's principle and that of virtual velocities.[38]

As far as I can see, there was no major difference with Mach's approach in his "Perspicuous criticism of Newton's presentation" in *Die Mechanik,* only Julius's treatment had been much more precise, surely more so than Mach's original exposition in his short article "Definition of Mass". Perhaps Julius had known the reprint of this paper in *Die Erhaltung der Arbeit*. For there were still more parallels with this work in Uncle Julius's book as I shall show presently. In his general philosophy of science, Julius distinguished, as usual, the "content" from the "form" of a science: the content being "the totality of facts which have come to our knowledge", while the form belonged to ourselves, whereas the process of choosing was only

---

37 V.A. Julius had been Professor of Physics, Physical Geography, and Meteorology at Utrecht from 1888 to 1896 before taking the chair of Mathematical Physics and Theoretical Mechanics, leaving the chair in experimental physics to his nephew W.H. Julius. For a biography and bibliography see Molhuysen and Blok 1912: 644-646, written by W.H. Julius.

38 V.A. Julius, *Beschouwingen over de grondslagen der naturkunde,* Breda, 1880, pp. 31-57.

governed by the desire for one aim: *explaining the largest possible number of phenomena in the simplest way.*[39] Julius, however, like other Dutch physicists, conidered hypotheses to be indispensable, hence allowed for "explanation" as opposed to the mere "description", which both Mach and Kirchhoff had thought primarily characterized scientific conclusions. Still, Julius pointed out that hypotheses came and went. One of his examples was especially interesting, namely, the supercedence of the emission-hypothesis of light by the undulation-hypothesis. Julius gave as the reason that "more and more phenomena were known that necessitated different hypotheses in the first theory, whereas the undulation-hypothesis could account for it in a much simpler way". But this was not yet the end of the story: "who can safeguard us from the possibility that tomorrow an intellectual son of Clerk Maxwell will arise in order to show us that the undulation-hypothesis apparently lacking an explanation for the influence of magnets on polarized light etc. is also played out in physics and should itself be replaced by another theory...?[40]

The philosophical background of this position was akin to that of Mach, as set forth in the latter's "Die okonomische Natur der physikalischen Forschung" of 1882. Let us study the following short quotation from Uncle Julius:

> I do not contend that the choice of our contemporary foundations [in physics] is not the outcome of the nature of the noumenal world. I only maintain that we do not know the noumenal world and must stay within the phenomenal world, in the world of [sensory] phenomena, if we don't want to slide our slippery way down the greasy path of metaphysics.[41]

This was no casual remark. It enabled Julius to criticize William Thomson and P.G. Tait who believed that they could solve the problem of time measurement by claiming that the rotation of the earth was a case of movement which satisfied the condition better than any other case we could observe well that it was not driven by a force to change its velocity. The question was of course, how could one know that the earth satisfied this condition, since one normally judged the forces that worked on a body according to their movement. According to Uncle Julius, in general one could never prove the existence of a phenomenon merely by reasoning: "Such proofs can only be given by observation and experimentation; the use of reasoning

[39] Ibid., p. 4.

[40] Ibid.

[41] Ibid.

can only prove that the phenomenon is or is not in accordance with hypotheses which have been put forward."[42]

The critical considerations on the measurement of time which Uncle Julius had worked out in his book in 1880 were well ahead of his time. Even though Ernst Mach has often been seen as a forerunner of Albert Einstein for his remarks on time from the year 1882 on, it was Victor August Julius who treated time measurement in 1880 and even earlier in his article in the English journal *Nature* in 1877 and who had done so in a more detailed way.[43] Mach's target was Newton, but Julius also brought Maxwell, Thomson, and Tait into his sights, who seemed to believe that "there must exist one absolutely good method of time measurement arising from the nature of things".

Julius maintained, needless to say, that what counts is "a method which enables us to erect the building of science better than other methods; we do not know the noumenal world and can judge a method of time measurement only by means of the outcomes it gives".[44] "As soon as it can be demonstrated that a time measurement on a different basis than the contemporary approach is feasible, makes it possible to explain a larger amount of phenomena in simpler terms, then one should abandon the earlier time measurement."[45] In *Nature* Julius had already pointed out some consequences which resulted from using methods of measuring time which deviated from the then customary one. For example, while the so-called law of attraction seemed to be correct in terms of the method then used of contemporary time measurement, it was not so with the other methods. The same odd fact held for the law of conservation of energy or for any other principle where reference to acceleration was made.

A central focus for Victor August Julius, our Uncle Julius, was his criticism of an overly strong tendency to generalize, since what we observe in a local context, we often immediately apply to the universe, without asking ourselves whether the case doesn't lose its entire meaning for the universe. According to Julius, this not only holds for the law of the conservation of energy, but also for the second law of the mechanical theory of heat - "unless we see in it nothing more than a new form of Carnot's principle; but it is not feasible to comprise the whole universe on this point, as Clausius and Thomson do."[46] In other words, Julius could be as critical of the Second Law as was Mach, nor were his reasons utterly dissimilar.

[42] Ibid., p. 7. See also W. Thomson & P.G. Tait, *Treatise on Natural Philosophy,* I, Oxford, 1867, § 247.

[43] V.A. Julius, "On Time", *Nature,* **16** (1877), 122; 420.

[44] V.A. Julius (1880), op. cit., p. 14.

[45] Ibid., p. 8.

[46] Ibid., p. 84.

On the other hand, the agreement with Mach's remarks about the First Law of Thermodynamics in the latter's publication of 1872 was so conspicuous that one might conclude that Julius had consulted that book. For Mach had indeed touched on time there: "*Für das Weltall gibt es keine Zeit*" (For the universe as a whole, there is no time).[47] Furthermore, the view that traditional formulations of so-called laws have an indefinite character until they are made relative[48] to actual measurement is repeated by Julius. But the important thing was the difference in elaboration. Mach's brief book merely hinted and suggested, almost never going into extensive detail, but Julius spelled out his argumentation, doing all he could to give chapter and verse. This was especially clear when Julius stressed that no concept should be introduced without a proper development of its historical origin *and* the communication of the method of measurement that had been chosen, including the unit of measure. (One is also reminded of Percy Bridgman's much later theory of operationism.) As an example of what he was talking about, Julius called an expression such as "change of movement" (*mutatio motus*) in Newton's second (so-called) law, indeterminate so long as it could not be spelled out how one could measure this change of movement.

Let us emphasize the difference within basic agreement again. Whereas Mach tended to be satisfied with minimal remarks such as: "Fruitful as the principle of excluded perpetuum mobile is in the hands of an experienced scientist, how useless it becomes in a domain of experience which has not yet been investigated in a precise way."[49] Julius went into much greater detail. Indeed, was Mach too "economical" in his description of *Denkökonomie?* Wasn't Uncle Julius correct that clear and extensive description of *how* that application should take place to bring maximum advantage to science had to be at least one *exception* to "Denkökonomie"?

There is also another significant difference which may well be to the credit of Julius, He had already felt the need for an adequate philosophy of mathematics within the upbuilding of science. His view was that "the whole of mathematics is nothing else than the totality of consequences derived from posited definitions".[50] This could help explain how different geometries were possible. The only question for science was: Are the chosen definitions such that science can easily be constructed?

As the years passed after 1880, Julius occasionally began to refer to the recent publications of Wilhelm Ostwald and Ernst Mach. Indeed, just four weeks before his death in 1902, Julius returned to the problem of using hypotheses in physics - this time in connection with the ether hypothesis. He

---

47 E. Mach, *Erhaltung der Arbeit,* Prag, 1872, P. 37.

48 Ibid., p. 47.

49 Ibid., p. 91.

50 V.A. Julius (1880), op. cit., p. 92.

referred to Ostwald and to his own view that hypotheses have only a temporary existence. but he considered Ostwald's denial that scientific discoveries have often required hypotheses to be astonishing, since it would condemn the work of Ampere, Huygens, Fresnel, Clausius, and even Van der Waals.[51] I think that Mach would probably have taken the side of Julius here, for he too emphasized the heuristic function of hypotheses.

Speaking about Mach, at the end of his last lecture, Julius explicitly referred to his book on the foundations of physics, *Die Mechanik,* repeating some of its conclusions and quoting Mach's statement that the expression "absolute movement" has no sense. Julius endorsed it immediately: absolute motion was a *contradictio in terminis.* But it seemed to him that Mach had overlooked how Newton had understood "absolute movement" as "movement with respect to absolute space". Mach's attack should have been directed to the acceptance of an absolute space.[52] Finally, he repeats his thought that for each of the well-known fundamental theories of physics the environment *has to* be indicated with respect to which of these theorems are considered true. Nevertheless, Julius's firm conclusion that "everything drives us to choose the ether as this environment" - the ether replaces the *spatium absolutum* of Newton! - shows how far he stood away from the ideas that Einstein presented three years later.

V

After the death of Victor August Julius in 1902, Dutch philosophy of science went through a period of rapid evolution first under the influence of his nephew and successor Willem Henri Julius and then Johannes Diderik van der Waals (1873-1971), son of the better-known Johannes Diderik Van der Waals (1837-1923). As is well-known, a controversy between "atomists" and "energeticists" occured in physics from about 1892 to 1908, that is, until the reality of atoms appeared to be put beyond reasonable doubt. Most Dutch physicists were on the side of the atomists in so far as they considered the atomic theory indispensable in their work, but many were not yet ready to accept the reality of atoms, particularly because there was a strong suspicion during this period of history that a version of what was called "the electro-dynamic world picture" would prevail, a type of continuum theory. One tactic used by many atomists such as Ludwig Boltzmann to win supporters who opposed energeticism but were not yet ready to accept the reality of atoms was to allow that maybe atoms were just useful or indispensable "pictures" as Heinrich Hertz had suggested.

---

[51] V.A. Julius, *De ether,* Haarlem, 1902, pp. 10f.

[52] Ibid., pp. 53f.

It was the younger Julius who drew the attention of the Dutch physicists to the distinction between "facts" and "pictures" or "models". Indeed, Nephew Julius in his inaugural speech on the occasion of the succession to his uncle's position in Utrecht did his best to explicate the relevant parts of the Introduction to Hertz's book on Mechanics.[53] Hertz himself had been influenced by both Kant and Mach, and by supporting a version or interpretation of Hertz's *Bildtheorie* it could be maintained that Julius Jr. agreed with Mach's theory of economy and of course his rejection of the reality of atoms, but as suggested above, given the tactical nature of this "support", it could be very misleading to maintain that Julius Jr. was basically Machist. He was surely much less so than his uncle. Indeed, the nephew soon turned most of his attention back to physics and left philosophy of science in other hands.

Nevertheless, in August 1911, while still somewhat interested in philosophy, he made a serious effort to attract Albert Einstein to Utrecht. Professor Windt had just died and a physics chair was open.* Einstein had already begun working on his general theory of relativity and for some time he would allege that the only external support for his theory came from what he would soon call "Mach's principle", the notion that local inertia was determined by the universe as a whole. If Leyden had honored Planck briefly, could Utrecht do better than win Einstein permanently? Especially when he still seemed to be under Mach's influence? What a triumph! But the hope faded like leaves in the Fall when Einstein turned Julius Junior's feeler down and decided to return from Prague to Zürich. Shortly after Julius's failure, Hendrik Lorentz saw his chance and asked Einstein if he would like to come to Leyden, but he turned that down too.**

Concerning particles, it seems that Lorentz,[54] the leader of Dutch science who had become famous for his electron theory and transformations continued to be very cautious, not only about the reality of electrons but even concerning atoms and molecules, preferring like many scientists to speak about small particles as "pictures". But support for Hertz's terminology and way of speaking by no means meant that most physicists accepted either his idealized conception of mechanics or the Kantian or Machist background to many of his philosophical assumptions. *Bildtheorie* was simply a provisional way to protect oneself during a period of great apparent instability in the foundations of physics. Hendrik Lorentz's student Leonard Salomon Ornstein (1880-1941), Van der Waals successor at the University

53 W.H. Julius, *Kritiek in de naturkunde,* Utrecht, 1896, p. 19.

* R.W. Clark, *Einstein - The Life and Times,* New York, 1971, pp. 146-148. - Ed.

** Ibid., p. 149. - Ed.

54 H.A. Lorentz, "Nieuwe uitkomsten op het gebied der molukulaire theorieën, *Chemisch Weekblad,* **7** (1910), 1 ff.

of Groningen, was likewise attracted toward "pictures". Ornstein, however, was also intrigued by non-Hertzian aspects of Mach's philosophy. In particular, one of his key notions being the Machian concept of adaptation, as if pictures could be regarded as an adaptation of our mind to nature. However, reflecting on the question of why adaptations of mind to nature could take place, Ornstein alluded to Hertz's conviction that there were certain concurrencies between nature and our mind - a formulation which Lorentz was also very fond of[55] - and this brought him to the assumption that those elements in our pictures which we see as "necessary in thought" have "natural necessary reality". For example, we have to assume that the axioms of mathematics hold in nature.[56] This view seemed different from Mach's empiricism, but it was very well possible that Ornstein had just been stimulated to adopt his position by reading the chapter on hypotheses in Mach's *Erkenntnis und Irrtum,* especially those passages in which Mach spoke about "requirements for the thinkability of facts" - and conceptual characteristics which were necessarily attached to facts.[57]

The work of Einstein and Perrin on Brownian motion circa 1905 to 1908 plus the development of instruments such as the spinthariscope appeared to make the reality of atoms so certain shortly after 1911 that most Dutch physicists gradually began dropping Hertz's "picture" terminology and openly advocating the reality of atoms. This change was accompanied by the gradual realization that the "electromagnetic world picture" could not easily or readily be carried through. Both continuous and discontinuous physical phenomena had to be acknowledged as real. The new attitude put the position of Poincaré, Duhem, and especially Ernst Mach as the world's leading philosophers of science in considerable question given their long opposition to the reality of atoms. In The Netherlands, it was appropriate that the new leader of local philosophy of science should be Van der Waals Jr.,[58] son of a longtime atomist, contributor to the kinetic theory, and believer in the reality of atoms.

On the whole, the leading Dutch physicists were not very communicative about their "philosophical" views, let alone those of others. As far as I know, they refrained from direct comment after the guest-lectures in Holland of two opponents in the positivism-realism debate at the beginning of

---

55 H.A. Lorentz, *Electromagnetische theorieën van naturkundige verschijnselen,* Leiden, 1900, p. 10.

56 L. Ornstein, *De rol der wiskunde in de mathematische physica,* Leiden, 1909, p. 18 ff.

57 E. Mach, *Erkenntnis und Irrtum,* zweite Auflage, Leipzig, 1906, pp. 248 ff.

58 Johannes Diderik van der Waals (1873-1971) was Professor of Physics at the University of Groningen from 1903 to 1908 and Professor of Physics at the University of Amsterdam from 1908 to 1943.

the century: the one by Max Verworn,[59] the other by Max Planck.[60] Planck had said that Mach's theory of knowledge was widespread among the scientists, but there were few signs of it in the Netherlands. Perhaps the general opinion on Mach was best represented by Johannes Petrus Kuennen (1866-1922), professor of physics at Leyden, who, a year before Planck was to lecture there, called Mach "an author whose writings have a charm to which few readers can be insensitive, though many of these admirers find difficulty in sharing his dislike of molecular methods of description, which occasionally appear in his later works."[61]

While Mach was particularly influential in many parts of Europe as a philosopher and historian of science during the first decade of the 20th century, the Low Countries were different. Native positivists held no university chairs in either philosophy or physics. Even though some leading physicists used *Bildtheorie* language, they actually supported the atomic theory and strongly leaned toward accepting the reality of atoms. The timing of his potential impact did not work out.

Also, part of the philosophical problem was that "positivists" in general were often thought of as atheists or anti-religious with the result that they often had difficulty acquiring prominent positions, though this difficulty tended to be Europe-wide and was not just a Dutch problem.

Nevertheless, Mach the experimental scientist was not only respected by Dutch physicists, but two leaders, Hendrik Lorentz and Julius Junior, even proposed in a letter to the *Nobel Comité* in Sweden dated January 29, 1912 that Mach be offered a Nobel prize.[62] They cited his work in optics and acoustics and mentioned that all physicists were familiar with his books in the history and philosophy of science.

VI

But if blocked on the university level, apart from residual influence on science students at Utrecht, Mach did succeed in having a philosophical impact on some *Gymnasium* or high school teachers, particularly on two men who had presumably been "conditioned" by the earlier works of Opzoomer and Uncle Julius. Let us now turn to J.E. Enklaar and Jacob Clay, who both taught science in secondary schools in the period between 1880

---

59 M. Verworn, "Prinzipienfragen in der Naturwissenschaft", *Handelingen van het Tiende Nederlandsche Natuur- en Geneeskundig Congress,* (1905), pp. 46-63.

60 M. Planck, "Die Einheit des physikalischen Weltbildes", *Physikalische Zeitschrift,* **10** (1909), 62-75.

61 J.P. Kuenen, *het noumenale wereld en de sinnelijke waarnemingen,* Haarlem, 1907, p. 22.

62 See J. Blackmore & K. Hentschel (eds.), *Ernst Mach als Aussenseiter,* Wilhelm Braumüller Verlag: Vienna, 1985, pp. 95-96.

and 1920 and who are important not only because they surpassed so many physicists in the number of essays on physical subjects, but also because of their profundity. Many Dutch scientists started their career as teachers in secondary schools. Even Lorentz[63] taught for some years at an evening-school before he was appointed at Leyden University. But of course few such teachers had the opportunity to get a higher position. Yet there were many who had a doctor's degree and continued to make contributions in their field, though the accents often shifted from mathematics and natural sciences to the history of mathematics and science, or even to foundational studies. Sometimes there was an opportunity to become an unsalaried (external) university teacher (*privaatdocent* in Dutch), usually after having published one or more remarkable publications. The best-known example of such an author would probably be E.J. Dijksterhuis who taught 34 years at a secondary school, before he became extra-ordinarius professor in History of Mathematics and Natural Science. Where? Utrecht of course.[64]

Dr. Johannes Eliza Enklaar (1847-1926) was a teacher of chemistry at a secondary school at Deventer from 1885 to 1902 and from 1902 until 1913 in Utrecht. He wrote extensively on scientific subjects, and it is no surprise that he took Mach's writings into account. Two of his essays bear the title "Ernst Mach"; the first was published in 1905,[65] shortly after Koster's criticism, and was explicitly intended to draw attention to Mach's books on mechanics and physics; the second was written after Mach's death,[66] and contained an overall picture of Mach's contributions.

In 1905, Enklaar's emphasis was on clarifying Ernst Mach's endeavour "primarily to dispel all logical contradictions and all useless appendices from the positive area"[67] - notably the remnants of Medieval scholasticism. This was especially interesting to Enklaar, who was involved in polemics with Roman Catholic natural philosophers, who had tried to put new life into Aristotelian-Thomistic doctrine.[68] Yet, he did not regard Mach as a positivist `a outrance such as Ostwald, whose philosophical ideas had come

---

63 G.L. De Haas-Lorentz, *H.A. Lorentz - Impressions of His Life and Work,* Amsterdam, 1957, p. 27.

64 E.J. Dyksterhuis, *Doel en methode van de geschiedenis der exacte wetenschappen,* Amsterdam, 1953, p. 23.

65 J.E. Enklaar, "Ernst Mach", *Album der Natuur,* (1905), 353-364.

66 J.E. Enklaar, "Ernst Mach", *De Tijdspiegel,* (1917), 203-237.

67 Enklaar (1905), op. cit., p. 354.

68 J.E. Enklaar, "Tweeërlei natuurverklaring", *De Tijdspiegel,* (1894), 304-338.

off worst in The Netherlands[69] vis `a vis the energeticism-atomism conflict with Ludwig Boltzmann which had started a decade earlier and resurfaced in lectures given at the St. Louis World's Fair in 1904. In support of this view, Enklaar quoted from *The Principles of the Theory of heat,* in which Mach had openly acknowledged "the lasting achievement of mechanical physics".[70] He observed that Mach's criticism had been especially directed against the ideas of the psychologist Wilhelm Wundt. What disturbed Mach was Wundt's axiom that all physical causes were motional ones (*Bewegungsursachen*). Mach pointed out that such a view might lead to the opinion that our "most familiar world of sense" could suddenly appear as "the greatest riddle of the world".[71]

Mach's greatest merit according to Enklaar was his "critique of science", which had become in his hands "a science in itself", a rigorous logical and historical investigation into the concepts and laws of science, in view of their factual content.[72] As a paradigmatic example, he summarized Mach's critical considerations of the concept of temperature and concluded that it gave complete clarity. It was well-known that Mach had considered Black to have been a most excellent scientist and a worthy successor of Newton. It was typical of Mach's historical approach to try to show that some scientists were at least in a sense far ahead of their time. Galileo's anticipated the principle of inertia later clarified by Descartes and especially Newton, and may have had a more completė understanding of it than he has been given credit for. Similarly, Robert Mayer deserved as much credit for his factual discovery of the conservation of energy as Helmholtz and Joule for their later refinements, a discovery he supported in a stronger and less mystical way than was widely believed.

But Enklaar, who had done scholarly research into the history of physics, also expressed some scepticism about Mach's strong tendency to modernize scientists of the past. In particular, Enklaar believed that Black's concept of the "latent heat of fusion" rested on a scholastic idea: Heat could alternatively change from actual into potential existence. But he agreed with Mach that Black's idea of equivalence of heat was extremely fruitful. He also took Mach's side in his conception of the history of Black's idea of

69 See Enklaar's paper "Het wereldbeeld der naturwetenschap", *Onze Eeuw,* **5.1** (1905), 438-477 in which he discussed Ostwald's Energetics critically and where he admitted that Mach stood close to Ostwald in his anti-materialism - in the sense that "matter" was conceived as a natural thought symbol for a relatively stable complex of sensory elements. But he also pointed out that Mach acknowledged the value of artificial hypothetical atoms and molecules for modeling physico-chemical experiences - not of course for the explanation of psychical phenomena.

70 E. Mach, *Die Principien der Wärmelehre,* zweite Auflage, Leipzig, 1900, p. 318.

71 Ibid., p. 317.

72 J.E. Enklaar, "Ernst Mach", *Album der Natuur,* (1905), p. 354.

a heat substance: an hypothesis which went beyond the domain of known facts and thus could be the bearer of new ideas for some time. He observed that the "heat as substance" idea was only given up because it did not match well with discovered facts; but one must not say that it was destroyed when it was discovered that heat and energy were equivalent - this would only be the case for those scientists who believed that it was a settled matter that heat was indeed a motional state of matter and ether. *The statement that the incorporeality of heat has been proved is incorrect; it betrayed lack of insight into the meaning and use of physical concepts.*[73]

In the eyes of Enklaar, this was Mach's lasting contribution: the clarity which his considerations threw on the meaning of physical concepts. Mach had made clear that "they receive, like words, their characteristic mark by the use which is made of them. They are meant for use; they are not representations of reality."[74] But this judgment included that such concepts were only valid for a limited group of facts. In particular, it was not the case that the concept of energy was applicable to all material phenomena.[75]

A more favorable review of Mach's critical work hardly seemed possible. Yet, Enklaar thought that his historical treatment of the second law of thermodynamics had, even with its numerous strongpoints, some serious defects. Therefore, he referred the reader to the "excellent" *Thermodynamics* of Max Planck[76] for its acute formulation of this law and its criticism of less correct and less general formulations. More on Planck followed the event of Planck's lecture at Leyden at the end of 1908. Enklaar admitted that many professional physicists had come to differ with Mach, but he distinguished between lesser gods convinced of the completely independent reality of the mechanico-physical world picture and physicists of the first rank, who took the nature of our cognitive faculties into account.

Planck it seems belonged to the second group when he not only aimed at a physical world picture independent of the fluctuations of peoples and times and unconnected with the individuality of the active, constructive mind but also acknowledged that an absolute elimination of sensations was not possible, as if direct knowledge of the absolute were out of the question.[77] Enklaar remarked drily that elimination of the anthropomorphic element from the physical world picture was surely desirable since "after all even we [Mach and Enklaar?] don't want to be solipsists", but this did not have to mean that the "psychical in general" had to disappear from it.

---

73 Ibid., p. 359.

74 Ibid., p. 360.

75 Ibid., p. 361, with a reference to Mach (1900): 318-346.

76 M. Planck, *Vorlesungen über Thermodynamik* (appeared first in 1897; the second edition dates from 1905.)

77 J.E. Enklaar, "Levende en levenlooze stof", *Onze Eeuw,* **10.4** (1910), p. 413.

But time did not stand still. In a later essay from 1914, Enklaar acknowledged that there had been enormous progress in and about the atomic view since 1905. The statement that the material world consisted of individual particles with a diameter of about one tenmillionth millimeter had won the authority of sensory perception. Atomism had come off triumphant.[78] Did this mean that every physical theory should stand on the basis of atomistics?

Not at all. Enklaar insisted that the historical development of natural science gave a negative answer. Natural science looked for new points of view from which the largest domain of phenomena could be surveyed, and which would enable us to include as many and as diverse phenomena under the same concept. History showed that scientists had sometimes taken the atomistic point of view and sometimes that of continuity, but of course not by unconstrained choice, but under the influence of new theories and great discoveries, as if it were more *useful* to accomodate ourselves to what currently seemed to be most plausible. Utility was a major factor, though utility *within* science could be very different from the utility of *applied* science. Perhaps atomism was no more than a first, superficial look at what really existed, for there could well be "something artificial" in its very limitations, its discontinuity. Why should nature stick with limited divisibility? *L'infini nous tormente*.[79]

When Enklaar recapitulated Mach's treatment of the concept of force in his 1917 paper, he quoted Mach's statement that the modern concept was not something hidden in the processes but was a measurable, factual case of movement. Indeed, the facts, directly observed and experienced and the direct expression of facts were all that mattered. For Mach the relationist, the highest stage of science was to seek out relations.

This position was Enklaar's clue to Mach's attitude toward Einstein's relativity theory: "It is entirely different from the older doctrine of relativity which Mach proposed; Mach stands on the foundation of classical mechanics (Galilei, Newton). In many respects, the principles of Einstein and Minkowski do not fit this theory; they want to revise the old foundations."[80] Enklaar mentioned the constancy of the velocity of light, whereas "the old mechanics does not know maximal absolute velocities, since every determination of velocity has only relative meaning". According to Minkowski, the consequent development of the theory of relativity required the assumption of an absolute world of four dimensions (. . .). What would Mach have said about such a world, wholly independent of an observer, a *Ding an sich*, for which there was for us no spur of visualizability left? He

---

78 J.E. Enklaar, "De bouw der stof", *Onze Eeuw*, **14.1** (1914), 265.

79 Ibid., p. 268.

80 Enklaar (1917), op. cit., p. 226.

could have appreciated it in terms of mathematical contemplation, but he would never have assigned it real physical meaning.[81]

Returning to Planck's attack on Mach, J.E. Enklaar repeated his earlier comments and even tried to give more support to Mach's position by translating some passages from Mach's answer: Planck's statements were wholly in conflict with the first principles of the theory of knowledge. He had completely misjudged the relation of each experience, each act of knowing to the human mind. Planck had unjustly reproached Mach for his "subjective and changeable world picture". Mach's elements really existed, not only for particular individuals but for everyone. Mach in his own way had also eliminated the subjective, the particular quality of sensations for different observers, since they could be excluded if we made appropriate corrections. "Mach's world picture was as constant as such a picture could and had to be."[82]

VII

Jacob Clay (1882-1955) had also reviewed Planck's criticisms of Mach, indeed, as early as 1910.[83] Clay like Enklaar was long a secondary school teacher, but in 1920 became professor of physics in Bandung in the Dutch East Indies. Rather late (1929), he finally obtained a university post in The Netherlands itself, indeed, a fine one at Amsterdam. Clay, as a student of both Lorentz and the Hegel-oriented Leyden philosopher Gerardus Bolland (who's hostile book review of Mach's *Analyse* we briefly mentioned near the beginning of this article) tried to interpret the developments of modern physics in Hegelian terms. But unlike Bolland he tended to favor Mach, at least when compared to Planck. In particular, he found numerous "naiveties" and "illusions" on the part of Max Planck. Ignoring the latter's many qualifications, Clay insisted that it was an illusion that physics could disengage itself completely from anthropomorphism,[84] and his ideal of forming "a constant world picture, independent of every human, or more generally, every intellectual individuality" belonged to the world of castles in the air, if not of pure ether.[85] Max Planck was even criticized for maintaining the "naive-realist" position that one's world picture of the future and the world were one and the same - which was allegedly a sure sign of defective logi-

[81] Ibid., p. 227.

[82] Ibid., p. 234.

[83] J. Clay, "Eenheid en eenzijdigheid in de natuurwetenschap", *Tijdschrift voor Wijsbegeerte*, **4** (1910), 70-82. See also Clay's earlier paper: "Naturphilosophie en Atomistiek" in the same journal, **1** (1907), 64-84.

[84] Ibid. (1910), p. 74.

[85] Ibid., p. 79.

cal and metaphysical development. (In fact, Planck's "future identity thesis" was apparently at least in part an attempt to win the support of Neo-Kantians in his struggle against Mach's phenomenalism, and was later abandoned in favor of indirect realism about the physical world.) Concerning how far a perceived process depended on the senses, Clay advised Planck to read Mach's *Analyse der Empfindungen* more carefully (an opinion which could scarcely have pleased Bolland who was still very much alive).

Clay gave an even more favorable account of some of Mach's ideas five years later in his *Sketch of a Critical History of the Concept of Law of Nature in the Newer Philosophy* (in Dutch), which was awarded a prize by a committee of the University of Leyden.[86] Clay considered Ernst Mach's conclusion about scientific laws being mere conceptual expressions of constancies in the connections or relations between sensations (elements) as an example where he went "beyond empiristic prejudices". He also saw Mach's opinion as an indication of his own that "proper laws as mathematico-functional relationships have only an (ideal) existence in our thought".[87] Clay regretted that Mach had not emphasized this point enough.

Despite the appearance of a positive assessment of Mach's philosophy of science, Clay continued to hold a kind of idealistic world view in which the goal was the development of a "world consciousness" which would "liberate the finite human being from his narrow individualism". This notion came out rather strongly in his 1920 book *The Development of Thought* (in Dutch).[88]

One consequence of the popularization of Einstein's special and general theories of relativity during the 1920's was the development of the view that if Mach's criticisms of Newton's notions of absolute space and time had influenced Einstein, then perhaps relativity theory was not just compatible with positivist philosophy but was an exemplification of it. Joseph Petzoldt in Germany and Philipp Frank in Austria, Czechoslovakia, and later America helped propagate this opinion. In reality, Machists were split. Many if not most were upset by Einstein's constancy of light axiom, as if a new absolute had been introduced into physics, but others became forerunners of or participants in the Vienna Circle and Ernst Mach Verein of the 1920's which evolved into the Logical Positivism of later decades.

---

[86] J. Clay, *Schets eener kritische geschiedenis van het begrip natuurwet in de nieuwere wijsbegeerte,* Leiden, 1915, pp. 336-342.

[87] Ibid., p. 341, with reference to E. Mach, *Analyse der Empfindungen,* dritte Auflage, Jena, 1902, p. 276. Clay's exposition of Mach's views on the content and value of laws of nature is mainly based on the relevant chapters of *Erkenntnis und Irrtum.*

[88] J. Clay, *De ontwikkeling van het denken,* Arnhem, 1920. (For comments on Mach see p. 126 ff.)

Jacob Clay in afteryears seems to have gravitated from Hegelianism toward Logical Positivism. (One is reminded of the German-American philosopher J.B. Stallo who decades earlier had done something similar in terms of the positivism of his era.) It cannot be said that Logical Positivism was a major force in The Netherlands before the Second World War or even for some time afterwards (though it should be mentioned that Otto Neurath, an important member of the Vienna Circle did stay in Holland for a time during the late 1930's after political opponents had taken over in Austria), but the influence of Kantian and Hegelian philosophy did show signs of weakening, though much of it would yield to Phenomenology and Existentialism before any significant turn to "scientific philosophy" would occur but while the rest of the Mach-influenced writers we will consider in this article were also associated during at least part of their lives with Utrecht, eventually, very late, it would finally cease to be the almost exclusive center of what would develop in philosophy of science, at least in a sense sympathetic to Mach. Indeed, the establishment of new universities would make the situation in philosophy much more complex. But since these changes cannot be brought under the head of the direct reception of Mach in The Netherlands it would be inappropriate to go into great detail about them.

VIII

The last Dutchmen I would like to consider here are the historian of science, E.J. Dijksterhuis, and my own former teacher, the logician and philosopher of science, Evert W. Beth. Both men have exercized extensive influence on intellectual thought both within and well-beyond The Netherlands, and strongly supported a number of views favorable to Mach and his philosophy of science.

While still a secondary school teacher, Dijksterhuis published a number of works on the history of mechanics. His special interest was Simon Stevin,[89] concern which naturally drew him to Mach's observations on Stevin in his book on mechanics. In matters of detail, Dijksterhuis was critical of Mach, and concerning his views on Galileo very critical, especially about Mach's underqualified attribution of the discovery of inertia to the Tuscan scientist. Already in his first historical book, *Fall and Throw, a Contribution to the History of Mechanics from Aristotle to Newton* (1n Dutch,1924), Dijksterhuis criticized Mach's treatment of Galileo's views on movements where something is cast or thrown, notably his statement that a toss oblique or diagonal (presumably to the followthrough of the thrower) did not

[89] See E.J. Dijksterhuis, "Simon Stevin", *Archives du Musée Teyler,* III, 9 (1941), 268-298. and his book *Simon Stevin,* 's-Gravenhage, 1943.

trouble Galileo, even though nothing could be found in Galileo's writings to confirm this. Mach's accusation of a *Fehlschluss* in Galileo's discussion on the law of falling bodies in the *Discoursi* also annoyed Dijksterhuis. But of more fundamental importance were his criticisms of Mach when he attributed the discovery of the law of inertia to Galileo. Dijksterhuis thought that Galileo did not see that a movement arising from abstracting away all external influences had to be rectilinear. He also commented on Mach's handling of the law of inertia in Newton's work, but I think that the above summary may suffice to to make Dijksterhuis's conclusion[90] understandable that Mach was "not very particular about historical exactness."

On the other hand, where philosophy of science was concerned there appeared to be greater convergence than one might have expected given Mach's atheism and Dijksterhuis's sensitivity about science encroaching on religion. This came out clearly in the book which transformed Dijksterhuis into one of the best-known and most important historians of science of his era. It was originally published in Dutch under the title *De mechanisering van het wereldbeeld* in 1950 and appeared English in 1961 as *The Mechanization of the World Picture*. While his thinking was very careful and qualified, many readers have undoubtedly cut the Gordian knot and interpreted his book as support for the positivist thesis that not only has "the mechanical world view" been replaced by "modern physics" but that there has also been a philosophical shift from the mind-matter dualism and indirect realism espoused by Galileo, Descartes, Boyle, Locke, and very possibly by Newton to a positivist or phenomenalist approach rejecting "metaphysics". Many passages in this now classic book when taken out of context could easily be viewed as supporting this hypothesis and other aspects of Machism. For example:

> According to this conception [nominalism], not only are terms such as 'attraction' and 'tendency' words which can be dispensed with and which merely imply a superfluous specification of ...force, but the same holds also for 'force' itself. Properly speaking one is only justified in saying - for this alone is taught by mathematically treated experience - that two particles in each other's presence have accelerations in opposite directions along the line joining them and that these accelerations vary inversely as certain coefficients to be assigned to the particles, namely their masses; it is nothing but an abbreviation to say that they are acted upon by equal and opposite forces, whose magnitude for each of the two particles is the product of the mass of that particle and its acceleration.[91]

[90] See E.J. Dijksterhuis, *Val en worp. Een bijdrage tot de geschiedenis der mechanica van Aristoteles tot Newton*, Groningen, 1924, particularly pages 249, 268, 271, and 285.

[91] E.J. Dijksterhuis, *The Mechanization of the World Picture*, trans. C. Dikshoorn, Oxford, 1961, p. 483.

and:

> ...an entirely new standpoint towards nature had to be reached: 'substantial' thinking which inquired about the true nature of things, had to be exchanged for 'functional thinking', which wanted to ascertain the behavior of things in their interdependence....functional thinking with its essentially mathematical mode of expression has not only been maintained, but has even come to dominate science.[92]

A final point about Dijksterhuis and Mach as historians of science. While the Dutch scholar continued to criticize Mach for his "Galileo-myth" in his *The Mechanization of the World Picture,* Dijksterhuis did change his mind about Mach's analysis of Stevin's *clootcrans* proof.[93] According to Mach, Stevin's derivation was one of the most valuable guides in the history of mechanics, because it illuminated the origin of science from "instinctive insights". In his first publication on Stevin,[94] Dijksterhuis had not shared this view about Stevin's derivation, but afterwards he changed his opinion - (as he says) on the basis of discussions with a member of the Dutch patent office. After that (1943), he ranked Stevin among "those mechanists, who intuitively used the principle of conservation of energy in the gravitational field of the earth, long before its conscious formulation".[95] This was a remarkable turn in Dijksterhuis's understanding, and no doubt helped raise his appreciation for Mach as an historian, and possibly as a philosopher as well.

Nor should we fail to re-emphasize that both Dijksterhuis and Beth, like so many other Dutch scientists and philosophers who we have mentioned, were clearly associated with Utrecht. Dijksterhuis, as already touched on, finally obtained a university position at that perspicacious university and Beth completed his studies in mathematics and physics there in 1932. What remains somewhat mysterious are the reasons for or causes of this apparent connection, though the more lasting influence of C.W. Opzoomer and V.A. Julius on science students with an interest in philosophy as opposed to philosophy students per se seems to have been a factor.

Evert W. Beth (1908-1964), like the last three gentlemen we have discussed, was a secondary school teacher for many years (1935-1945), but also like the last two finally found a university position. His greatest interests were in mathematics, psychology, and philosophy. Though he is best recalled as a philosopher of logic and mathematics, he also co-authored an interesting but posthumous book *Mathematical Epistemology and Psycholo-*

---

92 Ibid., p. 501.

93 E. Mach, *Die Mechanik,* sechste Auflage, Leipzig, 1908, p. 29.

94 Dijksterhuis (1941), op. cit..

95 Dijksterhuis, (1943), op. cit., p. 123.

gy (1966) with the world-famous psychologist Jean Piaget. Concerning philosophical outlook, After early attraction toward Kant, he turned away from him, and, possibly under the influence of G. Mannoury[96] and A. Fraenkel,[97] became quite sympathetic toward several of Mach's ideas.

Being myself a former student of Beth, I remember his utmost negative attitude towards Husserl, whom he even accused of having stolen his most important ideas from Mach's lecture on the principle of comparison in physics, which was reviewed by Edmund Husserl in 1897.[98] In the book mentioned above, Beth elaborated:

> It is not surprising that Husserl should have had so much interest and admiration for Mach's lecture. In the *Philosophie der Arithmetik* [1891] he had alluded to the possibility of an influence due to Mach. In the same book he also talks of the "description of a phenomenon". . . . His [Husserl's] anti-psychologism is thus not a revolt against psychology or its influence in different spheres; in origin it is only a revolt against the application of certain methods in psychology and is shown in an attempt to establish a descriptive or phenomenological psychology, analogous to phenomenological physics or chemistry as propounded by Duhem, Mach, Ostwald, and Voigt. Later the need to maintain the pretension of a hegemony over other scientific field for this refined psychology, compelled Husserl to make his descriptive or phenomenological psychology into an actual philosophy. As for its content, Husserl's psychology is a speculative psychology which differs from the systems of Mill and Ziehen only by the admission of different principles.[99]

It is apparently still little-known that Beth had attempted to elaborate a mathematical theory of apparent Mach systems in his *Philosophy of Nature* (in Dutch,1948).[100] I also found during some research of my own an anonymous manuscript on a generalization of Mach's definition of mass in the Beth archives. Beth's judgment on Mach was, in general, very favorable, but it must not be forgotten that the relevant publications had been written

[96] A relativism akin to Mach's, was defended by the Dutch mathematician and leader of the Dutch signific movement, Gerrit Mannoury (1867-1956). Moreover, his own contribution to the Dutch translation of Kern's *Manuel of Buddhism* also shows remarkable similarities with Mach's reflections on Buddhism. Beth praised Mannoury in *Mathematical Epistemology and Psychology*, Dordrecht, 1966, p. 35 & 48.

[97] Beth praised him as follows: "From 1933 onwards, I was fortunate enough to be able to attend a series of lectures on the foundations of mathematics given by A. Fraenkel at the University of Utrecht." (*Mathematical Epistemology and Psychology*, p. 4)

[98] E. Husserl, a review of E. Mach's "Über das Princip der vergleichung in der Physik", *Verhandlungen der Gesellschaft Deutscher Naturforscher und Ärzte,* Theil 1 (1894), 44-56 in *Archiv für systematische Philosophie,* N.F. **3** (1897), 241-244. [Please see chapter 12 for Keiichi Noe's comment on Husserl's admission in Holland of his debt to Mach. - Ed.]

[99] Beth (1966), pp. 32-33.

[100] E.W. Beth, *Natuurphilosophie,* Gorinchem, 1948.

under the influence of the Vienna Circle.[101] That many of Evert W. Beth's students, in their turn, should have paid attention to Mach's writings, may thus become more understandable.

Another development ensuring that Mach's ideas will retain some influence in the Netherlands has been the establishment of the Vienna Circle Archive in Amsterdam and the associated work of A. Kox and H. L. Mulder. Perhaps even more encouraging for the future has been the interest of Dutch publishers in Mach and Logical Positivism. In particular, D. Reidel Publishing Company of Dordrecht, which has now become part of the Kluwer Group, has come out with several books in its "Vienna Circle Collection", including two translations of books by Ernst Mach, titled *Knowledge and Error* (1976) and *Principles of the Theory of Heat* (1986). Brian McGuinness edited both. Robert S. Cohen, managing editor of the series of which this anthology is a part has also written articles and edited a book on Mach. [See bibliography for titles.]

In closing, let me mention that in 1980, S.J. Doorman, Professor of Philosophy at the Technical University of Delft since 1972, another former student of Beth, prepared an anthology of available translations of central texts by Ernst Mach.

What about the future? The sharp expansion of publications by and on Mach during the 1980's [See the bibliography to this volume] may help answer the question. And what about Mach's influence in The Netherlands? After so much opposition for so long, even from some parties at Utrecht, perhaps the timing is finally appropriate for a more general appreciation of his contributions to science, history, and philosophy.

101 Beth (1966), p. 4: "Later, the study of R. Carnap's work in the *Genootschap*, and then personal contacts with R, Freys, H. Scholz, and A. Tarski led me to turn towards logic rather than towards psychology."

ERNST MACH BIBLIOGRAPHY
(1971 - 1991)

(plus some useful earlier items)

I

OTHER BIBLIOGRAPHIES

All recent bibliographies of the writings of Ernst Mach are based on the work of Joachim Thiele, in particular his detailed article "Ernst Mach-Bibliographie" in *Centaurus,* **8** (1963), 189-237, which has been republished in part in numerous other books. There are also several recent bibliographies of articles and books *about* Ernst Mach starting with Thiele's article. For English-speaking readers the current bibliography may be understood as a continuation of volume 6 of *Boston Studies in the Phikosophy of Science* which came out in 1970 and this writer's own biography of Mach which appeared in 1972:

Mach, Ernst, *Kultur und Mechanik* . . . [with foreword and extension of 1963 Mach-Bibliography by J. Thiele] , Bonset: Amsterdam, 1969.

Blüh, Otto and W.F. Merzkirch, "Ernst Mach Bibliography", in *Ernst Mach - Physicist and Philosopher,* edited by Robert S. Cohen and Raymond J. Seeger, (*Boston Studies in the Philosophy of Science,* vol. 6), D. Reidel: Dordrecht, 1970, pp. 274-290.

Blühdorn, Jürgen and Joachim Ritter (eds.), *Positivismus im 19, Jahrhundert: Beiträge zu seiner geschichtlichen und systematischen Bedeutung,* (*Studien zur Philosophie und Literatur des 19. Jahrhunderts,* vol. 16, [with a bibliography of Ernst Mach pp. 183-201]), Klostermann: Frankfurt, 1971.

Blackmore, John T., "A Select Bibliography", in *Ernst Mach - His Life, Work, and Influence,* University of California Press: Berkeley, 1972, pp. 361-375.

D'Elia, Alfonsina, "Recente Bibliografia Machiana", in *Rivista Critica di Storia della Filosofia,* **3** (1975), 189-203. [This article includes commentary and judgment on numerous articles and books written about Mach between 1964 and 1972.]

Thiele, Joachim, "Bibliographie", *Wissenschaftliche Kommunikation - Die Korrespondenz Ernst Machs,* Aloys Henn Verlag: Kastellaun, 1978, pp. 302-309.

Stadler, Friedrich, "Ernst-Mach Bibliographie" and "Quellen- und Literaturverzeichnis", in *Vom Positivismus zur 'wissenschaftlichen Weltauffassung' - Am Beispiel der Wirkungsgeschichte von Ernst Mach in Österreich von 1895 bis 1934,* [Especially useful is the list of letters to Ernst Mach held by the Ernst-Mach-Institut], Löcker Verlag: Vienna & Munich, 1982, pp. 283-300 & 307-334.

Wolters, Gereon, "Nachweise", *Mach I, Mach II, Einstein und die Relativitätstheorie - Eine Fälschung und ihre Folgen,* Walter de Gruyter: Berlin and new York, 1987, pp. 448-461.

Information on the thousands of books in Mach's own library can be found in the following two catalogs published by Theodor Ackermann Antiquariat in Munich, Germany, Promenadeplatz 11.

Bibliothek Ernst Mach, Teil I: Alte Drucke-Aeronautik-Astronomie-Optik-Photographie-Physik-Mathematik-Uhren, Theodor Ackermann Antiquariat: München, 1959, Katalog 634, ninety-three pages.

Bibliothek Ernst Mach, Teil II : Philosophie-Psychologie-Medizin-Chemie-Naturwissenschaften-Technik, Theodor Ackermann Antiquariat: München, 1960, Katalog 636, ninety-nine pages.

II

SOME RECENT WORKS BY ERNST MACH

A. Works in English:

Space and Geometry - in the Light of Physiological, Psychological, and Physical Inquiry, translated by T.J. McCormack, 2nd edition, [based on articles appearing in the *Monist* from 1901 to 1903], Open Court: La Salle, 1960.

The Science of Mechanics - A Critical and Historical Account of its Development, translated by T.J. McCormack, introduction by K. Menger, sixth edition, based on the German sixth edition with revisions through the ninth edition, Open Court: La Salle, 1960.

The Analysis of Sensations, chapters I & XV, in *The European Philosophers from Descartes to Nietzsche,* edited by and introduction from M.C. Beardsley, The Modern Library, New York, 1960, pp. 768-200.

"The Significance and Purpose of Natural Laws", translated from the last chapter of *Erkenntnis und Irrtum,* Leipzig, 1905, and "Newton's Views on Time, Space, and Motion" from *Science of Mechanics,* in *Philosophy of Science,* editors: A. Danto and S. Morgenbesser, World Publishing Company: Cleveland, 1960 & 1964, pp. 266-273 & 335-348.

"Newton's Views on Time, Space, and Motion", from the *Science of Mechanics,* in *Problems of Space and Time,* ed. J.J.C. Smart, New York, 1964, pp. 126-132.

"The Guiding Principles of my Scientific Theory of Knowledge and Its Reception by my Contemporaries", in *Physical Reality: Philosophical Essays on 20th century Physics,* edited by Stephen Toulmin, Harper & Row: New York, 1970.

Science of Mechanics - A Critical and Historical Account of its Development, 3rd paperback edition, [reprint of the 1960 edition] , Open Court, La Salle, 1974.

Knowledge and Error - Sketches on the Psychology of Enquiry, edited by B. McGuinness, introduction by E.N. Hiebert, translators: T.J. McCormack & P. Foulkes from the 5th German edition (1926), (*Vienna Circle Collection,* vol. 3), D. Reidel: Dordrecht, 1976.

"Three Autobiographical Manuscripts by Ernst Mach", edited by J. Blackmore, *Annals of Science,* **35** (1978), 401-418. [For a Japanese translation of the longest manuscript see: Kazuo Shibuya, *Il Saggiatore - kagaku gijutsu to sono jidai,* Nr 17, 1988, pp. 43-59.]

The Analysis of Sensations and the Relation of the Physical to the Psychical, translated by C.M. Williams, 4th edition in English, [reprinted from the second English edition, 1914], Open Court: 1984.

Space and Geometry - In the Light of Physiological, Psychological and Physical Inquiry, third edition in English [based on articles appearing in The Monist from 1901-1903], Open Court: La Salle, 1984.

Principles of the Theory of Heat - Historically and Critically Elucidated, edited by B. McGuinness, introduction by M.J. Klein, translated by T.J. McCormack, P.E.B. Jourdain, and A.E. Heath from the second German edition, 1900, (*Vienna Circle Collection,* volume 17), D. Reidel: Dordrecht & Boston, 1986.

Popular Scientific Lectures, translation by T. J. McCormack, introduction by J. Bernstein, sixth English edition, [the text is reprinted from the third English edition, 1898], Open Court: La Salle, 1986.

B. Works in Other Languages:

Helmholtz, Bernard, Haeckel & Mach Chosakushu (A Collection of Writings by . . .), [The Mach selection is from *Knowledge and Error*], ed. S. Tanabe, Tokyo, 1961.

Kankaku no bunseki (Analysis of Sensations), edited by W. Hiromatsu and G. Sudo, introduction by W. Hiromatsu, Tokyo, 1963.

Die Mechanik in ihrer Entwicklung. Historisch-kritisch dargestellt, [reprinted from the 7th edition, 1912], Wissenschaftliche Buchgesellschaft: Darmstadt, 1963.

Arbeiten über Erscheinungen an fliegenden Projektilen, edited by J. Thiele, [reprinted from the *Sitzungsberichten der kaiserlichen Akademie der Wissenschaften* in Vienna (1887 to 1892)], Verlags-Vertriebe-Gesellschaft m.b.H., Hamburg-Altona, 1966.

Ninshiki no bunseki (Analysis of Knowledge), edited by W, Hiromatsu and N. Kato, introduction and afterward by W. Hiromatsu, [This book is a translation of three articles from Mach's *Populär-wissenschaftliche Vorlesungen* plus two articles from *Erkenntnis und Irrtum*] Tokyo, 1966.

Grundlinien der Lehre von den Bewegungsempfindungen, [reprint from 1st edition, 1875], Bonset: Zandoort, 1967.

Erkenntnis und Irrtum - Skizzen zur Psychologie der Forschung, [reprinted from the 5th edition, 1926], Wissenschaftliche Buchgesellschaft: Darmstadt, 1968.

Kultur und Mechanik, reprinted from the 1915 edition, with a foreword and extension of 1963 Mach-Bibliography by J. Thiele, plus *Die Geschichte und die Wurzel des Satzes von der Erhaltung der Arbeit,* reprinted from the second edition (1909), plus "Zur Geschichte des Arbeitbegriffes", reprinted from the *Sitzungsberichte der kaiserlichen Akademie der Wissenschaften,* Math.- Natur. Classe in Vienna, **68** (1873), Abt. 2, pp. 479-488, Bonset: Amsterdam, 1969.

Machrikigaku: rikigaku no hihanteki hattenshi (Mach's Mechanics - The Critical Historical Growth of Mechanics), ed. Y. Fushimi, Tokyo, 1969.

Kankaku no bunseki (Analysis of Sensations), edited by W. Hiromatsu & G. Sudo, introduction by W. Hiromatsu, Hosei University Press: Tokyo, 1971.

Ninshiki no bunseki (Analysis of Knowledge), edited by W, Hiromatsu and N. Kato, introduction and afterward by W. Hiromatsu, [This book is a translation of three articles from Mach's *Populär-wissenschaftliche Vorlesungen* plus two articles from *Erkenntnis und Irrtum*], Hosei University Press: Tokyo, 1971.

Rikigakushi: koten rikigaku no hatten to hihan (History of Mechanics - Growth and Criticism of Classical Mechanics), trans. H. Iwano, Tokyo, 1973.

L'analisi delle sensazioni e il raporto fra fisico e psichico, translated by L. Sosio, Feltrinelli: Milan, 1975.

Rikigakushi: koten rikigaku no hatten to hihan (History of Mechanics - Growth and Criticism of Classical Mechanics), translated by H. Iwano, Tokyo, 1976.

Jikan to kukan (Time and Space), [The book consists of five articles from Mach's *Erkenntnis und Irrtum* and one from his *Populär-wissenschaftliche Vorlesungen*], edited, introduction & afterword by Keiichi Noé, Hosei University Press: Tokyo, 1977.

La meccanica nel suo suillupo storico-critico, Boringhieri: Turin, 1977.

Vorträge über Psychophysik, ed. J. Thiele, reprinted from the *Öesterreichische Zeitschrift für Praktische Heilkunde,* vol. 9, 1863, privately published: Hamburg, 1978.

"Vorwort", in Pierre Duhem, *Ziel und Struktur physikalischer Theorien,* trans. F. Adler, ed. Lothar Schäfer, [first edition, 1908], Hamburg, 1978.

Netsugaku no shogenri (Various Principles of Heat Theory), translated by S. Takata, Tokai University Press: Tokyo, 1978.

Erkenntnis und Irrtum - Skizzen zur Psychologie der Forschung, [reprinted from the 5th edition, 1926], Wissenschaftliche Buchgesellschaft: Darmstadt, 1980.

Die Principien der Wärmelehre. Historisch-kritisch entwickelt, [reprinted from the 2nd edition, 1900], Minerva: Frankfurt, 1981.

Die Mechanik in ihrer Entwicklung historisch-kritisch dargestellt, [reprinted from the 9th edition, 1933], Wissenschaftliche Buchgesellschaft: Darmstadt, 1982.

Die Prinzipien der physikalischen Optik, [reprinted from the 1st edition, 1921], Minerva: Frankfurt, 1982.

Conosenza ed errore. Abbozi per una psicologia della recerca, introduction and translation by A. Gargani, Einaudi: Turin, 1982.

Einleitung in die Helmholtzsche Musiktheorie - Populär für Musiker dargestellt, [reprinted from 1866], Sändig Reprint: Schaan, 1985.

Optisch-akustische Versuche - Die spektrale und stroboskopische Untersuchung tönnender Körper, [reprinted from 1873] Sändig Reprint: Schaan, 1985.

Die Analyse der Empfindungen und das Verhältnis des Physischen zum Psychischen, with a foreword by G. Wolters [text reprinted from the 9th ed., 1922], Wissenschaftliche Buchgesellschaft: Darmstadt, 1985.

Erkenntnis und Irrtum - Skizzen zur Psychologie der Forschung, [reprinted from the 5th edition, 1926], Wissenschaftliche Buchgesellschaft: Darmstadt, 1987.

Populär-wissenschaftliche Vorlesungen, introduction by A. Hohenester and Foreword by F. Herneck, [text reprinted from the 5th edition, 1923], Böhlau Verlag: Vienna-Cologne-Graz, 1987.

Die Analyse der Empfindungen und das Verhältnis des Physischen zum Psychischen, with a foreword by G. Wolters, [text reprinted from the 9th edition, 1922], Wissenschaftliche Buchgesellschaft: Darmstadt, 1987.

La Mécanique, exposé historique et critique de son développment, translated by E. Bertrand, introduction by E. Picard, [reprinted from the 1904 edition], J. Gabay: Paris, 1987.

Die Mechanik in ihrer Entwicklung. Historisch-kritisch dargestellt, ed. E. Wahsner & H.-H. v. Borzeszkowski with appendices by L. Mach, E. Mach, M. Planck, and A. Einstein, plus 4 letters from Einstein to Mach, [text reprinted from the 7th edition, 1912], Akademie-Verlag: Berlin, 1988.

"Die Leitgedanken meiner naturwissenschaftlichen Erkenntnislehre und ihre Aufnahme durch die Zeitgenossen", [reprinted from *Scientia,* **7** (1910), 225-240], in *Die Mechanik in ihrer Entwicklung,* Akademie-Verlag: Berlin, 1988, pp. 653-669.

Die Mechanik in ihrer Entwicklung. Historisch-kritisch dargestellt, [reprinted from the 9th edition, 1933], Wissenschaftliche Buchgesellschaft: Darmstadt, 1988.

C. A Chronological List of Mach Book Publications:

Edition:	1	2	3	4	5	6	7	8	9			
Mechanik	1883	1888	1897	1901	1904	1908	1912	1921	1933	1963	1982	1988 (2)
English	1893	1902	1907*	1919	1942	1960	1964'	1974				
Japanese	1931	1969	1973	1976								
French	1904	1925	1987									
Italian	1908	1977										
Russian	1909											

* A supplement was published in 1915.
' Some chapters published.

Analyse...	1886*	1900	1902	1903	1906	1911	1918	1919	1922	1985
English	1897'	1914	1959	1984						
Russian	1904	1907	1908							
Italian	1903	1975								
Hungarian	1910	1922 (circa)								
Japanese	1963	1971								
Spanish	1925	1987								

* The original title was:
Beiträge zur Analyse der Empfindungen
' The original title was:
Contributions to the Analysis of Sensations

Erkenntnis...	1905 1906 1917 1920 1926 1968 1980 1987	
English	1906* 1960* 1976+ 1984*	
Japanese	1961' 1966' 1971' 1977'	* Three chapters published under the title: *Space and Geometry*
French	1908	
Italian	1982	' One chapter was published in 1961, two in 1966 & 1971, and five in 1977.
Russian	1909	
Spanish	1948	+ This is the first full edition in English of *Knowledge and Error,* Kluwer: Dordrecht.
Turkish	1935	

Populär...	1896 1897 1903' 1910' 1923' 1987
English	1895 1897 1898 1910 1943 1986
Russian	1901 1909 1920
Japanese	1971* 1977+
Italian	1900

' The German edition of 1903 (but not foreign ones) was increased to 19 chapters, the edition of 1910 to 26 chapters, and that of 1923 to 33 chapters.
* Three chapters were published. + One chapter published.

Erhaltung der Arbeit	1872 1909 1969
English	1911

"Die Leitgedanken meiner naturwissenschaftlichen Erkenntnislehre..."	1910 (2)	1919 1988
English	1970	1992 (abridged)

Wärmelehre	1896 1900 1919 1923 1981
English	1900' 1902* 1903* 1986+
Japanese	1978

Die Optik	1921 1982
English	1926 1953 1956

' Five chapters published. * One-and-a-half chapters published in 1902 and again in 1903.
+ This is the first full edition in English of *Principles of Heat Theory,* Kluwer: Dordrecht.

III

BOOKS ON ERNST MACH

A. Anthologies Devoted to Mach:

(Article titles are listed under the author's name in sections IVA & IVB.)

(Merzkirch, Wolfgang, ed.), *Symposium aus Anlass des 50. Todestages von Ernst Mach veranstaltet am 11/12 Marz 1966 vom Ernst-Mach-Institut Freiburg i. Br.,* Ernst-Mach-Institut: Freiburg, 1967-68. Contributors: G. v. Békésy, M. Bunge, F. v, Hayek, E. Hiebert, H. Jochim, F. Herneck, D. Herrmann, H. Hönl, R. Jung, R. Kutterer, E. Lesky, J. Mayerhöfer, W. Merzkirch, J. Pachner, R. Ratliff, M. Strauss, J. Thiele.

Hintikka, Jaakko, (ed.), "A Symposium on Ernst Mach", *Synthese - An International Journal for Epistemology, Methodology and Philosophy of Science,* **18** (1968), pp. 132-301. Contributors: S.G. Brush, M. Capek, R.S. Cohen, A. Koslow, M. Strauss, J. Thiele, W. Yourgrau & A. v. der Merve.

Cohen, Robert S. and Raymond J. Seeger, (eds.) *Ernst Mach - Physicist and Philosopher,* (*Boston Studies in the Philosophy of Science,* vol. 6), D. Reidel Publishing Company. Dordrecht-Holland, 1970. Contributors: P. G. Bergmann, O. Blüh, R.S. Cohen, P. Frank, H. Goenner, E.N. Hiebert, G.J. Holton, K. Menger, W.F. Merzkirch, R. v. Mises, F. Ratliff, R.J. Seeger.

Haller, Rudolf and Friedrich Stadler, (eds.), *Ernst Mach - Werk und Wirkung,* Verlag Hölder-Pichler-Tempsky: Vienna, 1988. Contributors: J.T. Blackmore, J. Dvorak. P.K. Feyerabend, R. Haller, A. Hohenester, W. Leinfellner, P. Mahr, H. Motz, M. Sommer, W. Swoboda, F. Stadler, G. Wolters.

Hoffmann Dieter and Hubert Laitko, (eds.), *Ernst Mach - Studien und Dokumente zu Leben und Werk,* Deutscher Verlag der Wissenschaften: Berlin, 1991. Contributors: M. Diersch, H. Hecht, F. Herneck, D. Hoffmann, A. Laab, H. Laitko, W. Manthei, E. Schmuzer, G. Tembrock.

B. Books on Ernst Mach in English:

Barbour, Julian B., *Absolute or Relative Motion? - A Study from a Machian Point of View of the Discovery and the Structure of Dynamical Theories,* Vol. I: *The Discovery of Dynamics,* Cambridge University Press: Cambridge/New York, 1989.

Blackmore, John, *Ernst Mach - His Life, Work, and Influence,* 2 vols., doct. diss.; University of California at Los Angeles, 1970. [available in xerox and microfilm versions from University Microfilms, no. 71-589, Ann Arbor, Michigan].

Ernst Mach - His Life, Work, and Influence, [This is an abridged version of the doctoral dissertation. Due to a printing error, the title page reads "Ernst Mach - His Work, Life, and Influence"], University of California Press: Berkeley, 1972.

Bradley, J., *Mach's Philosophy of Science,* [translated into German, 1974], The Athlone Press of the University of London: London, 1971.

Broda, Engelbert, *The Intellectual Quadrangle: Mach - Boltzmann - Planck -Einstein,* Cern Publication 81-10: Geneva, July 22, 1981, 20 pages.

Ludwig Boltzmann: Man - Physicist - Philosopher, trans. L. Gay & E. Broda, [original in German 1955 and later in 1986], Ox Bow Press: Woodbridge Conn., 1983.

Carnap, Rudolf, *The Logical Structure of the World and Pseudoproblems in Philosophy,* translated by Rolf. A. George from the German edition of 1928, University of California Press; Berkeley and Los Angeles, 1969.

Cohen, Robert S. and Raymond J. Seeger (eds.), *Ernst Mach - Physicist and Philosopher,* (*Boston Studies in the Philosophy of Science,* vol. 6), D, Reidel: Dordrecht, 1970.

D'Espagnat, Bernard, *Conceptual Foundations of Quantum Mechanics,* Benjamin: Menlo Park, Calif., 1971, ch. 17.

Reality and the Physicist, trans. J.C. Whitehouse & author, Cambridge University Press: Cambridge, 1990, especially Part I.

Feuer, Lewis S., *Einstein and the Generations of Science,* Basic Books: New York, 1974.

Fine, Arthur, *The Shaky Game: Einstein, Realism, and the Quantum Theory,* University of Chicago Press: Chicago and London, 1986, ch. 7.

Hiebert, Erwin N., *The Conception of Thermodynamics in the Scientific Thought of Mach and Planck,* Ernst-Mach-Institut Bericht Nr. 5, [106 pages], Ernst-Mach-Institut: Freiburg im Breisgau, 1968.

Holton, Gerald, *Thematic Origins of Scientific Thought,* [original edition, 1973], revised edition, [includes an article on Mach as well as remarks about G. Wolters's forgery thesis concerning the preface of Mach's *Optik*], Harvard University Press: Cambridge, 1988.

Johnston, William M., *The Austrian Mind - An Intellectual and Social History 1848-1938,* University of California Press: Berkeley, 1972.

Lenin, V.I., *Materialism and Empiriocriticism - Critical Comments on a Reactionary Philosophy,* [trans. from 1908 original], International Publishers: New York, 1927.

Murad, Hazim B., *Mach Revisited: A Reinterpretation of Mach's Philosophy of Science, and of his Opposition to Atomism,* University of Pittsburgh diss., 1990, (Univ. Microfilms, Ann Arbor, Michigan, no. 91-20161.)

Musil, Robert, *On Mach's Theories,* [trans. by K. Mulligan from Musil's doct. diss.], intro. by G.H. v. Wright, Catholic U. of America Press & Philosophia: Munich, 1982.

Pinkava, J. (ed.), *The Correspondence of the Czech Chemist Frantisek Wald with W. Ostwald, E. Mach, P. Duhem, J.W. Gibbs, and Other Scientists of That Time,* Prag, 1987.

Ratliff, Floyd, *Mach Bands - Quantitative Studies on Neural Networks in the Retina,* [with translations of six early Mach articles], Holden-Day: San Francicsco, 1965.

Sachs, M., *Einstein versus Bohr,* foreword by J. Agassi, La Salle, 1988, ch. 2 and 9.

Sewell, William C., *Einstein, Mach and the General Theory of Relativity,* doct. diss., Case Western Reserve University: Cleveland Ohio, 1975.

Shapere, Dudley, *Galileo - A Philosophical Study,* [many remarks on Mach's treatment of Galileo], University of Chicago Press: Chicago & London, 1974.

Swoboda, Wolfram, *The Thought and Work of the Young Ernst Mach and the Antecedents to his Philosophy,* Univ. of Pittsburgh doct. diss., University Microfilms #74-19, 440, Xerox Corporation: Ann Arbor, 1974.

Zahar, Elie, *Einstein's Revolution - A Study in Heuristik,* Open Court: La Salle Ill., 1989.

C. Books on Ernst Mach in Other Languages:

Adler, Friedrich, *Ernst Mach e il materialismo,* edited and translated by Antimo Negri, [from *Ernst Machs Überwindung des mechanischen Materialismus,* 1918], Armando Editore: Rome, 1978.

Blackmore, John, & Klaus Hentschel (eds.), *Ernst Mach als Aussenseiter - Machs Briefwechsel über Philosophie und Relativitätstheorie mit Personlichkeiten seiner Zeit - Auszug aus dem letzten Notizbuch (Faksimile) von Ernst Mach,* (Philosophica 3, Hrsg, F. Wallner), Wilhelm Braumüller: Vienna, 1985.

Blühdorn, Jürgen and Joachim Ritter (eds.), *Positivismus im 19, Jahrhundert: Beiträge zu seiner geschichtlichen und systematischen Bedeutung,* [with a bibliography of Ernst Mach pp. 183-201], (*Studien zur Philosophie und Literatur des 19. Jahrhunderts,* vol. 16), Klostermann: Frankfurt, 1971.

Boltzmann, Ludwig, *Ludwig Boltzmann 'Principien der Naturfilosofi' - Lectures on Natural Philosophy 1903-1906,* ed. Ilse M. Fasol-Boltzmann, Springer: Berlin, 1990.

Boniolo, Giovanni, *Mach e Einstein - Spazio e massa gravitante,* preface by Bruno Bertototi, Armando Editore: Rome, 1988.

Bradley, J. *Machs Philosophie der Naturwissenschaft,* [original in English, 1971], Stuttgart, 1974.

Brentano, Franz, *Über Ernst Machs 'Erkenntnis und Irrtum',* edited by and with an introduction from R. Chisholm & J. Marek, (*Studien zur Österreichischen Philosophie,* vol. 6), [This book also includes the extant Mach-Brentano correspondence.] Rodopi: Amsterdam, 1988.

Broda, Engelbert, *Ludwig Boltzmann: Mensch - Physiker -Philosoph,* first edition, Berlin, 1955, translated into English, 1983, Franz Deuticke: Vienna, 1986.

D'Elia, Alfonsina, *Ernst Mach,* La Nuova Italia: Florence, 1971.

Diersch, Manfred, *Empiriokritizismus und Impressionismus. Über Beziehungen zwischen Philosophie, Ästhetik und Literatur um 1900 in Wien,* [largely on Mach's influence on H. Bahr and A. Schnitzler], Rutten & Loening: Berlin, 1973.

Feuer, Lewis S., *Einstein to kagaku kakumei* (Einstein and the Scientific Revolution), Hosei University Press: Tokyo, 1991.

Fistetti, Francesco, *Lenin i il machismo: da Materialismo ed empiriocriticismo al quaderni filosofici*, Feltrinelli: Milan, 1977.

Frank, Philipp, *Einstein - Sein Leben und seine Zeit*, [reprinted from first German edition, 1949, which has material not in American edition of 1947, including a foreword by A. Einstein. Both contain data on Mach.], Braunschweig-Wiesbaden, 1979.

Haller, Rudolf & Friedrich Stadler (eds.) *Ernst Mach - Werk und Wirkung*, Verlag Hölder-Pichler-Tempsky: Vienna, 1988.

Hentschel, Klaus, *Interpretationen und Fehlinterpretationen der speziellen und der allgemeinen Relativitätstheorie durch Zeitgenossen Albert Einsteins*, (Science Networks: Historical Studies, 6), Birkhäuser: Basel, 1990.

(See John Blackmore for co-authored book.)

Herneck, F., *Die heilige Neugier*, [includes three earlier articles on Ernst Mach: "Wiener Physik vor Hundert Jahren", pp. 143-152; "Ernst Mach als Mensch", pp. 152-165; "Zum Briefwechsel Albert Einsteins mit Ernst Mach", pp. 165-175], Berlin, 1983.

Hirche, J.A., *Das Ich des Empiriokritizismus, eine erkenntnistheoretische Studie*, Annaberg, 1913.

Hiromatsu, Wataru, *Sekai no kyodoshukanteki sonsaikozo* (The Intersubjective Ontic Structure of the World), Keiso Shobo: Tokyo, 1972. See pages 3-20.

Kototeki sekaiken e no zensho (Outposts of the Relational Worldview), Keiso Shobo: Tokyo, 1975, See pages 51-75.

Hoffmann, Dieter, *Studien zu Leben und Werk von Ernst Mach*, doctoral dissertation Humboldt University, Berlin, 1989.

Hoffmann, Dieter & Hubert Laitko (eds.), *Ernst Mach - Studien und Dokumente zu Leben und Werk*, Deutscher Verlag der Wissenschaften: Berlin, 1991.

Holton, Gerald, *Thematische Analyse der Wissenschaft: Die Physik Einsteins und seiner Zeit*, trans. H. Huber, Frankfurt, 1981.

Ishiwara, Jun, *Shizen kagaku gairon* (Outline of Natural Science), Tokyo, 1929.

Johnston, William M., *Österreicische Kultur- und Geistesgeschichte. Gesellschaft und Ideen im Donauraum 1848 bis 1938*, [original in English, 1972] , Wien, 1974.

Kaizik, Jürgen, *Versuch einer Begegnung - Mach gegen Boltzmann*, Franz Deuticke: Vienna, 1986.

Kiss, Endre, *Der Tot der K.U.K. Weltordnung in Wien*, Böhlau: Wien-Graz-Köln, 1986.

Kleinpeter, Hans, *Der Phänomenalismus*, Leipzig, 1913.

Kuwaki, Ayao, *Butsurigaku to ninshiki* (Physics and Cognition), Tokyo, 1922.

Einstein-den, Tokyo, 1934. [a biography of Einstein with some references to Mach]. Also the same book but revised by T. Kuwaki and S. Nsio and republished in 1979.

Lübbe, Hermann, *Bewusstsein in Geschichten: Studien zur Phänomenologie der Subjektivität: Mach, Husserl, Schlapp,Wittgenstein*, Freiburg, 1972; [also in Japanese: *Rikishi ni okeru igi* (Meaning in History), trans. H. Kawashima, Kyoto, 1988.]

Merzkirch, Wolfgang. (ed.), *Symposium aus Anlass des 50. Todestages von Ernst Mach veranstaltet am 11/12 Marz 1966 vom Ernst-Mach-Institut Freiburg i Br.*, intro. F. Kerkhof, Ernst-Mach-Institut: Freiburg im Breisgau, 1967/68.

Musil, Robert, *Beitrag zur Beurteilung der Lehren Machs und Studien zur Technik und Psychotechnik*, [reprinted from doct. diss., Berlin, 1908], Reinbek bei Hamburg, 1980; and in French: *Pour une evaluation des doctrines de Mach*, ed. P.-L. Assoun, trans. from German by M.-F. Demet, Presse universitaires de France: Paris, 1985.

Noé, Keiichi, *Shinrigaku no tetsugaku* (Philosophy of Psychology), vol. 2, Tokyo, 1988.

Petronijevic, Branislaw, *Principien der Metaphysik*, vol, II, Heidelberg, 1912.

Onovi teirije saznanja (Foundations of Theory of Knowledge), Beograd, 1923.

Schulzki, E., *Der Mensch als Elementenkomplex und als denkökonomische Einheit - Zur Anthropologie Ernst Machs*, phil. diss., Münster, 1979.

Sommer, Manfred, *Husserl und der frühe Positivismus*, Frankfurt, 1985.

Evidenz im Augenblick - Eine Phänomenologie der reinen Empfindung, Suhrkamp: Frankfurt, 1987.

Stadler, Friedrich, *Positivismus als Lebensform. Zur Wirkungsgeschichte von Ernst Mach in Österreich-Ungarn 1895-1918,* phil. diss., Salzburg, 1981.

Vom Positivismus zur 'wissenschaftlichen Weltauffassung' - Am Beispiel der Wirk-Wirkungsgeschichte von Ernst Mach in Österreich von 1895 bis 1934, Löcker Verlag: Vienna and Munich, 1982.

(See Rudolf Haller for co-edited book.)

Stegmüller, Wolfgang, *Der Phänomenalismus und seine Schwierigkeiten,* Munich, 1958.

Taketani, Mitsuo, *Benshoho no shomondai* (The Problems of Dialectic), Keiso Shobo: Tokyo, 1968.

Thiele, Joachim, *Die Bedeutung Ernst Machs für die Wende von der klassischen zur modernen Physik,* doct. diss., Hamburg, 1959.

Wissenschaftliche Kommunikation - Die Korrespondenz Ernst Machs, A. Henn Verlag: Kastellaun, 1978.

Thiele, Rudolf, *Charakteristik von Mach's Erkenntnislehre,* [reprinted from 1914 edition], Olms: Hildesheim & New York, 1981.

Tretera, Ivo, *E. Mach na Prazske Universite - E. Mach at the University of Prague,* [a long booklet subtitled "International Conference on Ernst Mach in the Development of Physics" which consists largely of documents by and on Mach], Prague 1988, thirty-five pages.

Wolters, Gereon, *Mach I, Mach II, Einstein und die Relativitätstheorie - Eine Fälschung und ihre Folgen,* Walter de Gruyter: Berlin & New York, 1987.

Zagoda, Claudio, *Prinzipien in Machs Methodologie,* doct. diss., Graz, 1989, 230 pages.

IV

ARTICLES ON ERNST MACH

A. Articles on Ernst Mach in English:

Agassi, Joseph, "Mach on the Logic of Enquiry - Taste Maker Philosopher of Science", in *The Gentle Art of Philosophical Polemics - Selected reviews and Comments,* Open Court: La Salle, 1988, pp. 21-32.

Alpert, A.M., "Knowledge and Cosmos in the Philosophies of Mach and Ch'eng I: An Analysis of the Cognitive Structures of Empiricism in Two Cultures", *Philosophy East West,* **30** (1980), 163-179.

Arens, Katherine, "Mach's 'Psychology of Investigation'", *Journal of the History of the Behavioral Sciences,* **21** (1985), 151-168.

Barbour, Julian B., "Relative Distance Machian Theories", *Nature,* **249** (1974), 328-329.

Barbour, Julian B., & B. Bertotti, "Gravity and Inertia in a Machian Framework", *Nuovo Cimento,* **38B** (1977), 1-27.

"Mach's Principle and the Structure of Dynamical Theories", *Proceedings of the Royal Society of London,* **A382** (1982), 295-306.

Bergman, Peter G., "Ernst Mach and Contemporary Physics", in *Ernst Mach - Physicist and Philosopher,* edited by R. Cohen and R. Seeger, (*Boston Studies in the Philosophy of Science*, vol. 6), D. Reidel: Dordrecht, 1970, pp. 69-78.

Bernstein, Jeremy, "Ernst Mach and the Quarks", *Amer. Scholar,* **53** (1983-84), 7-14.

"Introduction", to Ernst Mach, *Popular Scientific Lectures,* trans. T. J. McCormack, Open Court: La Salle, 1986.

Bertotti, B. (See J.B. Barbour as co-author for two papers.)

Blackmore, John, (ed.), "Three Autobiographical Manuscripts by Ernst Mach", *Annals of Science,* **35** (1978), 401-418.

"Boltzmann's Concessions to Mach's Philosophy of Science" in *Ludwig Boltzmann Ausgewählte Abhandlungen,* edited by R. Sexl & J. Blackmore, (*Ludwig Boltzmann Gesamtausgabe,* vol. 8), Friedrich Vieweg & Sohn: Braunschweig & Wiesbaden, 1982, pp. 155-190.

"A Historical Note on Ernst Mach", *British Journal for the Philosophy of Science,* **36** (1985), 299-305.

"Ludwig Boltzmann as a Major Source of 20th Century Philosophy", *Berichte des 11. Internationalen Wittgenstein-Symposiums* (1986), eds. P. Weingartner & G.Schurz, Hölder-Pichler-Tempsky: Vienna, 1987, pp. 105-112.

"Mach Competes with Planck for Einstein's Favor", *Historia Scientiarum,* **35** (1988), 45-89.

"Evidence that Mach Rejected Einstein's Theory of Relativity", *Berichte des 13. internationalen Wittgenstein Symposiums,* (1988), Hölder-Pichler-Tempsky: Vienna, 1989, pp. 87-90.

"Ernst Mach Leaves 'The Church of Physics'", *British Journal for the Philosophy of Science,* **40** (1989), 519-540.

Blüh, Otto, "Ernst Mach - His Life as a Teacher and Thinker", in *Ernst Mach - Physicist and Philosopher,* eds. R. Cohen and R. Seeger, Reidel: Dordrecht, 1970, pp. 1-22.

Blüh, Otto & Wolfgang F. Merzkirch, "Ernst Mach Bibliography", in *Ernst Mach - Physicist and Philosopher,* edited by. R.S. Cohen & R.J. Seeger, (*Boston Studies in the Philosophy of Science,* vol, 6) D. Reidel: Dordrecht, 1970, pp. 274-290.

Boltzmann, Ludwig, "On the Indispensability of Atomism in Natural Science" and "On the Question of the Objective Existence of Processes in Inanimate nature" (both 1897), in *Theoretical Physics and Philosophical Problems,* edited by Brian McGuinness, translated by Paul Foulkes, (*Vienna Circle Collection,* vol. 5), D. Reidel: Dordrecht, 1974, pp. 41-53 & 57-76.

Brush, Stephen, "Mach and Atomism", *Synthese,* **18** (1968), 192-215.

"Mach", in *The kind of Motion We Call Heat - A History of the Kinetic Theory of Gases in the 19th Century, Book I: Physics and the Atomists,* North Holland Publishing Company: Amsterdam, New York, Oxford, 1976, pp. 274-299.

"Ludwig Boltzmann and the Foundations of Natural Science", in *Ludwig Boltzmann Principien der Naturfilosofi' - Lectures on Natural Philosophy 1903-1906,* ed. Ilse M. Fasol-Boltzmann, Springer Verlag: Berlin, 1990, pp. 43-61.

Bunge, Mario, "Mach's Critique of Newtonian Mechanics", *American Journal of Physics,* **34** (1966), 585-596; and "On Mach's Non-Concept of Mass (L)", **36** (1968), 167.

Capec, Milic, "Mach's Biological Theory of Knowledge", *Synthese,* **18** (1968), 171-191.

Cohen, I. Bernard, "History and the Philosopher of Science", [partly on Mach's treatment of Galileo], in *The Structure of Scientific Theories,* edited by F. Suppe, 2nd edition, University of Illinois Press: Urbana and Chicago, 1977, pp. 308-373.

Cohen, Robert S., "Ernst Mach: Physics, Perception and the Philosophy of Science", *Synthese,* **18** (1968), 132-170; also in *Ernst Mach - Physicist and Philosopher,* edited by R.S. Cohen & R.J. Seeger, D. Reidel: Dordrecht, 1970, pp. 126-164.

Dilworth, Craig, "Empiricism vs. Realism: High Points in the Debate during the Past 150 Years", *Studies in the History and Philosophy of Science,* **21** (1990), 431-462.

Dipert, Randall R., "Peirce on Mach and Absolute Space", *Transactions of the C.S. Peirce Society,* **9** (1973), 79-94.

Disalle, Robert, "Critical Notice: Gereon Wolters Mach I, Mach II . . .", [a review article of Gereon Wolters's 1987 book on Mach] in *Philosophy of Science,* **57** (1990), 712-723.

Feyerabend, Paul K. "Zahar on Mach, Einstein, and Modern Science", *British Journal for the Philosophy of Science,* **31** (1980), 273-282.

"Mach's Theory of Research and Its Relation to Einstein", in *Studies in History and Philosophy of Science,* **15** (1984), 1-22. This article was reprinted with a few changes in *Farewell to Reason,* Verso: London & New York, 1987, pp. 192-218.

Flis, Andrzej, "Cracow Philosophy of the Beginning of the Twentieth Century and the Rise of Malinowski's Scientific Ideas", [largely about Mach's influence on Malinowski], in *Malinowski Between Two Worlds,* eds. R. Ellen, E. Gellner, G. Kubica, J. Mucha, Cambridge University Press: Cambridge, 1988, pp. 105-127.

Frank, Philipp, "The Importance of Ernst Mach's Philosophy of Science for Our Times", in *Ernst Mach - Physicist and Philosopher,* eds. R.S. Cohen and R.J. Seeger, [first appeared in *Die Naturwissenschaften,* **5** (1917), 65-72], D. Reidel: Dordrecht, 1970, pp. 219-234; see also "Ernst Mach and the Unity of Science", pp. 235-244.

Gampier, P., "Mach and Freud - A Comparison", *Zeitgeschichte,* **17** (1990), 291-310.

Gargani, Aldo, "Remarks on Mach's Theoretical Physics and Epistemology", in *Berichte des 13. Internationalen Wittgenstein-Symposiums* (1988), Eds. P. Weingartner and G. Schurz, Hölder-Pichler-Tempsky: Vienna, 1989, pp. 95-97.

"The Good Austrian - Ernst Mach, Scientist and Philosopher", *Traditionen und Perspektiven der analytischen Philosophie - Festschrift für Rudolf Haller,* eds. W. Gombocz, et al, Hölder-Pichler-Tempsky: Vienna, 1989, pp. 135-148.

Gellner, Ernst, " Zeno of Cracow...", *Malinowski Between Two Worlds,* [largely on Mach's impact on Bronislaw Malinowski], eds. R. Ellen, E. Gellner, G. Kubica, J. Mucha, Cambridge University Press: Cambridge, 1988, pp. 164-194.

Goe, George, "On a Criticism by Mach on Galileo", *Scientia,* **116** (1981), 93-99.

Goenner, Hubert F., "Mach's Principle and Einstein's Theory of Gravitation", in *Ernst Mach - Physicist and Philosopher,* edited by R.S. Cohen & R.J. Seeger, (*Boston Studies in the Philosophy of Science*, vol. 6, Reidel: Dordrecht, 1970, pp. 200-215.

Grattan-Guinness, Ivor, [a review article of J. Blackmore's 1972 book on Mach], *British Journal for the History of Science,* **7** (1974), 75-79.

Haller, Rudolf, "New Light on the Vienna Circle", *Monist,* **65** (1982), 25-37.

"Poetic Imagination and Economy: Ernst Mach as Theorist of Science", in *Scientific Philosophy Today: Essays in Honor of Mario Bunge,* edited by Joseph Agassi and Robert S. Cohen, Reidel: Dordrecht and Boston, 1982, pp. 71-84.

Hamilton, A., "Ernst Mach and the Elimination of Subjectivity", *Ratio,* **3** (1990), 117-135.

Heilbron, J.L., "Defending the World Picture Against Mach", in *The Dilemmas of an Upright Man - Max Planck as Spokesman for German Science,* University of California Press: Berkeley, Los Angeles, London, 1986, pp. 47-60.

Heller, Michael, "The Influence of Mach's Thought on Contemporary Relativistic Physics" *Organon,* **11** (1975), 271-283.

"Between Newton and Einstein: Mach's Reform of the Newtonian Mechanics", in *Newton and the New Direction in Science,* edited by G.V. Coyne, M. Heller, and J. Zycinski, Specola Vaticana: Vatican City, 1988, pp. 155-173.

Hentschel, Klaus, "On Feyerabend's Version of 'Mach's Theory of Research and Its Relation to Einstein'", *Studies in History & Phil. of Science,* **16** (1985), 387-394.

Hiebert, Erwin N., "The Conception of Thermodynamics in the Scientific Thought of Mach and Planck", *Symposium aus Anlass des 50. Todestages von Ernst Mach,* ed. W. Merzkirch, [For the much longer monograph with the same title, see the book section], Ernst-Mach-Institut: Freiburg, 1967-68, pp. 186-187.

"The Genesis of Mach's Early Views on Atomism", *Ernst Mach - Physicist and Philosopher,* eds. R.S. Cohen & R.J. Seeger, D. Reidel: Dordrecht, 1970, pp. 79-106.

"Mach's Philosophical Use of the History of Science" in *Historical and Philosophical Perspectives of Science,* ed. R. Stuewer, (*Minnesota Studies in the Philosophy of* Science, vol. 5), University of Minnesota Press: Minneapolis, 1970, p. 184-213.

"The Energetics Controversy and the new Thermodynamics" in D.H.D. Roller, (ed.), *Perspectives in the History of Science and Technology,* University of Oklahoma Press: Norman, 1971, pp. 67-86.

"Ernst Mach", *Dictionary of Scientific Biography,* ed. C.C. Gillispie, vol 8., Charles Scribner's Sons: New York, 1973, pp. 595-607.

"An Appraisal of the Work of Ernst Mach: Scientist-Historian-Philosopher", in *Motion and Time Space and Matter*, eds. P.K. Machamer and R.G. Turnbull, Ohio State University Press: Columbus, 1976, pp. 360-389.

"The Influence of Mach's Thought on Science", *Philosophia Naturalis*, **21** (1984), 598-615.

Hintikka, Jaako (ed.), "A Symposium on Ernst Mach" in *Synthese*, vol. **18** (1968), pp. 132-301.

Hirosige, Tetu, "The Ether Problem, The Mechanistic World View, and the Origins of the Theory of Relativity" in *Historical Studies in the Physical Sciences*, **7** (1976), 3-82. Also included in *Einstein Studies*, ed. S. Nishio, Tokyo, 1977.

Holton, Gerald, "Influences on Einstein's Early Work in Relativity Theory", *American Scholar*, **37** (1967-68), 59-79.

"Mach, Einstein, and the Search for Reality", *Daedalus*, **97** (1968), 636-673. Later versions of this article can be found in *Ernst Mach - Physicist and Philosopher*, eds. R. Cohen & R. Seeger, D. Reidel: Dordrecht, 1970, pp. 165-199, and Gerald Holton, *Thematic Origins of Scientific Thought - Kepler to Einstein*, Harvard University Press: Cambridge Mass., 1972, pp. 219-259.

"More on Mach and Einstein": Investigation of an Important Phase in the Rise of Logical Positivism/Empiricism", [an essay review of books by G. Wolters (1987) and J. Blackmore/ K. Hentschel (1985)], *Methodology and Science*, **22.2** (1989), 67-81.

"Ernst Mach's Heritage in America", *Isis*, **83** (1992), forthcoming.

Howard, Don, "Einstein and Duhem", *Synthese*, **83** (1990), 363-384.

Itagaki, Ryoichi, "Why Did Mach Reject Einstein's Theory of Relativity?", *Historia Scientiarum*, 1982, Nr. 22, pp. 81-95.

Janko, Jan., "Mach's Critique of Knowledge and the Physiology of the Senses", *Filosoficky Casopis*, **38** (1990), 118-131.

Jungnickel, Christa and Russell McCormmach, "Training 'Scientific' Physicists in Austria", *The Intellectual Mastery of Nature - Theoretical Physics from Ohm to Einstein*, [The above section is partly on Mach as a *Privatdozent* in Vienna], Vol. I, The University of Chicago Press: Chicago & London, 1986, pp. 202-209.

Kartsev, V., "The Mach-Boltzmann Controversy", in *Thermodynamics and Physical Reality*, (Pisa Conference on the History and Philosophy of Science, 1978), *Proceedings*, Vol. 2: *Probabilistic Thinking,Thermodynamics, and the Interaction of the History and Philosophy of Science*, Reidel: Dordrecht, 1981.

Katsumori, Makoto, "Einstein's Philosophical Turn and the Theory of Relativity", *Berichte des 13. Internationalen Wittgenstein-Symposiums* (1988), eds. P. Weingartner and G. Schurz, Hölder-Pichler-Tempsky: Vienna, 1989, pp. 98-101.

Kiefer, G.W., "Alfred Stieglitz and the 'Steerage' - An Empirio-Critical Correlation", *Word & Image*, **7** (1991), 58-70.

Koslow, A., "Mach's Concept of Mass: Program and Definition", *Synthese*, **18** (1968), 216-233.

Laudan, Larry, "The Methodological Foundation of Mach's Anti-Atomism and Their Historical Roots", in *Motion and Time Space and Matter*, eds. P.K. Machamer and R.G. Turnbull, Ohio State University Press: Columbus, 1976, pp. 390-417.

"Ernst Mach's Opposition to Atomism", *Science and Hypothesis - Historical Essays in Scientific Methodology*, ed. L. Laudan, D. Reidel: Dordrecht-Boston-London, 1981, pp. 202-225.

Leegwater, A., "The Development of Wilhelm Ostwald's Chemical Energetics", *Centaurus*, **29** (1986), 314-337.

Lezhneva, O.A., "Boltzmann, Mach, and Russian Physicists of the Late 19th Century", in *Thermodynamics and Physical reality*, (Pisa Conference on the History and Philosophy of Science, 1978), *Proceedings*. Vol. 2: *Probabilistic Thinking, Thermodynamics, and the Interaction of the History and Philosophy of Science*, Reidel: Dordrecht, 1981.

Loparic, Zeljko, "Problem-solving and Theory Structure in Mach", *Studies in the History and Philosophy of Science,* **15** (1984), 23-49.

Lucas, J., "The Influence of Ernst Mach on Modern Theories of Cosmology", *Journal of the British Astronomical Association,* **91** (Feb., 1981), no. 2, pp. 158-166.

Martin, Ronald, "Duhem & the Origin of Statics - Ramifications of the Crisis of 1903-04", *Synthese,* **83** (1990), pp. 337-355.

Mathews, M., "Ernst Mach and Contemporary Science-Education Reforms", *International Journal of Science Education,* **12** (1990), 317-325.

McCrea, W.H., "Doubts about Mach's Principle", *Nature,* **230** (1971), 95-97.

McGinn, Colin, "Mach and Husserl", *Journal of the British Society for Phenomenology,* **3** (1972), 146-157.

McGuinness, Brian F., "Ernst Mach and His Influence on Austrian Thinkers", in *Traditionen und Perspektiven der analytischen Philosophie - Festschrift für Rudolf Haller,* eds. W.L. Gombocz, H. Rutte, W. Sauer, Verlag Hölder-Pichler-Tempsky: Vienna, 1989, pp. 149-156.

McNulty, Michael, "James, Mach, and the Problem of Other Minds", *Transactions of the C.S. Peirce Society,* **18** (1982), 244-254.

Menger, Karl, "Introduction", to Ernst Mach, *Science of Mechanics,* 6th edition, Open Court: La Salle, 1960, pp. v-xxi.

"Mathematical Implications of Mach's Ideas: Positivistic Geometry, The Clarification of Functional Connections", *Ernst Mach - Physicist and Philosopher,* eds. R.S. Cohen & R.J. Seeger, D. Reidel: Dordrecht, 1970, pp. 107-125.

Merzkirch, Wolfgang, "Mach's Contribution to the development of Gas Dynamics", in *Ernst Mach - Physicist and Philosopher,* eds. R.S. Cohen & R.J. Seeger, D. Reidel: Dordrecht, 1970, pp. 42-59.

Mises, Richard von, "Ernst Mach and the Empiricist Conception of Science", *Ernst Mach - Physicist and Philosopher,* eds. R.S. Cohen & R.J. Seeger, (*Boston Studies in the Philosophy of Science,* vol. 6), and as "Ernst Mach and the Scientific Conception of the World", *Unified Science,* ed. B. McGuinness, (*Vienna Circle Collection,* vol. 19) [both translated from Heft 7: *Ernst Mach und die empirische Wissenschaftsauffassung,* 1938, from the *Verein Ernst Mach* series], and both published by Kluwer (Reidel): Dordrecht and Boston, the first in 1970, pp. 245-270, and the second in 1987, pp.166-190.

Mulligan, Kevin and Barry Smith, "Mach and Ehrenfels: The Foundations of Gestalt Theory", in *The Foundations of Gestalt Theory,* ed. B. Smith, Munich, 1986.

Niiniluoto, Ilkka, "Ernst Mach as a Philosopher of Science", in *Essays in Honor of Kalvero Vihtori Laurikainen...* (*Annales Academise Scientiarum Fennicae,* series A, VI: *Physica,* 431), *Suomalainen Tiedeskatemia,* 1986, pp. 190-203.

Norton, John, "Einstein, Mach, and the Origins of the General Theory of Relativity", *Proceedings of the AAHPSSS,* **12.2** (1980/81), 277-288.

Nyhof. J., "Philosophical Objections to the Kinetic-Theory", *British Journal for the Philosophy of Science,* **39** (1988), 81-109.

Ono, Y.A., "Einstein's Speech at Kyoto University, December 14, 1922", [Einstein's Kyoto speech was originally published in the magazine *Kaizo* in 1923. See Jun Ishiwara for this source.], *NTM. Schriftenreihe für Geschichte der Naturwissenschaften, Technik, und Medizin,* **20** (1983), 25-28.

Planck, Max, "The Unity of the Physical Worldpicture", in *Physical Reality: Philosphical Essays on 20th Century Physics,* ed. Stephen Toulmin, Harper & Row: New York, 1970.

"On Mach's Theory of Physical Knowledge", in *Physical Reality: Philosophical Essays on 20th Century Physics,* ed. Stephen Toulmin, Harper & Row: New York, 1970 .

Popper, Karl, "A Note on Berkeley as Precursor of Mach and Einstein", [reprinted from *British Journal for the Philosophy of Science,* **4** (1953), 26-36], in *Conjectures and Refutations,* Harper and Row: New York, 1968, [numerous editions since the middle 1960's], pp. 166-174.

Putnam, Hilary, "Quinean Positivism", *Realism and Reason - Philosophical Papers,* Vol. 3., Cambridge University Press: Cambridge, 1986, pp. 240-244.
Raine, D.J., "Mach's Principle and Space-Time Structure", *Reports of Progress in Physics,* **44** (1981), 1151-1195.
Ratliff, Floyd, "On Mach's Contribution to the Analysis of Sensations", *Ernst Mach - Physicist and Philosopher,* eds. R.S. Cohen & R.J. Seeger, D. Reidel: Dordrecht, 1970, pp. 23-41.
Reichenbach, H., "Contributions of Ernst Mach to Fluid Dynamics", [with biographical information], Ernst-Mach-Institut: *Annual Review of Fluid Mechanics,* **15**, 1983, booklet 1.
Seeger, Raymond, J., "On Mach's Curiosity about Shockwaves", in *Ernst Mach - Physicist and Philosopher,* eds. R.S. Cohen & R.J. Seeger, Reidel: Dordrecht, 1970, pp. 60-68.
Smith, Barry, (See Kevin Mulligan.)
Smutny, Frantisek, "Ernst Mach and Wolfgang Pauli's Ancestors in Prague", *Gesnerus,* **46** (1989), 183-194. [Also in *European Journal of Physics,* **11** (1990), 257-261]
Sorensen, Roy, "Thought Experiments", *American Scientist,* **79.3** (1991), 250-263.
Stein, Howard, "Some Philosophical Prehistory of General Relativity". in *Foundations of Space-Time Theories,* eds. Earman, Glymour, Stachel, (*Minnesota Studies in the Philosophy of Science,* vol. 8), University of Minnesota Press: Minneapolis, 1977, pp. 3-39.
Strauss, M. "Einstein's Theories and the Critics of Newton. An Essay in Logical Historical Analysis", *Synthese,* **18** (1968), 251-284.
Swoboda, Wolfram, "Physics, Physiology, and Psychophysics: The Origins of Ernst Mach's Empiriocriticism", *Rivista di Filosofia,* **73** (1982), 234-274.
Takata, Seiji, "A Textual Comparison of All Four Editions of Mach's *Wärmelehre* ", *Historia Scientiarum,* **32** (1986), 21-31.
Taketani, Mitsuo, "Methodological Approaches in the Development of the Meson Theory of Yukawa in Japan", in *Science and Society in Modern Japan,* eds. Nakamura, S. et al., University of Tokyo Press: Tokyo, 1974, pp. 24-28.
Thornton, R.J., "'Imagine yourself set down...': Mach, Frazer, Conrad, Malinowski and the Role of Imagination in Ethnology", *Anthropology Today,* **1.5** (1985), 7-14.
Uebel, T., "Scientific Racism in the Philosophy of Science - Some Historical Examples", *Philosophical Forum,* **22** (1990), 1-18.
Visser, Henk, "Wittgenstein's Debt to Mach's *Popular Scientific Lectures* ", *Mind,* **91** (1982), 102-105.
"Mach's Method in Wittgenstein's Later Philosophy", *8. Internationalen Wittgenstein Symposiums,* eds. R. Haller and W.L. Gombocz, Kirchberg am Wechsel, (1983), Verlag Hölder-Pichler-Tempsky: Vienna, 1984.
Wheeler, J.A., "Mach's Principle as Boundary Condition for Einstein's Field Equations", in *Lectures in Theoretical Physics,* v, eds. W.E. Brittin, B.W. Downs, J. Downs, New York-London, Sydney, 1963, pp. 528-578.
Wolters, Gereon, "Ernst Mach and the Theory of Relativity", in *Philosophia Naturalis,* **21** (1984), 630-641.
"Phenomenalism, Relativity, and Atoms: Rehabilitating Ernst Mach's Philosophy of Science", in Logic, Methodology, and Philosophy of Science VIII: *Proceedings of the Eighth International Congress of Logic, Methodology, and Philosophy of Science,* 1987, North-Holland: Amsterdam, 1989, pp. 641-660.
Yan, Kangnian, "A Re-examination into Newton's Definition of Mass and Mach's Criticism", *Historia Scientiarum,* Nr. 40, June 1990, 29-40.
Yourgrau, Wolfgang & Alwyn van der Merve, "Did Ernst Mach 'Miss the Target'?", *Synthese,* **18** (1968), pp. 234-250.
Zahar, Elie, "Mach, Einstein, and the Rise of Modern Science", *British Journal for the Philosophy of Science,* **28** (1977), 195-213.

"Second Thoughts about Machian Positivism - A Reply to Feyerabend", *British Journal for the Philosophy of Science,* **32** (1981), 267-276.

B. Articles on Mach in Other Languages:

Aler, Jan, "Als Zögling zwischen Maeterlinck und Mach - Robert Musils literarisch-philosophische Anfänge", in *Problem des Erzählens in der Weltliteratur,* ed. F. Martini, Ernst Klett Verlag: Stuttgart, 1971, pp. 234-290.

Apa, H., "Tra Rokitansky e Mach - analogie con il Sezessionsstil", in *Le Arta A Vienna,* ed. M. Marchetti, La biennale de Venezia, 1984, pp. 465-467.

Baatz, Ursula, "Der Traum der Sehnerven", in *Debuts eines Jahrhundert - Essays zur Wiener Moderne,* ed. W. Pircher, Falter: Vienna, circa 1985, pp. 93-109.

"Ernst Mach und der Buddhismus", *Conceptus - Zeitschrift für Philosophie,* Nr. 56, **22** (1988), 19-33.

Bayertz, Kurt, "Wissenschaftsentwicklung als Evolution? Evolutionäre Konzeptionen wissenschaftlichen Wandels bei Ernst Mach, Karl Popper, and Stephen Toulmin", *Zeitschrift für allgemeine Wissenschaftstheorie,* **18** (1987), 61-91.

Békésy, Georg von, "Die Erweiterung der Existenz der Mach-Bänder auf andere Sinnesgebiete", *Symposium aus Anlass des 50.Todestages von Ernst Mach,* [includes diagrams], ed. W. Merzkirch, Ernst-Mach-Institut: Freiburg, 1967, pp. 161-170.

Bitbol, Michel, "Mach et Schrödinger: De l'element an tout", *Fundamenta Scientiae,* **8** (1987), 345-372.

Blackmore, John, "Mach über Atome und Relativität - Neuste Forschungsergebnisse", *Ernst Mach - Werk und Wirkung,* ed. R. Haller & F. Stadler, translated by H. Pauer-Studer, Verlag Hölder-Pichler-Tempsky: Vienna, 1988, pp. 463-483.

Boltzmann, Ludwig, "Über die Unentbehrlichkeit der Atomistik in der Naturwissenschaft" (abridged) and "Über die Frage nach der objektiven Existenz der Vorgänge in der unbelebten Natur (slightly abridged), *Populäre Schriften,* edited by R. Sexl, introduction by E. Broda, Vieweg: Braunschweig/Wiesbaden, 1979, pp. 78-119.

Broda, Engelbert, "Das intellektuelle Viereck: Mach-Boltzmann-Planck-Einstein", in E. Broda, *Wissenschaft Verantwortung Frieden,* posthumously prepared and edited by P. Broda et al, Franz Deuticke: Vienna, 1985, pp. 129-150.

Bunge, Mario, "Mach y la teoria atomica", *Boletin del Quimico Peruano,* **3** (1951), 12.

"Machs Kritik an der Newtonschen Mechanik", *Symposium aus Anlass des 50. Todes-Ttages von Ernst Mach,* edited by W. Merzkirch, [also in English, 1966], Freiburg im Breisgau, 1967/68, pp. 227-248.

"Mach's Beitrag zur Grundlegung der Mechanik", *Philosophia Naturalis,* **11** (1969), 189-213.

Buttemeyer, W., "Roberto Ardigo and Ernst Mach", *Rivista di Storia della Filosofia,* **46** (1991), 109-126.

Chisholm, R., and J. C. Marek, "Einleitung der Herausgeber", in Franz Brentano, *Über Ernst Machs 'Erkenntnis und Irrtum',* Rodopi: Amsterdam, 1988, pp. 1-12.

Costa de Beauregard, O., [a review article in French on J. Blackmore's 1972 book on Mach], *Revue de Synthèse,* 1974, pp. 271-282.

Dambmann, Holger, "Die Bedeutung des Machschen Prinzips in der Kosmologie", *Philosophia Naturalis,* **27** (1990), 234-271.

D'Elia, Alfonsina, "Recente Bibliografia Machiana", *Revista Critica di Storia della Filosofia,* **3** (1975), 189-203.

Desportes, Yvon, "Vergleichende Untersuchung eines Stils und seiner Philosophie: Ein Werk Musils aus der Sicht Machs", in *Robert Musil,* ed. R. v. Hydebrand, Wissenschaftliche Buchgesellschaft: Darmstadt, 1982, pp. 281-295.

Diersch, Manfred, "Spiegelungen. Ernst Machs Subjekt-Objekt-Verständnis als Anregung für die Kunst im 20. Jahrhundert.", in *Ernst Mach - Studien und Dokumente zu Leben und Werk,* eds. D. Hoffmann & H. Laitko, Deutscher Verlag der Wissenschaften: Berlin, 1991, pp. 328.

Dvorak, Johann, "Ernst Mach ein 'moderner Materialist'? - Zur wissenschaftlichen Methode bei Marx, Engels und Mach", in *Ernst Mach - Werk und Wirkung,* eds. R, Haller & F. Stadler, Verlag Hölder-Pichler-Tempsky: Vienna, 1988, pp. 329-341.

Einstein, Albert, "Ernst Mach", [reprinted from *Physikalische Zeitschrift,* **17** (1916), 101-104], in Ernst Mach, *Populär-wissenschaftliche Vorlesungen,* Böhlau: Wien-Köln-Graz, 1987, pp. 641-646. Also in *Die Mechanik* together with "Vier Briefe an Ernst Mach", Akademie Verlag: Berlin, 1988, pp. 679-689.

Feyerabend, Paul, "Mach's Theorie der Forschung und ihre Beziehung zu Einstein", in *Ernst Mach - Werk und Wirkung,* edited by R. Haller & F. Stadler, Verlag Hölder-Pichler-Tempsky: Vienna, 1988, pp. 435-462.

Gargani, Aldo, "'La buona austriacita' di Ernst Mach", in Ernst Mach, *Conosenza ed errore. Abbozzi per una psicologia della recerca,* Turin, 1982, pp. vii-xxxii.

Goe, George, "Su una critica Mach a Galileo", *Scientia,* **116** (1981), 83-92.

Goenner, Hubert F., "Machsches Prinzip und Theorien der Gravitation", in *Grundlagenprobleme der modernen Physik. Festschrift für Peter Mittelstadt zum 50.Geburtstag,* edited by. J. Nitsch, J. Pfarr, and E. Stachow, Mannheim, 1981, pp. 85-101.

Haller, Rudolf, "Der erste Wiener Kreis", *Erkenntnis,* **22** (1985), 341-358; also in R. Haller, *Fragen zu Wittgenstein und Aufsatze zur Österreichische Philosophie,* Rodopi: Amsterdam, 1986, pp. 89-107.

"Poetische Phantasie und Sparsamkeit - Ernst Mach als Wissenschaftstheoretiker", in R. Haller, *Fragen zu Wittgenstein und Aufsätze zur österreichischen Philosophie,* Rodopi: Amsterdam, 1986, pp. 70-85; Also in *Ernst Mach - Werk und Wirkung,* edited by R. Haller & F. Stadler, Vienna, 1988, pp. 342-355.

"Grundzüge der Machschen Philosophie", in *Ernst Mach - Werk und Wirkung,* eds. R. Haller & F. Stadler, Verlag Hölder-Pichler-Tempsky: Vienna, 1988, pp. 64-86.

"Ernst Mach - Das unrettbare Ich", in *Grundprobleme der Grossen Philosophen,* ed. Josef Speck, Vandenhoek & Reprecht in Göttingen, 1991, pp. 210-244.

Haubelt, J., "K Machove zadosti o profesoru na przske polytechnice (Mach as candidate for the Prague Polytechnic)," *Dejiny ved a techniky,* **5** (1972), 52-55.

Hayek, Friedrich von, "Diskussionsbemerkungen über Ernst Mach und das sozialwissenschaftliche Denken in Wien", *Symposium aus Anlass des 50.Todestages von Ernst Mach,* Ernst-Mach-Institut: Freiburg, 1967-68, pp. 41-44.

Hecht, Hartmut, "Facetten des Briefwechsels zwischen Ernst Mach und Joseph Petzoldt", in *Ernst Mach - Studien und Dokumente zu Leben und Werk,* eds. D. Hoffmann & H. Laitko, Deutscher Verlag der Wissenschaften: Berlin, 1991, pp. 224-242.

Heintel, E., "Die Dialektik des Ich bei Mach", in *Jahrbuch der Österreichische padegogische Gesellschaft,* Vienna, 1969, pp. 62-71.

Hentschel, Klaus, "Ernst Mach", *Neue Deutsche Biographie,* vol. 15, Duncker & Humblot: Berlin, 1987, pp. 605-609.

"Die Korrespondenz Duhem-Mach: Zur 'Modellbeladenheit' von Wissenschaftsgeschichte", *Annals of Science,* **45** (1988), 73-91.

Herneck, Friedrich, "Ernst Mach und Albert Einstein", *Symposium aus Anlass des 50. Todestages von Ernst Mach,* ed. W. Merzkirch, Ernst-Mach-Instiut: Freiburg, 1867-68, pp. 45-61.

"Die Beziehungen zwischen Einstein und Mach, dokumentarisch zusammengestellt", in *Einstein und sein Weltbild,* Der Morgan: Berlin, 1976, pp. 109-155.

"Ernst Mach und Albert Einstein", in *Wissenschafts-Geschichte - Vorträge und Abhandlungen,* Berlin, 1984, pp. 108-119.

"Über eine unveröffentlichte Selbstbiographie Ernst Machs", [reprinted from *Physilali-*

scher Blätter, **14** (1958), 385-390 with an afterword dated Fall 1988], in *Ernst Mach - Studien und Dokumente zu Leben und Werk,* eds. D. Hoffmann & H. Laitko, Deutscher Verlag der Wissenschaften: Berlin, 1991, pp. 107-140.

Herrmann, Dieter, "Ernst Mach und seine Stellung zur Doppler-Theorie", *Symposium aus Anlass des 50.Todestages von Ernst Mach,* ed. W. Merzkirch, Ernst-Mach-Institut: Freiburg, 1967-68, pp. 171-184.

Hiromatsu, Wataru, "Mach no tetsugaku" (Mach's Philosophy), in Ernst Mach, *Kankaku no bunseki* (Analysis of Sensations), eds. W. Hiromatsu and G. Sudo, Hosei University Press: Tokyo, 1971, pp. 332-355.

"Mach no tetsugaku to sotaisei riron", (Mach's Philosophy and the Theory of Relativity) in Ernst Mach, *Ninshiki no bunseki* (Analysis of Knowledge). [This book is largely a translation of three articles from Mach's *Populär-wissenschaftlicheVorlesungen* plus two articles from *Erkenntnis und Irrtum*], eds. W. Hiromatsu and N. Kato, Hosei University Press: Tokyo, 1971, pp. 136-173.

"Mach no tetsugaku to sotaisei riron" (Mach's Philosophy and the Theory of Relativity) in *Sotaisei riron no tetsugaku* (The philosophy of the Theory of Relativity), [The article on Mach published in 1971 is doubled in size, and a new section is added. The third section of the book is by Ryoiichi Itagaki], first edition, Nippon Britannica: Tokyo, 1981, pp. 89-178. [In the second edition published by Keiso Shobo: Tokyo, 1986, the article by Itagaki is replaced by one by Makoto Katsumori.]

Hoffmann, Dieter, "Sozial- und Gesellschaftspolitisches bei Ernst Mach", *NTM,* **26.2** (1989), 61 ff; also "Gesellschafts- und Sozialpolitisches bei Ernst Mach", 86-90.

"Ernst Mach in Prag", in *Ernst Mach - Studien und Dokumente zu Leben und Werk,* eds. D. Hoffmann & H. Laitko, Berlin, 1991, pp. 141-178.

Hoffmann, Dieter & Hubert Laitko, "Ernst Mach (1838-1916)" in *Wissenschaftliche Welt,* **32.4** (1988), 2-6.

"Ernst Mach: k 150-letiju so dnja rozdanija" (150th anniversary of Ernst Mach's birth) in *Voprosy istorii estestvoznanija i techniki,* 1988, H. 4, pp. 45-57.

Hoffmann, Dieter & Wolfgang Manthei, "Ernst Mach als Schulphysiker, Didaktiker, und Bildungstheoretiker", in *Ernst Mach - Studien und Dokumente zu Leben und Werk,* eds. D. Hoffmann & H. Laitko, Berlin, 1991, pp. 279-314.

Hohenester, Adolf, "Die Aussenwelt und die Innenwelt - Ernst Mach und die Physikdidaktik", *Zeitschrift für Didaktik der Philosophie,* **3** (1981), 215-219.

"Einleitung zum Neudruck von Ernst Mach", *Populär- wissenschaftliche Vorlesungen,* Böhlau Verlag: Wien-Köln-Graz, 1987, pp. xv- xxxvii.

"Ernst Mach als Didaktiker, Lehrbuch- und Lehrplanverfasser", in *Ernst Mach - Werk und Wirkung,* edited by R. Haller & F. Stadler, Verlag Hölder-Pichler-Tempsky: Vienna, 1988, pp. 138-163.

Holton, Gerald, "Où est la réalité? Les réponses d'Einstein", *Science et Synthèse,* trans. S. Bricianer & F. Lot, intro. R. Maheu, Gallimard: Paris, 1967, pp. 97-140.

"Mach, Einstein et la recherche du réel", *L'invention scientifique,* Press Universitaires de France-Croisees: Paris, 1982, pp. 233-287.

Honda, Shuro, *Gendai butsurigakusha no sei to tetsugaku Mach kara Einstein* (The Life and Philosophy of Recent Physicists from Mach to Einstein), Tokyo, 1981.

Hönl, Helmut, "Entwicklungsstufen des Mach'schen Prinzips", *Symposium aus Anlass des 50.Todestages von Ernst Mach,* ed. W. Merzkirch, Ernst-Mach-Institut: Freiburg, 1967-68, pp. 62-84.

Imai, Michio,:A Perspective on Mach" (in Japanese), *Hokkaido Daigaku Tetsugakai Tetsugaku',* **8** (1971), 83-110 and "Logic and Psychology of Research Related to E. Mach's *Erkenntnis und Irrtum*", *Hokkaido Daigaku Tetsugakai Kaiho,* **28** (1981, July issue), 1-7.

"Ernst Mach's Epistemology - Theory of Elements and *Denkökonomie* ", (in Japanese) *Higashi Nihon Gakuin Daigaku Kyoyoobu Ronshuu,* **1** (1975, June issue), 1-24.

"On Robert Musil's Doctoral Dissertation", *Hokkaido Daigaku Bungakubu Kiyoo,* **31**

(1983, issue 2), 1-32; "On the Origins of Mach's Philosophy", **35** (1988, issue 2), 1-30; and "An Aspect in the Development of Mach's Ideas", **37** (1988, issue 1), 35-87. (all in Japanese).

Ishiwara, Jun, "Vutsurigakujo no gainen ni tai suru yuibutsusei no imi ni tsuite", (The Meaning of Materiality with Respect to Physical Concepts), in *Shizen kagakuteki sekaizo* (A Natural Scientific World View), Hyoronsha: Tokyo, 1948, pp. 259-265.

Ishiwara, Jun, (ed.) "Einstein's Speech at Kyoto University (in Japanese)", in *Einstein Ko-en Roku,* (The Records of Einstein's Addresses), [See the article by Y.A. Ono for an English translation of Einstein's Kyoto speech.], Tokyo, 1971, pp. 78-88.

Itagaki, Ryoichi, "Einstein to Mach" (Einstein and Mach), in *Sotaisei riron no tetsugaku* (The philosophy of the Theory of Relativity), [The first two sections of the book are by W. Hiromatsu], Nippon Britannica: Tokyo, 1981, pp. 179-243.

Janich, Peter, "Hat Ernst Mach die Protophysik der Zeit kritisiert?" *Philosophia Naturalis,* **22** (1985), 51-60.

Jochim, H., "Phänomenologie und Modalität", *Symposium aus Anlass des 50. Todestages von Ernst Mach,* Ernst-Mach-Institut: Freiburg, 1967-68, pp. 189-202.

Jung, R. "Ernst Mach als Sinnesphysiologe", *Symposium aus Anlass des 50. Todestages von Ernst Mach,* Ernst-Mach-Institut: Freiburg, 1967-68, pp. 132-137.

Kaiser, Walter, "Ernst Mach und das Mechanismusproblem in der Biologie", *Medizinhistorische Journal,* **17** (1982), 179-194.

Katsumori, Makoto, "Sotaisei riron to Einstein no tetsugakuteki tenkai" (The Theory of Relativity and Einstein's Philosophical Turn) as Section 3 of *Sotaisei riron no tetsugaku,* (The Philosophy of the Theory of Relativity). [The first two sections are by W. Hiromatsu], Keiso Shobo: Tokyo, 1986, pp. 179-253.

Kühn, Karl, "Ernst Machs Herkunft und Abstammung", *Heimatbildung,* Reichenberg, **19** (1938), 268-269.

Kutterer, R., "Die Beitrage Ernst Mach zur Ballistik", *Symposium aus Anlass des 50. Todestages von Ernst Mach,* edited by W. Merzkirch, Ernst-Mach-Institut: Freiburg, 1967-68, pp. 96-112.

Kuwaki, Ayao, "Ryugaku Zakki", (Notes on my Study Abroad), [These notes describe Kuwaki's visits to Mach, Planck, Einstein, and other scientists.], *Toyo Gakugei Zasshi,* 1910, No. 345 P. 283, No. 347 p. 369, No. 348 p. 427, and No. 349 p. 481.

"Butsurigakuju no nishiki no mondai" (Epistemological Problems in Physics), *Rigakukai,* **9** (1912), 1-8.

"Mach to Sotaisei Genri" (Mach and the Theory of Relativity) (1922) and "Mach no sekaikan", (Ernst Mach's World View) (1938) in *Kagakushi-ko* (Essays in History of Science), Tokyo, 1944, pp. 385-392 & 450 ff.

Laas, Andreas, "Vom Sinn des Machschen Philosophierens", in *Ernst Mach - Studien und Dokumente zu Leben und Werk,* edited by D. Hoffmann & H. Laitko, Deutscher Verlag der Wissenschaften: Berlin, 1991, pp. 206-223.

Laitko, Hubert, "Mach I und Mach II? Annäherung an einen Widersprüchlichen", in *Sonntag,* **42** (1988), Nr. 27, p. 9.

"Ernst Mach und der Ökonomie in der Wissenschaft", in *Ernst Mach - Studien und Dokumente zu Leben und Werk,* Deutscher Verlag der Wissenschaften: Berlin, 1991, pp. 243-278.

(See Dieter Hoffmann for co-authored works.)

Leinfellner, Werner, "Physiologie und Psychologie - Ernst Machs *Analyse der Empfindungen*", in *Ernst Mach - Werk und Wirkung,* eds. R. Haller & F. Stadler, Verlag Hölder-Pichler-Tempsky: Vienna, 1988, pp. 113-137.

Lesky, Erna, "Kompensationslehre und denkökonomisches Prinzip", *Symposium aus Anlass des 50.Todestages von Ernst Mach,* edited by W. Merzkirch, Ernst-Mach-Institut: Freiburg, 1967-68, pp. 26-40.

Lübbe, Hermann, "Positivismus und Phänomenologie: Mach und Husserl", in *Bewusst-*

sein in Geschichten: Studien zur Phänomenologie der Subjektivität: Mach, Husserl, Schlapp,Wittgenstein, Freiburg, 1972, pp. 33-62. [This article has the same title as one published in *Beiträge zu Philosophie und Wissenschaft* in 1960]

Mach Ludwig, "Vorwort zur neunten Auflage", [reprinted from Ernst Mach, *Die Mechanik,* Brockhaus: Leipzig, 1933], in *Die Mechanik,* Akademie-Verlag: Berlin, 1988, pp. 650-652.

Mahr, Peter, "Ernst Mach - Gestaltwahrnehmung - Minimal Art", in *Ernst Mach - Werk und Wirkung,* eds. R. Haller & F. Stadler, Verlag Hölder-Pichler-Tempsky: Vienna, 1988, pp. 404-431.

Manthei, Wolfgang, (See Dieter Hoffmann for co-authored article.)

Marek, Johann C. (See Franz Brentano for a co-edited book on Mach and Roderick M. Chisholm for a co-authored introduction to it.)

Mayerhöfer, Josef, "Ernst Machs Berufung an die Wiener Universität 1895", *Symposium aus Anlass des 50. Todestages von Ernst Mach,* ed. W. Merzkirch, Ernst-Mach-Institut: Freiburg, 1967-68, pp. 12-25.

Merzkirch, Wolfgang, "Die Beiträge Ernst Machs zur Entwicklung der Gasdynamik", in *Symposium aus Anlass des 50. Geburtstages von Ernst Mach,* ed. W. Merzkirch, Ernst-Mach-Institut: Freiburg, 1967-68, pp. 114-131.

(See Otto Blüh for co-authored article in the above book.)

Motz, Hans, "Ernst Mach als Physiker und Historiker der Physik mit Zukunft", *Ernst Mach - Werk und Wirkung,* edited by R. Haller & F. Stadler, Verlag Hölder-Pichler-Tempsky: Vienna, 1988, pp. 87-112.

Mulligan, Kevin and Barry Smith, "Mach und Ehrenfels: Über Gestaltqualitäten und das Problem der Abhängigkeit", in *Christian von Ehrenfels - Leben und Werk,* ed. R. Fabian, Amsterdam, 1985.

Negri, Antimo, "Mach e il Materialismo", *Giovanale Critico della Filosofia Italiana,* **8** (1977), 167-181; and in French: "Mach et le matérialisme", *Revue philosophique de la France et de l'Etranger,* **171** (1981), 67-86.

Nelson, Leonard, (book review of E. Mach, *Erkenntnis und Irrtum,* 2nd edition, 1906), [reprinted from *Göttingsche gelehrte Anzeigen,* **169** (1907), 636-658], in *Gesammelte Schriften,* Bd. 2, Hamburg, 1973, p. 31 ff.

Noé, Keiichi, "Mach tetsugaku no shatei" (The Range of Mach's Philosophy), in Ernst Mach, *Jikan to kukan* (Time and Space), ed. K. Noé, Hosei University Press: Tokyo, 1977, pp. 201-246.

"Kindai Gorishugi no chokoku: Hiromatsu tetsugaku to gendai butsurigaku" (Overcoming Modern Rationalism: Hiromatsu's Philosophy and Modern Physics), in *Hiromatsu-ron* (Reviews of Hiromatsu), eds. K. Ohba et al., Unite: Nagoya, 1982, pp. 197-246.

Pachner, J., "Machs Kritik an der Newtonschen Mechanik und ihre Weiterentwicklung zum Machschen Prinzip", *Sympsium aus Anlass des 50. Todestages von Ernst Mach",* ed. W. Merzkirch, Ernst-Mach-Institut: Freiburg, 1967-68, pp. 216-226.

Pauletta, Gian Carlo, "Relazione e relativitá nelle conoscenza psicoanalitica: Da Mach a Freud a Einstein", *Rivista di Psicoanalisi,* **39** (1983), 334-351.

Petronijevic, Branislav, "Iz mojih uspomena na nemacke filosofe" (From My Recollections of German Philosophers), *Srpski narod* (Serbian Volk), Beograd, 2/1943, br. 30, 7 Avgust, str. 9.

Planck, Max, "Zur Machschen Theorie der Physikalischen Erkenntnis - Eine Erwiderung", [reprinted from *Physikalische Zeitschrift,* **11** (1910), 1180-1190], an appendix added to Ernst Mach, *Die Mechanik,* Akademie-Verlag: Berlin, 1988, pp. 670-678.

Ratliff, Floyd, "Mach Bänder in der Physik, der Physiologie und der Psychologie", *Symposium aus Anlass des 50.Todestages von Ernst Mach,* edited by W. Merzkirch, Ernst-Mach-Institut: Freiburg, 1967-68, pp. 138-160.

Ritter, Joachim (See Jürgen Blühdorn.)

Rossi, Arcangelo, "Mach ed Einstein: Influenza della *Meccanica* di Mach sul pensiero di

Einstein", *Physis,* **22** (1980), 279-292.
"Ernst Mach e l'epistemologia della fisica nel primo '900", *Physis,* **25** (1983) 481-496.
Sajner, Josef, "Ernst Machs Beziehungen zu seinem Heimatort Chirlitz (Chrlice) und zu Mähren", *Bohemia,* **24.2** (1982), 358-368.
Shibuya, Kazuo, "Ernst Mach jiden - 1913 nen soko no honyaku", (Ernst Mach's Autobiography - Translation of a Manuscript from 1913), *Il Saggiatore - kagaku gijutsu to sono jidai,* [This journal is the organ of a group of historically-minded researchers who call themselves "Il Saggiatore". The group seems to have originally developed at the Tokyo Institute of Technology. The journal started in 1975 with two issues a year but since 1980 has issued only one a year.], Nr. 17, 1988, pp.43-59.
Schmutzer, Ernst., "Ernst Mach und das relativistische Denken in der gegenwärtigen Phy-Physik", in *Ernst Mach - Studien und Dokumente zu Leben und Werk,* edited by D. Hoffmann & H. Laitko, Deutscher Verlag der Wissenschaften: Berlin, 1991, pp. 179-189.
Sommer, Manfred (ed.), "Aus den Notizbüchern [Ernst Machs] 1871-1910", in *Ernst Mach - Werk und Wirkung,* eds. R. Haller & F. Stadler, Verlag Hölder-Pichler Tempsky: Vienna, 1988, pp. 157-211.
"Denkökonomie und Empfindungstheorie bei Mach und Husserl - Zum Verhältnis von von Positivismus und Phänomenologie", in *Ernst Mach - Werk und Wirkung,* eds. R. Haller & F. Stadler, Verlag Hölder-Pichler-Tempsky: Vienna 1988, pp. 309-328.
Stadler, Friedrich, "Aspekte des gesellschaftlichen Hintergrunde und Standorts der Wiener Kreises am Beispiel der Universität Wien", *Akten des 3. International Wittgenstein-Symposiums,* eds. H. Berghel, A. Hübner, E. Köhler, Hölder-Pichler-Tempsky: Vienna, 1979, pp. 41-59.
"Popularisierungsbestrebungen im Wiener Kreis und 'Verein Ernst Mach'", in Philosophie, *Wissenschaft, Aufklärung - Beiträge zur Geschichte und Wirkung des Wiener Kreises,* ed. Hans-Joachim Dahms, Walter de Gruyter: Berlin & New York.
"Ernst Mach - Zu Leben, Werk und Wirkung", in *Ernst Mach - Werk und Wirkung,* eds. R. Haller & F. Stadler, Hölder-Pichler-Tempsky: Vienna, 1988, pp. 11-63.
Strauss, M., "Zur Logik der Begriffe 'Inertialsystem' und 'Masse'", *Symposium aus Anlass des 50. Todestages von Ernst Mach,* ed. W. Merzkirch, Ernst-Mach-Institut: Freiburg, 1967-68, pp. 203-215.
Stubenberg, Leopold, "Parfit und Mach über den Tod", *Conceptus - Zeitschrift für Philosophie,* Nr. 53/54, **21** (1987), 201-216.
Svatek, J. "Zapomenuty prirodovedec F.X. Vesely (1819-1904)" [about Mach's favorite teacher] , *Zpravy Vlastivedneho ustavu v Olomouci,* Nr. 108 (1963), 1-7.
Swoboda, Wolfram, "Physik, Physiologie und Psychophysik - Die Wurzeln von Ernst Machs Empiriokritizismus", in *Ernst Mach - Werk und Wirkung,* eds. R. Haller & F. Stadler, Verlag Hölder-Pichler-Tempsky: Vienna, 1988, pp. 356-403.
Takata, Seiji, "A Comparative Study of the Editions of Mach's *Die Principien der Wärmelehre* " (in Japanese), *Kagakushi Kenkyu,* **20** (1981), 110-116.
"Unexplored Corners of E. Mach's Achievements - at the 150th Anniversary of his Birth", (in Japanese), *Kagakushi Kenkyu,* (Journal of History of Science, Japan), Series II, **27** (1988), 179-182.
Tambrock, Günter, "Ernst Mach und die Sinnesphysiologie", *Ernst Mach - Studien und Dokumente zu Leben und Werk,* eds. D. Hoffmann & H. Laitko, Deutscher Verlag der Wissenschaften: Berlin, 1991, pp. 190-205.
Tanabe, Hajime, *Saikin no shizen kagaku* (Recent Natural Science), Iwanami Shoten: Tokyo, 1915, pp. 199-320.
Thiele, Joachim, "Bemerkungen zu einer Äusserung im Vorwort der 'Optik' von Ernst Mach", *Schriftenreihe für Geschichte der Naturwissenschaft, Technik und Medizin,* **2** (1965), 10-19.
"Zur Wirkungsgeschichte der Methodenlehre Ernst Machs", *Symposium aus Anlass*

des Todestages von Ernst Mach, ed. W. Merzkirch, Ernst-Mach-Institut: Freiburg, 1967-68, pp. 85-95.

"Briefe Robert Lowies an Ernst Mach", *Isis,* **59** (1968), 84-87.

"Paul Carus und Ernst Mach: Wechselbeziehungen zwischen deutscher und amerikanischer Philosophie um 1900", *Isis,* **62** (1970), 208-219.

"Zur Kritik des Empiriomonismus . . .", *Zeitschrift für philosophische Forschung,* **24** (1970), 412-427.

"Briefe Albert Einsteins an Joseph Petzoldt", *NTM Zeitschrift für Geschichte der Naturwissenschaft, Technik und Medizin,* **8** (1971), 70-74.

"Franz Walds Kritik des theoretischen Chemie [with unpublished letters to Ernst Mach]", in *Annals of Science,* **30** (1973), 417-433.

"Natur und Geschichte in den wissenschaftlichen Schriften Ernst Machs", in *Natur und Geschichte,* 10 Deutscher Kongress für Philosophie, Kiel, 1972, (Hamburg, 1973), pp. 449-452.

"Zur Analyse der Bewegungsempfindungen - Briefe von Josef Breuer an Ernst Mach", *Centaurus,* **19** (1975), 40-53.

"Briefe deutscher Philosophen an Ernst Mach aus den Jahren 1865 bis 1915", *Philosophia Naturalis,* **16** (1976), 64-84.

"Briefe Hugo von Seeligers an Ernst Mach" in *Philosophia Naturalis,* **17** (1979), 391-399.

Tretera, Ivo, "Casus Stepan Doubrava (K neznamym dopism Ernsta Machs Augustu Seydlerovi z roku 1882)", [on Mach's Criticism of his former student Doubrava], *Acta Universitis Carolinae,* **XXV** (1985), 89-94.

Wolters, Gereon, "Topik der Forschung. Zur wissenschaftlichen Funktion der Heuristik bei Ernst Mach" in *Technische Rationalität und rationale Heuristik,* edited by C. Burrichter, R. Inherveen, and R. Kötter, Paderborn, 1986.

"Atome und Relativität - Was meinte Mach?", in *Ernst Mach - Werk und Wirkung*, eds. R. Haller & F. Stadler, Hölder-Pichler-Tempsky: Vienna, 1988, pp. 484-507.

"Il positivismo visionario: La filosofia della scienza di Mach", *Interprezioni,* **10**, (1990) 319-343.

NAME INDEX

Abbe, Cleveland, 65.
Abraham, Max 299, 301, 306, 310, 326.
Adler, Friedrich 118n, 133n, 159-161, 223n, 266-267, 274n, 277, 282, 290-291.
Adler, Max, 391-392.
Adler, Viktor, 159n.
Agassi, Joseph, 215n.
Aichi, Keiichi, 322.
Amano, Kiyoshi, 331.
Ampère, Andre Marie, 310, 416.
Anderson, J.S., 69, 282n.
Arago, D.F.J., 65, 68.
Archimedes, 20, 63-65, 327.
Aristotle, 197, 244, 254, 303, 327, 343-345, 387.
Avenarius, Richard, 117, 134, 274, 335, 342, 387, 389, 391-392.

Baatz, Ursula, 183-199.
Babbage, Charles, 197.
Bach, Alexander, 2.
Bacon, Francis, 154, 255, 327, 365.
Baege, M.H., 107.
Barth zu Barthenau, Ludwig, 150n.
Bauer, Otto, 367, 370-371, 375-376.
Becquerel, Antoine Henri, 162.
Benedetti, Vincent Comte, 62.
Berghel, H., 377.
Bergson, Henri, 162, 390.
Berka, K., 37n.
Berkeley, George, 22, 115, 215n, 218n, 239, 250, 261, 341-342, 382, 387, 393.
Bernstein, Jeremy, 374.
Bertalanffy, Ludwig von, 367, 371.
Besso, Michelle, 264, 278, 326, 358.
Beth, Evert W., 237n, 421, 428-430.
Biermann, Otto, 48.
Biot, Jean Baptiste, 136.
Bismarck, Otto Prinz von, 4.
Black, Joseph, 58, 421-422.
Bloch, Ernst, 66-67, 211.
Bogdanov, Alexander, 119.
Bohr, Niels, 327, 345-349.
Bolland, Gerardus, 403n, 407, 411, 424-425.
Boltzmann, Ludwig, xi, xiv, 17n, 26, 58n, 71n, 127n, 137, 151, 152n, 192, 219, 238, 261, 197-299, 304, 327-328, 363, 379n, 416, 421.
Bolyai, Janos, 205n.
Bonitz, Hermann, 21n.
Boscovic, Rudjera, 385n, 390.
Boutroux, Émile, 104.
Bouvier, Robert, 194.
Boyle, Robert, 427.
Bradley, J., 208n.
Brahe, Tycho, 343.
Brauer, Eberhard, 90.
Brauner, Bohuslav, 49, 152n, 164.
Brentano, Franz, 221.
Breton [Brethon], Baron Klaudius Waldfried, 1, 17.
Breton [Brethon], Josefine (née Lanhaus), 17.
Breuer, Josef, 24, 148.
Bridgman, Percy, 415.
Brodetsky, S., 68-69.
Brown, Crum, 24, 148.
Brücke, Ernst, 24, 148, 198.
Brühl, Levy, 18.
Brunsvicg [Brunschvicg?] 104.
Brunschvicg, Léon, 162.
Brunswik, E., 368, 371-372, 374.
Brush, Stephen, 238n.
Bucherer, Alfred, 301.
Buddha, Gautama, 183-199.
Bunge, Mario, 215n, 243-261.
Burmester, Ludwig, 95n, 97, 107-108.
Burrichter, C., 264n.
Buys Ballot, C.H.D., 406.

Cahan, D. 31n.
Cajori, F., 244n, 254n.
Cantor, Mathias, 74-75.
Cantor, M.B., 97-98.
Cardan [Cardano, Girolamo], 62.
Carnap, Rudolf, 92n, 215, 220, 239, 327, 363n, 364-377, 396, 430.
Carnot, N.L. Sadi, 58, 144.
Cartan, Elié Joseph, 162.
Carus, Paul, 88, 191-192, 195.
Cassidy, David, C., 312.
Cauchy, A.L.B., 261.
Cassirer, Ernst, 211, 212n, 235.
Cekic, Miodrag, 201-214, 379-402.
Chapran, J., 31n, 32n.
Chisholm, Roderick M., 221n.
Clark, R.W., 417n.
Classen, Johannes, 266.
Clausius, Rudolf, 58, 128, 144, 414, 416.
Clay, Jacob, 419, 424-426.
Cohen, 410.
Cohen, Hermann, 335, 338-340.
Cohen, I.B., 63n, 152n, 153n.
Cohen, R.S., xvii, 215n, 219n, 220n, 314-315, 363n, 373, 430.
Coleman, Samuel, 334n.
Comte, Auguste, 116, 310, 320, 328-329, 333, 404, 411.
Comton, Arthur Holly, 326.
Condillac, Étienne Bonnet de, 126, 134.
Conze, E., 189n.
Copernicus, Nicolaus, 131, 249.
Cornelius, Hans, 117.
Coulomb, Charles Augustin de 144.
Couturat, Louis, 78.
Crooke, William, 151.
Curie, Marie Sklodowska, 310.

D'Abro, A., 177n.
D'Alembert, Jean, 63, 234, 250.
Dahms, Hans-Joachim, 363n.
Dalton, John, 58, 119.
Darboux, Gaston, 311.
Darwin, Charles, xii-xv, 122, 133-134.
De Broglie, Louis, 323, 327.
Deffenhart, Leo, 2.
De Haas-Lorentz, G.L., 420n.
Democritus, 302.
Descartes, René, xii, 62, 125, 254-255, 310, 339, 341, 351, 387, 406, 421, 427.
Dick, Georg, 48.
Dijksterhuis, E.J., 63n, 420, 426-428.
Dingler, Frau Hugo, 92n, 110.
Dingler, Hugo, xvi, 91-110, 170-174, 206, 209, 216n, 217n, 219n, 266, 272, 280-281, 265-286, 290, 294, 327.
Djuric, Dragisa, 382-383.
Dollfuss, Engelbert, 375-377.
Donders, Frans Cornelis, 406.
Doorman, S.J., 430.
Dostal, J., 21.
Dostoevsky, Fedor Mikhailovich, 324.
Doubrava, Stefan, 49-50, 53.
Drake, Stillman, 63n.
Du Bois-Reymond, Emil, 130, 198.
Duhem, Pierre, 62, 81n, 117, 135, 223, 232, 249, 318, 418, 429.
Dvorak, Anton, 37.
Dvorak, Cenek [Vincenz], 25, 33, 48, 50-51.

Eddington, Sir Arthur stanley, 322.
Edel, A., 374.
Einstein, Albert, xi-xvi, 67, 71n, 125, 127n, 139, 147, 152-177, 181, 184, 206-209, 214, 233, 247-248, 263-332, 334-335, 352-361, 393-395, 401, 414, 416-418, 423, 425.
Einstein, Elsa [second wife], 322, 324, 326.
Einstein, Hans Albert, 313-315.
Einstein, Liserl [Lieserl], 309, 312-315.
Einstein, Mileva [first wife, née Maric], 309, 312-315.
Eisenbach, Georg Michael, 191.
Eisenbud, L., 259.
Elster, J.L.L.J., 151.
Engelhart, Josef, 2.
Engels, Friedrich, 392, 395.
Engelmeyer, Peter Klementitsch von, 97.
Enklaar,Johannes Eliza, 419-424.
Escherich, Theodor von, 150n,
Ettinghausen, Andreas von, 23.
Euclid, 65.
Eugénie [Bonaparte], Empress, 8.
Euler, Leonhard, 60, 244n.
Exner, Franz, 21n, 149n.
Exner, Sigmund, 149n.

Faraday, Michael, 131, 144, 327.
Fasol-Boltzmann, Ilse, 238n, 379n.
Fechner, Gustav, 24, 111-114, 116, 192, 196, 203n, 404.
Feddersen, W., 144.
Feigl, Herbert, 367, 370-371.
Fenalosa, Ernst, 323.
Fermi, Enrico, 346-347.
Feuerbach, Ludwig, 391.
Feyerabend, Paul, 217n, 219, 240, 241.
Fichte, Johann Gottlieb, 213, 338.
Finley-Freundlich, E., 374.
Finsterwalder, Sebastion, 107-109.
Fischer, Georg, 49.
Flexner, Abraham, 326.
Folia, J., 40n.
Folse, Henry, 349n.
Föppl, August, 267, 278, 299.
Fohringer, Anton, 2.
Foucault, Michel, 179n.
Foulkes, Paul, 373.
Frank, Philipp, 152-153, 206, 208, 215n., 249, 267-269, 279, 280n, 364, 367-369, 371-374, 425.
Franz Joseph [Habsburg], 30-31, 41, 91.
Freed, J & L., 373.
Frege, Georg, 98, 232, 233n.
Frenkel-Brunswik, Else, 371.
Fresnel, Auguste Jean, 136, 138, 261, 354, 416.
Freys, R., 430.
Friedjung, Josef Karl, 367, 370.
Fritsch, Alois, 165n.
Fujino, Seigyo, 305.
Fürst, P. Leo, 1-3.

Galilei, Galileo, xii, 25, 60, 122-123, 202, 207, 249-250, 254, 343-345, 421, 426, 427.
Galison, Peter L., 269.
Gargani, Aldo, 217n.
Garibaldi, Giuseppe, 4.
Gauss, J.K.F., 59-60, 163.
Gay-Lussac, Joseph Louis, 50.
Gehlen, A., 197n,
Geiger, Hans, 164.
Geiringer, Hilda, 373.
Geisser, Karl Friedrich, 163.
Geitel, Hans, 151.
Geitler-Armingen, Joseph Ritter von, 48.
Gerber, Paul, 268.
Gibbs, Josiah Willard, 58, 81n, 327.
Gicklhorn, R., 368.
Gleich, G., 294.
Goethe, Johann Wolfgang von, 15, 69, 197-198.
Goldschmidt, Heinrich, 50.
Goll, J., 38n.
Goltz, F.L., 24.
Gomperz, Heinrich, 183, 366, 369, 372.
Gomperz, Theodor, 150n.

Goodman, Nelson, 215, 249.
Goto, Makita, 298.
Grailich, Josef, 23.
Grassmann, Hermann, 119.
Grillparzer, Franz, 16.
Grimaldi, Franciscus Maria, 96-97.
Grobben, Karl, 147n.
Grossmann, Marcel, 270n, 271, 273, 275, 328.
Gruber, Jacob, 1,2.
Grünwald, Anton Karl, 48.
Gruss, Gustav, 48.

Haberditzl, Andreas, 48.
Haberlandt, Gottlieb, 147.
Hackel, O., 39n, 192.
Hadamard, Jacques, 162.,
Haga, Hermanus, 403.
Hahn, Hans, 363n, 364n, 365, 367-369, 372.
Hahn, Otto, 305.
Haldane, Sir Richard Burdon, 326.
Halla, Joseph, 35.
Haller, Friedrich, 325.
Haller, Rudolf, 195n, 198n, 215-228, 265, 266n, 272n, 280n, 281n, 287n, 291n, 366n.
Halpern, George, 368.
Halsted, G.B., 250n.
Hamilton, Sir William Rowan, 325.
Hanson, N.R., 240.
Hartmann, Eduard, 303.
Hartmann, Heinz, 368-370.
Hartmann, Ludo, 80.
Haubelt, J., 30n, 40.
Havranek, J., 41n.
Hegel, Georg, 333, 335, 338, 343, 348, 390, 424.
Hegeler, Alfred, 192.
Hegselmann, Rainer, 370n, 372n.
Heidegger, Martin, 183, 199, 214, 298, 350.
Heilmann, Friedrich, 2.
Heinzel, Richard, 159n.
Heisenberg, Werner Karl, 323-324, 327, 346-347, 395, 401.
Heisig, James, 338.
Heller, K.D., 36n, 39n, 46n, 216, 243n.
Hempel, Carl G., 374-375.
Helmholtz, Hermann von, 24, 51, 135, 154, 198, 235, 286, 297-298, 302-303, 327, 421.
Henkin, C., 258n, 260n.
Henning, Hans, 91.
Hentschel, Klaus, 45n, 71n, 92n, 118n, 147n, 192n, 241n, 263, 265, 266n, 273, 277n, 279n, 283n, 284n, 419n.
Heraclitus, 349.
Herbart, Johann F., 116, 195-196, 199, 386.
Hering, Ewald, 42-43, 116, 118, 125, 147.
Hermann, Emanuel, 133, 218.
Hermes, H., 258, 259.
Herneck, Friedrich, 17n, 36n, 194n, 268, 281,n, 292.
Heymans, Gerardus, 403-404, 407n.
Hertz, Heinrich, xi, 59, 127n, 135, 144, 154, 219, 298-299, 310, 321, 327, 416, 418.
Hesse, Mary, 240.
Hiebert, Erwin, 214, 222n.
Hirokawa, Shunkiti, 335n.
Hiromatsu, Wataru, 278n, 333, 349n, 350n, 351-361.
Hirosige, Tetsu, 241, 279n, 349, 351.
Hitler, Adolf, 375-377.
Höffding, Harald, 152n.
Höfler, Alois,
Hoffmann, Dieter, 29-55, 71n, 84n, 85n, 266n, 293n.
Hofmannsthal, Hugo von, 191.
Hodge, William, 373.
Hölder, O., 60.
Hollingdale, R.J., 203n.
Holton, Gerald, 23n, 82, 184, 209, 263, 275, 279, 282n, 286, 289n, 292, 294n, 358.
Holzapfel, Herr, 14.
Holzapfel, Mieze, 8-9, 15.
Holzapfel, Frau, 15.
Horstmann, August Friedrich, 58.
Hostinsky, Otakar, 50-52.
Huber, K.A., 43n.
Hübner, Adolf, 377.
Huet, Josué L'Ange, 404-406.
Humboldt, Alexander von, 6.
Hume, David, 115, 154, 156, 193, 195-196, 221, 261, 303, 329, 341, 358.
Hürbe, Johanna, 3.
Hus, John, 37.
Husserl, Edmund, 230, 237, 351, 429.
Huygens, Christiaan, 25, 59, 131, 136, 292, 416.

Inagaki, Morikatsu, 322.
Inhetveen, R., 264n.
Ishiwara, Jun, 163n, 294, 319, 322, 333, 335-337.
Itagaki, Ryoichi, 184n, 277-295.
Ivanov, N., 46n.

Jackson, C.T., 192n.
Jacob, Joseph, 91.
Jäger, Gustav, 267.
Jäger, Herr von, 3.
Jahoda, Maria, 370n, 371.
James, William, 195, 264, 273, 287, 319.
Jaumann, Gustav, 33, 48-50, 54-55, 307.
Jeffery, G.B., 171n, 208n, 278n.
Jenks, W.A., 37n.
Jerusalem, Wilhelm, 27n, 159, 363.
Jiracek, J., 34n, 35.
Jodl, Friedrich, 363.

Joergensen, J., 372, 374-375.
Johnston, William M., 21n.
Jones, R., 127n.
Joseph II [Hapsburg], 21, 38.
Joule, James Preston, 58, 421.
Jourdain, P.E.B., 62-63, 102.
Julius, Victor August, 412-417, 419.
Julius, Willem Henri, 412, 416-417, 419.
Jungnickel, Christa, 165n.

Kaal, Hans, 370n, 372n.
Kalidasa, 192n.
Kalupahana, D., 190n.
Kaluza, Theodor, 322.
Kamerlingh-Onnes, M. Heike, 312, 404.
Kant, Immanuel, xii, 18, 22, 26, 100, 115, 125, 132, 162, 185, 191, 193-194, 211n, 230, 235, 302, 308, 310, 327, 333, 335, 337, 341, 387, 407, 417, 429.
Kant, O., 373.
Kaufmann, W., 203n, 326.
Kautsky, Karl, 210n.
Kavka, F., 46n.
Kayser, Rudolf, 324.
Kelson, Hans, 374.
Kelvin, Lord, 58-59, 144, 234, 381, 413-414.
Kemmerich, Max, 21.
Kepler, Johannes, 125, 131, 249, 294n, 343.
Kern, 429.
Kessel, J. 25.
Kienzl, W., 33n, 192.
Kirchhoff, Gustav, 116, 135-136, 154, 255, 258, 261, 299, 321, 413.
Kittler, Friedrich, 197n.
Klebs, E., 39.
Klein, Martin, 57n, 221n, 225n.
Kleinpeter, Hans, 97, 192.
Knezevic, Bozidar, 383.
Köhler, Eckehart
Kolacek, F., 40, 48.
Kolakowski, Leszek, 212n.
Koster, Willem, 407-411.
Kotliar, J., 117.
Kotter, R., 204n.
Kowarik, W., 1.
Koy, A., 430.
Koyré, Alexandre, 63n.
Kraft, Victor, 216, 373.
Kretschmann, E., 167.
Krupka, Cölestin, 48.
Külpe, Oswald, 26.
Kuennen, Johannes, Petrus, 419.
Kuhn, Thomas, 240, 374.
Kuipers, Annie, xvii.
Kunz, Anton, 2, 13-14.
Kusaka, Shiichi, 152n, 208n.
Kus-Ninolajev, Mirko, 391-392.
Kuwaki, Ayao, 280, 290, 297-332, 334, 335.
Kuwaki, Gen-yoku, 298, 304, 320, 323.
Kuwaki, Tsutomu, 280n, 298, 313.

Laas, Ernst, 383.
Lagrange, Joseph Louis, 136, 234, 316.
Laitko, Hubert, 29n, 33n, 71n, 84n, 85n, 183, 293n.
Lamarck, Jean Baptiste, xiv-xv, 20, 133.
Lampa, Anton, 33, 185, 268.
Lampe, E., 59-61.
Lanczos, Cornelius, 271.
Land, Pieter Nicolaus, 403n.
Lang, Victor von, 149n.
Lange, Ludwig, 62.
Langevin, Paul, 162.
Lanhaus, Wenzel, 3.
Laplace, Pierre Simon, 208, 235.
Larmor, Joseph, 327.
Lasic, Stanko, 398.
Laue, Max von, 65-66, 325-326.
Lausch, Johann, 2.
Lazarsfeld, Paul, 371.
Lederer, C., 36n.
Leibniz, G.W., 59, 180, 196, 211n, 244, 248, 254-255, 299, 385.
Leipen, Robert, 50.
Lenin, Vladimir, 210, 239, 335, 341-342, 393, 396-397, 399, 401.
Lenzen, V.F., 63-65, 181n, 374.
Leonardo da Vinci, 62, 327.
Le Roy, Edouard, 162, 249.
Levi-Civita, Tullio, 260.
Lichtenberg, Georg Christoph, 193, 195.
Lieben, Adolf, 150n.
Liebig, Justus von, 23.
Linnemann, Professor, 49-50.
Lippich, Ferdinand, 48-49, 149n.
Lobachevsky, Nikolai Ivanovich, 205n, 232-233, 380.
Locke, John, xii, 203, 213, 427.
Lodge, Sir Oliver, 286-287.
Loeb, Jacques, 264.
Lorentz, H.A., xi, 136-137, 139, 158, 168, 171n, 172, 261, 269, 278-279, 300-301, 304, 312, 316, 327, 354, 404, 417-420, 424.
Lotze, Rudolf, 111.
Louis Philippe [Bourbon], 4.
Lovejoy, Arthur, 174-182.
Löwy, Heinrich, 207n, 292n, 365.
Luce, B.R.D., 224n.
Lucretius, 244, 302.
Ludwig, Carl, 24, 148.
Lueger, Karl, 4, 16.
Luggin, Hans, 49.
Luther, Martin, 191.

Mach, Anna Karma, 291-293.
Mach, Carolina, 13, 27, 36.

Mach, Felix, 28.
Mach, Heinrich, 28, 87.
Mach, Johann Nepomuk, 1-20, 195, 329.
Mach, Joseph, 3.
Mach, Josepha, 3, 5-7, 11-12, 14, 17.
Mach, Ludwig, xvi, 10, 54, 75, 108, 152n, 159-161, 265, 272, 281-282, 288, 290, 293-294.
Mach, Luise [née Ludovica Marussig], 27-28.
Mach, Marie, 2-16, 87, 329n.
Mach, Octavie [Octavia], 3, 6, 8, 10-11, 13-14.
Mach, Victor, 28.
März, Eduard, 371n.
Mainx, F., 374.
Mannoury, Gerrit, 372, 429.
Marek, J. Christian, 221n.
Maric, Mileva (See Mileva Einstein.)
Marie Louise [2nd wife of Napoleon I], 7.
Marinelli, W., 367, 371.
Markovic, Sima, 391, 393-395, 402.
Marx, Karl, 210n, 341-342, 351, 391-392, 395.
Maslow, Abraham, 195.
Massieu, Francois-Jacques-Dominique, 58.
Mathias, Fabian, 21.
Mathiesen, Jochmann, 51.
Mauer, Clementine, 10, 12.
Mauer, Uncle, 12.
Mauthner, Fritz, 192-193.
Mayer, J.R., 136, 421.
Maxwell, James Clerk, 135, 158, 238, 261, 282, 303, 327, 413-414.
McCormack, T.J., 57n, 156n, 354n.
McCormmach, R., 165n.
McGuinness, Brian, 57n, 221n, 225n, 366n, 370n, 373n, 430.
Meitner, Lise, 305.
Menger, Karl, 366.
Merkl, Manon, 13, 14.
Mertens, Franz, 150n.
Meyer, Stefan, 151-152.
Meyer, W.H., 373.
Meyerson, Émile, 162, 327, 330.
Michelson, Albert Abraham, 290-291, 293-294, 300, 312.
Miki, Kiyoshi, 349.
Milankovic, Uros, 387.
Milanov, Kajica, 401.
Mill, John Stuart, 154, 303, 333, 404, 406.
Miller, Arthur I., 293n.
Miller, Dayton C., 293-294.
Minkowski, Hermann, 125, 139, 169, 171-172, 177, 266-270, 274-275, 277-280, 285, 316, 326, 423.
Misar, W., 367.
Mises, Richard von, 364n, 372-374.
Möbius, A.F., 25.
Monet, Claude, 205n.
Moos, Meister von, 7.
Morley, Edward Williams, 290-291, 294n, 300, 312.
Morris, Charles, 372, 374.
Moszkowski, Alexander, 324.
Motoki, Jindayu, 321.
Müller, David Heinrich, 150n.
Müller, Johannes, 125, 197-198.
Müller, Max, 192.
Mulder, G.J., 406.
Mulder, Henk L., 365n, 373, 430.
Murphy, A.E., 176n.
Mussolini, Benito, 375, 377.

Nachtikal, Frantisek, 36.
Nagel, E., 374.
Nagaoka, Hantaro, 297-298, 300, 316, 319, 322.
Nagarjuna, 190.
Nakamura, S., 333n, 342n.
Napoleon I, 7.
Napoleon II [See Duke of Reichstadt]
Napoleon III, 8.
Narasu, Lakshmi, 86-88.
Nedeljekovic, Dusan, 382, 390.
Neumann, C. 60, 248.
Neumann, Franz, 24, 299.
Neumann, Karl Eugen, 191.
Meumann, Miromil, 48, 50.
Neurath, Marie, 363n, 373.
Neurath, Otto, 239, 264-265, 363n, 364-377, 426.
Neurath, Paul, 370n.
Newton Sir Isaac, 25, 60, 63, 68-69, 121, 123, 131, 136, 138, 156-157, 169, 172n, 173, 202, 207-209, 234, 237, 238, 243-261, 278, 283, 289, 297-299, 303, 325, 327, 329, 334, 343-345, 352-358, 382, 385, 393-394, 412, 414-416, 421, 425, 427.
Newton, Roger, G., 374.
Neyman, J., 373.
Nietzsche, Friedrich, 179, 203n.
Nishida, Kitaro, 320, 333, 337-339, 349, 350.
Nisio, Sigeko, 241n, 280n, 313.
Noé, Keiichi, 229-242, 350n, 429n.
Noll, W., 260.
Nomorarius, Jordanus, 62.
Nordmann, Johannes, 9.
Novak, V., 40n.
Novy, L. 40n.

Oettingen, A. von, 51, 58.
Ogawa, Tsuyoshi, 294n.
Ohba, Ken, 350n.
Oldenberg, Hermann, 191, 194.
Ono, Y.A., 163, 289n.
Opzoomer, C. W., 404, 406, 409, 411, 419, 428.
Ordnung, Hugo, 13.
Ornstein, Salomon, 417-418.

Ostwald, Walter, 77n, 87.
Ostwald, Wilhelm, xi, xvi, 17n, 26, 58, 61, 71-91, 95, 103, 120, 135, 152n, 241n, 281n, 283n, 308, 324, 380, 408, 415-416, 420, 421n, 429.

Padoa, Alessandro, 261.
Painlevé, Paul, 162.
Pais, Abraham, 287n, 346n, 347n.
Panum, Peter Ludwig, 125.
Pappenheim, M., 369.
Pascal, Blaise, 390.
Pauli, Wolfgang, 346-347.
Pearson, Karl, 61, 249.
Pegram, Professor, 305-306.
Pendse, C.G., 252n, 253n, 256.
Perrett, W., 171n, 208n, 278n.
Perrin, J.B., xi, 71n, 162, 418.
Petronijevic, Branislav, 211, 212n, 379-382, 384-390.
Petzval, Josef, 23, 148.
Petzoldt, Joseph, 82, 91, 107-109, 117, 134, 174-175, 207, 208n, 209, 266-267, 272-276, 283, 287-288, 366, 425.
Pfaundler, Leopold, 147, 212n.
Philipp, H., 20.
Philo of Byzantium, 19.
Piaget, Jean, 429.
Picard, Émile, 310-311.
Pierson, Allard, 403n, 407.
Piéron, H., 162.
Pinkava, J., 81n.
Pinter, Tomislav, 391-393.
Piovesana, Gino, 333, 338n, 339n.
Pircher, Wolfgang, 197n.
Pius X, 149.
Planck, Emma, 306.
Planck, Greta, 306.
Planck, Max, xi-xii, xiv, xvi, 17n, 26, 127-149, 152n, 206, 214, 219, 238-239, 266-267, 270n, 279, 284, 300-301, 303-310, 317, 326-328, 330-331, 334-336, 340, 360, 393, 395, 401, 404, 417, 419, 422, 424-425.
Plato, 197, 231, 244, 269, 387.
Pliva, E., 41n.
Ploetz, Karl, 21.
Plücker, J., 25.
Plutarch, 20.
Podhorsky, Richard, 391, 393, 395, 399-401.
Poe, Edgar Allen, 197.
Poincaré, Henri, 60, 92n, 97, 100, 117, 250n, 298-300, 304, 311, 318, 326-327, 335, 418.
Poess, O., 31n, 32n.
Polanyi, K., 368, 371.
Popovic, Milos, 391, 401-402.
Popper, Sir Karl, 215n, 218n, 224n, 228, 239n, 373.
Popper-Lynkeus, Josef, 45, 88, 90, 192.
Poser, H., 223n.
Prica, Ognjen, 391, 397-400, 401.
Przibram, Karl, 151.
Ptolemy, Claudius, 125, 249.

Quine, W.V., 234, 241.

Rabinowitsch, E., 373.
Raiffa, H., 224n
Rankine, W.J., 58, 135.
Ratliff, Floyd, 403n.
Rayleigh, John William Strutt, 3rd Baron, 326.
Rehmke, Johannes, 387, 389.
Reich, Wilhelm, 368, 371.
Reichenbach, Hans, 209, 244, 249, 366.
Reichstadt, Duke of [Napoleon II], 7.
Reimann, Dora, 63n, 232.
Reininger, Robert, 367.
Reinisch, Leo, 150n.
Reiser, Anton, 324.
Renn, Jürgen, 312, 314.
Renoir, Pierre Auguste, 205n.
Ribot, Theodule Armand, 192.
Richtmann, Zvonimir, 391, 395-402.
Rickert, Heinrich, 335, 338.
Riecke, Eduard, 316.
Riemann, Georg Friedrich Bernhard, 163, 205n, 207, 232-233, 271, 329, 380.
Riess, P.T., 144.
Ritz, Walther, 311.
Roberval, Gilles Personne, 62.
Roozeboom, Bakhuis, 410.
Rosen, George, 152n, 208n.
Rosicky, Wenzel, 48.
Rudas, L., 399.
Russell, Bertrand, 62, 192, 299, 327, 363.
Rutherford, Ernest, 164, 327.
Rynin, D., 373.

Saigusa, Hiroto, 341.
Sajner, Josef, 17n, 36n.
Sakata, Shoichi, 342, 346n, 347.
Santillana, G. de, 374.
Santone, Michael A. Jr., 333-361.
Sarton, George, 331.
Sartre, Jean Paul, 350.
Sartres, 398.
Sassen, F., 411.
Schachter, J., 373.
Schedling, Anton, 13.
Scheler, Max, 213n.
Schelling, Friedrich Wilhelm, 382.
Scherer, Wilhelm, 205n.
Schiffner, Ludwig, 15.
Schiller, Friedrich, 4, 6.
Schilpp, P.A., 215n.
Schlick, Moritz, 239, 264, 327, 364-377, 393.
Schlosser, Anton, 49.
Schmidt, Rudiger, 185n.

Schneewind, E.H., 373.
Scholl, D., 373.
Scholz, H., 430.
Schopenhauer, Arthur, 69, 184, 191, 192n, 194-195, 197.
Schrödinger, Erwin, 323.
Schubert-Solden, R. von, 117.
Schütte-Linetsky, Margarete, 371n.
Schulmann, Robert, 312, 314n.
Schuppe, Wilhelm, 117, 134, 387, 389.
Schuster, A., 69.
Schuster, Captain, 5.
Schuster, Sir Arthur, 299.
Schwädelbach, H., 197n.
Schwarzenberg, Friedrich Fürst von, 43, 45.
Scott, Sir Walter, 8.
Seeger, Raymond J., 220n.
Seelig, Carl, 287.
Seeliger, H. von, 45n, 107-108.
Segré, Emilio, 346.
Seibt, F., 43n.
Seuret, Georges Pierre, 205n.
Seydler, A., 33n, 40, 48, 50, 53.
Shankland, R.S., 294.
Shapere, Dudley, 63n.
Shapeley, 326.
Shaw, George Bernard, 326.
Siegert, M., 377n.
Signac, Paul, 205n.
Simon, H.A., 259.
Simonides, Jaroslav, 48.
Skimunt, S., 117.
Smeaton, Amethe, 373.
Smetena, Bedrich, 37.
Solovine, M., 310.
Sommer, Jan, 48.
Sommer, Manfred, 280n, 287n.
Sommerfield, Robert, 312n.
Sommerfeld, A., 171n, 278n, 301, 335.
Spencer, Herbert, 333, 404.
Spinoza, Benedictus de, 154, 387.
Spruyt, C.B., 403n, 405.
Stachel, John, 312n.
Stadler, Friedrich, 195n, 198n, 265n, 266n, 272n, 280n, 281n, 287n, 291n, 363-377.
Stadion, Gräfin Kunigunde, 3, 11.
Stadion, Graf Philipp von, 1-3, 11.
Stallo, John Bernard, 78, 303, 426.
Stark, Johannes, 67-68.
Stevinus, Simon, 62, 426, 428.
Stöhr, Adolf, 363.
Stojkovic, Andrija, 382n.
Strigl, Richard, 368, 371.
Strouhal, Cenek, 33, 40.
Stuewer, R., 214n, 217n, 222n.
Stumpf, Carl, 26, 121-126, 279.
Suess, Edward, 147.
Suppe, F., 63n.
Suppes, P., 258n, 260, 261n.
Svatos, M., 42n.
Swabey, M.C., 176n.
Swenson, Lloyd, C., 293n.
Swoboda, W., 195n.

Taaffe, Count von, 37.
Tagore, Rabindranath, 192.
Tait, P.G., 298, 302, 413-414.
Takakasu, J., 189n.
Takata, Seiji, 331.
Taketani, Mitsuo, 333, 342-350.
Tanabe, Hajime, 320, 333, 339, 349-350.
Tanaka, Setsuko, 297-332.
Tanakadate, Aikitsu, 297-298.
Tandler, Julius, 370.
Tanikawa, Yasutaka, 346n, 347n.
Tarski, A., 258n, 260n, 430.
Terada, Torahiko, 318-319, 323.
Tesinska, E., 40n.
Thiele, Joachim, 30n, 45n, 71n, 265n, 281n, 404n.
Thiele, Frau Joachim, 71n, 92n, 111n.
Thirring, Hans, 370.
Thomsen, Anton, 273n.
Thomson, J.J., 287, 298, 302.
Thucydides, 153n.
Thun, Count Leo, 21n, 23.
Tintner, Ivo, 374.
Tosaka, Jun, 341, 349.
Toulmin, S., 240.
Tretera, Ivo, 47-55.
Trevor, J.E., 57-59.
Truesdell, C., 244, 260.
Tschermak, Gustav von, 147n.
Tumlirz, Ottokar, 48-49.
Twyman, F., 69.

Uhlig, Viktor, 147n.

Vaihinger, Hans, 255.
Vanacek, J., 41n.
Van der Waals, J.D., 404, 416.
Van der Wyck, Bernard, 403n, 406, 409, 411.
Van't Hoff, J.H., 404, 410.
Verworn, Max, 419.
Visser, Henk, 63, 237n, 403-430.
Vitruvius, 20.
Voet, Gisbert, 406.
Voigt, Woldemar, 429.
Vokolek, H, 367.
Volkelt, Johannes, 211, 212n.
Volta, Alessandro Giuseppe, 327.
Voss, A., 60.
Vucic, Radmilo, 382-383.

Wahrig-Schmidt, Bettina, 185n.
Waismann, Friedrich, 366, 368-369.

Wald, Frantisek, 81, 117, 152n.
Waldhofen, Adalbert von, 30.
Walzel, Oskar, 205n.
Wanka, Joseph, 48-49.
Warburg, E., 294n.
Ward, James, 3.
Weber von Ebenhof, Philipp Freiherr, 41.
Webster, 326.
Weinek, L., 49.
Wessely, Peter F.X., 12, 20, 133.
Weyl, Hermann, 171n, 278n, 322.
Whitehead, Alfred North, 174, 176n, 327.
Whittaker, Sir Edmund, 260.
Wiechert, Emil, 165-170, 292.
Weinburg, Carlton B., 152n, 153n.
Wien, Willy, 136.
Wiener, Otto, 31, 293n.
Wiener, Philip, 223n.
Wilkinson, John, 393
Williams, D.H., 127n.
Willy, Rudolf, 408.
Wind, C.H., 403.
Wingler, H.M., 370n.
Wirtinger, Wilhelm, 150n.
Wittgenstein, Ludwig, 220-221, 239, 263, 366.
Wohlwill, Emil, 62.
Wolff, Christian, 387.
Wolters, Gereon, 23n, 36n, 92n, 152n, 184, 263-276, 280n, 281, 290.
Woodger, J.H., 372, 374.
Wosycka (Wosyka), Jaromir, 48, 50, 53-54.
Wundt, Wilhelm, 192, 235, 308, 387, 421.

Yajima, Suketoshi, 331.
Yamamoto, Mr. [from Kaizosha Pub. Co.], 322.
Young, A.F.A., 69, 282n, 325.
Young, Thomas, 69-70, 136, 138.
Yukawa, Hideki, 333, 339, 342-350.

Zahar, Elie, 289n.
Zeeman, Pieter, 404.
Zeisel, Hans, 368, 371.
Zermelo, Ernst, 98.
Ziehen Theodor, 117, 408-409, 429.
Zilsel, Edgar, 365, 367-368, 370-371, 374.
Zotz, V., 191.

SUBJECT INDEX

Absolute Motion, Space, and Time (See also Mechanics and Relativity Theory): xii, 62, 156-157, 207, 236-237, 244-250, 267, 274, 298-301, 310, 317, 353-355, 360-361, 382, 385-386, 393-394, 416, 425.
Acoustics: 24-25, 31, 47, 51-52, 122, 148, 158, 201, 204, 419.
Analysis of Sensations: 26-27, 29, 33, 45-46, 79, 96, 102, 115-117, 126, 129n, 134, 185-188, 191, 193, 197, 201n, 203n, 204n, 212n, 229n, 236n, 318, 351, 407, 409, 411, 424-425.
Anti-Semitism: 4, 16.
Atomism: xi, 66, 81, 112-114, 129, 137-139, 145, 149n, 151-153, 162, 164, 210-211, 236-239, 243, 254, 265n, 278, 280, 283-284, 290, 300-303, 307, 310, 327-330, 352, 369, 393, 403, 409-410, 416-419, 423.

Ballistics (See Shockwaves.)
Bewegungsempfindungen (See *Perceptions of Movement.*)
Biology (See also Evolution Theory): 122-123, 138, 148, 187, 196, 201, 229, 236, 240, 367, 374, 383, 390.
Buddhism: 26, 87, 183-199, 338, 349, 351, 429.
Burschenschaften: 39.

Calorimetry: 57-58.
Catholic Church: 1-3, 43-46, 80, 375, 404.
Causality (See also Dependent Co-Arising and Mathematical Functions): 104, 204, 235, 237, 249, 289, 330, 344-345, 352, 366-368, 373-374, 393-396, 399, 405.
Chemistry: 49-50, 71, 74-76, 124, 152, 164, 206, 410, 420.
Color: 69, 111-112, 118-119, 121-122, 131, 204, 230, 236, 356.
Compendium der Physik für Mediziner: 329.
Consciousness: 125-126, 187-188, 196, 351-352, 387-389, 410.
Conservation of Energy: 25, 35, 134, 170, 222, 267, 277, 407, 412.
Conservation of Energy (See also Thermodynamics): 129-130, 143, 145, 235, 269, 385, 421, 428.
Continuum Theory: 58, 206-207, 257, 416.
Conventionalism: 100, 249, 303.
Copernican Theory, 63, 145, 249.

Darwinism (See Evolution Theory).
Denkökonomie (See Economy of Thought).
Dependent Co-Arising: 186-187, 190.
Ding-an-sich (See Thing-in-itself).
Discontinuity Theory (See Atomism).
Dogmatism: 145, 174, 397-402.
Doppler Effect: 23, 249.
Doppler Theory: 29.

Economy, Theory of: xi, xiii-xv, 115, 131-139, 142-143, 146, 149n, 153-154, 167, 186, 212, 215-228, 249, 308, 317, 383-384, 390, 394, 415, 417.
Education (See also Tutoring): 18-23.
Ego or Self: 26-27, 185-188, 197, 405, 410.
Electricity: 47-49, 111-112, 123, 144.
Electrodynamics: 61, 136, 300-301, 308.
Electromagnetic World View: xi, 61, 416, 418.
Electromagnetism: xi, 65, 261, 287, 293, 322.
Electrons: 128-129, 137, 210, 261, 290, 299-302, 326, 337, 417.
Elements, Theory of: xii-xiii, xv-xvi, 24, 70, 111-114, 116, 118-126, 129, 137, 158, 162, 185, 187-188, 202, 204, 210, 212-213, 215, 218-219, 221, 236, 278, 286, 338, 340, 353, 355, 387-388, 393, 407, 410, 424.
Empirio-Criticism (See also Phenomenalism): 29, 134, 335, 341, 393.
Energeticism: xi, 26, 58, 120, 135, 280, 302, 304, 308, 416.
Energy (See Thermodynamics, Energeticism, and Conservation of Energy).
Epistemology: xiii, 22, 24-26, 74, 78, 93, 100-101, 115, 117-127, 130-131, 133-139, 141-146, 148, 154-157, 160, 168, 174n, 183-199, 201-214, 229-241, 243-261, 265, 267, 272, 280, 284, 287-290, 303, 308, 316-320, 328, 339-341, 347, 351-352, 356-358, 379-430.
Erhaltung der Arbeit (See *Conservation of Energy*).
Erkenntnis und Irrtum (See *Knowledge and Error*).
Esperanto: 78.
Ether: 66, 112, 165-170, 292-295, 311, 329, 353-354, 405, 415-416.
Evolution Theory: xiv, 66, 116, 133-134.
Experiment: 10, 23-25, 29, 31, 33, 47, 74, 158, 239, 252, 57-258, 272, 284, 290-295, 312, 324, 343.

Farming: 1-2, 8, 329.
Fits, Newton's Theory of: 69.
Flowers: 7, 16.

Geometrical Optics: 65, 67, 97.
Geometry: 95-97, 125, 163, 207, 230-234, 241, 268, 277, 285-286, 316, 354-356, 380-382, 423.
Gesellschaft für positive Philosophie (and its journal): 102-103, 105-106, 109.

Haidinger Fringes: 123.
Heat Conduction: 57-58.

Heat Theory (See Thermodynamics).
History of Ideas: 68, 195, 307, 316, 317-320, 333-430.
History of Science: 25, 57-70, 74, 97, 115, 154-157, 240, 299, 302, 304, 310, 319, 320-327, 330-332, 342, 345, 386, 421-423, 425-428.

Idealism: 211n, 218, 308, 319, 338, 347, 352, 389, 393, 399-402, 408, 425.
Ido: 78.
Imagination: 215, 224-228.
Interferometer: 69, 282, 288, 290, 293.

Kantianism and Neo-Kantianism: 22, 141, 213, 333, 335, 338-342, 345, 351, 366-367, 401, 403, 405, 409, 411, 426.
Kinematics: 145, 259-260.
Kinetic Theory: 74, 128, 130, 136-137, 172, 238, 279.
Knowledge and Error: 26-27, 102, 134, 171, 186, 201n, 203n, 204n, 211, 217, 220-221, 222n, 226n, 229n, 231n, 234n, 245n, 285n, 351, 380, 430.

Labyrinth Function: 24,29, 148.
Laws (See Natural Laws).
Leading Thoughts of my Scientific Epistemology: 133-139, 141-146, 218n, 219n, 269n, 279n, 284n.
Lectures on Psycho-Physics: 111-114.
Leitgedanken (See *Leading Thoughts...*).
Lever, Theory of: 63-65.
Light (See Optics).
Light Constancy Axiom: 207, 274, 288, 329, 423, 425.
Logic: xiv, 78, 104, 109, 180, 225, 239-240, 254, 260-261, 342, 368-369, 372-374, 395-396, 400-401, 421.
Logical Positivism (See Positivism).
Lokal Lorentz System: 288.
Lorentz-Fitzgerald Contraction: 175, 319.
Lorentz Transformations: 417.

Mach Bands: 29, 403.
Mach number (See Shockwaves).
Mach Principle: 63, 157-158, 163, 170, 208, 248, 264, 289, 417.
Magnetism: 49, 123.
Mass: 208-209, 214, 235-236, 244, 250-258, 287, 317, 328, 332, 353-354, 356, 358, 412, 427, 429.
Materialism: 111, 308, 342, 352, 389, 391, 393.
Mathematical Functions: xiv, 289, 336, 410, 425.
Mathematics (See also Geometry and Set Theory): 19, 23-25, 92, 94, 96, 102, 109, 122, 124, 207, 225, 230, 266-271, 273, 302, 307, 322, 334, 339.
Matter: 119-120, 139, 170, 210, 278-279, 315, 336-337, 340, 353, 366, 368-369, 372, 374, 380-382, 384, 393-394, 408-410, 415, 420, 428-429.
Mechanics: xii, 47, 57, 59-65, 135, 145, 156-158, 201, 230, 234, 236-241, 243-261, 264, 297-298, 303, 308, 310, 315, 329, 334, 344-345, 352, 381-382, 385, 412, 417, 421, 426-428.
Mechanik (See *Science of Mechanics*).
Mechanistic World View: 61, 229, 234-241, 372, 394, 427-428.
Medicine and Illness: 24-25, 35, 75, 89, 96, 102, 109, 110-111, 113, 126, 152, 186, 189.
Meson Theory: 342-350.
Metaphysics (See also Ontology): xii, xv, 22, 66, 104, 151-152, 209-210, 214, 219, 239-240, 251, 254-255, 307-308, 317, 337-338, 345, 349, 355, 367, 379, 381-382, 384-385, 387-388, 394, 397, 399, 404, 407, 425, 427.
Monism: 83, 85-88, 126, 286, 329, 408, 410-411.
Motion: 60, 120, 123, 143-145, 251, 254-259, 298-300, 308, 337, 344, 356, 385, 404.
Music: 7, 12-13, 15, 51-52, 97n, 122n.

Nationalism: 16, 30, 35-46, 375-378.
Natural Law: xiii, 121, 126, 163, 166, 221, 223, 229, 234, 251, 253, 259, 275, 339, 343, 356, 358, 367, 393-394, 396, 410, 414-415, 425.
Nature Philosophy: 76, 103, 106, 142, 197.
Negative Mass: 61.
Newton's Bucket Experiment (See also Mach Principle): 63, 157, 289, 357.

Observation: xiv, 31, 107-108, 121-123, 130, 136, 147, 152, 219, 237, 251, 254, 256-259, 318.
Ontology (See also Metaphysics, World Picture, and World View): xiii, 114, 174n, 203, 211, 239, 341, 357-358, 387-388, 410-411.
Operationism (Operationalism): 201, 357-358, 415.
Optics: 31, 47-48, 65-70, 112, 136, 159, 165-170, 273, 292-295, 357, 359, 382, 413, 417.
Optische-Akustische Versuche: 25, 31.
Optik (See *Principles of Physical Optics*).
Optik Preface: 23, 65-70, 89-91, 159-161, 173, 184, 263, 271-272, 281-283, 291-295.

Pali Canon, 183, 191.
Pan-Slavism: 37.
Perceptions of Movement, 25.
Phenomenalism: xi, xvii, 22, 173, 183-200, 201-214, 215-228, 229-241, 243-261, 290, 302, 338, 345, 357-359, 361, 269, 381-430.
Phenomenological Physics: 237-238, 344-346, 429.

Phenomenology (Husserl): 237-238, 426, 429.
Philosophy (See particular branches and movements).
Philosophy of Science: 74, 99, 114, 127-132, 133-139, 146, 162-163, 201-214, 215-229, 239-241, 307, 316-320, 330-430.
Photography: 29, 158, 167, 205n, 212.
Physics (See particular branches of).
Physics, Church of: 91, 138, 148-149, 173, 272, 283.
Piarist Order: 12, 20.
Poetry: 13-15, 50-51.
Polarization of Light: 47, 67-68, 124.
Popular Scientific Lectures: 27, 34, 101, 223n, 231n, 232n, 245n, 407-408.
Positivism and Logical Positivism: xv, xvii, 115, 130, 139, 174, 215-216, 218, 225, 239, 249, 274, 283, 308, 310, 327, 329, 360, 363-377, 382, 392, 399-430.
Principles of Physical Optics: 23, 65-70, 97, 118-121, 159-161, 173, 184, 263, 271-272, 281-283, 291-295.
Principles of the Theory of Heat: 23, 26, 57-59, 98, 116, 134, 143-144, 160-161, 170-171, 206-207, 224, 225n, 237n, 238n, 288n, 292, 381-382, 407, 421, 430.
Psychophysics: 24, 111-114, 116, 203.

Quantum Theory: 206-207, 210-211, 238-239, 254, 309, 322-323, 326-327, 330-331, 334, 336-37, 342-350, 392, 394-395.

Radiometers: 47.
Realism: 211, 219, 236, 249, 261, 271, 302, 340, 357-361, 381, 384, 387-390, 409, 424, 427.
Rectorships, Mach's: 30, 37-46.
Reflection: 67, 138.
Refraction: 67.
Relationism: 174-182, 352-361.
Relativism, Objective: 174-182.
Relativity Theory, Einstein's: 66-68, 125, 139, 149n, 153, 156-158, 160-163, 165-182, 184, 207-211, 233, 239, 249, 263-276, 277-295, 298-299, 303, 305, 308-310, 316-317, 319, 322-323, 326, 328, 334, 353-361, 394-395, 417, 423, 425.
Relativity Theory, Mach's: 60, 90, 125, 139, 145, 165-175, 186, 229-241, 243-261, 274-277, 279, 285, 289, 299, 301, 303, 308, 320, 354, 385, 394, 423, 429.

Science of Mechanics: 25, 29, 42, 57, 59-65, 76-77, 98, 100-101, 116, 134, 144, 156-158, 169-172, 202, 230n, 236n, 241, 244n, 249n, 250n, 282, 285n, 286-287, 288n, 294, 298-299, 330-331, 334, 381-382, 412, 416.
Sensations (See Elements, Sense Physiology, and Phenomenalism).
Sensory Elements: 118-126, 212n.
Sense Physiology: 107-108, 115, 118-125, 148, 201, 230, 274-275, 284.
Set Theory: 97-98, 105.
Shock Waves: 29, 31-32, 46-47, 158-159, 201-202, 212-213.
Sinnliche Elemente (See *Sensory Elements*).
Soul: 113-114, 196.
Spectroanalysis: 49, 113.
Spinthariscope: 151-152, 164, 284.
Stroboscopy: 47.
Substance: 113, 196, 211, 236-237, 344-352, 367, 428.

Temperature (See Thermodynamics and Thermometry).
Thermodynamics: xi, 57-58, 116, 136, 142-144, 160, 235, 238-239, 288, 308, 414-415, 421-422.
Thermometry: 57-58, 144.
Thing-in-itself: 26, 115, 307, 338, 423.
Truth: 16, 190, 204, 214, 226, 229, 232-233, 236, 320.
Tutoring: 1, 7, 24.

Unity of Science Movement: 241, 364-377.
Universals: 181.
Utraquism: 38, 40.

Verein Ernst Mach: 183, 239-240, 363-377.
Vienna Circle: 183, 215, 220, 239-240, 327, 363-377, 430.

Wärmelehre (See *Principles of the Theory of Heat*).
Weltbild (See World Picture).
Wiener Kreis (See Vienna Circle).
World Picture: 127-132, 135n, 136-137, 217, 236, 279, 299, 302, 416-419, 422, 424, 426-428.
World View (See also Ontology): 26-27, 337, 364-365, 367, 372, 383, 416.